Plasma Physics via Computer Simulation

Plasma Physics Series

Series Editor: **Professor E W Laing**, University of Glasgow

Advisory Panel

> **Dr J Lacina**, Czechoslovak Academy of Sciences
> **Professor M A Hellberg**, University of Natal
> **Professor A B Mikhailovskii**, I V Kurchatov Insitute of Atomic Energy
> **Professor K Miyamoto**, University of Tokyo
> **Professor H Wilhelmsson**, Chalmers University of Technology

Other books in the series

Radiofrequency Heating of Plasmas
R A Cairns

An Introduction to Alfven Waves
R Cross

Transition Radiation and Transition Scattering
V L Ginzburg and V N Tsytovich

Tokamak Plasma: A Complex Physical System
B B Kadomtsev

MHD and Microinstabilities in Confined Plasma
W M Manheimer and C N Lashmore-Davies

Electromagnetic Instabilities in an Inhomogeneous Plasma
A B Mikhailovski

Physics of Intense Beams in Plasmas
M V Nezlin

Plasma Physics Series

Plasma Physics via Computer Simulation

C K Birdsall

*Electrical Engineering and Computer Sciences Department,
University of California, Berkeley*

A B Langdon

*Physics Department, Lawrence Livermore Laboratory,
University of California, Livermore*

Institute of Physics Publishing
Bristol and Philadelphia

British Library Cataloguing in Publication Data

CIP catalogue recored for this book is available from the British Library

ISBN 0 7503 0117 1

Library of Congress Cataloging-in-Publication Data

Birdsall, Charles K.
    Plasma physics via computer simulation.

    Bibliography: p.
    Includes Index.
    1. Plasma (Ionized gases)-Simulation methods.
2. Computer simulation. I. Langdon, A. Bruce.
II. Title.
QC718.4.B57   530.4'4   81-8296
ISBN 0 7503 0117 1   AACR2

Reprinted 1995

Published by Institute of Physics Publishing, wholly owned by the Institute
of Physics, London

Institute of Physics Publishing, Techno House, Redcliffe Way, Bristol BS1
6NX, UK

US Editorial Office: Institute of Physics Publishing, The Public Ledger
Building, Suite 1035, Independence Square, Philadelphia, PA 19106, USA

Printed in Great Britain by Galliard (Printers) Ltd, Great Yarmouth, Nor-
folk

Contents

6   A 1d Electromagnetic Program EM1   133

7   Projects for EM1   145

Part 2   Theory

Plasma Simulation Using Particles in Spatial Grids with Finite Time Steps — Warm Plasma   **153**

8   Effects of the Spatial Grid   155

Part 3 Practice
Programs in Two and Three dimensions:
Design Considerations 303

16 Particle Loading, Injection; Boundary Conditions and External Circuit 387

Foreword

The complex nature of the problems encountered in plasma physics has motivated considerable interest in computer simulation, which has played an essential role in the development of plasma theory. In addition, computer simulation is also becoming an efficient design tool to provide accurate performance predictions in plasma physics applications to fusion reactors and other devices, which are now entering the engineering phase.

Computer simulation of plasmas comprises two general areas based on kinetic and fluid descriptions, as shown in Figure a. While fluid simulation proceeds by solving numerically the magnetohydrodynamic (MHD) equations of a plasma, assuming approximate transport coefficients, kinetic simulation considers more detailed models of the plasma involving particle interactions through the electromagnetic field. This is achieved either by solving numerically the plasma kinetic equations (*e.g.* Vlasov or Fokker-Planck equations) or by "particle" simulation, which simply computes the motions of a collection of charged particles, interacting with each other and with externally applied fields. The pioneering work of Dawson and others in

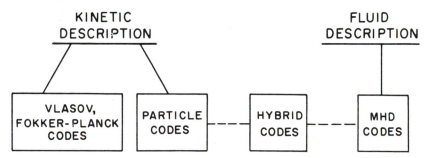

Figure a Classification of computer simulation models of plasmas.

the early 60's has shown that, when appropriate methods are used, relatively small systems of a few thousand particles can indeed simulate accurately the collective behavior of real plasmas. Since then, the development of new algorithms and the availability of more powerful computers has allowed particle simulation to progress from simple, one-dimensional, electrostatic problems to more complex and realistic situations, involving electromagnetic fields in multiple dimensions and up to 10^6 particles. Kinetic simulation has been particularly successful in dealing with basic physical problems in which the particle distributions deviate significantly from a local Maxwellian distribution, such as when wave-particle resonances, trapping, or stochastic heating occur. MHD simulation, on the other hand, has generally been applied to large-scale problems directly related to the behavior of experimental devices. However, this simple distinction between kinetic and MHD simulations is becoming more complex through the increasingly-common use of "hybrid" codes, in which for example, fluid and particle treatments are applied to different components of a given plasma, and through the introduction of particle-hydrodynamic codes, in which fluid equations are solved by particle methods. The recent development of implicit algorithms is also expected to allow application of particle, Vlasov or Fokker-Planck codes to long-time-scale transport problems, which have generally been treated by fluid simulation.

Both MHD and kinetic simulations, including particle simulation, are well-developed disciplines, which have become an integral part of plasma physics, and the need for textbooks and basic references in these areas has been felt for some time. "Plasma Physics via Computer Simulation," by C. K. Birdsall and A. B. Langdon, provides a clearly-written and competent answer to this need in the area of particle simulation. The book consists of three parts, Part One presenting elementary particle simulation methods, while Parts Two and Three deal with the fundamental numerical analysis problems of particle simulation, and with more advanced particle simulation techniques applicable to electromagnetic fields and to problems in several dimensions. The elementary description of Part One is supported by the electrostatic code ES1 and electromagnetic code EM1, and proceeds with a number of physically interesting projects. This is a very appropriate approach in a field which can best be learned by actual practice, and as implied in the title of the book, computer simulation can itself be a useful pedagogical tool. For example, phase-space plots from a particle simulation can be worth "a thousand equations" to illustrate nonlinear wave-particle interactions. Thus, this part of the book is not limited to specialists in computer simulations, but should be of interest to plasma physicists in general, and should provide a better understanding of the capabilities and limitations of these methods. Part Two and Three are addressed to readers who need a deeper knowledge of particle simulation. These parts provide a knowledgeable presentation of advanced subjects and guidance through the extensive literature of this field.

Ned Birdsall and Bruce Langdon are recognized authorities in plasma physics subjects associated with particle simulation. Both have made important contributions to the development and to the applications of particle simulation in plasma physics problems related to magnetic- and inertial-confinement. The book has evolved from class notes, compiled over a decade of teaching of the subject at Berkeley, which have already served a generation of students, many of whom are now well established in the field. The authors have also called on several contributors to present certain subjects, and the present work represents a comprehensive text on the fundamental aspects of particle simulations.

J. Denavit
Lawrence Livermore Laboratory

Preface

Our book is on particle simulation of plasmas, aimed at developing insight into the essence of plasma behavior. Major current applications are to magnetically- and inertially-contained fusion plasmas. However, particle simulations are also being used to gain understanding of plasmas in space, and man-made plasmas such as occur in electron and ion guns, plasma propulsion and microwave devices, and in nuclear explosions. Our title notwithstanding, we make no pretense of covering all of plasma physics or all of computer simulation.

Plasma is the fourth state of matter, consisting of electrons, ions and neutral atoms, usually at temperatures above 10^4 degrees Kelvin. The sun and stars are plasmas; the earth's ionosphere, Van Allen belts, magnetosphere, etc., are all plasmas. Indeed, plasma makes up much of the known matter in the universe.

Plasma is the medium for magnetically or inertially-confined controlled thermonuclear fusion. A plasma of deuterium and tritium ions heated to a temperature of 10^8 degrees Kelvin undergoes thermonuclear burn, producing energetic helium ions and neutrons from fusion reactions. Such plasmas could be used as sources of heat which may be used to produce steam to run turbines to drive electric generators. Current studies of fusion plasmas include theory, experiment, and computation using large and fast computers. The latter is very extensive because it has produced results useful to theory and experiment and because current fusion plasma experiments cost considerable time (years) and money (hundreds of millions of dollars). Included in computation is plasma simulation; part is done using fluid models; part is done using many-particle models (meaning 10^3 to 10^6 particles) in order to obtain detailed kinetic behavior; part is done using hybrid models with both fluids and particles.

Plasma simulation using particles has grown from art to science in the past twenty years, and is now in world-wide use. The cost has diminished as computers have improved to the point where the simulation codes presented in the text may be run on small computers or run very quickly on large computers. Hence, particle simulation is used not only by the large laboratories, but is well within reach of small university groups. And, very ambitious programs may be run on large computers, as through the network connecting universities and laboratories to the National Magnetic Fusion Energy Computer Center at Livermore.

The text has four major Parts:

Part One, *Primer*. This part is intended for the novice who wishes to do plasma simulation with particles by actually simulating. The computer code ES1 [an electrostatic one dimensional (1d) code] is described, with a listing. EM1 (an electromagnetic 1d code) is also described, along with the numerical methods needed to do the computational physics. The Primer is useful for an introductory course, complete with problems, or for the beginning of a longer course. Projects are provided, with suggested initial values and selected results.

Part Two, *Theory*. This part provides most of the current theory on electrostatic particle simulation, helping to explain the effects of finite time steps and of the spatial grid on the plasma physics. Part Two provides the mathematical and physical foundations for the algorithms used in Part One, and their effects.

Part Three, *Practice*. This part covers more complicated simulations in two dimensions, both electrostatic and electromagnetic; it is close to current plasma research, and is intended for use by research people and by students.

Part Four, *Appendices*. These provide many of the details essential to running computer simulations.

These features (programs/projects/problems/theory) were developed for a course which has been taught over the past decade in Berkeley.

A result rewarding to us in doing plasma simulations has been the fun of obtaining thorough physical understanding of plasma behavior. There is considerable excitement in making a program produce good physics. We recommend that the projects be started during the first week of class so that students are doing simulations as the theory is being covered in lectures and reading. The quiz questions in Chapter 3 may be used to ensure reading and understanding of the program. Simulation is almost always the only *direct* experience students have with plasma oscillations, streaming instabilities and the kinetic behavior of warm plasmas; almost all of their other experiences (in theory and lab) are *indirect*. Our experience has been that those students willing to work through the projects in Part One and add the theoretical understanding of Part Two will have deepened their insight into plasma behavior, whether they continue in theory or simulation or experiment.

We quote from the introduction to "Elements of Color," by Johannes Itten: Learning from books and teachers is like travelling by carriage, so we are told in the Veda. The thought goes on, "But the carriage will serve only when one is on the highroad. He who reaches the end of the highroad will leave the carriage and walk afoot." Doctrine and theory are best for weaker moments. In moments of strength, problems are solved intuitively, as if of themselves.

We have emphasized mathematical tools with which to construct algorithms with desired properties and analyze algorithms. Many of the algorithms were developed without use of such tools by people whose style and intuition leads to successful algorithms. Many useful codes have been assembled in an *ad hoc* manner and (often) work well, even though it may be impractical to calculate analytically the effects of finite Δx and Δt on the outcome.

This book gathers information which is valuable to simulators, some of which is scattered through published journals, and some of which is unpublished. Hence, it is intended also for reference use.

We are delighted to recommend "Computer Simulation Using Particles" by R. W. Hockney and J. W. Eastwood (1981, McGraw-Hill) as a complementary text. Where our emphasis is on plasma simulation, they extend the techniques developed primarily for plasmas to simulations of semiconductor devices, of gravitational problems, and of solids and liquids.

Charles K. (Ned) Birdsall
A. Bruce Langdon

Acknowledgments

We owe a special debt to J. M. Dawson for many helpful discussions from the early 1960's on. The idea of finite-size particle interactions and the understanding of such physics was shared with him as well as the importance of understanding the statistical or noisy behavior of simulation plasmas. To O. Buneman and R. W. Hockney go thanks for leading the way with particle integrators in magnetic fields and Poisson solvers in two dimensions. To C. W. Barnes, J. P. Boris, J. Denavit, J. W. Eastwood, M. R. Feix, D. W. Forslund, B. B. Godfrey, H. R. Lewis, E. L. Lindman, R. L. Morse, C. W. Nielson, K. V. Roberts, and K. R. Simon go thanks for many discussions. H. Okuda joined us in Berkeley in 1968-1970, helping produce the initial theory and verification for gridded particle models. We are especially indebted to J. Denavit for his Foreward and his general counsel on the book.

Birdsall thanks W. B. Bridges for our work with electron diode simulations begun in 1959, which in turn benefited from the pioneering work on electron device simulations in the 1950's of T. G. Mihran and S. P. Yu and of P. K. Tien. A. Hasegawa introduced me to 1½d models in 1962. J. A. Byers helped me on 2d Poisson solvers and linear weighting in 1964. T. Kamimura helped me with 2d and 3d gridded simulations in 1966 in Osaka, as did D. Fuss in 1967-1970, and N. Maron in 1972-75, at LLNL, Livermore. Collaboration with Langdon began in 1967 and has been most challenging and productive.

Langdon began plasma simulation with J. M. Dawson, who sets an exceptional example of the symbiosis of theory, simulation, and intuition. Much of the theoretical understanding of simulation methods and applications was done with, or stimulated by, Birdsall, who has catalyzed many successful projects and careers. Many collaborations have been instructive, especially the Astron simulations with J. Byers, J. Killeen, and others, and the development and extensive application of the ZOHAR code with B. Lasinski and other members of the plasma physics group, led by W. L. Kruer in support of the Livermore inertial-confinement fusion project.

We are especially grateful to B. I. Cohen and M. Mostrom for Chapters 6 and 7, and to W. M. Nevins for Chapter 11 and Appendix E.

It is a real pleasure to acknowledge the contributions of students in class and in research who have used particle simulations in their studies. The feedback from them made for better notes and better programs. Special thanks are due to L. Anderson, L. Chen, B. Cohen, R. Gordon, R. Littlejohn, C. McKee, W. Nevins, D. Nicholson, G. Smith, and D. Wong.

We gratefully acknowledge the support given to our effort by the Department of Energy. In particular, we wish to acknowledge the encouragement given by B. Miller, D. B. Nelson, D. Priester, and W. L. Sadowski in Washington, D. C., and to T. K. Fowler, W. L. Kruer, B. McNamara, and L. D. Pearlstein at LLNL, Livermore. Birdsall was both directly and indirectly supported in various periods in Berkeley for the express purpose of developing and producing the notes for use within the Magnetic Fusion Energy effort of the Department of Energy. We have used the National Magnetic Fusion Energy Computer Center at Livermore and wish to express appreciation for their operation, and especially to the NMFECC director, J. Killeen, and to his associates, H. Bruijnes, and D. Fuss. Birdsall is grateful to the British Science Research Council for support for part of the summer in 1976 at Reading University for time to work on Chapters 14 and 16 and to his host R. W. Hockney and his colleague J. W. Eastwood. Birdsall is grateful to the Japanese Ministry of Education for support at the Institute of Plasma Physics, Nagoya University, and to his host T. Kamimura during fall and winter 1981-2 when many corrections were made.

Our book originated as a set of class notes intended for use by graduate students who were learning to simulate using ES1. The first set was written about 1973; the second set, then in two parts (Primer and Theory) was finished in 1975; the third set, with the theory part rewritten and Practice and Appendices added, was completed in 1978. The current text thus contains sections written over most a decade during which our secretaries struggled in typing from pretty rough notes, namely Paula Bjork, Pamela Humphrey, and Michael Hoagland in Berkeley and Jill Dickinson in Reading, to whom we are most grateful.

The production team for putting the final version into camera-ready form during 1980-1984, was lead by Douglas W. Potter, who developed the macros and was responsible for the final photocopies. H. Stephen Au-Yeung and Carolyn Overhoff typed much of the book and the corrections. Thomas King and Fiona E. O'Neill, did all of the drawings, usually starting from rough hand-drawn sketches. Ginger Pletcher located references, handled correspondence and other tasks. Thomas L. Crystal coordinated the production during the last three years and was responsible for much of the typing and final corrections in Chapters 12-16. Our team performed admirably through continual changes, adding immeasurably to the appearance of the book, for which we are most grateful. The errors, of course, are our responsibility.

We acknowledge, with thanks, permission from authors and editors and publishers to draw on material published in journals and in books and modified to conform to our text style. A list of publishers follows:

Academic Press: *Journal of Computational Physics, Methods in Computational Physics, Computational Techniques*
American Institute of Physics: *Journal of Applied Physics, Physics of Fluids, Physical Review Letters*
Gordon and Breach
McGraw-Hill
Springer Verlag
Conferences on Numerical Simulation of Plasmas (starting in 1967, with the tenth held in 1983)
Government and University Laboratories

We have been most fortunate to work with patient and professional editors and associates at McGraw-Hill, notably B. J. Clark, David Damstra, Madelaine Eichberg, Diane Heiberg, and T. Michael Slaughter, who suffered through our pioneering task of producing camera-ready copy in Berkeley.

Charles K. (Ned) Birdsall
A. Bruce Langdon

ONE

PRIMER

ONE DIMENSIONAL
ELECTROSTATIC AND
ELECTROMAGNETIC CODES

This part of the book is truly a primer for plasma simulation using particles and is intended for use by those with some knowledge of plasmas and some ability in numerical methods and programming. However, the reader with no prior plasma or numerical experience may still profit from this part, using additional texts on plasmas.

These seven chapters have been used in teaching particle simulation for roughly a decade. The lectures follow the chapters as written. The student homework, however, begins in the first week with assignment of the projects of Chapter 5. Hence, students are actively involved in running a one-dimensional electrostatic code from the first day of class.

Chapter 1 is introductory, and is intended to make the reader feel comfortable with using a few hundred to a few thousand particles to simulate a laboratory plasma of perhaps 10^{14} to 10^{24} particles.

Chapter 2 presents some details of the one-dimensional electrostatic program ES1 in terms of the differential and difference equations to be solved. The algorithms, initial values, and diagnostics are described briefly.

Chapter 3 presents ES1 in detail and includes the listings of most parts of the program. Some of the problems are used as classroom quizzes.

Chapter 4 goes into the numerical details of the particle mover, the meaning of particle shapes as inferred from weighting from particle to grid and grid to particle, and the Poisson equation solver.

Chapter 5 presents a number of introductory projects on plasma oscillations, waves, and instabilities for students to run using ES1. Details are presented on choice of initial conditions, development of linear analysis to check against, and approximate nonlinear analysis. These projects start with cold plasmas which are very easy and inexpensive to run and to understand. Successful completion of most or all of these projects is essential as an introduction to professional plasma particle simulation. This chapter (and Chapter 7) are the laboratory parts of the course.

Chapter 6 describes a one-dimensional electromagnetic program EM1 which uses Maxwell's equations for the transverse fields. Chapter 7 presents several projects for the student to run using EM1.

While Part One may stand alone for a course in particle simulation, teachers and students are encouraged to read ahead in Part Two for more complete development of the theory of particle simulation, especially for insight into some of the results they obtain in the projects.

ONE

WHY ATTEMPTING TO DO PLASMA PHYSICS VIA COMPUTER SIMULATION USING PARTICLES MAKES GOOD PHYSICAL SENSE

The idea of obtaining more or less valid plasma physics by using a computer to follow charged particles occurred to a number of people, notably John Dawson at Princeton and Oscar Buneman at Cambridge, in the late 1950's and early 1960's. After all, there had been simulations following particles of electron beams in vacuum tubes all during the 1950's, so why not simulate plasmas? It simply was not clear that taking the step from cold beams with charges all of one sign to thermal distributions with essentially equal density of charges of opposite signs and greatly different masses would be successful.

Why not?

In simulating an electron beam, it is reasonable to think that the computations would be valid if some small number of particles like 16 or 32 is used per wavelength. These particles are usually disks with diameter of the beam and are followed by the computer (like buttons on a string, except that they are tenuous and allowed to overtake each other) during their interaction with microwave circuits (say, resonant cavities or slow-wave structures) over some five or ten wavelengths. A total of 10 times 16 or 32 particles are used and followed for, say, 10 to 20 cycles (a few hundred time steps) from linear modulation through to nonlinear saturation. Once computers became easily

3

programmable, these simulations became rather straightforward. These simulations succeed with few particles because the field of one electron acts on a large fraction of the electrons; the collective interaction length is comparable to the dimensions of the electronic device.

In considering simulating *neutral* plasmas, one first turns to the early pages of numerous plasma texts, where the Debye length is introduced. This length is numerically related to the *plasma frequency* $\omega_p \propto$ (density)$^{1/2}$, by

$$\lambda_D \equiv \frac{v_{thermal}}{\omega_p} \propto \left[\frac{\text{temperature}}{\text{density}} \right]^{1/2} \tag{1}$$

This is the distance traveled by a particle at the thermal velocity in $1/2\pi$ of a plasma cycle, the shielding distance around a test charge and the scale length inside which particle-particle effects occur most strongly and outside of which collective effects dominate. We also read the assertion that "plasmas of interest are many λ_D across, $L \gg \lambda_D$." In order to have many particles within a collective interaction length, as in the electronic devices, we must have a much larger number of particles in simulating a neutral plasma.

(It is assumed that our readers will have experience with plasmas through lecture and/or laboratory courses and will turn to plasma texts for help when needed. Although some elementary statements or definitions are given in the text, not all of the plasma physics required for our text will be developed; our text is aimed at complementing plasma texts.)

It may seem worse. In the texts we often find a chart of density (from about $n = 10^6/cm^3$ to $n = 10^{22}/cm^3$) versus energy (from about 0.01 eV to 10^5 eV) with lines of constant N_D, the number of particles in a Debye cube,

$$N_D \equiv n\lambda_D^3 \tag{2}$$

Although very dense plasmas can have small N_D, we see that alkali vapor plasmas have $N_D \approx 10^2$, the earth's ionosphere has $N_D \approx 10^4$, and magnetically confined fusion experiment plasmas have $N_D \approx 10^6$! A literal simulation of the latter experiments is inconceivable in the foreseeable future.

Hence, we might be tempted to stop and say that simulation is hopeless, requiring many orders of magnitude more particles than can be handled by any existing computer.

If we succumb to this argument, then we miss the general character of much of plasma behavior. First, we are often going to be interested in the *collective behavior of collisionless plasmas at wavelengths longer than the Debye length, $\lambda \gtrsim \lambda_D$.* Second, we can obtain much useful information in one and two dimensions where we need not satisfy the three-dimensional requirement, say, $N_D = 10^6$, but only the one-dimensional $N_D = n\lambda_D \approx (10^6)^{1/3} = 10^2$ or the two-dimensional $N_D = n\lambda_D^2 \approx (10^6)^{1/2} = 10^3$. Third, we can alter the simulation so that we may use even smaller N_D but keep the essential plasma behavior. *The models we will use are intended to produce the essence of the plasma, without all of the details.*

Let us examine more closely the rough statements just made. To be sure, a collisionless laboratory plasma is characterized by $N_D \gg 1$ and $L \gg \lambda_D$. However, the physical behavior of a plasma is one of electrons and ions moving in their Coulomb fields with sufficient kinetic energy to inhibit recombination. Hence, another characterization of a plasma is that

$$\frac{\text{Thermal kinetic energy (KE)}}{\text{Microscopic potential energy (PE)}} \gg 1 \qquad (3)$$

This ratio, for laboratory plasmas, is indeed N_D. However, the fundamental physics only requires KE $\gg$ PE, which may be quite satisfactory at N_D as low as 10, or which may be achieved by means other than requiring $N_D \gg 1$, if necessary. The second characterization, collisionless, is also tied to N_D through

$$\frac{\nu}{\omega_p} \approx \frac{1}{N_D}\ln N_D \ll 1$$

and may likewise either be acceptable for $N_D = 10$ or achieved by means other than $N_D \gg 1$. The third description, $L \gg \lambda_D$, can be realized, in many problems, by use of periodic models which look at a slice of an infinite plasma, and use, for example, $L = 50\lambda_D$. This paragraph says that we may choose to simulate warm plasmas at small values of N_D or small L/λ_D. Examination of the physics of cold plasmas (*e.g.*, Langmuir oscillations at $\omega = \pm\omega_p$) and of nonneutral plasmas (ion density not equal to electron density in zero order) also shows that simulation with particles is practical.

What are these "means other than $N_D \gg 1$"? We use several in our simulations. First, we seldom keep track of interactions down to infinitesimal charge separations. In 3d (shorthand for 3 dimensions) and 2d, the impact parameters are always relatively large even for $N_D \approx 10$. In 1d, collisions are of a different nature; also, in 1d, making $N_D = n\lambda_D$ large is not a difficulty. If we use a *spatial grid* in order to simplify the calculation of the fields, then fields and forces at lengths less than a grid cell are not observable and may be considered to be smoothed away; we can do even more *smoothing* by such means as dropping forces at short wavelengths (large wave vectors, $k = 2\pi/\lambda$), by dropping charge densities, potentials, and fields at large values of k.

Second, we may deliberately alter the physics in order to achieve either KE $\gg$ PE or $\nu \ll \omega_p$, or both. For example, laboratory particles are generally considered to be point particles, with forces between particles separated by distance r given by $1/r^2$, $1/r^1$, or $1/r^0$ in 3d, 2d, and 1d. These Coulomb singularities in 3d and 2d (there are none in 1d) may be removed by using *finite-sized particles*, a notion employed long before plasma simulation; for example, see *Vlasov* (1950), who calls his particles *clouds*, as we do also. The important results here are that:

(a) Finite-size particles occur naturally when we use a spatial grid.
(b) As the finite-size particle radius R is made comparable or larger than the Debye length $(R \geq \lambda_D)$, then the collision cross-section σ and collision frequency ν diminish rapidly relative to that of point particles, $R = 0$ in 2d and 3d.
(c) There is considerable latitude for invention in terms of particle and force weightings relative to the spatial grid in order to achieve results desired in any one simulation problem (applicable in one but perhaps not in another).

Nonetheless, we simulate with far fewer particles than are in the laboratory plasma of interest. As a result, we usually live with relatively higher collision rates, *e.g.*, $\nu \ll \omega_p$, but not $\nu \lll \omega_p$ (and altered collisional dependence on parameters), somewhat higher noise levels (our *superparticles* may have the same charge-to-mass ratio as in the lab but each q and m is much larger), and inability to examine all time and space scales (all frequencies ω and wave vectors k).

An alternative to simulation using particles is integration of the collisionless kinetic ("Vlasov") equation, which treats phase space as a continuum (which is also an approximation). This approach does avoid statistical errors present in particle simulation, and has been used successfully. Vlasov simulation has not so far proven to be as adaptable as particle simulation, especially in multidimensional problems, and untempermental, accurate, yet economical, representation of velocity space has been difficult in long-time simulations.

We simulate only over limited time and space and so can tolerate small errors. We even use mass ratios $m_{ion}/m_{electron}$ like 100 rather than 1836 (proton/electron) or sometimes freeze the motion of the neutralizing or background particles or put in a linearized susceptibility $\chi(\omega, k)$ say, for warm electrons. We find that the finite time and space gridding may itself produce waves, even instabilities, that are nonphysical and, hence, unwanted.

Indeed, we are accused of tomfoolery more than we deserve. We then simply admit to being in good company with the rest of plasma physics, with theorists and experimentalists who also have their kit bags of approximate (and occasionally inaccurate) tools.

The point is, within all three branches of plasma study (experiment, theory, computation), practitioners must exercise a great deal of care, enough to obtain the essence of the problem, but not so much as to inhibit achieving any result. We need only do well enough. This text is intended to provide insight into commonly used methods, to offer practice problems and projects and to give the reader a starting place for doing his own simulations.

Hockney and Eastwood (1981, chapter 1) provide similar and elegant reasoning for performing computer experiments using particle models as applied to plasma physics and to a variety of other physical areas.

TWO

OVERALL VIEW OF A ONE-DIMENSIONAL ELECTROSTATIC PROGRAM

2-1 INTRODUCTION

In this chapter, without becoming overly involved in the details of the various parts, we present a brief once-through of a particle simulation program. At the same time, we hope that the reader does not think that the whole program is too elementary; we will take up improvements, subtleties, and alternatives later.

Our procedure is to follow a widely-used program ES1, an ElectroStatic 1-dimensional program, complete with initial conditions and diagnostics to tell us about the physics, as well as check on the numerics. This is preparatory to running the program from tens to thousands of time steps on trial runs. We may find out that we were not as wise as we thought on the first run; hence, we will modify the program, improve the initial conditions, add some new diagnostics and try again. After a few such go-arounds as this, we may have what we started out after, the essence of the physics. Of course, along the way, we probably will find that we want more solutions to the dispersion equation for the waves that we were studying, or better estimates of the nonlinear behavior to be observed, or other information.

Our initial system for study has charged particles in both self and applied electrostatic fields and an applied magnetic field. This initial use of an electrostatic model follows the historical development of plasma particle simulation, is perhaps the easiest of all models to understand, and also leads directly into fully electromagnetic models.

We do this in a spirit of complementing theory or experiment, in order to observe new phenomena, or to understand what has been predicted or observed in the laboratory. It is wise to remain coupled to both theory and experiment.

2-2 THE ELECTROSTATIC MODEL: GENERAL REMARKS

The model (Figure 2-2a) consists of charged particles moving about due to forces of their own and applied fields. The physics comes from two parts, the fields produced by the particles and the motion produced by the forces (or fields). The fields are calculated from *Maxwell's equations* by knowing the positions of all of the particles and their velocities; the forces on the particles are found using the electric and magnetic fields in the *Newton-Lorentz* equation of motion. One calculates the fields from the initial charge and current densities, then moves the particles (small distances) and recalculates the fields due to the particles at their new positions and velocities; this procedure is repeated for many time steps.

The difference from a laboratory plasma is that simulations proceed discontinuously in time *step by step,* using *digital* rather than analog computation. We must show care in developing *numerical methods* that provide sufficient *accuracy and stability* to make the simulations useful through many characteristic cycles of the plasma, whether these be plasma, cyclotron, or hybrid (or whatever) periods of the ions or electrons. We use a *temporal grid*

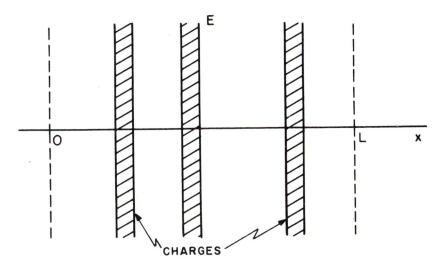

Figure 2-2a A one-dimensional model, consisting of many sheet charges, with self and applied electric fields **E** directed along the coordinate x. There are no variations in y or z.

which is sufficiently fine grained to follow the plasma with acceptable accuracy and stability.

A second difference is the use of a *spatial grid* on which the fields will be calculated. One might ask, why not use *Coulomb's law* directly (for forces between charges separated by distance r, where the force decays as $1/r^2$, $1/r^1$, $1/r^0$ in three, two, and one dimensions)? Consider the calculation of these forces in terms of both the numerical operations required and in terms of the actual physics; *i.e.*, are we interested in the details of close encounters among particles and are close encounters at all frequent? The answer to both parts of the question is almost always no. We recall our first course on electromagnetic theory where we met Coulomb's law and went quickly on to the notion of an *electric field* $\mathbf{E}$ and found that $\mathbf{E}$ was seldom to be found by summing the effect of each individual charge. Instead, we were introduced to the notion of a *charge density ρ*, and that $\mathbf{E}$ was to be obtained from this density, which was to be thought of as varying continuously in space. The idea of working with something like 10^{25} charges with as many calculations of $1/r^2$ was dismissed, with relief; the problem of what to do with 10^{25} singularities (at $r \to 0$) also vanished. Nearly all of the plasma physics which we will do requires knowledge only down to some scale length, with charge density (and current density) considered continuous; the finer grained behavior is omitted. Furthermore, in plasmas, with many particles in a characteristic length (which is the *Debye length),* there are relatively few close particle encounters, that is, few large angle deflections from single encounters; rather large deflections come mostly from the cumulative effect of many small deflections. Hence, we are encouraged by the nature of plasma problems to take advantage of the simplifications that come about in using a *mathematical spatial grid,* as shown in Figure 2-2b, usually fine enough to resolve a *Debye length,* in order to measure the charge density and, thence, calculate the electric field $\mathbf{E}$.

There are some exceptions to these generalities. For example, one may calculate the electric field in one-dimensional problems relatively easily without using a spatial grid. However, the grid provides a *smoothing effect* by not resolving spatial fluctuations that are smaller than the grid size; an exact field calculation would keep everything, which is usually more than we want.

The use of temporal and spatial grids, which are mathematical and not physical, causes concern about accuracy and may create what we will term *nonphysics.* However, we only touch on such in Part One, and develop this material later in Part Two. Suffice it to say: the possibility of nonphysical effects may restrict our choices of parameters on occasion, but generally these effects can be avoided; inaccuracies will always be with us and simply must be made small.

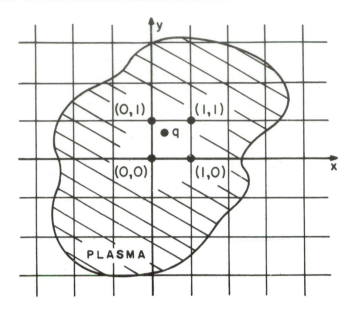

Figure 2-2b A mathematical grid is set into the plasma region in order to measure charge and current densities ρ, J; from these we will obtain the electric and magnetic fields $\mathbf{E}$, $\mathbf{B}$ on the grid. A charged particle q at (x,y) will typically be counted in terms of ρ at the nearby grid points $(0,0)$, $(1,0)$, $(1,1)$, $(0,1)$ and in terms of J at the faces between these points. The force on q will also be obtained from the fields at these nearby points.

PROBLEMS

2-2a Sketch the electric field $E(x)$ versus x for the 1d model of Figure 2-2a, for various boundary conditions, such as (a) potential equal to zero at $x = 0$, $x = L$, or (b) the system is periodic, with period of L, or (c) there is an applied potential difference [say, $\phi(0) = 0$, $\phi(L) = V_0$]. Let the charges have some thickness as shown in Figure 2-2a to show E within the charge. Consider neutral and nonneutral systems (equal or nonequal numbers of positive and negative charges). Consider a stationary uniform background of charge of one sign and sheet charges of another sign, with net neutrality; show $\rho(x)$, $E(x)$, and $\phi(x)$.

2-2b Let the charge density of a sheet extend from x_a to x_b, and be zero outside these values. Show that the electrostatic force on the sheet, for arbitrary charge distribution within the sheet [that is, $\rho(x)$ is arbitrary between x_a and x_b], is $q(E(x_a) + E(x_b))/2$, where q is the total charge per unit area of the charged sheet. *Hint*: Consider the integral $\int \rho E\,dx$ $= \int \frac{\partial}{\partial x}(E^2/2\epsilon_0)\,dx$ across the sheet. This simple and ancient result is a "gift" of the one-dimensional model, one of several that we will use to advantage. (See, *e.g.*, *Portis*, 1978, p. 68, for force on charged sheet and p. 324 for force on a current sheet.)

2-2c The text states that the fields can be found exactly analytically in one dimension so that a spatial grid is not needed solely for this purpose. Suppose that we used sheets of zero thickness and found the fields exactly at time t and then again at time $t + \Delta t$, etc.; however, these fields change only when two (or more) particles cross which may be during Δt, leading to error. What corrections would you suggest to account for such crossing(s) during Δt? (See *Dawson*, 1970, p. 4-10, for discussion of sheet crossings.) Explain why a gridded model with finite-size particles has less of a problem in accounting for crossings.

2-3 THE COMPUTATIONAL CYCLE: GENERAL REMARKS

At each step in time, the program solves for the fields from the particles and then moves the particles; this cycle is shown in Figure 2-3a. There may be tens of steps in a characteristic period of the plasma and there may be tens of periods in a typical run, which adds up to hundreds or thousands of time steps in a given run.

The cycle starts at $t = 0$, with some appropriate initial conditions on the particle positions and velocities. The computer runs to the number of time steps it is told. Various *diagnostics* are printed out at the end of the run; some are in the form of *snapshots* at particular times, such as densities or fields or velocity distributions; some are in the form of *time histories,* such as energy versus time. These graphs are the record from which one obtains the physics of the simulation. Numbers per se are very seldom the object of a plasma simulation.

The particles are processed through the boxes shown in Figure 2-3a, much as the fields and forces are created in the actual plasma. Let us follow a cycle by starting from initial values of positions and velocities. Keep in mind that hundreds or thousands (or 10^6 and up in two and three dimensions) of particles are being processed.

The particle quantities, such as velocity and position, are known at the particle and may take on all values in $\mathbf{v}$ and $\mathbf{x}$ space, called *phase space*. The name of the particle is given by index i, such as v_i and x_i. The field quantities will be obtained only on the spatial grid, known only at discrete points in space labeled with index j such as E_j. The ties from the particle position

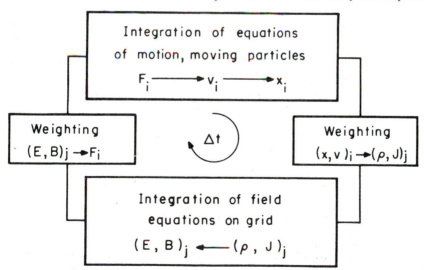

Figure 2-3a A typical cycle, one time step, in a particle simulation program. The particles are numbered $i = 1, 2, \ldots,$ NP; the grid indices are j, which become vectors in 2 and 3 dimensions.

and velocities to the field quantities are made by first calculating the charge and current densities on the grid; this step requires stating how to produce the grid densities from the particle positions and velocities. This process of charge and current assignment implies some *weighting* to the grid points that is dependent on particle position. Once the densities are established on the grid, then we will use various methods to obtain the electric and magnetic fields. With the fields known on the grid, but with the particles scattered around within the grid, we interpolate the fields from the grid to the particles in order to apply the force *at the particle* by again performing a weighting.

How do we distinguish particles? What information is stored for the particles and for the fields? The particles may be known by the way in which they are stored in the computer memory; they may be in some kind of order, with only their present velocity and position stored ($\mathbf{v}_i$, $\mathbf{x}_i$); their values of charge q_i and mass m_i may be put elsewhere (*e.g.*, with only two species, electrons and ions, q and m would change only once in running all of the particles through in a time step). The fields are known at the grid points and are stored, probably indexed in an array, so that they can be recalled readily. There will almost always be *many more particles than grid points,* and external storage allows use of more particles than might fit into the computer's fast memory, along with the field quantities. Since the particles are integrated independently, only a few need be in fast memory at a time. The field quantities probably will be retained in fast memory, as in most present methods they have to be recalled randomly.

There are many variations of the cycle shown. For example, the electrons (with relatively large ω_{pe}, ω_{ce}, ω_{uh}) may be advanced at a relatively small Δt_e; the ions (with much smaller ω_{pi}, ω_{ci}, ω_{lh}) may be advanced with a relatively large Δt_i; the fields may be obtained on yet a third scale, Δt_f, possibly relatively short for electromagnetic fields (waves move faster than particles) or relatively large for observing low frequency effects.

2-4 INTEGRATION OF THE EQUATIONS OF MOTION

A problem may call for 10,000 particles to be run for 1000 time steps. This means that the equations of motion must be integrated $10,000 \times 1000 = 10^7$ times. We want to use as fast a method as possible, and still retain acceptable accuracy. The time per particle per step currently is on the order of microseconds. Explicitly, a 10^4 particle program of 10^3 time steps running at 10 μsec/particle/step on a machine with a 720 dollar/hour charge would cost 20 dollars for the 100-second run.

In addition, our choice of method must take into account the storage capability of the computer we will use in terms of the number of quantities that may be kept for each particle. If we were to follow just the trajectory of one particle through given fields (so-called *trajectory calculations*), then we probably would choose to keep velocity and position information from several previous time steps and use a method of time integration with a high order of accuracy. However, the minimum information needed for integration is the particle velocity and position, two words per particle or 20,000 words for our problem. Storing the v_i, x_i for several previous time steps would multiply this number. Using a high-order method (*e.g.*, Runge-Kutta) would multiply the operations taken for each particle. Hence, we nearly always will choose to use the least information (storage) and the fastest method we can.

One commonly used integration is called the *leap-frog method*. The two first-order differential equations to be integrated separately for each particle are

$$m\frac{d\mathbf{v}}{dt} = \mathbf{F} \tag{1}$$

$$\frac{d\mathbf{x}}{dt} = \mathbf{v} \tag{2}$$

where $\mathbf{F}$ is the force. These equations are replaced by the *finite-difference equations*

$$m\frac{\mathbf{v}_{new} - \mathbf{v}_{old}}{\Delta t} = \mathbf{F}_{old} \tag{3}$$

$$\frac{\mathbf{x}_{new} - \mathbf{x}_{old}}{\Delta t} = \mathbf{v}_{new} \tag{4}$$

The flow in time and notation is shown in Figure 2-4a, which makes clear the *time centering*. The computer will advance $\mathbf{v}_t$ and $\mathbf{x}_t$ to $\mathbf{v}_{t+\Delta t}$, $\mathbf{x}_{t+\Delta t}$, even though $\mathbf{v}$ and $\mathbf{x}$ are not known at the same time. The user must show care in at least two ways: first, initial conditions for particle velocities and

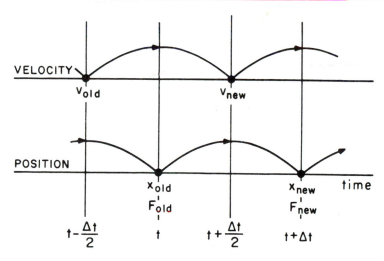

Figure 2-4a Sketch of leap-frog integration method showing time-centering of force $\mathbf{F}$ while advancing $\mathbf{v}$, and of $\mathbf{v}$ while advancing $\mathbf{x}$.

positions given at $t = 0$ must be changed to fit in the flow; we push $\mathbf{v}(0)$ back to $\mathbf{v}(-\Delta t/2)$ using the force $\mathbf{F}$ calculated at $t = 0$; second, the energies calculated from $\mathbf{v}$ (kinetic) and $\mathbf{x}$ (potential, or field) must be adjusted to appear at the same time.

The leap-frog method has error, with the error vanishing as $\Delta t \rightarrow 0$. We will use a leap-frog integrator in nearly all of our programs, because it is both simple (easy to understand, and with minimum storage) and surprisingly accurate (as shown later). Applying this method to integration of a simple harmonic oscillator of radian frequency ω_0 (in a later section), we will find that there is *no* amplitude error for $\omega_0 \Delta t \leqslant 2$ and that the phase advance for one step is given by

$$\omega_0 \Delta t + \frac{1}{24}(\omega_0 \Delta t)^3 + \text{higher-order error terms} \tag{5}$$

The error terms dictate a choice of $\omega_0 \Delta t \leqslant 0.3$ in order to observe oscillations or waves for some tens of cycles with acceptable accuracy.

The force $\mathbf{F}$ has two parts,

$$\mathbf{F} = \mathbf{F}_{\text{electric}} + \mathbf{F}_{\text{magnetic}} \tag{6}$$

$$= q\mathbf{E} + q(\mathbf{v} \times \mathbf{B}) \tag{7}$$

Here the electric field $\mathbf{E}$ and magnetic field $\mathbf{B}$ are to be calculated *at the particle*. Hence, using a spatial grid, we must interpolate $\mathbf{E}$ and $\mathbf{B}$ from the grid to the particle; we will do this in the same way as we determined the charge density (at the grid points) from the particle positions, a point discussed in Section 2-6. As we will see later, the electric force on a particle will depend not only on the distance to other particles (physical) but also on the position within the cell (nonphysical).

For our one dimensional purpose now, let us consider the particle displacement to be along x, and that we have velocities v_x and v_y, with a uniform static magnetic field B_0, along z. The $q(\mathbf{v} \times \mathbf{B})$ force, as shown in Figure 2-4b, is simply a *rotation* of $\mathbf{v}$; that is, $\mathbf{v}$ does not change in magnitude. However, the $q\mathbf{E} = \hat{x}qE_x$ force does alter the magnitude of $\mathbf{v}$ (v_x, that is); $E_y = 0$. Hence, a physically reasonable scheme which is centered in time is as follows (with t' and t'' as dummy variables, $t - \Delta t/2 < t' < t'' < t + \Delta t/2$):

Half acceleration

$$v_x(t') = v_x\left[t - \frac{\Delta t}{2}\right] + \left[\frac{q}{m}\right]E_x(t)\left[\frac{\Delta t}{2}\right] \tag{8}$$

$$v_y(t') = v_y\left[t - \frac{\Delta t}{2}\right]$$

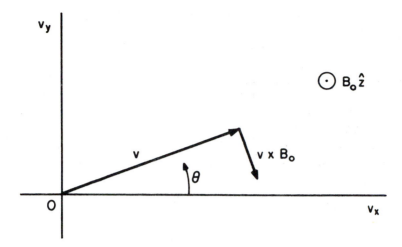

Figure 2-4b The v_x, v_y plane, showing the $q(\mathbf{v} \times \mathbf{B})$ force normal to $\mathbf{v}$, which results in a rotation of $\mathbf{v}$, with no change in speed with $\dot{\theta} < 0$ for $(q/m) > 0$, $B_0 > 0$.

Rotation

$$\begin{pmatrix} v_x(t'') \\ v_y(t'') \end{pmatrix} = \begin{pmatrix} \cos \omega_c \Delta t & \sin \omega_c \Delta t \\ -\sin \omega_c \Delta t & \cos \omega_c \Delta t \end{pmatrix} \begin{pmatrix} v_x(t') \\ v_y(t') \end{pmatrix} \tag{9}$$

Half acceleration

$$v_x\left(t + \frac{\Delta t}{2}\right) = v_x(t'') + \left(\frac{q}{m}\right) E_x(t) \left(\frac{\Delta t}{2}\right) \tag{10}$$

$$v_y\left(t + \frac{\Delta t}{2}\right) = v_y(t'')$$

The angle of rotation is

$$\Delta\theta = -\omega_c \Delta t \quad i.e. \quad \dot{\theta} = -\omega_c \tag{11}$$

as desired, where the _cyclotron frequency_ (radians/sec) as defined by

$$\omega_c \equiv \left(\frac{q}{m}\right) B_0 \tag{12}$$

which carries the sign of q and of B_0. This scheme (*Boris*, 1970b) is elaborated upon in Chapter 4.

One complication arises at $t = 0$ when the initial conditions, $\mathbf{x}(0)$ and $\mathbf{v}(0)$, are given at the same time. The main loop runs with $\mathbf{x}$ leading $\mathbf{v}$ by $\Delta t/2$. Hence, at the start, $\mathbf{v}(0)$ is moved backward to $\mathbf{v}(-\Delta t/2)$, by first rotating $\mathbf{v}(0)$ through the angle $\Delta\theta = +\omega_c \Delta t/2$ and then applying a half acceleration using $-\Delta t/2$ (really a deceleration?) based on $\mathbf{E}(0)$ obtained from $\mathbf{x}(0)$; see Sec. 2-7 and Figure 2-7a.

2-5 INTEGRATION OF THE FIELD EQUATIONS

Starting from the charge and current densities as assigned to the grid-points, we now obtain the electric and magnetic fields, in general, from Maxwell's equations, using ρ and J as sources. Here we take this step for an electrostatic problem (meaning $\nabla \times \mathbf{E} = -\partial \mathbf{B}/\partial t \approx 0$ so that $\mathbf{E} = -\nabla \phi$) in one dimension x.

The differential equations to be solved are

$$\mathbf{E} = -\nabla \phi \quad \text{or} \quad E_x = -\frac{\partial \phi}{\partial x} \tag{1}$$

$$\nabla \cdot \mathbf{E} = \frac{\rho}{\epsilon_0} \quad \text{or} \quad \frac{\partial E_x}{\partial x} = \frac{\rho}{\epsilon_0} \tag{2}$$

which are combined to obtain *Poisson's equation*

$$\nabla^2 \phi = -\frac{\rho}{\epsilon_0} \quad \text{or} \quad \frac{\partial^2 \phi}{\partial x^2} = -\frac{\rho}{\epsilon_0} \tag{3}$$

One approach is to solve the finite difference equations of (1) and (3), using the grid shown in Figure 2-5a, as

$$E_j = \frac{\phi_{j-1} - \phi_{j+1}}{2\Delta x} \tag{4}$$

$$\frac{\phi_{j-1} - 2\phi_j + \phi_{j+1}}{(\Delta x)^2} = -\frac{\rho_j}{\epsilon_0} \tag{5}$$

This last may be written in matrix form as

$$\mathbf{A}\phi = -\frac{(\Delta x)^2}{\epsilon_0}\rho \tag{6}$$

We are to use the ρ_j's known from the x_i's, to obtain the unknown ϕ_j's and then the E_j's, for j running from 0 to $L/\Delta x$ (roughly) where L is the length of the system of NG points. By using the known boundary conditions

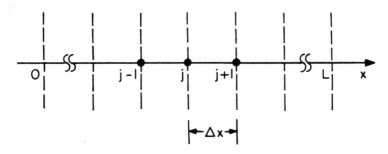

Figure 2-5a One dimensional numerical grid, with grid planes located at $X_j = j\Delta x$, uniformly spaced. The charge density ρ, the potential ϕ, and the electric field E_x will be obtained only at the X_j's.

at $x = 0$, L and all of the ρ_j there will be as many equations as unknowns (the ϕ); hence, the problem is solvable. There are numerous methods of solution. These are taken up in Appendix D.

A very powerful approach for *periodic systems* is to use a discrete *Fourier series* for all grid quantities. This approach also provides spatial spectral information on ρ, ϕ, and E which is useful in relating results to plasma theory, and which also allows control (smoothing) over the spectrum of field quantities. We now present this method in some detail; it is the basis for the field solver in our code ES1.

The ability to Fourier transform efficiently through use of the *fast Fourier transform FFT*, an invention of the 1960's, is a major enabling factor in this computation. The key to the solution is the assumption that, in the problems we attack, $\rho(x)$ and $\phi(x)$ have Fourier transforms, $\rho(k)$ and $\phi(k)$, where $\mathbf{k}$ is the wave vector in the Fourier transform kernel, $\exp(-i\mathbf{k}\cdot\mathbf{x})$. (We are implying certain boundary conditions by this step, such as periodic, which are considered in detail later.) This assumption allows us to obtain $\phi(k)$ from $\rho(k)$ in Poisson's differential equation directly as, in one dimension, $\partial^2/\partial x^2$ is replaced by $-k^2$; that is,

$$\phi(k) = \frac{\rho(k)}{\epsilon_0 k^2} \tag{7}$$

The next step is to take the inverse Fourier transform of $\phi(k)$ in order to obtain $\phi(x)$ and then $E(x)$ using (2). The overall sequence is given in Figure 2-5b. (The reader will quickly note that one may go directly to $E(k)$; we go into that possibility later.) The hitch in proceeding in this manner is that we have a finite Fourier series; we have $\rho(x)$ only at the X_j's, at NG points.

The solution using a finite Fourier series starts from the charge densities at the grid points, with values $\rho(X_j)$, $j = 0, 1, 2, \cdots$, NG $- 1$ for a total of NG values. Letting the grid functions $G(X_j)$ (standing for field or potential or charge density) be periodic, $G(X_j) = G(X_j + L)$, then the finite discrete Fourier transform is (sum on $X_j = j\Delta x$)

$$G(k) = \Delta x \sum_{j=0}^{NG-1} G(X_j) e^{-ikX_j} \tag{8}$$

The inverse transform is [the sum is on $k = n(2\pi/L)$]

$$G(X_j) = \frac{1}{L} \sum_{n=-NG/2}^{NG/2-1} G(k) e^{ikX_j} \tag{9}$$

which produces NG distinct values of $G(X_j)$. (The virtue of the FFT is that

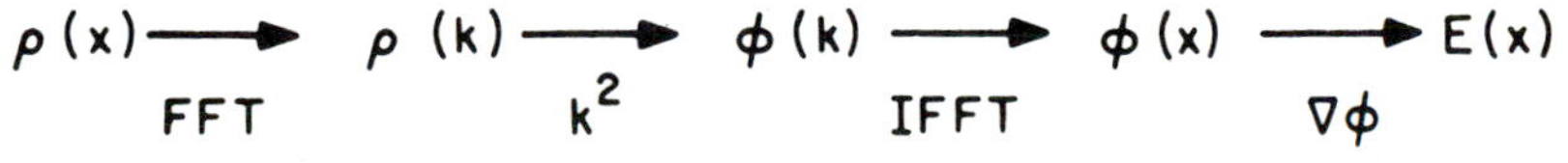

$$\rho(x) \xrightarrow[\text{FFT}]{} \rho(k) \xrightarrow[k^2]{} \phi(k) \xrightarrow[\text{IFFT}]{} \phi(x) \xrightarrow[\nabla\phi]{} E(x)$$

Figure 2-5b A possible sequence for solving Poisson's equation using the fast Fourier transform (FFT) and its inverse (IFFT).

it performs the sums rapidly.) Using the above series for $\rho(X_j)$, $\phi(X_j)$, and $E(X_j)$ with the particular set of finite difference equations chosen from (4) and (5), we obtain

$$E(k) = -i\kappa\phi(k) \tag{10}$$

where

$$\kappa = k\left[\frac{\sin(k\Delta x)}{k\Delta x}\right] = k\,\mathrm{dif}\,(k\Delta x) \tag{11}$$

where (from diffraction theory)

$$\mathrm{dif}\,\theta \equiv \frac{\sin\theta}{\theta}$$

and

$$\phi(k) = \frac{\rho(k)}{\epsilon_0 K^2} \tag{12}$$

where

$$K^2 = k^2\left[\frac{\sin\dfrac{k\Delta x}{2}}{\dfrac{k\Delta x}{2}}\right]^2 = k^2\mathrm{dif}^2(\frac{k\Delta x}{2}) \tag{13}$$

The finite difference terms, κ and K^2, approach the differential equation result, k and k^2, as the grid becomes finer, $k\Delta x \to 0$. The overall understanding of the role of the spatial grid, both in accuracy (*i.e.*, κ versus k, K^2 versus k^2) and in *aliasing* (the dynamics creates effects at $|k\Delta x| > \pi$ which are read falsely by the grid, being placed at $|k\Delta x| < \pi$) will be covered in detail in Part Two on the theory of simulation. A discussion on the difference between using k, k^2 and κ, K^2 is given in an appendix.

With the understanding that NG values of $\rho(X_j)$ are transformed to NG values of $\rho(k)$ and so on through to obtain NG values of $E(X_j)$, the solution sequence of Figure 2-5b still holds; this is commonly done by one Fourier transform, a division by K^2, an inverse transform and, lastly, a gradient operation [by finite differencing (4)] on $\phi(X_j)$ to obtain $E(X_j)$. There are NG values of k starting from $k_{\mathrm{minimum}} = 2\pi/L$ (L is the length of the system; see Figure 2-2a) up to $k_{\mathrm{maximum}} = \pi/\Delta x$ and their negatives; or, in terms of wavelengths, $\lambda_{\mathrm{maximum}} = L$ and $\lambda_{\mathrm{minimum}} = 2\Delta x$.

PROBLEMS

2-5a $\nabla\cdot\mathbf{E} = \rho/\epsilon_0$ may also be solved for $\mathbf{E}$ by Fourier methods. In 1d, obviously $E_x(k) = \rho(k)/ik\epsilon_0$. In 3d, electrostatic still, one must add $\nabla\times\mathbf{E} = 0$. Show that these equations produce solutions

$$E(\mathbf{k}) = \frac{\mathbf{k}\rho(\mathbf{k})}{ik^2\epsilon_0}$$

where $k^2 = k_x^2 + k_y^2 + k_z^2$.

2-5b Show that (9) is the inverse of (8). Show that $G(X_{j+NG})$, as given by (9) is equal to $G(X_j)$. Show that (9) is unchanged if the sum is taken from 0 to $NG - 1$.

2-5c When $G(X_j)$ is real, there is redundancy in the NG complex values of $G(k)$ in (8). Show that $G(-k) = G^*(k)$, and that there are $NG/2 + 1$ independent values of Re $G(k)$ (cosine coefficients) and $NG/2 - 1$ independent values of Im $G(k)$ (sine coefficients).

2-5d If $G(X_j)$ and $H(X_j)$ are both real, show that cosine and sine coefficients of G and H can be extracted from the transform [using (8)] of the complex sequence $G(X_j) + iH(X_j)$. This is accomplished by calls to subroutines CPFT and RPFT2, and the inverse by calls to RPFTI2 and CPFT (Appendix A).

2-5e Obtain κ in (11) and K^2 in (13) by inserting $E(X_j)$, $\phi(X_j)$, and $\rho(X_j)$ in the transform form (9) into the finite difference forms, (4), (5). Sketch (κ/k) and (κ^2/K^2) as functions of $k\Delta x$ from $k\Delta x = 0$ to π.

2-6 PARTICLE AND FORCE WEIGHTING; CONNECTION BETWEEN GRID AND PARTICLE QUANTITIES

It is necessary to calculate the charge density on the discrete grid points from the continuous particle positions and (after the fields are obtained), to calculate the force at the particles from the fields on the grid points. In Figure 2-3a, these calculations are called weighting, which implies some form of interpolation among the grid points nearest the particle. As shown later, it is desirable to use the same weighting in both density and force calculations in order to avoid a self-force (*i.e.*, a particle accelerates itself).

In *zero-order weighting,* we simply count the number of particles within distance $\pm\Delta x/2$ (one cell width) about the jth grid point and assign that number [call it $N(j)$] to that point; that is, the grid density (in one dimension) is simply $n_j = N(j)/\Delta x$. This is illustrated in Figure 2-6a(a). The common name for this weighting is *nearest-grid-point* or NGP. Computationally, the counting is fast, since only one grid look-up is done. The electric field to be used in the force is that at X_j, for all particles in the jth cell.

As a particle moves into the jth cell (through cell boundaries at $x = X_j \pm \Delta x/2$), the grid density due to that particle jumps up; as the particle moves out ($x > X_j + \Delta x/2$ or $x < X_j - \Delta x/2$), the grid density jumps down. The density observed at j is shown in Figure 2-6a(b). We see two effects here. One is that the particle appears to have a *rectangular shape* with a width of Δx. This leads us (*i.e.*, the grid) to think that we have a collection of *finite-size particles;* hence, the physics observed will be that of such particles rather than that of point particles. Because close encounters

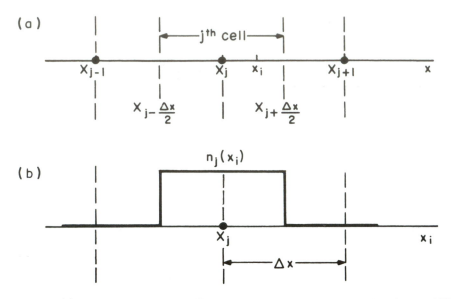

Figure 2-6a (a) Zero-order particle and field weighting, also called nearest-grid-point, or, NGP. Particles in the jth cell, that is, with positions $x_i \epsilon X_j \pm \Delta x/2$, are assigned to X_j to obtain grid density $n(X_j)$. All of these particles are acted on by the field at X_j, $E(X_j)$. (b) The density $n_j(X_i)$ at point X_j due to a particle at x_i, as the particle moves through the cell centered on X_j. This density may be interpreted as the effective particle shape.

between plasma particles are rare (*i.e.*, for many particles in a Debye length, $N_D \gg 1$, virtually all collisions are at large impact parameter), this new physics hardly alters the basic plasma effects to be studied. The second effect is that the jumps up and down as a particle passes through a cell boundary will produce a density and an electric field which are relatively *noisy* both in space and time; this noise may be intolerable in many plasma problems. Thus, we look for a better weighting.

First-order weighting smooths the density and field fluctuations, which reduces the noise (relative to zero-order weighting), but requires additional expense in accessing two grid points for each particle, twice per step. We may view this step either as an improvement in using finite-size particles or as one of better interpolation. The charged particles seem to be finite-size rigid *clouds* which may pass freely through each other. We call the model *cloud-in-cell* or CIC (*Birdsall and Fuss*, 1969). If we take the *nominal cloud* to be of uniform density and to be Δx wide as shown in Figure 2-6b(a) (the so-called square cloud), then the grid assignment is self-evident, using NGP for each element. That is, for total cloud charge of q_c, the part assigned to j is

$$q_j = q_c \left[\frac{\Delta x - (x_i - X_j)}{\Delta x} \right] = q_c \frac{X_{j+1} - x_i}{\Delta x} \tag{1}$$

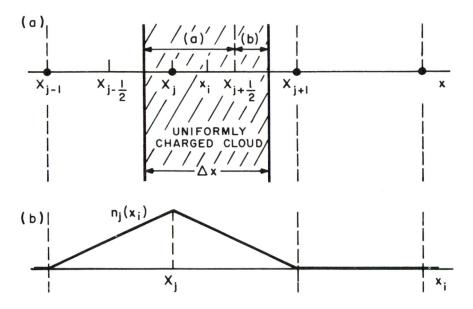

Figure 2-6b (a) First-order particle weighting, or cloud-in-cell model CIC. The nominal finite-size charged particle, or cloud, is one cell wide, with center at x_i. This weighting puts that part of the cloud which is in the jth cell at X_j, fraction (a), and that part which is in the $(j + 1)^{th}$ cell at X_{j+1}, fraction (b). This weighting is the same as applying NGP interpolation to each elemental part. (b) The grid density $n_j(x_i)$ at point x_i as the particle moves past X_j, again displaying the effective particle shape $S(x)$.

and the part assigned to $j + 1$ is

$$q_{j+1} = q_c \left[\frac{x_i - X_j}{\Delta x} \right] \tag{2}$$

The net effect is to produce a *triangular* particle shape $S(x)$ which has width $2\Delta x$. In computation, the nearest left-hand grid point j is located first, so that $x_i > X_j$ always; then the weights are calculated and the charges assigned. Note that assignment of a point charge at x_i to its nearest grid points by linear interpolation would produce the same result; this viewpoint is called *particle-in-cell,* or PIC modeling. As a cloud moves through the grid, it contributes to density much more smoothly than with zero-order weight, as seen from Figure 2-6b(b); hence, the resultant plasma density and field will have much less noise and be more acceptable for most plasma simulation problems.

Higher-order weighting by use of quadratic and cubic *splines* rounds off further the roughness in particle shape and reduces density and field noise, but at the cost of more computation. The use of splines for higher-order weighting is discussed later. Also, the effective particle shape may be altered during the field calculation after the charge density $\rho(x)$ is Fourier transformed to $\rho(k)$, *e.g.*, by cutting off $\rho(k)$ at some k_{last} or multiplying

$\rho(k)$ by, say, $\exp[-(k/k_{\text{last}})^n]$.

The *field or force weighting* operates in the same manner. The NGP force comes from the field at the nearest grid point. The *first-order force* (CIC-PIC) comes from linear interpolation, exactly as in the charge assignment; for a particle at x_i,

$$E(x_i) = \left[\frac{X_{j+1} - x_i}{\Delta x}\right] E_j + \left[\frac{x_i - X_j}{\Delta x}\right] E_{j+1} \tag{3}$$

The first-order weighting consumes more computer time per particle than does the zero-order weighting; however, for a given noise level, CIC allows both a coarser grid and fewer particles than NGP, and thus regains some of the additional computer time required per particle. We will also see that higher-order weighting and smoothing tend to reduce nonphysical effects.

PROBLEM

2-6a Show that second-order Lagrangian interpolation results in discontinuities. (Splines employ a piecewise-polynomial representation also, but the segments are joined smoothly.)

2-7 CHOICE OF INITIAL VALUES; GENERAL REMARKS

We have presented some ideas on how the computational cycle functions. Now we will say something about initiating the program. Let us suppose that the problem of interest has been looked at thoroughly so that we have done the design work of choosing:

(a) The numbers of particles and grid cells
(b) The weighting and smoothing
(c) The desired initial distribution function, $f(\mathbf{x}, \mathbf{v}, t = 0)$ including the initial perturbation (if any, random or ordered)

The next step is to place the particles in phase space $(\mathbf{x}, \mathbf{v})$ so that the problem desired is properly set up to run.

A cold, uniform periodic plasma of mobile electrons and immobile ions $(m_i/m_e \to \infty)$ is simplest. The electrons are put in uniformly, one or more per cell. The charge density and or field solver automatically puts in a uniform neutralizing ion background by having the $k = 0$ component of ρ and E vanish. The plasma wave may be excited by perturbing the uniform positions x_{i0} by

$$x_i(t = 0) = x_{i0} + x_{i1}\cos(k_s x_{i0})$$

where $k_{\text{min}} \leqslant k_s \leqslant k_{\text{max}}$ is some wave vector for which we want the plasma behavior. These x's and the appropriate velocities (here, all zero) are put into the $t = 0$ step. The program then finds the fields at $t = 0$ from which the velocities at $t = -\Delta t/2$ are found (only used once in a run) (Section 2-4 and Figure 2-7a). Then the cycle proceeds forward by advancing $v(-\Delta t/2)$ to $v(+\Delta t/2)$, then $x(0)$ to $x(\Delta t)$, and so on for as many steps as desired.

Changing to a warm plasma requires that each particle be given a velocity **v** such that over some specified region (perhaps several cells, the shortest λ kept), the desired velocity distribution of velocities is well approximated. Suppose that the desired velocity distribution is flat, from $-v_{max}$ to $+v_{max}$ as shown in Figure 2-7b(a). Then, where the cold model had, say, four particles per cell with zero velocity, we now may split each one (figuratively) into four parts, placed for example, as shown in Figure 2-7b(b). The overall results, cold and warm, are shown in Figure 2-7c. One can improve on these placements and usually must do so [e.g., to avoid the potential multibeam instability of the approximation in Figure 2-7b and Figure 2-7c to $f(v)$ of Figure 2-7b(a)]. Such improvements will be taken up later, as well as prescriptions for loading less-simple $f(\mathbf{x},\mathbf{v})$ such as Gaussian in $\mathbf{v}_i$ (a Maxwellian) and nonuniform in $\mathbf{x}_i$ are taken up in Chapter 16.

PROBLEMS

2-7a Show that the initial half-step backward has the same-order accuracy as the main program mover, that is, to order $(\Delta t)^2$.

2-7b Provide a sketch showing how the half-step sequence can be used after the main program has been started in order to obtain **x** and **v** at the same time; i.e., go from $\mathbf{x}(t)$ and $\mathbf{v}(t - \Delta t/2)$ to $\mathbf{x}(t)$ and $\mathbf{v}(t)$. Logic says half acceleration and then half rotation will work. Is this correct? With $\mathbf{x}(t)$ and $\mathbf{v}(t)$, the code may then be re-started, say, to run backward (as a check, by changing the sign of Δt) or to run forward with a new Δt.

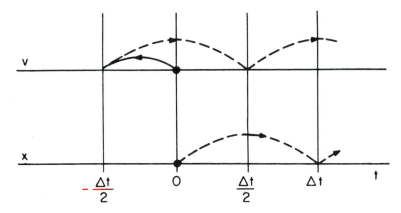

Figure 2-7a At $t = 0$, the x_i's ($t = 0$) and v_i's ($t = 0$) are put in, as shown. The very first step is to calculate the fields at $t = 0$ from the x_i's and move the v_i's back a half step, as shown by the solid line. Then the program advances the v_i's then the x_i's as shown earlier in Figure 2-4a.

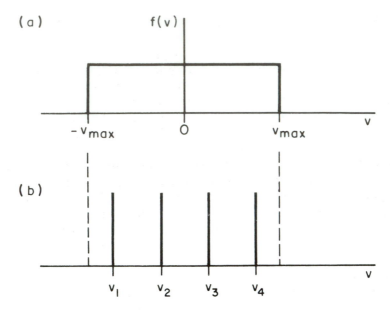

Figure 2-7b (a) The desired velocity distribution $f(v)$. (b) An elementary way of approximating $f(v)$ with four velocities.

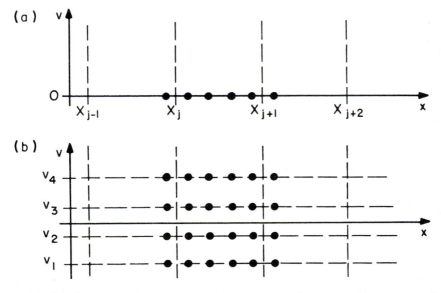

Figure 2-7c (a) Placement of particles in particle phase space for a cold uniform plasma, four particles per cell. (b) Placement of particles for a warm plasma with flat $f(v)$, using four velocities, making four beams.

2-8 CHOICE OF DIAGNOSTICS; GENERAL REMARKS

The object of simulation is to gain insight into the physics of plasmas. The object of any one simulation run or series of runs, is to study oscillations, or waves, or instabilities, or heating, or radiation, or transport, etc., but usually a limited number of these phenomena in any one set of runs. Hence, the simulation program output should consist of information that shows what physics is being done and should provide specific information about the phenomena under study.

One could ask that all of the information generated ($\mathbf{x}$, $\mathbf{v}$, ρ, ϕ, $\mathbf{E}$, etc., at every step) be put in storage (*e.g.*, magnetic tape) or printed out, and then diagnosed separately, by hand or by computer. This can and has been done for, say, a thermal (Maxwellian) plasma where the simulator wished to do his diagnostics at his leisure, after the run and more than once. However, the most common output is *graphical*, and consists of plots made during the run (*snapshots*) and at the end (*time histories*), complete with Fourier analyses in time and space, printed out on film and or paper. Since no other information will be saved, so that the plots chosen must contain the physics of interest. To accommodate hindsight, it is usually prudent to save more than the minimum perceived necessary.

The first page should contain the name of the program and the version being used; it must also have the initial conditions in considerable detail. The next pages may be snapshot plots made at intervals during the run, such as at $t = 0$, $t = 60\Delta t$, $t = 120\Delta t$. Information of interest in these snapshots might be, for particles:

(a) Phase space, v_x versus x
(b) Velocity space, v_y versus v_x
(c) Density in velocity, $f(v)$ versus v, or $f(v^2)$ versus v^2, or $\ln[f(v^2)]$ versus v^2

For grid quantities, one might plot:

(a) Charge density $\rho(x)$ versus x or particle density $n(x)$ versus x
(b) Potential $\phi(x)$ versus x
(c) Field $E(x)$ versus x
(d) Distribution of electrostatic energy $\frac{1}{2}\rho_k\dot{\phi}_k$ versus k

The result at the end of a run will consist of plots of histories of various quantities versus time, such as:

(a) Electrostatic energy $\sum_k \frac{1}{2}\rho_k\dot{\phi}_k$

(b) Particle kinetic energy by species $\sum_i \frac{1}{2} m_i v_i^2$

(c) Particle drift energy $\sum_i \frac{1}{2} m_i <v_i>^2$

(d) Particle thermal energy $\sum_i \frac{1}{2} m_i (<v_i^2> - <v_i>^2)$

(e) Total energy, electrostatic plus particle (usually with the zero suppressed, so as to magnify loss of energy conservation)

(f) Mode plots, $\frac{1}{2} \rho_k \phi_k^*$, for each k, possibly with Fourier analyses of each, with plots versus ω

These lists give an idea of a one-dimensional program output. In a typical run of 1000 steps, one may ask for snapshots every $60\Delta t$ of the seven particle and grid plots (126 plots), and at the end for the 5 energy plots, and, say, 32 mode plots; this is a total of $1 + 126 + 5 + 32 = 164$ plots which fits nicely on a 192 page 10×15 cm microfiche film.

Many other quantities may be of interest. In studying diffusion, one wishes to follow the mean square deviation of particle velocities or positions so that $[v_i - v_i(t = 0)]^2$ or $[x_i - x_i(t = 0)]^2$ must be stored, then plotted. In studying waves, one may wish to resolve frequencies, so that a Fourier analysis in time would be performed on some quantity like $\phi(k, t)$, perhaps following the run (called *postprocessing*), necessitating storage of the quantity. The dynamics of phase or velocity or distribution function space may be of interest in which case a *movie* might be made, probably from snapshots every Δt or $2\Delta t$.

In this way, the simulator, like an experimenter in the laboratory, will accumulate sufficient data to verify the correctness of his physics and to provide the insight desired into his particular plasma problem.

2-9 ARE THE RESULTS CORRECT? TESTS

The simulator, in demonstrating correctness, has many of the problems of a theorist or experimenter. The latter two may be questioned, for example, on their approximations and on their instruments. The simulator uses an unpublished program, with a restricted set of physics (*e.g.* forced to be in one dimension, or to be electrostatic or lacking collisions, radiation, etc.), with carefully chosen initial conditions and a limited amount of output. How can he tell himself and the world that his work is to be believed?

He can compare his results with those obtained in theory and or in experiment, obtain the desired results for problems with known answers, show invariance of his results as the nonphysical computer parameters (Δt, Δx, NP, NG, etc.) are changed, and so on. Even so, he may leave some doubters. Problems that are fundamental may be checked by simulators at different computer centers, using separate programs; an example of such a problem is plasma diffusion across a magnetic field.

The simulator himself must have confidence in his program and know the bounds within which it should work. The confidence must be real. First, all component parts of the program (particle mover, field solver, etc.) must be tested separately to produce predictable results.

The total program must then be run on *test problems* such as:

(a) Simple harmonic motion of a pair of test electrons in a uniform background; check frequency ω
(b) Cold plasma oscillations of many electrons, at long wavelength, in a uniform background; check ω as a function of k
(c) Instability of two opposing electron streams in a uniform background; check growth rate as a function of k
(d) Instability due to a ring velocity distribution in a magnetic field $f(v_\perp) \approx \delta(v_\perp - v_0)$; check complex ω

In any or all of these, try reducing the number of particles NP or the number of grid cells NG or increasing the time step Δt. See how far you can go before tell-tale signs of nonphysics, such as flagrant loss of energy conservation, show up. Confidence comes with experience.

THREE

A ONE-DIMENSIONAL ELECTROSTATIC PROGRAM ES1

Note

ES1 has changed considerably from the time of the first printing. The listing here is that of the first printing (1985); the current source is included on the diskette.

3-1 INTRODUCTION

We are ready to take a thorough look at a particular program. This look is in preparation for using this program (or any other) on one-dimensional electrostatic problems.

The program ES1 (ElectroStatic, 1-dimension) was written by A. B. Langdon for C. K. Birdsall's course in Berkeley, in 1972. Although ES1 was designed to be a teaching tool, it has formed a base for programs used professionally. ES1 is available to all users of the National Magnetic Fusion Energy Computer Center, and is easily adapted to other computer systems.

We go straight through the various parts, explain what each part does and present their Fortran listings. As in Chapter 2, we leave discussions of accuracy, alternative schemes, and comparisons with physics until later.

Some sections have a number of questions about statements in the program. These questions have been useful as classroom quizzes to ensure careful reading of the program by users.

3-2 GENERAL STRUCTURE OF THE PROGRAM ES1

The program flows very much like the scheme shown earlier in Section 2-3a, where the main loop steps were shown, without the initial step. Now we state the scheme in computer terms, use the names of *computer variables* and *subroutines*, bunch the steps together somewhat differently, and add the initial step.

First the program reads the input data, then initializes plotting routine HISTRY

Next, ES1 calls subroutines INIT, FIELDS (which calls the Fourier transform routine and its inverse), and SETV which calls ACCEL, in that order. These routines are used once to set up the proper initial conditions, *i.e.*, given $x(0)$, $v(0)$ for all particles, provide $x(0)$, $v(-\Delta t/2)$ as starting conditions for the main loop. Actually, more is done, as follows:

FIRST	Connect input file. Create output file for "tape 3" and plots.
Read data	
HISTRY	Later plots quantities (like the various energies) versus time.
INIT	Change input variables (like ω_p, q/m) into computer variables (like q, m). Calculate initial particle positions, including perturbations $x(0)$. Calculate initial particle velocities, such as cold or Maxwellian including perturbations $v(0)$.
SETRHO	Convert x to $x/\Delta x$, computer variable. Accumulate charge on the grid, with weighting.
FIELDS	Using charge on the grid, solve Poisson's equation for ϕ. Use CPFT, RPFT2, RPFTI2, CPFT, the Fourier transform subroutines. Calculate field energy from $\sum \rho_k \phi_k^*$. Difference ϕ to obtain E on grid. Make field "snapshots."
SETV	Change $v(0)$ to $v(-\Delta t/2)$, using $E(0)$ in ACCEL. Convert v to $v\Delta t/\Delta x$, computer variable.

Then the program continues for as many time steps as are requested in the main loop, which calls subroutines ACCEL, MOVE, and FIELDS in order, once each step. The details for each of these tasks are listed below, which is like the block diagram of Figure 2-3a, with the weighting moved into the subroutines ACCEL and MOVE.

Enter main loop from initializing steps, with $x(0)$, $v(-\Delta t/2)$, $E(0)$.	
PLOTXV, etc.	Particle diagnostics.
ACCEL	Convert E to $(q/m)E\Delta t^2/\Delta x$, called A, a computer variable. Advance velocity one time step, using weighted E (or A). Calculate momentum and kinetic energy. Repeat for each species.
MOVE	Advance position one time step. Accumulate charge density on grid, with weighting. Repeat for each species.
HISTRY	Saves quantities for plotting versus time.
Advance time step counter.	
FIELDS	(Already described.)
Return to start of main loop for total of NT steps.	

After running NT time steps we exit from the main loop through HISTRY and LAST.

HISTRY	Make plots of quantities versus time.
LAST	Close files and terminate execution.

More details as to what each step of each subroutine does are given in the sections following, more or less in the order used in the program.

There are few aspects of ES1 which are peculiar to the Livermore-developed operating systems for the Control Data Corporation 7600 and Cray Research CRAY-I computers. Most of these are confined to subroutines FIRST, LAST, and the subroutines which plot graphical output. We have included separate particle mover coding to gain some access to the vector capabilities of the CRAY-I computer; although this coding is specialized, it may be instructive. Finally, we have made use of a precompiler to facilitate changing many array dimensions and to replicate COMMON blocks in the subroutines which use them. PARAMETER statements define values which are fixed at compile time and remain constant (as compared to DATA statements which assign initial values to variables which may change during execution). CLICHE and ENDCLICHE delimit statements which are to be inserted in the source wherever cited by a USE statement. Thus, changes to array dimensions and COMMON blocks need only be made in one place.

The main program listing follows.

```
c   ES1 - a one-dimensional electrostatic plasma simulation code.
c   Written by A. Bruce Langdon, Livermore, 1972.
c   Revised 6/1976, 2/1977, 10/1977, 6/1978, 3-5/1979.
c
        cliche mparam
        common /param/ nsp, l, dx, dt, ntp
        real l
        endcliche mparam
        use mparam
c
c   ngmax=maximum number of cells.
        cliche mfield
        parameter(ngmax=256)
        parameter(ng1m=ngmax+1)
        common /cfield/ ng, ael, epsi, iw, vec, rho0, a1, a2,
     .    rho(ng1m), phi(ng1m), e(ng1m),
     .    e0, w0
        logical vec
        endcliche mfield
        use mfield
c
        cliche mptcl
c   particle coordinate and velocities.
        common x(8192), vx(8192), vy(8192)
        endcliche mptcl
        use mptcl
c   can declare vy to be length 1 if unmagnetized.
c
        cliche mcntrl
        common /cntrl/ it, time, ithl,
     .    irho, irhos, iphi, ie, ixvx, ivxvy, ifvx
        endcliche mcntrl
        use mcntrl
c
c   nth=number of time steps between history plots.
c   mmax=maximum number of different Fourier modes to plot.
c   nspm=maximum number of species.
c   allow up to nspm species.
        cliche mtime
        parameter(nth=500,mmax=10,nspm=3)
        parameter(nth1=nth+1,nth2=nth+2,nspm1=nspm+1)
```

```
      common /ctime/ ese(nth1), kes(nth1,nspm), pxs(nth2,nspm),
     *  nms(nspm), mplot(mmax), esem(nth1,mmax)
      real kes, nms
      endcliche mtime
      use mtime
c
      real ke, ms(nspm), qs(nspm), ts(nspm)
      integer ins(nspm1)
c
      data ntp/100/
      data rho0 /0./
      data it,time,ith,ithl/0,0.,0,0/
      data ins(1) /1/
c
c  input variables:
c  l       =physical length of system.
c  nsp     =number of particle species.
c  dt      =time step.
c  nt      =number of time steps to run (ending time=nt*dt).
c  ng      =number of spacial cells. *must* be power of 2.
c  iw      =mover algorithm selector. see subroutines accel and move.
c  vec     =.t. to select vectorized option where possible on cray.
c  epsi    =multiplier in poisson equation. 1 for rationalized units.
c  a1,a2   =field smoothing and mid-range boost. see fields.
c  e0,w0   =add uniform electric field e0*cos(w0*time).
c  irho    =plotting interval for rho (charge density). =0 for no plot.
c  irhos   =plotting interval for smoothed density.
c  iphi    =plotting interval for phi (potential).
c  ie      =plotting interval for e (electric field).
c  ixvx    =plotting interval for x vs. vx phase space.
c  ifvx    =plotting interval for f(vx) distribution function.
c             >0 gives linear, <0 gives semi-log.
c  ivxv,   =plotting interval for vx vs. vy phase space.
c  mplot   =fourier mode numbers to plot.
c
c  default input parameters:
      data nsp,l,dt,nt,epsi /1,6.28318530717958,.2,150,1./
      data ng,iw,a1,a2,vec /32,2,0.,0.,.true./
      data e0,w0 /0.,0./
      data irho,irhos,iphi,ie,ixvx,ivxvy,ifvx/7*0/, mplot/mmax*0/
      namelist /in/ l, nsp, nt, dt, epsi, ng, iw, vec, a1, a2, e0, w0,
     *  irho, irhos, iphi, ie, mplot, ixvx, ivxvy, ifvx
c
      call first
c
c  read namelist input. echo all values to output, including defaults.
c  empty output buffer in case some disaster occurs later.
      read(2,in)
      write(3,in)
      write(100,in)
c
      call histry
      dx=l/ng
      do 10 is=1,nsp
   10 call init( ins(is),ins(is+1),ms(is),qs(is),ts(is),nms(is),rho0 )
      call fields(0)
      do 11 is=1,nsp
   11 call setv( ins(is),ins(is+1)-1,qs(is),ms(is),ts(is),pxs(1,is) )
c
c
c  begin time step loop.
c  particle x is at time 0, vx and vy at -.5*dt.
  100 continue
      vl=0.
      vu=0.
```

```
c   phase space plot of all species.
        call plotxv(1,ins(nsp+1)-1,vl,vu)
c   distribution function for species 1.
        call pltfvx(1,ins(2)-1,vl,vu,qs(1),32)
        vmu=0.
        if( ts(1).ne.0. ) call pltvxy(1,ins(2)-1,vmu)
c   (if make restart file, here is where to save particles.)
        p=0.
        ke=0.
        do 101 is=1,nsp
c   advance velocities from it-.5 to it+.5.
        call accel( ins(is),ins(is+1)-1,qs(is),ms(is),ts(is),
          pxs(ith+2,is),kes(ith+1,is) )
        p=p+pxs(ith+2,is)
  101 ke=ke+kes(ith+1,is)
        do 102 is=1,nsp
c   advance positions from it to it+1.
  102 call move( ins(is),ins(is+1)-1,qs(is) )
        te=ke+ese(ith+1)
        write(3,950) time,ese(ith+1),p,(kes(ith+1,is),is=1,2),te
  950 format(" time,ese,p,ke1,ke2,te",f8.2,5e17.9)
        if( it.ge.nt ) go to 103
        if( ith.eq.nth ) call histry
        it=it+1
        time=it*dt
        ith=it-ithl
c   get fields at time step it.
        call fields(ith)
        go to 100
c
c   end of run.
c   particle x is at time nt*dt, vx and vy at (nt-.5)*dt.
  103 continue

        call histry
        call last

        end
```

3-3 DATA INPUT TO ES1

Data to be supplied to the main program includes:

L	Length of system
NSP	Number of species (1, 2, 3, ...)
DT	Time step
NT	Total number of steps to be run
NG	Number of grid points (power of 2)
IW	Weighting to be used:
	1 for zero order (NGP), momentum conserving
	2 for first order (CIC, PIC), momentum conserving
	3 for energy conserving (first-order for particles and zero-order for forces)

EPSI	$1/\epsilon_0$ (usually 1)
A1	Compensation factor (A1 = 0 means no compensation)
A2	Smoothing factor (A2 = 0 means no smoothing)
IPHI, etc.	Plotting frequencies

Data to be supplied to INIT for each species includes:

N	Number of particles (1, 2, 3, ...)
WP	ω_p (positive)
WC	ω_c (signed)
QM	q/m (signed)
VT1	Provides Gaussian velocity distribution of thermal velocity v_{t1} centered on $v_x = v_0$, $v_y = 0$, using random number routine; maximum velocity is $6v_{t1}$
VT2	Provides Gaussian (or other) velocity distribution of thermal velocity v_{t2} using inverse distribution functions, giving ordered velocities ("quiet start").
NLG	Number of sub-groups to be given the same velocity distribution; usually one.
NV2	Exponent of quiet start distribution $f(v) \propto (v/v_{t2})^{NV2}$ $\exp(-v^2/2v_{t2}^2)$, usually zero.
V0	Drift velocity in x direction (signed)
MODE	Number of the mode to be given an initial perturbation in x, v_x
X1	Magnitude of perturbation in x, generally less than half the uniform particle spacing, N/L; used as $X1\cos(2\pi x\,MODE/L + \theta_x)$
V1	Magnitude of perturbation in v; used as $V1\sin(2\pi x\,MODE/L + \theta_v)$
THETAX	θ_x
THETAV	θ_v

3-4 CHANGE OF INPUT PARAMETERS TO COMPUTER QUANTITIES

It is convenient to write the initial values in terms of plasma parameters, such as plasma and cyclotron frequencies (ω_p, ω_c) and charge-to-mass ratios (q/m) for each species. In ES1, ω_p^2 determines the average density and ω_c is uniform.

However, the program needs q and m. These are obtained from ω_p, q/m, and the particle density n; q is obtained from

$$\omega_p^2 = \frac{nq^2}{\epsilon_0 m} = \left(\frac{N}{L}\right)(\text{EPSI})(\text{Q})(\text{QM})$$

and m is obtained from

$$m = \frac{Q}{QM}$$

for each species, in the case of a uniform plasma where the density is given by the number of particles divided by the length $n = N/L$. It is convenient to avoid using ϵ_0 (or $1/4\pi$); hence, we set

$$\epsilon_0 = 1 \qquad \text{EPSI} = 1$$

in our program, a matter of units (not any new physics).

The user specifies all ω_c's and q/m's, so he must do this consistently $[\omega_c = (q/m)B_0]$. $T = -\tan(\omega_c \Delta t/2)$ is determined from ω_c, to be used in the particle mover for that species. If a species is meant to be unmagnetized then its $\omega_c = 0$. If $\omega_c \neq 0$, then $IW = 2$ (linear weighting) must be used.

3-5 NORMALIZATION; COMPUTER VARIABLES

The spatial grid spacing Δx and the time step Δt enter repeatedly into multiplication (or division) in several places where they may be *normalized* away.

In order to facilitate truncation, we change particle x to computer $x/\Delta x$. Next, writing the leap-frog particle mover of 2-4(3,4) as

$$v_{new} = v_{old} + \frac{F_{old}\Delta t}{m} \tag{1}$$

$$x_{new} = x_{old} + v_{new}\Delta t \tag{2}$$

we see that there are two multiplications by Δt for each particle each step, which are not necessary if we change

Particle v to computer $v\Delta t/\Delta x$
Grid $F/m = qE/m$ to computer $\left(q/m\right)E(\Delta t)^2/\Delta x$, called A

The Δx divisor in v and E is required by the x normalization. The equations now read

ep in TS/S

$$\left.\frac{v\Delta t}{\Delta x}\right|_{new} = \left.\frac{v\Delta t}{\Delta x}\right|_{old} + \frac{q}{m}\frac{E_{old}(\Delta t)^2}{\Delta x} \qquad \text{(ACCEL)} \tag{3}$$

$$\left.\frac{x}{\Delta x}\right|_{new} = \left.\frac{x}{\Delta x}\right|_{old} + \left.\frac{v\Delta t}{\Delta x}\right|_{new} \qquad \text{(MOVE)} \tag{4}$$

Hence, by forming A at the grid points (nearly always we have $NG \ll N$) means that ACCEL requires only an addition; MOVE also requires only an addition. No multiplications by Δt for each particle are needed each time step. For the first time step, each x is multiplied by $1/\Delta x$ (in SETRHO) and each v is multiplied by $\Delta t/\Delta x$ (in SETV). In obtaining kinetic energy for all particles, there is a multiplication by $(\Delta x / \Delta t)^2$ on the sum over all particle velocities, just once per time step (an *un*normalization).

[This normalization is a little questionable in 1d, as it clutters all diagnostic interpretation of x and v. In 2d, or on modern vector computers, little or no computer time is saved, so the variables should be normalized for ease in interpretation.]

3-6 INIT SUBROUTINE; CALCULATION OF INITIAL CHARGE POSITIONS AND VELOCITIES

This subroutine places every particle in x, v, phase space. This requires the simulator to know the equivalent of the initial density distribution, $f(x, v, t = 0)$, for each species.

There is no magic way of calculating initial particle densities and velocities. The simulator must specify *exactly* what he wants for *initial values* for every particle.

A spatially uniform cold plasma is simplest; the particles are spaced uniformly, with zero velocity, $v_{t1} = v_{t2} = v_0 = 0$. Adding a drift velocity v_0 to each particle creates a cold beam.

A uniform warm plasma requires finding a way to make $f_0(v)$ reasonably well filled out (enough velocities represented) for each x or range of x. This plasma obviously requires more particles since now, for each of the previous cold particles (which were sufficiently dense in x to be able to provide good spatial resolution), there must be a distribution in v as well. Some general techniques for solving $f_0(v)$ and $n_0(x)$ [really, $f_0(v, x, t = 0)$] for spacing in v and x space are given in Chapter 16. Here particles are created in NLG groups which are identical (before any perturbation is added) except for being displaced in x. In each group the velocities are any superposition of (a) simple drift, (b) random near-Maxwellian, and (c) nonrandom (*quiet start*) from any distribution. If (c) is used, then the positions are scrambled relative to the velocities in a manner that seems to fill v_x-x space smoothly. The v_{t2} (*quiet start*) is done first and the v_{t1} (random) is done second; this order allows a wee tweak to be made on all modes as the last step. The utility of the quiet start is seen in projects (Chapter 5) and is discussed in Chapter 16.

In addition, the program requires the simulator to specify the initial perturbations for the distribution [if small, those are related to $f_1(x, v, t=0)$], in order to initiate the interactions at a known level in a given mode (X1, V1, MODE).

PROBLEMS

3-6a If the perturbation in x is made first $(x_1 > 0)$ and then the perturbation in v is made $(v_1 > 0)$, is there cross coupling? Which perturbation should be made first to avoid cross coupling? How does ES1 avoid this problem?

3-6b Note that particles are first loaded at *half*integer multiples of L/N, in loops ending in statements numbered 10 and 41. What would happen after the perturbation is added if particles were loaded at integer multiples of L/N? Assume linear weighting (IW = 2), and that N/NG is an integer. *Hint*: first try N = NG = 4, MODE = 1, and check the charge density resulting from small values of X1 and values of THETAX such as 0 and $\pi/2$.

The subroutine INIT follows.

```
      subroutine init(il1,il2,m,q,t,nm,rho0)
c   loads particles for one species at a time.
      real m, nm
      use mparam
      real lg
      use mptcl
      data twopi/6.2831 85307 17958/
c
c   input variables:
c   n        =number of particles (for this species).
c   wp       =plasma frequency.
c   wc       =cyclotron frequency.
c   qm       =q/m charge:mass ratio.
c   vt1      =rms thermal velocity for random velocities.
c   vt2      =rms thermal velocity for ordered velocities.
c   nlg      =number of loading groups (sharing same ordered velocities).
c   nv2      =multiply maxwellian (for ordered velocities) by v**nv2.
c   v0       =drift velocity.
c   mode, x1, v1, thetax and thetay are for loading a sinusoidal perturbation.
c   velocity contributions thru vt1, vt2, v0 and v1 are additive.
c
c   default input parameters:
      data n,wp,wc,qm,vt1,vt2,nlg,nv2,v0,mode,x1,v1,thetax,thetav
     /128,1.,0.,-1.,2*0.,1,0,0.,1,4*0./
      namelist /in/ n, wp, wc, qm, vt1, vt2, nv2, nlg, v0,
     mode, x1, v1, thetax, thetav
      read(2,in)
      write(3,in)
      write(100,in)
      t=tan(-wc*dt/2.)
      il2=il1+n
      q=l*wp*wp/(n*qm)
      m=q/qm
      nm=n*m
c
c   load first of nlg groups of particles.
      ngr=n/nlg
      lg=l/nlg
      ddx=l/n
c   load evenly-spaced, with drift.
c   also does cold case.
      do 10 i=1,ngr
      i1=i-1+il1
      x0=(i-.5)*ddx
      x (i1)=x0
 10   vx(i1)=v0
      if(vt2.eq.0.) go to 50
c   load ordered velocities in vx ("quiet start", or at least subdued).
c   is set up for maxwellian*v**nv2, but can do any smooth distribution.
c   hereafter, ngr is preferably a power of 2.
```

```
c   first store indefinite integral of distribution function in x array.
c   use midpoint rule -simple and quite accurate.
        vmax=5.*vt2
        dv=2.*vmax/(n-1)
        vvnv2=1.
        x(il1)=0.
        do 21 i=2,n
        vv=((i-1.5)*dv-vmax)/vt2
        if(nv2.ne.0) vvnv2=vv**nv2
        fv=vvnv2*exp( -.5*vv**2 )
        il1=i-1+il1
  21    x(il1)=x(il1-1)+amax1(fv,0.)
c   for evenly spaced (half-integer multiples) values of the integral,
c   find corresponding velocity by inverse linear interpolation.
        df=x(il1)/ngr
        il1=il1
        j=il1
        do 22 i=1,ngr
        fv=(i-.5)*df
  23    if(fv.lt.x(j+1)) go to 24
        j=j+1
        if(j.gt.il2-2) go to 80
        go to 23
  24    vv=dv*( j-il1+(fv-x(j))/(x(j+1)-x(j)) )-vmax
        vx(il1)=vx(il1)+vv
  22    il1=il1+1
c   for ordered velocities, scramble positions to reduce correlations.
c   scrambling is done by bit-reversed counter -compare sorter in cpft.
c   xs=.000,.100,.010,.110,.001,.101,.011,.111,.0001.. (binary fractions).
        xs=0.
        do 41 i=1,ngr
        il1=i-1+il1
        x(il1)=xs*lg+.5*ddx
        write(3,99) x(il1), vx(il1)
  99    format(6f10.4)
        xsi=1.
  42    xsi=.5*xsi
        xs=xs-xsi
        if(xs.ge.0) go to 42
  41    xs=xs+2.*xsi
c   if magnetized, rotate (vx,0) into (vx,vy).
  50    if(wc.eq.0.) go to 80
        do 51 i=1,ngr
        il1=i-1+il1
        vv=vx(il1)
        theta=twopi*x(il1)/lg
        vx(il1)=vv*cos(theta)
  51    vy(il1)=vv*sin(theta)
c
c   copy first group into rest of groups.
  80    if(nlg.eq.1) go to 85
        j=ngr+1
        xs=0.
        do 81 i=j,n,ngr
        xs=xs+lg
        do 81 j=1,ngr
        il1=j-1+il1
        i2=il1+i-1
        x (i2)=x (il1)+xs
        vx(i2)=vx(il1)
  81    if(wc.ne.0.) vy(i2)=vy(il1)
c
c   add random maxwellian.
  85    if(vt1.eq.0.) go to 90
        do 86 i=1,n
        il1=i-1+il1
        do 86 j=1,12
        if(wc.ne.0.)
```

```
    .vy(i1)=vy(i1)+vt1*(ranf(dum)-.5)
86  vx(i1)=vx(i1)+vt1*(ranf(dum)-.5)
c
c   add perturbation.
c   loading x(t=0), v(t=0), remember so no dt/2 correction.
c   may want to perturb vy too.
    90 do 91 i=1,n
        i1=i-1+il1
        theta=twopi*mode*x(i1)/l
        x (i1)=x (i1)+x1*cos(theta+thetax)
    91 vx(i1)=vx(i1)+v1*sin(theta+thetav)
c
c   apply boundary conditions, collect charge density, etc.
        call setrho(il1,il2-1,q,n*q/l)
        return
        end
```

3-7 SETRHO, INITIALIZATION OF CHARGE DENSITY

This subroutine is called once for each species at $t = 0$. SETRHO converts x to $x/\Delta x$, enforces periodicity, and accumulates charges at the grid points, according to the weighting used.

The subroutine SETRHO follows.

```
        subroutine setrho(il,iu,q,rhos)
c   converts position to computer normalization and accumulates charge
c   density.
        use mparam
        use mfield
        use mptcl
c
        qdx=q/dx
c   if is first group of particles, then clear out rho.
        if( il.ne.1 ) go to 2
        do 1 j=1,ng
    1   rho(j)=rho0
        rho(ng+1)=0.
        dxi=1.0/dx
        xn=ng
c   add on fixed neutralizing charge density.
c   (not needed when all species are mobile -but harmless.)
    2   rho0=rho0-rhos
        do 3 j=1,ng
    3   rho(j)=rho(j)-rhos
c
        go to (100,200,200), iw
c
c ngp
    100 continue
        do 101 i=il,iu
        x(i)=x(i)*dxi
        if( x(i).lt.0. ) x(i)=x(i)+xn
        if( x(i).ge.xn ) x(i)=x(i)-xn
        j=x(i)+0.5
    101 rho(j+1)=rho(j+1)+qdx
        return
c
c   linear
```

```
200 continue
    do 201 i=il,iu
    x(i)=x(i)*dxi
    if( x(i).lt.0. ) x(i)=x(i)+xn
    if( x(i).ge.xn ) x(i)=x(i)-xn
    j=x(i)
    drho=qdx*( x(i)-j )
    rho(j+1)=rho(j+1)-drho+qdx
201 rho(j+2)=rho(j+2)+drho
    return
    end
```

3-8 FIELDS SUBROUTINE; SOLUTION FOR THE FIELDS FROM THE DENSITIES; FIELD ENERGY

The physics and numerics of field solving were given in Section 2-5. The subroutine FIELDS carries out the steps.

First, $K^2(k)$ is to be formed, the Fourier representation of the three-point finite difference form 2-5(5), which is 2-5(13). Since $1/K^2$ is used, this is what is formed, called KSQI(k), for each mode or harmonic. (We do not use k^2 for K^2 or k for κ, for reasons given in Appendix F.) Here KSQI(k) also allows a factor SM(k), called the *smoothing factor;* this factor is used to modify the k spectrum in going from $\rho(k)$ to $\phi(k)$ in any way desired, primarily in attenuating unwanted short wavelengths or in compensating (pre-emphasis) for unwanted distortions at long wavelengths. The details are discussed in Appendix B. As we see in Chapter 4, this smoothing also has the effect of broadening the particle shape $S(x)$.

Second, the charge density $\rho(x)$ is Fourier transformed to $\rho(k)$ through calls to subroutine CPFT (complex Fourier transform) and RPFT (an interface). These are listed in Appendix A.

Next, $\rho(k)$ is multiplied by KSQI(k) to produce $\phi(k)$. At this stage, we have both $\rho(k)$ and $\phi(k)$ so that we may form the electrostatic energy, k by k, as

$$\text{Electrostatic energy}(k) = \text{ESE}(k) \; \propto \; \frac{1}{2}\rho(k)\phi^*(k)$$

as well as accumulate ESE(k) to obtain

$$\text{ESE}(t) \; \propto \; \frac{1}{2}\sum_k \rho(k,t)\phi^*(k,t)$$

Next, both $\phi(k)$ and $\rho(k)$ are fed to the inverse Fourier transform, through calls to RPFTI2 and CPFT to produce $\phi(x)$ and the smoothed $\rho(x)$.

Last, $E(x)$ is to be obtained from $-\nabla\phi(x)$ by the differencing associated with the weighting to be used. For example, for zero- and first-order weighting, 2-5(4) is used which differences $\phi(x)$ over $2\Delta x$ to obtain $E(x)$. At this point E has not been renormalized.

PROBLEMS

3-8a Using the program names, starting with RHO(J) and ending with $E(J)$, show the steps in FIELDS and indicate by names what methods are used in each step. Be as symbolic as possible.

3-8b What is done by the statement
$$RHO(1) = RHO(1) + RHO(NG+1)$$
Use a sketch of the grid to be explicit.

3-8c RHO and RHOS are plotted. What is RHOS and why are both plotted?

3-8d What do the following statements do:
    DO 41 J = 1, NG
41   RHO(J) = RHO0
Why is RHO(NG1) left out of this loop and zeroed in the statement following?

3-8e What is the algebraic formula used to obtain the electrostatic energy ESE?

3-8f K = 1 in loop 20 is which laboratory harmonic?

3-8g Suppose you wished to compute for only one wave, one value of k. How would you do this? Will computing for only one value of k affect the physics at small amplitude? At large amplitude?

The subroutine FIELDS follows.

```
      subroutine fields(ith)
c  solves for phi and e, computes field energy, etc.
      use mparam
      use mfield
      use mtime
      use mcntrl
      real rhok(1), phik(1), scrach(1)
      equivalence (rho,rhok), (phi,phik), (e,scrach)
      parameter(ng2m=ngmax/2)
      real kdx2, ksqi(ng2m), li, sm(ng2m)
c
c  first time step duties.
      data ng2/0/
      if( ng2.ne.0 ) go to 2
      ng2=ng/2
      ng1=ng+1
c  set up ratio phik/rhok.
c  a2>0 gives a short-wavelength cutoff (smoothing).
c  a1>0 gives a mid-range boost to compensate for slight attenuation
c     in force calculation.
      data pi/3.1415 92653 58979/
      do 1 k=1,ng2
      kdx2=(pi/ng)*k
      sm(k)=exp(a1*sin(kdx2)**2-a2*tan(kdx2)**4)
    1 ksqi(k)=+epsi/((2.0*sin(kdx2)/dx)**2)*sm(k)**2
    2 continue
c
c  transform charge density.
      rho(1)=rho(1)+rho(ng+1)
      rho(ng+1)=rho(1)
      call plotf(rho,"charge density",irho)
      hdx=0.5*dx
      do 10 j=1,ng
      rhok(j)=rho(j)*hdx
```

```
   10 scrach(j)=0.
      call cpft (rhok,scrach,ng,1,1)
      call rpft2(rhok,scrach,ng,1)
      rhok(1)=0.
c
c  calculate phik and field energy.
      eses=0.
      phik(1)=0.
      do 20 k=2,ng2
      kk=ng+2-k
      phik(k )=ksqi(k-1)*rhok(k )
      phik(kk)=ksqi(k-1)*rhok(kk)
      eses=eses+rhok(k)*phik(k)+rhok(kk)*phik(kk)
      rhok(k )=sm(k-1)*rhok(k )
   20 rhok(kk)=sm(k-1)*rhok(kk)
      phik(ng2+1)=ksqi(ng2)*rhok(ng2+1)
      ese(ith+1)=(2.0*eses+rhok(ng2+1)*phik(ng2+1))/(2.0*l)
      rhok(ng2+1)=sm(ng2)*rhok(ng2+1)
c
c  save specified mode energies.
      do 21 km=1,mmax
      k=mplot(km)+1
      if( k.eq.1 ) go to 22
      kk=ng+2-k
      esem(ith+1,km)=( rhok(k)*phik(k)+rhok(kk)*phik(kk) )/l
   21 if( k.eq.kk ) esem(ith+1,km)=0.25*esem(ith+1,km)
c
   22 continue
c
c  inverse transform phi.
      li=1.0/l
      do 30 k=1,ng
      rho(k)=rhok(k)*li
   30 phi(k)=phik(k)*li
      call rpfti2(phi,rho,ng,1)
      call cpft (phi,rho,ng,1,-1)
      phi(ng+1)=phi(1)
      rho(ng+1)=rho(1)
      call plotf(rho,"smoothed density",irhos)
      call plotf(phi,"e potential",iphi)
c
c  uniform field.
      e0t=e0*cos(w0*time)
c
c  select electric field differencing.
      go to (100,100,200), iw
c
c  centered difference across 2 cells.
  100 continue
      hdxi=0.5/dx
      do 101 j=2,ng
  101 e(j)=( phi(j-1)-phi(j+1) )*hdxi+e0t
      e(1)=( phi(ng )-phi( 2) )*hdxi+e0t
      e(ng+1)=e(1)
      go to 40
c
c  centered difference across 1 cell.
  200 continue
      dxi=1.0/dx
      do 201 j=1,ng
  201 e(j)=( phi(j)-phi(j+1) )*dxi+e0t
      e(ng+1)=e(1)
c
   40 continue
      call plotf(e,"electric field",ie)
c  clear out old charge density.
      do 41 j=1,ng
   41 rho(j)=rho0
```

```
    rho(ng1)=0.
c   electric field has not been renormalized yet.
    ael=1.
    return
    end
```

3-9 CPFT, RPFT2, RPFTI2,
FAST FOURIER TRANSFORM SUBROUTINES

The coefficients of cosines and sines in a Fourier series may be calculated for a function which is given at a discrete number of evenly spaced points in a fashion similar to that for a function which is continuous. Since the middle 1960's it has been learned how to find these coefficients very rapidly (well beyond the gathering of coefficients of older texts) using methods commonly called *fast Fourier transforms* FFT.

We use an FFT called CPFT, for *complex Fourier transform,* which transforms $a(x) + ib(x)$ to $A(k) + iB(k)$. Since we have real quantities (ρ, ϕ) to be transformed, it pays us to set up a *pair* of real sequences $a(x)$, $b(x)$ and transform them together, gaining a factor of two in speed, if setting up the pairs is fast. It is; RPFT2 does this, in less than one tenth the time required for CPFT. The FFT calls are RPFT2, CPFT; the inverse FFT calls are RPFTI2, CPFT.

The listing of these subroutines is in Appendix A.

3-10 SETV, SUBROUTINE FOR INITIAL HALF-STEP IN VELOCITY

The simulator provides positions and velocities at time $t = 0$ for the initial conditions $x(0)$, $v(0)$. The program, however, with its leapfrog time-centered particle mover, wants x leading v by $\Delta t / 2$. Hence, at $t = 0$, we must operate on v to effect this separation.

$v(0)$ is moved to $v(-\Delta t / 2)$ using the fields at $t = 0$ by a half-step rotation (backward) and then half step acceleration, qE only (backward). The velocity advancer, ACCEL is used not with Δt replaced by $-\Delta t / 2$, but with q replaced by $-q / 2$, for this one call. (Do you see why?)

PROBLEMS

3-10a Consider a cold plasma and find the conditions on initial values of X and VX which excite a purely travelling wave. Now, if SETV were not used, how much of the (unwanted) oppositely-propagating wave would be excited?

3-10b C and S are cosine and sine of what angle?

3-10c Why is there the argument $-0.5*Q$ in the call to ACCEL?

3-10d When SETV is called, does the program have values for the electric field, E?

3-10e Sketch a CHANGV routine to take $x(t)$ and $v(t - \Delta t/2)$ to $x(t)$ and $v(t)$, the reverse of SETV. Now Δt may be changed, followed by another call to SETV. (Similar to Problem 2-7b.)

The subroutine SETV follows.

```
      subroutine setv(il,iu,q,m,t,p)
c   converts particle velocities at t=0 to computer normalization at
c   t=-dt/2.
      use mparam
      use mfield
      use mptcl
      dtdx=dt/dx
c   rotate v thru angle +0.5*wc*dt and normalize vy.
c   if t=0, there is no magnetic field. the rotation is omitted
c   and no references are made to vy.
      if( t.eq.0. ) go to 2
      c=1.0/sqrt(1.0+t*t)
      s=c*t
      do 1 i=il,iu
      vxx=vx(i)
      vx(i)= c*vxx+s*vy(i)
      vy(i)=-s*vxx+c*vy(i)
    1 vy(i)=vy(i)*dtdx
    2 continue
c   normalize vx.
      do 3 i=il,iu
    3 vx(i)=vx(i)*dtdx
c   electric impulse to go back 1/2 time step.
      data dum/0./
      call accel(il,iu,-0.5*q,m,0.,p,dum)
      return
      end
```

3-11 ACCEL, SUBROUTINE FOR ADVANCING THE VELOCITY

The first step is to normalize E at cell grid points to the form A as given in Section 3-5. Secondly, following choice of weighting, the velocity is advanced by additions (also shown in Section 3-5). Next, the particle momenta are calculated. Lastly, the particle kinetic energies are obtained, centered at time t (when the potential energy is known), by summing on

$$KE = \frac{m}{2} v_{old} v_{new}$$

This is simpler to calculate than $(v_{new}^2 + v_{old}^2)/4$ or $[(v_{new} + v_{old})/2]^2/2$, has the same value to order $(\Delta t)^2$, and has an obscure advantage to be seen in Problem 4-10e.

If there is a magnetic field, $\omega_c \neq 0$, then the procedure for advancing $\mathbf{v}$ is different, as outlined in Section 2-4(8,9,10) for a uniform magnetic field. *First,* INIT has calculated

$$T = \tan(-\omega_c \Delta t / 2)$$

Second, ACCEL tests whether $T = 0$ and, if not, sends the program to the half accel-rotation-half accel steps; also, the normalized value of **E** is halved, to use in the half-acceleration. (Only $IW = 2$ is used with a magnetic field.) *Third*, $\sin \omega_c \Delta t$ and $\cos \omega_c \Delta t$ are calculated from T. *Fourth*, the half accel in v_x is done; this produces $v_x^2 + v_y^2$ in the center of the step, at time t, so that this is the time to calculate the kinetic energy. *Lastly*, the rotation of **v** through angle $(-\omega_c \Delta t)$ is done, and the half accel in v_x added in.

Note that the kinetic energies and momenta are calculated from v_x^2 and v_x only if there is no magnetic field, $T = 0$. Otherwise there are both v_x and v_y so that the kinetic energy uses $v_x^2 + v_y^2$; no momenta are calculated, but they could be.

When logical variable VEC is *true*, the $IW = 2$ movers jump to special coding following labels 2000 and 2500, which is designed so that the Cray Research CFT compiler generates machine instruction sequences for the CRAY-I computer which are several times more efficient than it generates from the more straightforward Fortran. The form of this coding reflects both the vector architecture of the CRAY-I and the capability of the CFT compiler to recognize opportunities to use vector instructions. On a different vector computer, or as CFT becomes more cognizant, this coding should differ.

PROBLEMS

3-11a The statements following do what with the particle coordinates
    J = X(I) + 0.5
    J = X(I)

3-11b Show that KE from VN*V0 has the same order of accuracy as using the square of the average or the average of the squares of the old and new velocities. Why is VN*V0 to be preferred? Could there be a sign problem? At what time is KE obtained?

3-11c Linear weighting of the fields to the particles is given by 2-6(3) and the corresponding use in the program is
    VN = V0 + A(J+1) + (X(I)−J)*(A(J+2)−A(J+1))
or
    AA = A(J+1) + (X(I)−J)*(A(J+2)−A(J+1))
Explain all differences between the algebraic statement and that in the program.

3-11d The following statement does what:
    V1S = V1S + VN

3-11e The following statement does what:
    V2S = V2S + VN*V0

3-11f What form of electric field weighting is used in option $IW = 3$, energy conserving?

3-11g What is the purpose of the statement
           IF (T.NE.0) AE = 0.5*AE

3-11h In calculating momenta *P* why is DXDT there?

3-11i Following Problem 3-11b, compare the number of multiplies and adds per particle in obtaining KE in the three ways given. Ignore steps that are common to a given species.

The subroutine ACCEL follows.

```
      subroutine accel(ilp,iup,q,m,t,p,ke)
c   advances velocity one time step, computes momentum and kinetic energy
      real ke, m
      use mparam
      use mfield
      use mptcl
      real a(1)
      equivalence (a,e)
c   these arrays are used in vectorizing accel and move.
      common /scratch/ ji(64), al(64), ar(64), vni(64),
         v1si(64), v2si(64), aai(64)
      il=ilp
      iu=iup
c
c   renormalize acceleration if need be.
      dxdt=dx/dt
      ae=(q/m)*(dt/dxdt)
      if( t.ne.0. ) ae=0.5*ae
      if( ae.eq.ael ) go to 2
      ng1=ng+1
      tem=ae/ael
      do 1 j=1,ng1
    1 a(j)=a(j)*tem
      ael=ae
    2 continue
c
c   select acceleration weighting.
      go to (100,200,300), iw
c
c   ngp, grid points at j*dx.
  100 continue
      v1s=0.
      v2s=0.
      do 101 i=il,iu
      j=x(i)+0.5
      vo=vx(i)
      vn=vo+a(j+1)
      v1s=v1s+vn
      v2s=v2s+vn*vo
  101 vx(i)=vn
      p=p+m*v1s*dxdt
      ke=ke+0.5*m*v2s*dxdt*dxdt
      return
c
c   linear, momentum conserving.
  200 continue
      if( t.ne.0. ) go to 250
      v1s=0.
      v2s=0.
      if(vec) go to 2000
 2009 continue
      do 201 i=il,iu
      j=x(i)
      vo=vx(i)
      vn=vo+a(j+1)+( x(i)-j )*( a(j+2)-a(j+1) )
```

```
      v1s=v1s+vn
      v2s=v2s+vo*vn
201   vx(i)=vn
      p=p+m*v1s*dxdt
      ke=ke+0.5*m*v2s*dxdt*dxdt

      return
c
c   linear, energy conserving.
  300 continue
      v1s=0.
      v2s=0.
      do 301 i=il,iu
      j=x(i)
      vo=vx(i)
      vn=vo+a(j+1)
      v1s=v1s+vn
      v2s=v2s+vn*vo
  301 vx(i)=vn
      p=p+m*v1s*dxdt
      ke=ke+0.5*m*v2s*dxdt*dxdt
      return
c
c   linear, momentum conserving, uniformly magnetized.
  250 continue
      s=2.0*t/(1.0+t*t)
      v2s=0.
      if(vec) go to 2500
 2509 continue
      do 251 i=il,iu
      j=x(i)
      aa=a(j+1)+( x(i)-j )*( a(j+2)-a(j+1) )
      vyy=vy(i)
      vxx=vx(i)-t*vyy+aa
      vyy=vyy+s*vxx
      vxx=vxx-t*vyy
      v2s=v2s+vxx*vxx+vyy*vyy
      vx(i)=vxx+aa
  251 vy(i)=vyy
      ke=ke+0.5*m*v2s*dxdt*dxdt
      return
c
c
c***********************vectorized movers*****************************
c   These are functionally equivalent to the straightforward fortran movers
c   above, but are written in such a way that the cray 'cft'
c   compiler will use vector instructions for most operations.
c
c   Linear, momentum conserving, vectorized.
 2000 continue
      do 2004 i=1,64
      v1si(i)=0.
 2004 v2si(i)=0.
      do 2006 j=il,iu-63,64
      do 2001 i=0,63
 2001 ji(i+1)=x(i+j)
      do 2002 i=1,64,2
      al(i+1)=a( ji(i+1)+1 )
      ar(i+1)=a( ji(i+1)+2 )
      al(i )=a( ji(i )+1 )
 2002 ar(i )=a( ji(i )+2 )
      do 2005 i=0,63
      vni(i+1)=vx(i+j)+al(i+1)+( x(i+j)-ji(i+1) )*( ar(i+1)-al(i+1) )
      v1si(i+1)=v1si(i+1)+vni(i+1)
      v2si(i+1)=v2si(i+1)+vx(i+j)*vni(i+1)
 2005 vx(i+j)=vni(i+1)
```

```
2006  il=il+64
      do 2007  i=1,64
      v1s=v1s+v1si(i)
2007  v2s=v2s+v2si(i)
c     il is reset so that non-vector code will take care of remaining particles.
c     The number of particles is a multiple of 64, then il>iu and
c     CFT will not execute loop 201 at all.
      go to 2009
c
c     Linear, momentum conserving, vectorized.
c     See comments in unmagnetized coding above.
2500  continue
      do 2504  i=1,64
2504  v2si(i)=0.
      do 2506  j=il,iu-63,64
      do 2501  i=0,63
2501  ji(i+1)=x(i+j)
      do 2502  i=1,64,2
      al(i+1)=a( ji(i+1)+1 )
      ar(i+1)=a( ji(i+1)+2 )
      al(i  )=a( ji(i  )+1 )
2502  ar(i  )=a( ji(i  )+2 )
      do 2505  i=0,63
      aai(i+1)=al(i+1)+( x(i+j)-ji(i+1) )*( ar(i+1)-al(i+1) )
      vx(i+j)=vx(i+j)-t*vy(i+j)+aai(i+1)
      vy(i+j)=vy(i+j)+s*vx(i+j)
      vx(i+j)=vx(i+j)-t*vy(i+j)
      v2si(i+1)=v2si(i+1)+vx(i+j)*vx(i+j)+vy(i+j)*vy(i+j)
2505  vx(i+j)=vx(i+j)+aai(i+1)
2506  il=il+64
      do 2507  i=1,64
2507  v2s=v2s+v2si(i)
      go to 2509
c
      end
```

3-12 MOVE, SUBROUTINE FOR ADVANCING THE POSITION

This subroutine simply does 3-5(4) for each particle in a species, to obtain x_{new} from x_{old}. MOVE is called once for each species.

If 3-5(4) would place the particle outside the system, that is, if $x_{new} < 0$ or $x_{new} > L$, then MOVE replaces the particle in $0 \leqslant x \leqslant L$; this is done by moving it a full period L to the left or right.

With the x_{new}, MOVE now accumulates charge at the grid points, according to the weighting chosen.

As in ACCEL, if VEC = *true* and IW = 2, then we jump to special coding which exploits the vector capabilities of the CFT compiler on the Cray-I computer. (See discussion in Section 3-11.)

PROBLEMS

3-12a Using a sketch of the grid and location of a typical charge, show how linear weighting is obtained graphically, in at least three ways.

3-12b What are the units of RHO? State how you know, *e.g.*, where Q comes from.

3-12c Explain the following statement:
        IF (X(I).LT.0) X(I) = X(I) + XN

The subroutine MOVE follows.

```
      subroutine move(ilp,iup,q)
c   advances position one time step and accumulates charge density.
      use mparam
      use mfield
      use mptcl
c   these arrays are used in vectorizing accel and move.
      common /scratch/ ji(64), rhol(64), rhor(64)
      il=ilp
      iu=iup
      qdx=q/dx
      xn=ng
      go to (100,200,200), iw
c
c ngp
  100 continue
      do 101 i=il,iu
      x(i)=x(i)+vx(i)
      if( x(i).lt.0. ) x(i)=x(i)+xn
      if( x(i).ge.xn ) x(i)=x(i)-xn
      j=x(i)+0.5
  101 rho(j+1)=rho(j+1)+qdx
      return
c
c   linear
  200 continue
      if(vec) go to 2000
 2009 continue
      do 201 i=il,iu
      x(i)=x(i)+vx(i)
      if( x(i).lt.0. ) x(i)=x(i)+xn
      if( x(i).ge.xn ) x(i)=x(i)-xn
      j=x(i)
      drho=qdx*( x(i)-j )
      rho(j+1)=rho(j+1)-drho+qdx
  201 rho(j+2)=rho(j+2)+drho
      return
c
c
c*******************vectorized mover*****************************
c   This is functionally equivalent to the straightforward fortran
c   above, but is written in such a way that the cray 'cft' compiler
c   will use vector instructions for some operations.
c
c   Linear, vectorized.
c   See comments in accel.
 2000 continue
      do 2004 j=il,iu-63,64
      do 2001 i=0,63
      x(i+j)=x(i+j)+vx(i+j)
      x(i+j)=cvmgm(x(i+j)+xn,x(i+j),x(i+j))
      x(i+j)=cvmgp(x(i+j)-xn,x(i+j),x(i+j)-xn)
      ji(i+1)=x(i+j)
      rhor(i+1)=qdx*( x(i+j)-ji(i+1) )
      rhol(i+1)=qdx-rhor(i+1)
 2001 continue
      do 2002 i=1,64,2
      rho( ji(i  )+1 )=rho( ji(i  )+1 )+rhol(i  )
      rho( ji(i  )+2 )=rho( ji(i  )+2 )+rhor(i  )
      rho( ji(i+1)+1 )=rho( ji(i+1)+1 )+rhol(i+1)
 2002 rho( ji(i+1)+2 )=rho( ji(i+1)+2 )+rhor(i+1)
```

```
 2004  i l=i l+64
       go to 2009
c
       end
```

3-13 ADVANCE TIME ONE STEP

At the end of the time loop the velocity has been advanced, $v(t - \Delta t/2)$ to $v(t + \Delta t/2)$, the position advanced to $x(t + \Delta t)$, time advanced to $t + \Delta t$, and the field $E(t + \Delta t)$ is calculated. The time has been checked to see if NT steps have been run and, if so, the program goes to the end plots.

3-14 HISTRY SUBROUTINE; PLOTS VERSUS TIME

There is no physics in this subroutine. HISTRY is called to save quantities (like energies) to be plotted as a function of time [t from 0 to (NT)Δt] at the end of a run, or at intervals if NT > NTH. Some bookkeeping is also done, such as zeroing out the arrays (to be plotted) at $t = 0$.

The subroutine HISTRY follows.

```
       subroutine histry
c  plot energies etc. vs. time.
       use mparam
       use mcntrl
       use mtime
       real tim(nth1), pxsl(nspm)
       if( it.eq.0 ) go to 10
       tl=ithl*dt
       mth=it-ithl+1
       do 1 i=1,mth
    1  tim(i)=(i-1)*dt+tl
c
c  plot mode energies.
       do 50 km=1,mmax
       k=mplot(km)
       if( k.eq.0 ) go to 52
       call plthst(esem(1,km),tim,mth,tl,time,1)
   50  write(ntp,51) k, tl, time
   51  format("mode",i3," energy, time=",f10.4," to",f10.4)
   52  continue
c
c  plot field energy.
       call plthst(ese,tim,mth,tl,time,1)
       write(ntp,100) tl, time
  100  format("field energy, time=",f10.4," to",f10.4)
c
c  plot kinetic energies.
       do 200 is=1,nsp
       call plthst(kes(1,is),tim,mth,tl,time,0)
  200  write(ntp,201) is, tl, time
  201  format("kinetic energy",i2,", time=",f10.4," to",f10.4)
c
c  plot directed (drift) energies.
       do 300 is=1,nsp
       pxsl(is)=pxs(mth,is)
       do 299 i=1,mth
```

```
  299 pxs(i,is)=pxs(i,is)*pxs(i+1,is)/(2.*nms(is))
      call plthst(pxs(1,is),tim,mth,tl,time,0)
  300 write(ntp,301) is, tl, time
  301 format("drift energy",i2,", time=",f10.4," to",f10.4)
c
c  plot thermal energies
      do 400 is=1,nsp
      do 399 i=1,mth
  399 pxs(i,is)=kes(i,is)-pxs(i,is)
      call plthst(pxs(1,is),tim,mth,tl,time,1)
  400 write(ntp,401) is, tl, time
  401 format("thermal energy",i2,", time=",f10.4," to",f10.4)
c
c  plot total energy.
      do 499 is=1,nsp
      do 499 i=1,mth
  499 ese(i)=ese(i)+kes(i,is)
      call plthst(ese,tim,mth,tl,time,0)
      write(ntp,500) tl, time
  500 format("total energy, time=",f10.4," to",f10.4)
c
      ithl=it
c  last values now are first values for next time interval.
      ese(1)=ese(mth)
      do 2 is=1,nsp
      kes(1,is)=kes(mth,is)
      ese(1)=ese(1)-kes(1,is)
      pxs(1,is)=pxsl(is)
      pxs(2,is)=pxs(mth+1,is)
      do 2 i=2,mth
      kes(i,is)=0.
    2 pxs(i+1,is)=0.
      do 3 i=2,mth
    3 ese(i)=0.
      do 4 km=1,mmax
      esem(1,km)=esem(mth,km)
      do 4 i=2,mth
    4 esem(i,km)=0.
      return
c
c  at t=0 just zero arrays.
   10 mth=nth+1
      do 11 is=1,nsp
      pxs(1,is)=0.
      do 11 i=1,mth
      kes(i,is)=0.
   11 pxs(i+1,is)=0.
      do 12 i=1,mth
   12 ese(i)=0.
      do 13 km=1,mmax
      do 13 i=1,mth
   13 esem(i,km)=0.
      return
      end
```

3-15 PLOTTING AND MISCELLANEOUS SUBROUTINES

The following listings complete ES1: PLOTF, PLOTXV, PLTVXY, PLTFVX, and PLTHST. They make calls to plotting routines peculiar to Livermore; you can infer their function from context well enough to substitute your own. Subroutines FIRST and LAST were discussed already,

MYFRAME simply begins a new plot frame; these are not generally informative enough to list here.

```
      subroutine plotf(f,label,intrvl)
c  plot field at certain times.
      use mparam
      use mfield
      use mcntrl
      real xj(ng1m), f(1), label(2)
      data xj(2)/0./
      if( intrvl.le.0 ) return
      if( (it/intrvl)*intrvl.ne.it ) return
c
      if( xj(2).eq.dx ) go to 2
      ng1=ng+1
      do 1 j=1,ng1
    1 xj(j)=(j-1)*dx
    2 continue
c
      call myframe
      call cartmm(ng,rmin,rmax,f,1)
      call mapg(0.,l,rmin,rmax)
      call trace(xj,f,ng+1)
      call setch(10.,1.,1,0,1,0,1)
      write(ntp,3) label, time
    3 format(2a8," at time=",f10.4)
      return
      end
      subroutine plotxv(il,iu,vl,vu)
c  plot x-vx phase space at certain times.
      use mptcl
      use mparam
      use mcntrl
      if( ixvx.le.0 ) return
      if( (it/ixvx)*ixvx.ne.it ) return
      dxdt=dx/dt
c
c  set velocity range etc. if need be.
      if( vl.lt.vu ) go to 2
      vl=vx(il)
      vu=vl
      do 1 i=il,iu
      vl=amin1(vl,vx(i))
    1 vu=amax1(vu,vx(i))
      vl=vl*dxdt
      vu=vu*dxdt
    2 continue
c
      call myframe
      call maps(0.,l,vl,vu)
      call setpch(1,0,0,0,1)
c  if fewer than 2000 particles, plot pluses for best visibility.
      if( iu-il.gt.2000 ) go to 31
      do 3 i=il,iu
    3 call pointc( "+", (x(i)-.5*vx(i))*dx, vx(i)*dxdt, 1)
      go to 33
c  if more than 13000 particles, plot every 3rd, or 5th...
   31 int=2*((iu-il)/13000)+1
      do 32 i=il,iu,int
   32 call point( (x(i)-.5*vx(i))*dx, vx(i)*dxdt, 1)
   33 continue
c
c  label plot.
      call setch(10.,1.,1,0,1,0)
      tim=time-0.5*dt
      write(ntp,4) tim
    4 format("vx vs. x, time=",f10.4)
      return
      end
```

```
      subroutine pltvxy(il,iu,vmu)
c   plot vx-vy phase space at certain times.
      use mptcl
      use mparam
      use mcntrl
      if( ivxvy.le.0 ) return
      if( (it/ivxvy)*ivxvy.ne.it ) return
      dxdt=dx/dt
c
c   set velocity range etc. if need be.
      if( vmu.ne.0. ) go to 2
      vl=vx(il)
      vu=vl
      do 1 i=il,iu
      vl=amin1(vl,vx(i),vy(i))
    1 vu=amax1(vu,vx(i),vy(i))
      vmu=amax1( abs(vl),abs(vu) )*dxdt
    2 continue
c
      call myframe
      call maps(-vmu,vmu,-vmu,vmu)
      call setpch(1,0,0,0,1)
c   if fewer than 2000 particles, plot pluses for best visibility.
      if( iu-il.gt.2000 ) go to 31
      do 3 i=il,iu
    3 call pointc( "+", vx(i)*dxdt,vy(i)*dxdt, 1)
      go to 33
c   if more than 13000 particles, plot every 3rd, or 5th...
   31 int=2*((iu-il)/13000)+1
      do 32 i=il,iu,int
   32 call point( vx(i)*dxdt, vy(i)*dxdt, 1)
   33 continue
c
c   label plot.
      call setch(10.,1.,1,0,1,0)
      tim=time-0.5*dt
      write(ntp,4) tim
    4 format("vy vs. vx, time=",f10.4)
      return
      end

      subroutine pltfvx(il,iu,vl,vu,q,nbinsp)
c   plot distribution function f(vx) at selected times.
c   if plot interval ifvx is negative, plot is semilog.
      use mptcl
      use mparam
      use mcntrl
      parameter(maxbins=50)
      common /scratch/ vbin(maxbins), fbin(maxbins)
c   nbins must not exceed dimension of these arrays.
      if( ifvx.eq.0 ) return
      if( (it/ifvx)*ifvx.ne.it ) return
      nbins=amin0(nbinsp,maxbins)
      dxdt=dx/dt
c
c   set velocity range etc. if need be.
      if( vl.lt.vu ) go to 2
      vl=vx(il)
      vu=vl
      do 1 i=il,iu
      vl=amin1(vl,vx(i))
    1 vu=amax1(vu,vx(i))
      vl=vl*dxdt
      vu=vu*dxdt
      tem=(vu-vl)/(nbins-2)
      vl=vl-tem
      vu=vu+tem
```

```
      2 continue
        if( vl.eq.vu .or. nbins.le.1 ) return
        dvi=(nbins-1)/(vu-vl)
        dq=abs(q)*dvi
c
c  assign vx's to bins with linear weighting.
        do 3 i=1,nbins
        fbin(i)=0.
      3 vbin(i)=(i-1)/dvi+vl
        do 4 i=il,iu
        tem=(vx(i)*dxdt-vl)*dvi
        j=tem
        if( tem.lt.0. .or. j+2.gt.nbins ) go to 4
        tem=(tem-j)*dq
        fbin(j+1)=fbin(j+1)-tem+dq
        fbin(j+2)=fbin(j+2)+tem
      4 continue
c
        call myframe
        call cartmm(nbins,rmin,rmax,fbin,1)
        if( ifvx.lt.0 ) go to 5
        call maps(vl,vu,0.,rmax)
        go to 6
      5 rmin=amax1(rmin,1.e-5*rmax)
        call mapssl(vl,vu,rmin,rmax)
      6 call trace(vbin,fbin,nbins)
        call setch(10.,1.,1,0,1,0)
        tim=time-0.5*dt
        write(ntp,7) tim
      7 format("f(vx), time=",f10.4)
        return
        end

        subroutine plthst(rec,tim,mth,tl,tu,linlog)
c  plot time history, linear or log.
        real rec(mth), tim(mth)
        call myframe
        call cartmm(mth,rl,ru,rec,1)
        if( linlog.eq.1 .and. rl.ge.0. ) go to 1
        call mapg(tl,tu,rl,ru)
        go to 2
      1 rl=amax1(rl,1.e-5*ru)
        call mapgsl(tl,tu,rl,ru)
      2 call trace(tim,rec,mth)
        call setch(10.,1.,1,0,1,0,1)
        return
        end
```

FOUR

INTRODUCTION TO THE
NUMERICAL METHODS USED

4-1 INTRODUCTION

At this stage we have outlined, with some descriptions of the parts and overall structure, how a program is put together. We are now ready for a little seasoning to make the approaches used a little more palatable (although still not fully supported at this stage) prior to doing projects in the next chapter.

First, we look at the *particle mover*. The leap-frog method is examined for *accuracy*. The magnetic force method is looked at more generally.

Next, we consider the meaning of *particle shapes,* which we have already decided to be finite in size in contrast to point size particles. We look at the *shape factor* $S(x)$ and its Fourier transform $S(k)$. Other forms of weighting are mentioned.

Then we consider the *field solver,* in particular the *Poisson solver,* for *accuracy.* We mention alternative approaches as well as other boundary conditions (nonperiodic).

We say a little bit about the particle and field energies and how and when they are calculated.

4-2 PARTICLE MOVER ACCURACY; SIMPLE HARMONIC MOTION TEST

Let us try a simple model which illustrates the leap-frog method. The model is that of a simple harmonic oscillator, which is described by the second-order differential equation,

$$\frac{d^2x}{dt^2} = -\omega_0^2 x \tag{1}$$

This equation has solutions

$$x(t,t_0) = A(t_0)\cos\omega_0 t + B(t_0)\sin\omega_0 t \tag{2}$$

The finite-difference equations of the leap-frog method (Chapter 2) are

$$\frac{dx}{dt} \rightarrow \frac{x_t - x_{t-\Delta t}}{\Delta t} = v_{t-\Delta t/2} \tag{3}$$

$$\frac{dx}{dt} \rightarrow \frac{x_{t+\Delta t} - x_t}{\Delta t} = v_{t+\Delta t/2} \tag{4}$$

$$\frac{dv}{dt} \rightarrow \frac{v_{t+\Delta t/2} - v_{t-\Delta t/2}}{\Delta t} = \frac{x_{t+\Delta t} - 2x_t + x_{t-\Delta t}}{\Delta t^2} \tag{5}$$

In computation, one starts at some time and works forward, say, with x_t and $v_{t-\Delta t/2}$ given; first, knowing x_t, one finds the force and then solves for $v_{t+\Delta t/2}$ from the acceleration equation, then $x_{t+\Delta t}$ is solved for from the velocity equation. This gives a new x and the process is repeated time and time again. The equations are time centered for stability.

We are now interested in substituting these finite-difference approximations into a homogeneous equation of motion, and then solving it analytically and comparing the solution with the physical solution. The homogeneous equation to be solved is

$$\frac{x_{t+\Delta t} - 2x_t + x_{t-\Delta t}}{\Delta t^2} = -\omega_0^2 x_t \tag{6}$$

This is a standard finite-difference equation, which can be readily solved by assuming solutions of the form

$$x_t = A e^{-i\omega t} \tag{7}$$

A is an initial value and ω is the unknown. Substituting this (and $x_{t-\Delta t} = Ae^{-i\omega(t-\Delta t)}$, etc.) into (6), we obtain

$$\sin\left[\omega\frac{\Delta t}{2}\right] = \pm\omega_0\frac{\Delta t}{2} \tag{8}$$

(8) is plotted in Figure 4-2a. We see that for $\omega_0\Delta t/2 \ll 1$, $\omega \approx \omega_0$ as desired. However, we see that as $\omega_0\Delta t$ increases beyond 2, the (initially wholly) real solution for ω becomes complex, with growing and decaying roots, which indicates *numerical instability*.

Let us find the magnitude of the *phase error* for small $\omega_0\Delta t$. (The amplitude error is nil for $\omega_0\Delta t < 2$.) For $\omega_0\Delta t \ll 1$, one finds

$$\omega\frac{\Delta t}{2} \approx \omega_0\frac{\Delta t}{2}\left[1 + \frac{1}{6}\left[\omega_0\frac{\Delta t}{2}\right]^2 + \cdots\right] \tag{9}$$

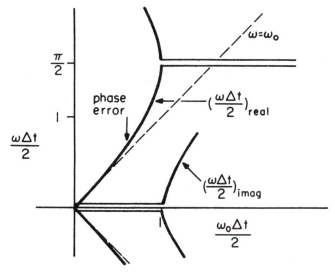

Figure 4-2a Solution for ω in terms of ω_0 for simple harmonic motion, from the leap-frog finite-difference equation. Frequency ω agrees with ω_0 for small $\omega_0\Delta t$, but is larger than ω_0 as $\omega_0\Delta t$ increases. For $\omega_0\Delta t/2 > 1$, the solution becomes complex with growing and decaying roots for ω (this is *numerical instability);* one trademark is the breakup into odd and even steps, the manifestation of the π phase shift. Note the double root at $\omega = 0$, $\pi/2$.

showing a *quadratic error term,* as desired. The cumulative phase after N steps is $\omega N\Delta t$, so that the cumulative

$$\text{Phase error} \approx \omega_0 N\Delta t\frac{1}{6}\left[\omega_0\frac{\Delta t}{2}\right]^2 = \frac{1}{24}N(\omega_0\Delta t)^3 \tag{10}$$

For this error to equal, say, $1/24$ radian, then

$$N = \frac{1}{(\omega_0\Delta t)^3} \quad \text{which} \begin{cases} \text{for } \omega_0\Delta t = 0.1 \text{ is 1000 steps, about 16 cycles} \\ \text{for } \omega_0\Delta t = 0.3 \text{ is 37 steps, about 2 cycles} \end{cases}$$

For this error to equal, say, 1 radian, then

$$N = \frac{24}{(\omega_0\Delta t)^3} \quad \text{which} \begin{cases} \text{for } \omega_0\Delta t = 0.1 \text{ is 24,000 steps, 384 cycles} \\ \text{for } \omega_0\Delta t = 0.3 \text{ is 890 steps, 42 cycles} \end{cases}$$

The points are: restricting the allowable error to small values limits the number of steps (and cycles); increasing the step increases the error as the cube of the step size. A common compromise is $\omega_0\Delta t \approx 0.2$; common usage is 1000 to 10,000 steps, with useful physics sometimes extending to even more steps.

PROBLEMS

4-2a Sketch the locations of a, v, and x for the alternative integrator,

$$\frac{v_{t+\Delta t} - v_t}{\Delta t} = \frac{1}{2}(3a_t - a_{t-\Delta t}) \tag{11}$$

$$\frac{x_{t+\Delta t} - x_t}{\Delta t} = \frac{1}{2}(v_{t+\Delta t} + v_t) \tag{12}$$

Show that this integrator produces a frequency for the simple harmonic oscillator (1) of

$$\omega \approx \omega_0 \left[1 + \frac{1}{6}(\omega_0 \Delta t)^2 + i\,\frac{(\omega_0 \Delta t)^3}{8} + \cdots \right] \tag{13}$$

Hint: Use $x_{t \pm n\Delta t} = z^{\pm n} x_t$, $z \equiv \exp(-i\omega\Delta t)$ and write (11) and (12) as a matrix. Set the determinant of the coefficients equal to zero to produce an equation in z (cubic) to be solved. The phase error is four times larger than that of the leap-frog scheme, and there is mild growth. Show how (with a sketch, as in Figure 2-4a) Δt can be halved or doubled in one step. This method has the disadvantage of requiring storage of the previous acceleration $a_{t-\Delta t}$ in addition to previous velocity v_t and position x_t.

4-2b Discuss the possible use of *integer arithmetic* (no floating point) in the mover, with some care as to the number of bits needed for reasonable accuracy. Keep in mind that small changes are lost; that is, nothing happens unless $v\Delta t$ exceeds the least step in x and $a\Delta t$ exceeds the least velocity. *Hint*: 17 bits is marginal or fatal in quiet start. This coding was exploited by *Estabrook and Tull* (1980) for very high speed, almost twice as fast on the CDC-7600 (in machine language) as ES1 is on the CRAY-I (in FORTRAN)!

4-3 NEWTON-LORENTZ FORCE; THREE-DIMENSIONAL v × B INTEGRATOR

The particle equations of motion to be integrated are

$$m\frac{d\mathbf{v}}{dt} = q(\mathbf{E} + \mathbf{v} \times \mathbf{B}) \tag{1}$$

$$\frac{d\mathbf{x}}{dt} = \mathbf{v} \tag{2}$$

We desire a centered-difference form of the Newton-Lorentz equations of motion. The magnetic term is centered by averaging $\mathbf{v}_{t-\Delta t/2}$ and $\mathbf{v}_{t+\Delta t/2}$, following *Buneman* (1967). The other terms are treated as before. Hence, (1) becomes

$$\frac{\mathbf{v}_{t+\Delta t/2} - \mathbf{v}_{t-\Delta t/2}}{\Delta t} = \frac{q}{m}\left[\mathbf{E} + \frac{\mathbf{v}_{t+\Delta t/2} + \mathbf{v}_{t-\Delta t/2}}{2} \times \mathbf{B} \right] \tag{3}$$

This vector equation for $\mathbf{v}_{t+\Delta t/2}$ can be solved as three simultaneous scalar equations, one for each component. Instead, we choose to obtain a simpler solution using several steps.

The first method (*Buneman*, 1967) is to subtract the drift velocity $\mathbf{E} \times \mathbf{B} / B^2$ from $\mathbf{v}$, as

$$\mathbf{v}'_{\text{old}} = \mathbf{v}_{t-\Delta t/2} - \frac{\mathbf{E} \times \mathbf{B}}{B^2} \tag{4}$$

$$\mathbf{v}'_{\text{new}} = \mathbf{v}_{t+\Delta t/2} - \frac{\mathbf{E} \times \mathbf{B}}{B^2} \tag{5}$$

Similar to (1), this leaves just a rotation of $\mathbf{v}_\perp$ and free acceleration of $v_\parallel$,

$$\frac{\mathbf{v}'_{\text{new}} - \mathbf{v}'_{\text{old}}}{\Delta t} = \frac{q}{m}\left[\mathbf{E}_\parallel + \frac{\mathbf{v}'_{\text{new}} + \mathbf{v}'_{\text{old}}}{2} \times \mathbf{B}\right] \tag{6}$$

We discuss the rotation in Problem 4-3a and Section 4-4.

Another method separates the electric and magnetic forces completely (*Boris*, 1970b). Substitute

$$\mathbf{v}_{t-\Delta t/2} = \mathbf{v}^- - \frac{q\mathbf{E}}{m}\frac{\Delta t}{2} \tag{7}$$

$$\mathbf{v}_{t+\Delta t/2} = \mathbf{v}^+ + \frac{q\mathbf{E}}{m}\frac{\Delta t}{2} \tag{8}$$

into (3); then $\mathbf{E}$ cancels entirely (not just $\mathbf{E}_\perp$), which leaves

$$\frac{\mathbf{v}^+ - \mathbf{v}^-}{\Delta t} = \frac{q}{2m}(\mathbf{v}^+ + \mathbf{v}^-) \times \mathbf{B} \tag{9}$$

which is a rotation (see Problem 4-3a). The steps to compute are: add half the electric impulse to $\mathbf{v}_{t-\Delta t/2}$ using (7) to obtain $\mathbf{v}^-$; rotate according to (9) to obtain $\mathbf{v}^+$, and add the remaining half of the electric impulse (8) to obtain $\mathbf{v}_{t+\Delta t/2}$. These are the same steps, motivated differently, as in Section 2-4. Separation of parallel and perpendicular components is not needed with Boris' method, and the relativistic generalization is straightforward.

Finally, we check the angle of rotation θ which we expect to be close to $\omega_c \Delta t = qB\Delta t/m$. By inspection of Figure 4-3a, we see that

$$\left|\tan\frac{\theta}{2}\right| = \frac{|\mathbf{v}^+_\perp - \mathbf{v}^-_\perp|}{|\mathbf{v}^+_\perp + \mathbf{v}^-_\perp|} = \frac{qB}{m}\frac{\Delta t}{2} = \frac{\omega_c \Delta t}{2} \tag{10}$$

where we have used (9) in the last step. Hence, the difference equation (9) produces a rotation through angle

$$\theta = 2\arctan\left(\frac{qB}{m}\frac{\Delta t}{2}\right) = \omega_c \Delta t\left(1 - \frac{(\omega_c \Delta t)^2}{12} + \cdots\right) \tag{11}$$

which has less than one percent error for $\omega_c \Delta t < 0.35$.

PROBLEMS

4-3a Show that (9) is only a *rotation* of $\mathbf{v}$. *Hint*: take the scalar product of (9) with $(\mathbf{v}^+ + \mathbf{v}^-)$.

4-3b Consider a model with $\mathbf{B}_0$ uniform and $\mathbf{E}_\perp = 0$, in which a particle at speed v has circular motion in the $\mathbf{x}_\perp$ plane. Let the orbit be followed by (subscripts refer to time steps)

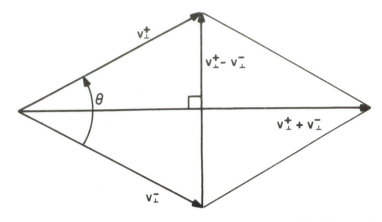

Figure 4-3a Knowing that (9) represents a rotation, we construct this diagram, from which $\tan(\theta/2)$ is readily obtained.

$$\frac{\mathbf{x}^+ - \mathbf{x}^0}{\Delta t} = \mathbf{v}^+ \qquad \frac{\mathbf{x}^0 - \mathbf{x}^-}{\Delta t} = \mathbf{v}^-$$

with $\mathbf{v}^+$ obtained from

$$\frac{\mathbf{v}^+ - \mathbf{v}^-}{\Delta t} = \lambda \left(\frac{\mathbf{v}^+ + \mathbf{v}^-}{2}\right) \times \left(\frac{q\mathbf{B}}{m}\right)$$

as shown in Figure 4-3b. Using $\tan\alpha$ from this figure [*Hint*: see (10)], show that

$$\lambda = \frac{\tan\alpha}{\tfrac{1}{2}\omega_c\Delta t}$$

which is $(\tan\alpha)/\alpha$ if we ask that the mover produce the correct gyrophase, $\alpha = \omega_c\Delta t/2$. Next, requiring that the gyroradius and period be reproduced correctly, show that

$$|\mathbf{v}^+| = |\mathbf{v}^-| = v\left(\frac{\sin\alpha}{\alpha}\right)$$

Last, show that when $\mathbf{E}_\perp$ is included, the λ multiplier appears as $\lambda(\mathbf{E}_\perp + \mathbf{v} \times \mathbf{B})$ in order to produce the correct $\mathbf{E} \times \mathbf{B}/B^2$ drift. [These ideas came from *Hockney* (1966), *Buneman* (1967), and R. H. Gordon (unpublished Berkeley seminar, 1971).]

4-3c (Due to R. H. Gordon). Using any of the $\mathbf{v} \times \mathbf{B}$ integrators given here, for uniform $B_0\hat{z}$, plot particle orbits in the x, y plane:
(a) Without the $\tan\alpha/\alpha$ correction, supposedly with a integral number of steps per cyclotron period; note where points on the second and succeeding cycles are with respect to those on the first.
(b) Repeat (a) with $\tan\alpha/\alpha$ correction; try $0 < \alpha < \pi$.
(c) Like (b), with $\omega_c\Delta t = 2\pi$; compare with true orbit; explain the motion; is this instability?
(d) Like (b), with $\omega_c\Delta t = 2\pi - \delta$; compare with true orbit; explain the motion.

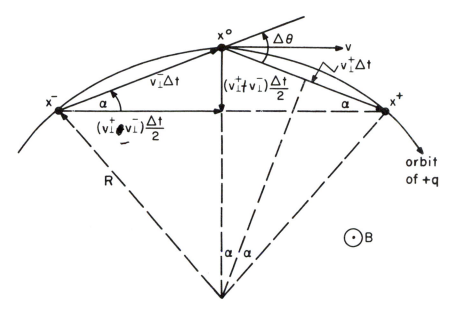

Figure 4-3b Velocities and positions in the plane normal to the uniform magnetic field $\mathbf{B}_0$, with $\mathbf{E}_\perp = 0$ in which the particle orbit is a circle (cyclotron motion). The computer or finite difference orbit is made up of straight line segments connecting old and new positions.

4-4 IMPLEMENTATION OF THE v × B ROTATION

First consider the case in which **B** is parallel to the z axis. In the x-y plane the rotation is through an angle θ where

$$\tan \frac{\theta}{2} = - \frac{qB}{m} \frac{\Delta t}{2} \tag{1}$$

This gives a good approximation to rotation angle θ when θ is not too large [4-3(11)], and is convenient when B is not constant. In ES1, B is fixed so we evaluate $\tan \theta / 2 = - \tan (qB\Delta t / 2m)$; obtaining the correct rotation angle costs nothing more.

Now we use this value of $\tan \theta / 2$ in the half-angle formulas to obtain $\cos \theta$ and $\sin \theta$ for the velocity rotation. Letting

$$t = - \tan \frac{\theta}{2} \tag{2}$$

we have

$$s \equiv - \sin \theta = \frac{2t}{1 + t^2} \tag{3}$$

$$c \equiv \cos \theta = \frac{1 - t^2}{1 + t^2} \tag{4}$$

The rotation becomes

$$v_x^+ = c v_x^- + s v_y^- \tag{5}$$

$$v_y^+ = -s v_x^- + c v_y^- \tag{6}$$

The mover requires no evaluation of transcendental functions, which is a significant time saving when B is not constant. Equations (3) to (6) require 7 multiplies, 1 divide, and 5 adds. *Buneman* (1973) reduces this to:

$$v_x' = v_x^- + v_y^- t \tag{7}$$

$$v_y^+ = v_y^- - v_x' s \tag{8}$$

$$v_x^+ = v_x' + v_y^+ t \tag{9}$$

with 4 multiplies, 1 divide, and 5 adds. The saving of 3 multiplies per particle per time step is desirable.

When the directions of $\mathbf{B}$ and $\mathbf{v}$ are arbitrary, a convenient rotation in vector form is described by *Boris* (1970). First $\mathbf{v}^-$ is incremented to produce a vector $\mathbf{v}'$ which is perpendicular to $(\mathbf{v}^+ - \mathbf{v}^-)$ and $\mathbf{B}$ (see Figure 4-4a).

$$\mathbf{v}' = \mathbf{v}^- + \mathbf{v}^- \times \mathbf{t} \tag{10}$$

The angle between $\mathbf{v}^-$ and $\mathbf{v}'$ is just $\theta/2$, therefore the vector $\mathbf{t}$ is seen from Figure 4-4a to be given by

$$\mathbf{t} \equiv -\hat{\mathbf{b}} \tan \frac{\theta}{2} = \frac{q\mathbf{B}}{m} \frac{\Delta t}{2} \tag{11}$$

Finally, $\mathbf{v}^+ - \mathbf{v}^-$ is parallel to $\mathbf{v}' \times \mathbf{B}$, so

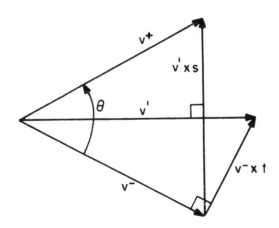

Figure 4-4a Velocity space showing the rotation from $\mathbf{v}^-$ to $\mathbf{v}^+$. The velocities shown are projections of the total velocities onto the plane perpendicular to $\mathbf{B}$.

$$v^+ = v^- + v' \times s \tag{12}$$

where **s** is parallel to **B** and its magnitude is determined by the requirement $|v^-|^2 = |v^+|^2$

$$s = \frac{2t}{1 + t^2} \tag{13}$$

Boris' algorithm is readily made relativistic; see Chapter 15.

PROBLEMS

4-4a Verify that (7) - (9) has the same result as (5) and (6).

4-4b Using $|v^+| = |v^-|$ (a rotation), obtain s (13).

4-4c Show that Boris' rotation satisfies the equation of motion 4-3(9), if $t = q\mathbf{B}\Delta t / 2m$.

4-5 APPLICATION TO ONE-DIMENSIONAL PROGRAMS

In programs with one dimension x and with *two velocities*, v_x and v_y, that allow a magnetic field B_z, the motion is all perpendicular to **B**. Hence, using the vector equations of the previous sections, we obtain the 1d2v accel-rotate-accel algorithm in Section 2-4.

A program with one dimension and with *three velocities* 1d3v may be set up as shown in Figure 4-5a. Let $\mathbf{B}_0$ (constant, uniform) be in the x-z plane, and make an angle θ with the z axis. The self-consistent **E** and **k** must be along x, normal to the sheets. If we stopped here, then the perpendicular motion in z could be ignored ($F_z = 0$). However, on occasion we may be interested in applying an electric field E_{ext} along y which produces a v_y which in turn produces an $F_z = -qv_yB_x$ and a z drift, $(v_E)_z = -(E_{ext})_y / B_x$. The point of the exercise is to include both $v_\parallel$ and $v_\perp$ as well as $k_\parallel$ and $k_\perp$ in this model. It is convenient to have the motion solved for in the parallel and perpendicular directions, which leads to the invention of the x' coordinate ($\perp$ to $\mathbf{B}_0$, at angle θ with respect to x). Call the field due to the charges, $\mathbf{E}_{\text{self-consistent}} \equiv \mathbf{E}_{sc}$. Then the fields are:

$$\mathbf{E}_{sc} = \hat{\mathbf{x}}E_{sc} = \hat{\mathbf{x}}'E_{sc}\cos\theta + \hat{\mathbf{b}}_0 E_{sc}\sin\theta \tag{1}$$

$$\mathbf{E}_{ext} = \hat{\mathbf{y}}E_{ext} \tag{2}$$

$$\mathbf{B} = \mathbf{B}_0 \tag{3}$$

The equations of motion in the (x', y, B_0) coordinates are integrated by the accel-rot-accel method, as follows

$$v_{x_1'} = v_{x'}(t - \Delta t/2) + \frac{q}{m}\frac{\Delta t}{2}E_{sc}\cos\theta \tag{4}$$

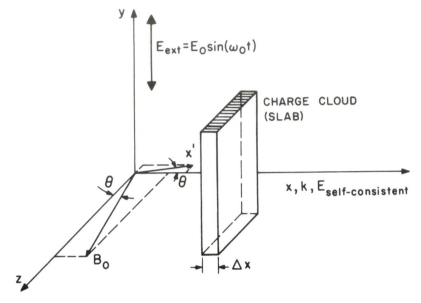

Figure 4-5a One-dimensional sheet model, with x displacement and v_x, v_y, v_z velocities 1d3v. The self-consistent field, that due to the sheets, is along x, as is **k**. There may be an applied electric field along y. The magnetic field is in the x-z plane. *(From Chen and Birdsall, 1973.)*

$$v_{y_1} = v_y \left(t - \Delta t/2 \right) + \frac{q}{m} \frac{\Delta t}{2} E_{\text{ext}} \tag{5}$$

$$v_{x'} (t + \Delta t/2) = v_{x'_1} \cos \left(\omega_c \Delta t \right) + v_{y_1} \sin \left(\omega_c \Delta t \right) + \frac{q}{m} \frac{\Delta t}{2} E_{\text{sc}} \cos \theta \tag{6}$$

$$v_y (t + \Delta t/2) = -v_{x'_1} \sin \left(\omega_c \Delta t \right) + v_{y_1} \cos \left(\omega_c \Delta t \right) + \frac{q}{m} \frac{\Delta t}{2} E_{\text{ext}} \tag{7}$$

$$v_{\hat{B}_0} (t + \Delta t/2) = v_{\hat{B}_0} (t - \Delta t/2) + \frac{q}{m} \Delta t \, E_{\text{sc}} \sin \theta \tag{8}$$

In many problems the magnetic field is uniform and constant, $B_z = B_0$, and is independent of x and t. In this situation $\omega_c \Delta t$ can be calculated once at the start of the problem, and hence also $\cos \omega_c \Delta t$ and $\sin \omega_c \Delta t$.

In some problems B_z may vary in t or x so that one may be forced to obtain $\cos \omega_c \Delta t$ and $\sin \omega_c \Delta t$ each step and for each particle. Time and space centering for such models are treated by *Langdon and Lasinski* (1976) and *Nevins et al.* (1979).

The 1d models of this section have sometimes been called 1½d models due to use of more than the x component of velocity. Early use was given by *Auer et al.* (1961, 1962) for sheet currents and sheet charges and by *Hasegawa and Birdsall* (1964). The work given here on 1d3v is from *Chen and Birdsall* (1973).

PROBLEMS

4-5a With uniform **B** parallel to the z axis and $\mathbf{E}_y = \mathbf{E}_z = 0$, as in ES1, the y component of the equation of motion 4-3(1) can be integrated to find a constant of motion

$$\omega_c X_{gc} \equiv v_y + \omega_c x = \text{constant}$$

where X_{gc} is the location of the *guiding-center* and is related to the canonical momentum p_y. Show that the difference equation 4-3(3) *also* has the analogous exact constant of motion

$$\omega_c X_{gc,t+\Delta t/2} = v_{y,t+\Delta t/2} + \frac{\omega_c}{2}(x_t + x_{t+\Delta t})$$

i.e., show that

$$X_{gc,t+\Delta t/2} \equiv X_{gc,t-\Delta t/2}$$

4-5b Show that the equation of motion

$$\frac{x_{t+\Delta t} - 2x_t + x_{t-\Delta t}}{\Delta t^2} = \frac{q}{m}E_x + \omega_c^2 \left[X_{gc} - \frac{x_{t-\Delta t} + 2x_t + x_{t+\Delta t}}{4} \right]$$

is *equivalent* to 4-3(3) in one-dimension. Here, X_{gc} (defined in Problem 4-5a) is stored instead of v_y. An even simpler equation of motion due to *Byers* (1970) reduces the magnetic term to $\omega_c^2(X_{gc} - x_t)$

4-6 PARTICLES AS SEEN BY THE GRID; SHAPE FACTORS $S(x)$, $S(k)$

The use of a spatial grid, with interpolation to obtain the charge density, leads to the appearance of charges that are at least one cell wide as already shown in Section 2-6, Figure 2-6a, b. This appearance comes from *making observations at the grid points* and from the fact that the fields are calculated from these observations. The corollary is that the particles never behave as if they had zero thickness. Hence, it is wise to be concerned with the *effective shape* of the particles, through the *particle shape factor* $S(\mathbf{x})$ (as already implied in Figures 2-6a(b), 2-6b(b)) and the Fourier transform of the shape factor $S(\mathbf{k})$.

In this section, we present a short analysis of finite-size particles *without grid effects*. In Part Two, we include grid effects in great detail. Here, we simply say that the particles have finite size. This approach is given in detail by *Langdon and Birdsall* (1970) and *Okuda and Birdsall* (1970).

The particles have a spread-out charge distribution and move rigidly without rotation or internal change and pass freely through one another. It seems natural to call these particles *clouds*. The interactions of the system of clouds is a straightforward generalization of the point particle interaction; in fact, as noted in Chapter 1, certain divergences in the kinetic theory and in classical electromagnetic theory are removed.

The charge density at laboratory coordinate $\mathbf{x}'$ of a cloud whose center is at $\mathbf{x}$ is changed from $q\delta(\mathbf{x}' - \mathbf{x})$ for a point particle, to $qS(\mathbf{x}' - \mathbf{x})$ for a cloud, where q is the total charge given by $q \int dx' S(x' - x)$. Let $\mathbf{J}_p$ and ρ_p be the current and charge densities of a system of point charges located at the (x')'s; then the densities $\mathbf{J}_c$ and ρ_c for a system of clouds, whose centers coincide with the point particles, are

$$\begin{bmatrix} \rho_c(\mathbf{x},t) \\ \mathbf{J}_c(\mathbf{x},t) \end{bmatrix} = \int dx' \, S(x' - x) \begin{bmatrix} \rho_p(x',t) \\ \mathbf{J}_p(x',t) \end{bmatrix} \tag{1}$$

These cloud densities are to be used in Maxwell's equations to find the fields $\mathbf{E}$ and $\mathbf{B}$. The Newton-Lorentz force on one cloud of total charge q, with (center) position $\mathbf{x}$ and velocity $\mathbf{v}$ is then

$$\mathbf{F}(\mathbf{x},\mathbf{v},t) = q \int dx' S(x' - x) \cdot [\mathbf{E}(x',t) + \mathbf{v} \times \mathbf{B}(x',t)] \tag{2}$$

These relations are convolutions and, therefore, take on a very simple form when Fourier transformed in space:

$$\begin{bmatrix} \rho_c(\mathbf{k},t) \\ \mathbf{J}_c(\mathbf{k},t) \end{bmatrix} = S(\mathbf{k}) \begin{bmatrix} \rho_p(\mathbf{k},t) \\ \mathbf{J}_p(\mathbf{k},t) \end{bmatrix} \tag{3}$$

$$\mathbf{F}(\mathbf{k},\mathbf{v},t) = qS(-\mathbf{k}) [\mathbf{E}(\mathbf{k},t) + \mathbf{v} \times \mathbf{B}(\mathbf{k},t)] \tag{4}$$

where

$$S(\mathbf{k}) = \int d\mathbf{x} \, S(\mathbf{x}) \exp(-i\mathbf{k} \cdot \mathbf{x}) \tag{5}$$

Our transform convention is such that in the point particle limit, (*particle radius* $R \to 0$) or long-wavelength limit $k \to 0$, $S(k) \to 1$. The size of the cloud by some criterion is denoted by R; then, $S(k)$ becomes small for $|k| \geqslant R^{-1}$.

The shape factor S need not be isotropic (and has not been in practice, *e.g.*, squares and cubes) or symmetric (but usually is). However, in this section we assume that $S(\mathbf{x})$ is isotropic. Therefore, $S(\mathbf{k})$ is isotropic and real valued. (For asymmetrical clouds, the only change in most results is to replace $S^2(k)$ with $|S^2(\mathbf{k})|$.)

Using (3) and a little care, one can now redo most plasma theory with few changes by the replacement of the charge q by $qS(\mathbf{k})$. For example, the dielectric tensor for a uniform Vlasov gas of clouds, and therefore, dispersion relations, are unchanged except that the plasma frequency squared ω_p^2 must everywhere be multiplied by $S^2(k)$ (one S from the equation of motion, another from relating position to density) as,

$$\omega^2 \approx S^2(k) \, \omega_p^2 \tag{6}$$

This result may be viewed as a k-dependent plasma frequency or charge when adapting linear stability analyses, etc., to cloud plasmas. Two shapes have been met so far, those with zero-order and first-order weightings (Figure 2-6b, d) for which the shape factor transforms are, in 1d,

$$S_0(k) = \frac{\sin \dfrac{k\Delta x}{2}}{\dfrac{k\Delta x}{2}} \tag{7}$$

$$S_1(k) = \left(\frac{\sin \dfrac{k\Delta x}{2}}{\dfrac{k\Delta x}{2}} \right)^2 \tag{8}$$

Hence, our first guesses for cold plasma dispersion are as shown in Figure 4-6a. Later, in adding a spatial grid, the finite differencing adds further k-dependence to the dispersion.

In the presence of a uniform imposed magnetic field, the correct k to use in the zero-order cyclotron frequency $[\omega_{c0} = qB_0 S(k)/m]$ is 0, so that ω_{c0} is unchanged from the point-particle value. This is an example of the care that must be used if there are several spatial dependences in the system, which also occurs if there are several waves interacting nonlinearly or if a spatial grid is used.

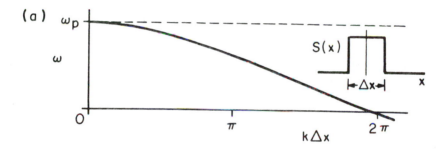

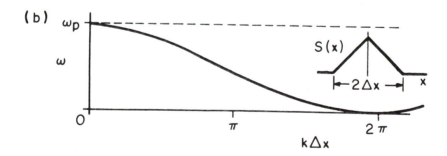

Figure 4-6a (a) Cold plasma dispersion expected for finite-size charges, with shape factor $S(x)$ is shown for zero-order weighting NGP. The Langmuir result, $\omega = \omega_p$, is shown dashed. The drop in ω is due to the smoothing effect of the clouds, as the wavelength becomes comparable to the cloud radius. There is no grid. (b) Similarly, for first-order weighting CIC.

PROBLEMS

4-6a What could happen if we applied the cloud notion inconsistently, *e.g.*, we used a different S in (1) than in (2)? For example, if we do *no* convolution on the fields, then what does the dispersion relation (6) become? The resulting instability, at wavevectors for which $S(k) < 0$, is like that of a system of gravitationally *attracting* particles. Show that the potential energy is lower when there is a density perturbation at such a k than if the density is uniform. See *Langdon and Birdsall* (1970).

4-6b Can different shapes S be used for different species?

4-7 A WARM PLASMA OF FINITE-SIZE PARTICLES

Here we drop the magnetic field and look only at the longitudinal plasma oscillations to see how the simulation differs from the physics for point particles.

The longitudinal dielectric function for a cloud plasma is

$$\epsilon(k,\omega) = 1 + S^2(k)\frac{\omega_p^2}{k^2}\int \mathbf{k} \cdot \frac{\partial f_0}{\partial \mathbf{v}} \frac{d\mathbf{v}}{\omega - \mathbf{k}\cdot\mathbf{v}} \tag{1}$$

with the standard symbol definitions. Space-time dependence $\exp(i\mathbf{k}\cdot\mathbf{x} - i\omega t)$ is assumed, and the usual remarks about analyticity apply. For a Maxwellian velocity distribution with no drift and thermal velocity, $v_t = [v_{av}^2/3]^{1/2}$, the dielectric function becomes

$$\epsilon(k,\omega) = 1 - \frac{1}{2}\left(\frac{S\omega_p}{kv_t}\right)^2 Z'\left(\frac{\omega}{\sqrt{2}kv_t}\right) \tag{2}$$

where Z' is the derivative of the plasma dispersion function of *Fried and Conte* (1961).

The dispersion relation for longitudinal waves is $\epsilon = 0$. When $kv_t/S\omega_p = k\lambda_D/S \ll 1$, we can use the large argument asymptotic expansion for Z' and find an approximate solution for ω which shows weak Landau damping of the oscillations

$$(\text{Re}\,\omega)^2 \approx S^2(k)\omega_p^2 + 3k^2v_t^2 \tag{3}$$

$$\text{Im}\,\omega \approx -\left(\frac{\pi}{8}\right)^{1/2} S\omega_p\left(\frac{S}{k\lambda_D}\right)^3 \exp\left[-\frac{1}{2}\left(\frac{S}{k\lambda_D}\right)^2 - \frac{3}{2}\right] \tag{4}$$

With small clouds (of radius $R < \lambda_D$) and weak damping we have $kR < k\lambda_D \ll 1$ so that $S \approx 1$. Thus, the weakly damped oscillations are little affected with small clouds, as we would hope. See the exact solution for uniform density clouds in Figure 4-7a. For large clouds ($R \geqslant \lambda_D$) and weak damping, $\text{Re}\,\omega$ can be very different from the point-particle result when $kR \geqslant 1$; see Figure 4-7b.

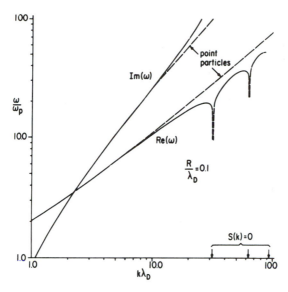

Figure 4-7a Dispersion relation roots (ω vs. k) for small clouds, $R = 0.1\lambda_D$. There is no grid. *(From Langdon and Birdsall, 1970.)*

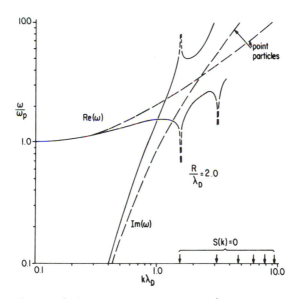

Figure 4-7b Same as Figure 4-7a for large clouds, $R = 2\lambda_D$. *(From Langdon and Birdsall, 1970.)*

When $k\lambda_D/S \gtrsim 1$, the oscillations are strongly damped. Thus, in a cloud plasma the onset of damping as k increases occurs when $k\lambda_D \approx 1$ *or* when kR is large enough. For some cloud shapes, such as cubes, $S \rightarrow 0$ for finite k. Where this happens the asymptotic solutions for strong damping show that $\mathrm{Im}\,\omega \rightarrow \infty$, $\mathrm{Re}\,\omega \rightarrow 0$, which can be seen in Figure 4-2a. Of course, when S is very small the electric interaction is disabled and the clouds free stream. The dispersion roots then do not describe the time evolution of a density perturbation; the free-streaming time evolution goes as $\exp(-\frac{1}{2}k^2v_t^2t^2)$. Where $S(k) \to 0$, a perturbation could be made in density which would produce no E, hence be undamped; this perturbation might recur and cause trouble nonlinearly. At such wavelengths where the clouds strongly affect the plasma, nothing destructive occurs; in fact, very little happens.

This section is intended as an introduction to finite-size particle physics, without the grid. More detail is given on potentials, shielding, energies, collisions, etc., in the articles by *Langdon and Birdsall* (1970) and *Okuda and Birdsall* (1970). These issues are studied in detail in Part Two with a spatial grid.

4-8 INTERACTION FORCE WITH FINITE-SIZE PARTICLES IN A GRID

Having observed the density produced at the grid points, for zero and first-order density weighting, let us now look at the force between two particles, with these weightings, in one dimension. Coulomb's law for sheets says that the force $\propto 1/r^0$, that is, the force is independent of the separation, although it jumps and may change sign as two particles pass; hence, the force with no grid is a step function.

Adding a spatial grid modifies the $1/r^0$ law at short range (charge separation less than the particle thickness) for zero-order and first-order weighting, as follows. (In both weightings, the three-point finite-difference expression for $\partial^2\phi/\partial x^2$ and two-point expression for $\partial\phi/\partial x$ are used, as noted earlier.) The interaction force of a particle at x_1 on a particle at x_2 is taken as (*Langdon and Birdsall*, 1970)

$$F(x_1,x_2) = F\left[\bar{x} - \frac{1}{2}x, \bar{x} + \frac{1}{2}x\right] \tag{1}$$

F depends on the *separation*

$$x \equiv x_2 - x_1 \tag{2}$$

as well as the *mean position in the grid,*

$$\bar{x} \equiv \frac{x_1 + x_2}{2} \tag{3}$$

When nominal point particles are used ($R = 0$ of Section 4-7) with zero-order-weighting (the NPG scheme), then the force is the expected Coulomb step function only for $\bar{x} = \Delta x/2$ (mean position is midway between grid points) but has two steps for other values of $\bar{x}$ as is shown in Figure 4-8a(a). The particles indeed now know the presence of the grid. When nominal

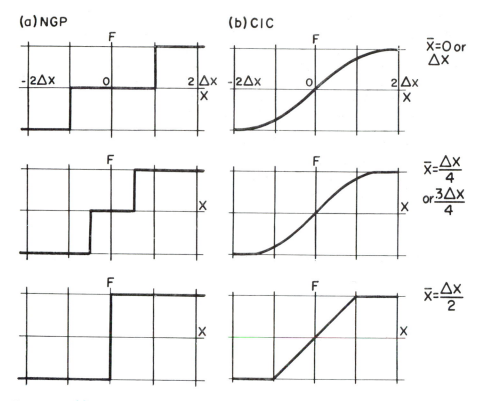

Figure 4-8a (a) Interaction force F for nominal zero-size particles ($R \to 0$, sheets) using the zero-order or nearest-grid point (NGP) density and force weighting. (b) F for nominal clouds of width Δx ($R = \Delta x/2$, slabs), using the first-order or cloud-in-cell(CIC), particle-in-cell(PIC) weightings. *(From Langdon and Birdsall, 1970.)*

clouds of width $2R = \Delta x$ are used with first-order weighting (CIC, PIC), then the force (though not its derivative) is continuous with x as shown in Figure 4-8a(b) and the variation with mean grid position $\bar{x}$ is much less pronounced than with zero-order weighting.

It is informative to separate the interaction force into two parts: an *averaged part* which is invariant under displacement of the grid,

$$\bar{F}(x_1, x_2) = \bar{F}(x_1 - x_2) \tag{4}$$

$$= \frac{1}{\Delta x} \int_{\Delta x} F\left[\bar{x} - \frac{x}{2}, \bar{x} + \frac{x}{2}\right] d\bar{x} \tag{5}$$

and the remainder, which is a *nonphysical grid force,*

$$\delta F = F - \bar{F} \tag{6}$$

If the effects of δF may be neglected, then the system can be analyzed by the methods discussed earlier. For example, in the dispersion relation for plasma oscillations, one must multiply S^2 by $(k\kappa/K^2)$, the finite-difference terms anticipated in Section 2-5, which shows that one effect of the grid is to smooth the interaction, which we have argued is beneficial.

A detailed formulation of the grid force problem shows that the principal consequence of the grid force δF is that perturbations at wavevector k are coupled to perturbations at wavevectors differing from k by integer multiples of $2\pi/\Delta x$ (the *aliasing problem*, taken up in Chapter 8). This coupling is not important when $\lambda_D \gg \Delta x$; then the approximate description in terms of $\bar{F}$ is quite accurate.

We see that finite-size particles in a spatial grid have lost their short-range interactions and pass smoothly through one another with relatively small noise. These effects carry over nicely to two and three dimensions, and leads to much reduced cross sections and no large-angle scattering; *i.e.*, collisional and short-wavelength fluctuations are much reduced. In addition, for cloud size $R \leqslant \lambda_D$, longitudinal waves and Debye shielding are nearly the same as for a laboratory plasma. All of this is covered in more detail later.

PROBLEMS

4-8a Copy Figure 4-8a(a) and overlay the $\bar{x} = 0$, $\Delta x/4$, $\Delta x/2$ graphs. This makes the dependence of F on grid location more emphatic.

4-8b Repeat (a) for Figure 4-8a(b).

4-8c Construct a graph for second-order weighting using a quadratic spline, supporting your sketches with analysis. (See Section 8-8 for formulation of the spline.) Repeat the overlay given in (a) and observe that the position of the grid changes F very little, much less than with zero and first order weighting.

4-9 ACCURACY OF THE POISSON SOLVER

As with the particle mover where we found an error term $\propto (\Delta t)^2$, we would like to have some idea of the error in the field solutions, and we want an error term to be $\propto (\Delta x)^2$ or smaller.

Let us assume that we know the ρ_j and are now ready to find the ϕ_j using Poisson's equation,

$$\frac{d^2\phi(x)}{dx^2} = -\frac{\rho(x)}{\epsilon} \tag{1}$$

This must be put in finite-difference form as we know ρ only at the grids and so find ϕ only at the grids, ϕ_j. Fortunately, there is a great store of information and practice on writing and solving such equations.

From *Collatz* (1966) (see also the error analysis of *Forsythe and Wasow*, 1960), we find

$$\frac{\phi_{j-1} - 2\phi_j + \phi_{j+1}}{(\Delta x)^2} + \left[\frac{1}{12}(\Delta x)^2 \phi_\epsilon^{IV}\right] = -\rho_j \tag{2}$$

with Collatz's $h \to \Delta x$, $y_0'' \to -\rho_0$, $j - 1 < \epsilon < j + 1$. This form is the usual 3 points, which goes to 5 points in two dimensions 2d and 7 points in 3d. The error is the $[(\Delta x)^2 \phi^{IV}]$ term, which is $\propto (\Delta x)^2$, as desired, making it quite small for most of the usable range (say, $0 < k\Delta x < \pi/2$, with $\pi/2 < k\Delta x < \pi$ smoothed away). This form is widely used.

A second form, of higher accuracy, is,

$$\frac{\phi_{j-1} - 2\phi_j + \phi_{j+1}}{(\Delta x)^2} + \left[\frac{1}{240}(\Delta x)^4 \phi_\epsilon^{VI}\right] = -\frac{\rho_{j-1} + 10\rho_j + \rho_{j+1}}{12} \tag{3}$$

which goes to 9 points in 2d and 19 in 3d. While the accuracy of the second form (and there are others) appears to recommend it (and it requires only a small amount more effort, still using information only at the j, $j \pm 1$ planes), remember the earlier hint: *it is the over-all system error that counts.* Indeed, while the second form above is superior in the Poisson step, it may not aid as well in compensating for errors elsewhere in the force calculation as well as does the first.

PROBLEMS

4-9a In order to convince yourself of the statement on accuracy, assume that

$$\phi(x) = A \cos\frac{2\pi mx}{L} \quad m = 1, 2, \ldots, m_{\max}$$

The largest allowable k $(=2\pi m/L)$ is $\pi/\Delta x$ so that $m_{\max} = L/2\Delta x = NG/2$. Devise a relative error term and plot versus m in the range $(1 \leqslant m \leqslant NG/2)$ for (2) and (3). At what $m/m_{\max}$ is the error becoming large (say, 5 percent)? A smoothing factor is generally used beyond some $m/m_{\max}$ (like the 5 percent error value), attenuating the potentials at short wavelengths.

4-10 FIELD ENERGIES AND KINETIC ENERGIES

In all simulation methods, we generally keep track of quantities like field or potential (PE), kinetic (KE), and total (TE) energies (while being careful with energy sources and sinks, if any) and of particle momenta. These quantities are plotted versus time at the end of the program, and usually display considerable interesting physics (*e.g.*, growth rates, exchange of PE with KE, saturation of instabilities, changes of state). In addition, independent of such program names as energy conserving or momentum conserving etc. (usually valid names only in some limit, like $\Delta t \to 0$), we wish to see how much these quantities vary from expected values as a measure of trustworthiness of the over-all program. Hence, there should be some care in calculation of PE, KE, and TE as well as in display (log plots, suppressed zeros, etc.).

The electrostatic field energy (ESE) in our ES1 model is obtained from summing the $\rho_k \phi_k$ product, as

$$\text{ESE} = \frac{1}{L} \sum_{k=k_0}^{k_{max}} \rho_k \phi_k^* \qquad (1)$$

drawing on the Parseval equality

$$\frac{1}{2} \int_0^L \rho(x)\phi(x)\,dx = \frac{1}{L} \sum_k^\infty \rho_k \phi_k^* \qquad (2)$$

As $\rho(x)$, $\phi(x)$, ρ_k, ϕ_k are known at t, $t + \Delta t$, etc., the ESE is found at those times. those times.

The field energy, one might argue, may also be obtained from the electric field, as

$$\frac{1}{L} \sum_k |E_k|^2 \qquad (3)$$

Several points need to be noted if one were to use $|E_k|^2$. First, there is a real difference between $\rho_k \phi_k^*$ and $|E_k|^2$. If we take ρ_k as given and find ϕ_k and E_k as in 2-5(10) and 2-5(12), then

$$\frac{|E_k|^2}{\rho_k \phi_k^*} = \frac{\kappa^2}{K^2} \qquad (4)$$

which, using the differencing of 2-5(4), 2-5(5) is

$$\frac{\kappa^2}{K^2} = \cos^2\left(\frac{k\Delta x}{2}\right) \qquad (5)$$

Hence, $|E_k|^2$ drops off at large $k\Delta x$ relative to $\rho_k \phi_k^*$ so that the sum of $|E_k|^2$ over k is appreciably smaller than that of $\rho_k \phi_k^*$, especially when there is relatively large energy at short wavelengths. The real point is that the basic potential energy calculation is that of $q\phi$, summed over all charges, which is generalized to $\frac{1}{2}\int \rho\phi\, d(\text{volume})$; then, via Coulomb's law ($E \propto q/r^2$), this integral is transformed to $\frac{1}{2}\int E^2 d(\text{volume})$. Hence, in our model, where we have modified Coulomb's law (attenuated the short-range forces, as shown in Figure 4-8a), we cannot expect the $\frac{1}{2}\int E^2 d(\text{volume})$ form to agree with the $\rho\phi$ calculation which is correct for any force law. That is, we take $\frac{1}{2}\int \rho\phi\, d(\text{volume})$ to be correct. Furthermore, even with the grid, the $\rho\phi$ calculation is correct for the energy conserving algorithm (IW = 3). We say more on the subject in Part Two.

The kinetic energy of the particles (KE), is simply a sum over $\frac{1}{2}mv^2$ of all particles; or is it? If we obtained $\frac{1}{2}\sum_{p=1}^{NP} mv^2$, then we would produce KE at times $t + \Delta t/2$, $t + 3/2\Delta t$, etc., interlaced but not simultaneous with ESE.

We can do better, as already noted in Section 3-11, by averaging KE in some way. Between old and new times $(t - \Delta t / 2, t + \Delta t / 2)$, we may choose the mean square velocity from among

$$\frac{1}{2}(v_{new}^2 + v_{old}^2) \tag{6}$$

$$\left[\frac{v_{new} + v_{old}}{2}\right]^2 \tag{7}$$

$$v_{new} v_{old} \tag{8}$$

On the basis of which is quickest to calculate, the last form is obviously the winner. All of the forms have the same value through order Δt, differing only in $(\Delta t)^2$ terms, which is the order of accuracy of the leap-frog integrator which produces v_{new} and v_{old}; hence, a better KE form (better interpolation) would imply an accuracy that is not there.

PROBLEMS

4-10a Obtain the results given in (4) and (5).

4-10b Suppose that ρ_k is smoothed by a factor $SM(k)$ (usually to attenuate the noise at large $k\Delta x$) and that ϕ_k is obtained from $\rho_k SM(k) \equiv \rho_{k\ smoothed}$. Show that (4) still holds.

4-10c List the number of adds and multiplies for each of the forms (6), (7), and (8). Calculate each of the forms, assuming v_{new} expressed in terms of v_{old}, to prove the statement in the text that (6), (7), and (8) are the same through order Δt and obtain the differences through order $(\Delta t)^2$.

4-10d Show that, if $\frac{1}{2}\sum mv_{new}^2$ is used as the kinetic energy, then in a cold plasma oscillation the total energy (kinetic plus electric) oscillates with amplitude $\propto (\omega_p \Delta t)$.

4-10e Look ahead to Part Two and show whether or not the $(v_{new}v_{old})/2$ form makes KE + PE exact for the energy conserving algorithm. For cold oscillations only? For a warm plasma?

4-11 BOUNDARY CONDITIONS FOR CHARGE, CURRENT, FIELD, AND POTENTIAL

A one-dimensional *periodic system* is usually thought of as a part of an infinite system. However, there are several variations worth considering, including open circuit, short circuit, and driven systems.

First, let us consider the implications of a periodic field. Integrating $\partial E / \partial x = \rho / \epsilon$ over a period L produces

$$\int_x^{x+L} \frac{\partial E}{\partial x} dx = E(x + L) - E(x) = \frac{1}{\epsilon} \int_x^{x+L} \rho \, dx = \frac{L}{\epsilon} <\rho> \tag{1}$$

That is, the average charge density vanishes, $<\rho> = 0$, if the field is periodic. In ES1 there is no net charge in any period, $\rho(k = 0) = 0$.

Second, if the potential is taken to be periodic, then integrating $\partial\phi/\partial x = -E$ over a period L,

$$\int_{x}^{x+L} \frac{\partial\phi}{\partial x} dx = \phi(x + L) - \phi(x) = -\int_{x}^{x+L} E\, dx = -L <E> \tag{2}$$

forces $<E> = 0$, which, if calculated in ES1, would mean $E(k = 0) = 0$. ES1 has provision for adding $E_0 \cos \omega_0 t$ uniform field, a *driven* system, which implies a non-periodic potential. Where we do use a periodic potential, with $\phi(0) = \phi(L)$, then we call the system a *short circuit*.

Third, we need to consider the total current density, convection plus displacement,

$$\mathbf{J}_{total} = \rho\mathbf{v} + \frac{\partial\mathbf{E}}{\partial t} \tag{3}$$

In one dimension, this must be independent of x, a result which comes from

$$\nabla \times \mathbf{H} = \mathbf{J}_{total}, \qquad \nabla \cdot \nabla \times \mathbf{H} \equiv 0 = \nabla \cdot \mathbf{J}_{total} \tag{4}$$

This is $\partial J_{total,x}/\partial x = 0$; hence, $J_{total,x}$ is dependent of x, $J_{total,x}(k = 0, t)$ may exist, dependent on time, simply $J(t)$.

Hence, in contrast with ES1 with periodic E and ϕ, the *driven* one-dimensional plasma model, Figure 4-11a, may have net values of field or charge or potential difference; *i.e.*, there may be $k = 0$ components of E, ρ, or ϕ, in addition to J_{total}. These components may be induced by various methods, which may or may not be specified. We need additional means to obtain the $k = 0$ components, as follows. Integrate (3) over the region, of length L, using $J_{total} = J_{total}(t)$, $E = -\partial\phi/\partial x$, and, for a model with N sheets replace $\int_{0}^{L} \rho v\, dx$ with $\sum_{i=1}^{N} \rho_{si} v_i$ (as $\rho dx = \rho_s$), to obtain

$$J(t) = \frac{1}{L}\sum_{i=1}^{N}\rho_{si} v_i - \frac{1}{L}\frac{\partial}{\partial t}[\phi_b(t) - \phi_a(t)] \tag{5}$$

where $x = 0$ is the a plane and $x = L$ is the b plane. The first term is the spatial average of the convection current, the $k = 0$ component, also called the *induced current* or *driving current* in the parlance of devices. The second term is the displacement current calculated as if *no* charges were present $0 < x < L$, the $k = 0$ value. (For more details on devices, see *Birdsall and Bridges,* 1966, especially Section 1-3 on the steps above.)

For a region driven by an *external current source* (a current generator equivalent circuit is an open circuit) which is given as a function of time, the $k = 0$ value of E may be advanced in time, from E^n to E^{n+1}, using (3) or (5) in the form

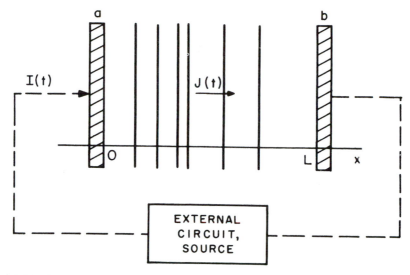

Figure 4-11a Driven plasma model, bounded by planes a and b which may be emitters, absorbers, reflectors, or transparent (*i.e.*, ends of a periodic system). $J_{total}(t)$ is the total current density in the region (independent of x) and times the area forms the current I in the external circuit.

$$J_{external}^{n+\frac{1}{2}} = J_{plasma,k=0}^{n+\frac{1}{2}} + \frac{E_{k=0}^{n+1} - E_{k=0}^{n}}{\Delta t} \qquad (6)$$

where

$$J_{plasma,k=0}^{n+\frac{1}{2}} = \frac{1}{L}\sum_{i=1}^{N} q_i\, v_i^{n+\frac{1}{2}} \qquad (7)$$

as the v_i are already known. The periodic $E(k \neq 0)$ are obtained as in ES1, keeping overall neutrality, but allowing for nonperiodic ϕ.

A region connected to an *external circuit*, through grid planes a and b is called a *device*. In such a model which may have nonzero net charge, the *external circuit equation* must also be supplied, such as

$$\phi_0(t) - \phi_L(t) = I(t)R \qquad (8)$$

for an external resistor (see *Birdsall and Bridges*, 1966, sec. 3-12); for external L and C, use the usual Kirchhoff circuit equations as given in detail in Chapter 16. The potential within the device is now due to the charges within the region $0 < x < L$ *and* to charges on the electrodes at $x = 0, L$. Whereas in the periodic ES1, the fields and particles are treated similarly, here the fields and particle boundary conditions may be specified separately. Thus, we need both the particular and homogeneous solutions of Poisson's equation, with the former due to the free charges inside the device (much as obtained in the periodic model) and the latter due to charges at $x = 0, L$, as

$$\phi_{\text{homogeneous}}(x,t) = A(t)x + B(t) = [\phi_L(t) - \phi_0(t)]\frac{x}{L} + \phi_0(t) \qquad (9)$$

$$E_{\text{homogeneous}}(t) = -\frac{\phi_L(t) - \phi_0(t)}{L} \qquad (10)$$

These are the $k = 0$ solutions to be added to the $k \neq 0$ solutions which are obtained from the Poisson solver. However, the appropriate Poisson solver boundary conditions are now the short-circuit conditions

$$\phi_0(t) = 0 = \phi_L(t) \qquad (11)$$

These require changing the periodic code solutions for $\phi(x)$ with both $\cos kx$ and $\sin kx$ terms to one with $\sin kx$ terms only. One easy way to do this (minimal code changes in ES1) is to *invert* the record of $\rho(x)$ from $x=0$ to $x=L$ into the region $L < x < 2L$, as shown in Figure 4-11b; this makes $\rho(x)$ into an odd function $0 < x < 2L$, producing $\sin kx$ terms in ρ and ϕ, as desired, with $<\rho> = 0$ over $0 \leqslant x \leqslant 2L$. The inverse of $\phi(k)$ is used, of course, only in the space, $0 < x < L$ to produce the $k > 0$ solutions; the $k = 0$ homogeneous solutions must be added. Another method of solution where there is net charge and biased electrodes, is to use a direct solution, as in Appendix D.

Another model obtainable from periodic ES1 is that used to model the sheath around a floating probe at $x = L/2$ (*Birdsall*, 1982). The probe is simply the center grid plane, programmed to accumulate any active charge passing it, deleting such charges from the list of active particles. The boundaries at $0 = x = L$ are biased to zero potential. Particles are used only in the left half of the region, with the weighted density mirrored into the right half (but not inverted), taking advantage of the symmetry. Particles passing $x = 0$ are reflected specularly. The initial conditions are a thermal plasma with net charge of zero, which remains zero as the particles drain slowly to

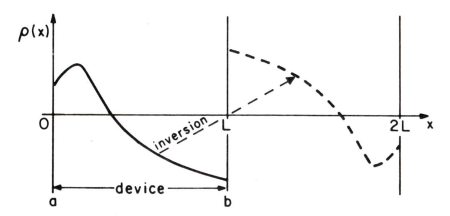

Figure 4-11b Charge density $\rho(x)$ in the device, $0 < x < L$. This is inverted through $x = L$ into the interval $L < x < 2L$, in order to make a density in $0 < x < 2L$ which is odd and has $<\rho> = 0$, producing only $\sin kx$ terms in $\rho(k)$ and hence in $\phi(k)$.

the probe. After a charging transient (about three plasma cycles) the floating potential drops to the value predicted from time-independent kinetic theory by *Emmert et al.*, (1980). With no sources and no collisions, the faster electrons are all absorbed early on and the probe potential slowly rises. Using a pair creating source (to maintain neutrality) over some region in x, presumably would result in reaching a steady state, with fluctuations.

PROBLEM

4-11a Sketch the nominal charge and invent the first-order charge assignment to the grid for $x_i = -\Delta x/4$, 0, $\Delta x/2$ for an emitting wall at $x = 0$. Do the same for the last four x_i's for a reflecting wall at $x = 0$. Repeat using zero-order weighting. Compare your inventions with the algorithm proposed in Figure 16-9b.

FIVE

PROJECTS FOR ES1

5-1 INTRODUCTION

We are ready to have fun with plasma projects. In this chapter, we become quite intimate with a plasma by following in detail *oscillations*, *waves*, and *instabilities*. We provide theory to go with the projects, such as the amplitudes needed for particle crossing and the peak value of electric field expected in various instabilities. Our object in running these simulations is to learn basic "laboratory plasma behavior" starting with cold plasmas.

This chapter consists of relatively detailed sections connected with *projects* requiring one or more computer runs of a few tens or hundreds of time steps. In sections where computer simulations are to be done using ES1, we put *project* in the section heading. These sections have details on to how to do the simulation, what to observe, and what to calculate in order to help explain what is observed. The work assigned or implied varies from trivial to difficult. Students benefit from doing these projects and writing them up in a format which includes statement of the problem, theory or analysis, choice of parameters, simulation results, and comparison with theory.

5-2 RELATIONS AMONG INITIAL CONDITIONS; SMALL AMPLITUDE EXCITATION

Theories usually give us an idea of the initial charge or particle *density*. However, to start the code, we need the initial *position* and *velocity* of all the particles. Hence, we derive particle positions from the particle densities and then display all quantities, with proper phase and amplitude relations, at $t = 0$, for small-amplitude sinusoidal excitation.

Particle positions are the x's. Let there be two neighboring particles, with positions x_a and x_b, as shown in Figure 5-2a; the uniform plasma has the particles at x_{0a} and x_{0b}, with zero-order uniform density (one dimensional),

$$n_0 = \frac{1}{x_{0b} - x_{0a}} \tag{1}$$

We now perturb the zero-order positions, to

$$x_a = x_{0a} + \delta x_a ; \quad x_b = x_{0b} + \delta x_b \tag{2}$$

so that, at some x, $x_a < x < x_b$, the new density is

$$n(x) = n_0 + n_1(x) = \frac{1}{x_b - x_a} \tag{3}$$

$$= n_0 \frac{1}{1 + \dfrac{\delta x_b - \delta x_a}{x_{0b} - x_{0a}}} \tag{4}$$

Another point of view is to call the zero-order quantity of fluid between x_{0b} and x_{0a}, Δf. Then $n(x) = \Delta f / (x_b - x_a)$. As the order of a and b is immaterial, replace $(x_b - x_a)$ with $|x_b - x_a|$. In order to account for more than one element of fluid appearing at x, then $n(x)$ is to be a sum over $\Delta f / |\Delta x|$. Let the displacement from uniformity $\delta x(x_0)$ be a continuous variable which may be considered as the displacement of an element of fluid,

$$\delta x(x_0) \equiv x - x_0 \tag{5}$$

then we have

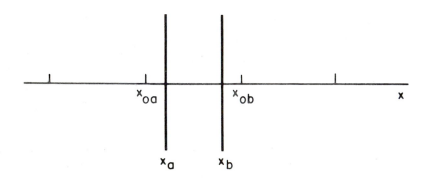

Figure 5-2a Unperturbed adjacent particle positions, x_{0a} and x_{0b}, and their perturbed positions, x_a and x_b.

$$n(x) = n_0 \frac{1}{1 + \dfrac{\partial \delta x}{\partial x_0}} \tag{6}$$

The solution is [assuming that n_0 and $n_1(x)$ are given, and δx is to be found],

$$\frac{\partial \delta x}{\partial x_0} = \frac{-n_1(x)}{n_0 + n_1(x)} \tag{7}$$

We can identify some measure of smallness by noting that particles touch at

$$x_b = x_a, \quad \text{or} \quad \frac{\partial \delta x}{\partial x_0} = -1 \tag{8}$$

producing a peak in density, $n(x)$ [and $n_1(x)$] $\to \infty$. See Figure 5-2b.

If we are interested in small-amplitude excitation, $|n_1(x)| \ll n_0$, then $|\partial \delta x / \partial x| \ll 1$ and

$$\frac{\partial \delta x}{\partial x} \approx \frac{-n_1(x)}{n_0} \tag{9}$$

In three dimensions, this would read $\nabla \cdot \delta \mathbf{x} = -n_1(\mathbf{x}) / n_0$; the derivation is given in Section 9-2.

Let the excitation be periodic, sinusoidal in x,

$$n_1(x) = A \sin kx \tag{10}$$

with perturbed charge density

$$\rho_1(x) = q n_1(x) = q A \sin kx \tag{11}$$

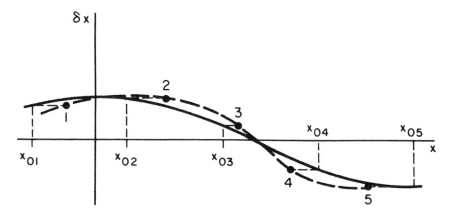

Figure 5-2b Uniform particle positions x_{0i} with their sinusoidally perturbed positions (heavy dots) obtained from $x_i = x_{0i} + \delta x(x_{0i})$. Note that the first particles to touch (and cross) are 3 and 4, where $\partial \delta x / \partial x_0$ has its largest negative value; they touch at $\partial \delta x / \partial x_0 = -1$ (for the x_{0i} very closely spaced, that is, many particles in a wavelength of the perturbation). Or take $x_{02} - x_{03}$ to be an element of fluid, displaced to $x_2 - x_3$.

Presumably either A or qA is known (as well as k). Then the displacement δx is obtained by integration of (9), as

$$\delta x(x) = x - x_0 = \frac{A}{n_0} \frac{1}{k} \cos kx \tag{12}$$

The perturbed electric field is obtained from $\nabla \cdot \mathbf{E}_1 = \rho_1/\epsilon_0$, also by integration in x, to produce

$$E_1(x) = -\frac{qA}{\epsilon_0} \frac{1}{k} \cos kx \tag{13}$$

Lastly, the potential is obtained from the field ($\nabla \phi_1 = -\mathbf{E}_1$), by integration or from $\nabla^2 \phi = -\rho_1/\epsilon_0$ (take your pick),

$$\phi_1(x) = \frac{qA}{\epsilon_0} \frac{1}{k^2} \sin kx \tag{14}$$

These relations are shown in Figure 5-2c, with amplitudes relative to the initial density perturbation, $|n_1| = A$, and proper phases. Working backward from a small initial displacement ($x_1 \ll x_{02} - x_{01}$),

$$x = x_0 + x_1 \cos kx \tag{15}$$

we obtain a *peak particle density* of

$$|n_1| = A = (n_0 k) x_1 \tag{16}$$

with a *field peak* of

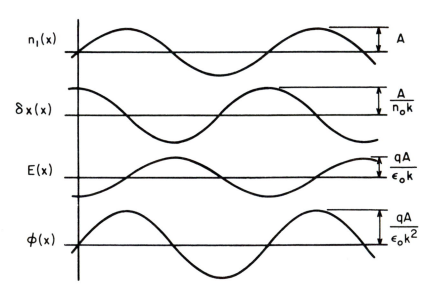

$$
\begin{array}{ll}
n_1(x) & A \\
\delta x(x) & \dfrac{A}{n_0 k} \\
E(x) & \dfrac{qA}{\epsilon_0 k} \\
\phi(x) & \dfrac{qA}{\epsilon_0 k^2}
\end{array}
$$

Figure 5-2c Starting from a small perturbation in density $n_1(x)$ with amplitude A, the perturbations in displacement δx, field E, and potential ϕ are shown, with relative amplitudes.

$$|E_1| = \frac{qA}{\epsilon_0 k} = \frac{qn_0}{\epsilon_0} x_1 \tag{17}$$

and a *peak acceleration* of

$$\frac{q}{m}|E_1| = \frac{q}{m}\frac{qn_0}{\epsilon_0}x_1 = \omega_p^2 x_1 \tag{18}$$

and a *peak potential* of

$$|\phi_1| = \frac{qn_0}{\epsilon_0 k} x_1 \tag{19}$$

This goes with a *peak electrostatic energy density* of

$$(W_E)_{\max} = \frac{1}{2}[\rho_1(x)\phi_1(x)]_{\max} = \frac{1}{2}q[n_1(x)\phi_1(x)]_{\max} = \frac{1}{2}\frac{q^2 n_0^2}{\epsilon_0}x_1^2 \tag{20}$$

If we have a thermal (warm) plasma with zero-order kinetic-energy density of

$$W_K = \frac{1}{2}n_0 m v_t^2 \tag{21}$$

then the ratio of $(W_E)_{\text{average}}$ (due to the first-order perturbations) to W_K is

$$\frac{W_E}{W_K} = \frac{1}{2}\frac{q}{m}\frac{qn_0}{\epsilon_0 v_t^2}x_1^2 = \frac{1}{2}\frac{\omega_p^2}{v_t^2}x_1^2 = \frac{1}{2}\left(\frac{x_1}{\lambda_D}\right)^2 \tag{22}$$

The answer is neat, but should be interpreted with care.

In following an instability, the field grows to some E_{peak}, corresponding to a $(W_E)_{\text{peak}}$; usually

$$(W_E)_{\text{peak}} \lesssim f W_K \tag{23}$$

where f is some fraction like 1/10 or 1/100. If we wish to follow the growth from small amplitudes (where linear analysis applies and theoretical growth rates can be compared with the simulation), then the initial field energy $(W_E)_{\text{initial}}$ must be, say, 10^{-3} of the final value, that is,

$$(W_E)_{\text{initial}} \lesssim 10^{-3}(W_E)_{\text{peak}}; \qquad (W_E)_{\text{peak}} \lesssim 10^{-2}W_K \tag{24}$$

or, roughly,

$$(W_E)_{\text{initial}} \lesssim 10^{-5}W_K \tag{25}$$

This says, for a thermal plasma [stretching the meaning of (22)],

$$\left(\frac{x_1}{\lambda_D}\right)^2 \lesssim 10^{-5}, \quad x_1 \lesssim 3 \times 10^{-3}\lambda_D \tag{26}$$

For a cold plasma, say, a cold beam of velocity v_0, and $W_K = \frac{1}{2}nmv_0^2$, then the same relation exists with λ_D replaced by v_0/ω_p; this presumes that inequalities (23) and (24) still hold.

The manner of estimating how to choose x_1, (21) to (26), is crude, but it is best to make some estimate rather than none.

PROBLEMS

5-2a Using (6) for $n(x)$ and $\delta x = B \sin kx$, sketch $n(x)$ for $0 < x < 2\pi$ for $kB = 0.1$, 0.5, and 0.9; sketch particle positions as in Figure 5-2b also. At $kB = 1.0$, $n(x) \to \infty$ at $kx = \pi$. What about using $kB > 1$ in (6), producing $n(x) < 0$, nonphysical? Change the formulation to allow for $\partial \delta x / \partial x_0 < -1$, particles having crossed.

5-2b Show that the perturbation procedure in subroutine INIT produces an initial distribution

$$f(x, v, 0) = [1 - kx_1 \sin(kx + \theta_x)]^{-1} f_-[v - v_1 \sin(kx + \theta_v)]$$

where $f_-(v)$ is the spatially-uniform distribution before the perturbation.

5-3 COLD PLASMA (OR LANGMUIR) OSCILLATIONS; ANALYSIS

A "cold plasma" is both a self-contradiction (as most plasmas have at least one velocity component with thermal energy greater than about $1 \, \text{eV} \approx 10^4 \text{K}$), and a bit singular (vanishing Debye length, $\lambda_D \to 0$; addition of thermal energy to make $\lambda_D > 0$ alters the behavior markedly). Nonetheless, we can learn much by using the simplification of $v_{\text{thermal}} \to 0$, which allows us to use simple initial conditions and easily understood diagnostics. Initially we restrict our calculations to regions where $k\lambda_D \lesssim 0.1$.

From your earlier exposure to uniform plasmas, you are well aware that the charged particles have *simple harmonic motion* about their equilibrium positions (the x_0's), as described by

$$\delta x \equiv x - x_0, \quad \delta \ddot{x} = -\omega_p^2 \delta x \tag{1}$$

with solutions

$$\delta x(t - t_0) = A(t_0) \cos \omega_p(t - t_0) + B(t_0) \sin \omega_p(t - t_0) \tag{2}$$

The common derivations make assumptions, *e.g.* $m_i \to \infty$, or that particles are distributed continuously in space, or that particles do not cross. Chances are, because the results came out so neatly (so obviously correct), that you never seriously questioned the assumptions or results. Let us do some questioning now.

First, there is no problem with letting both electrons and ions move. The result is the same: both species oscillate harmonically, at the same frequency ($\omega^2 = \omega_p^2 = \omega_{pe}^2 + \omega_{pi}^2$), but with the perturbed ion density and velocity smaller than that of the electrons by the factor m_e / m_i. Hence, we treat the ions as stationary and take $m_i / m_e \to \infty$ for high-frequency oscillations.

Second, for our computational purposes, we need to decide how to put in the ions, or we may decide to ignore their motion and follow only that of the electrons.

Suppose that we ignore the spatial grid for a moment and look at a uniform distribution of electrons and ions in one dimension (sheets), Figure 5-3a. Let each sheet have charge q or $-q$; the system is neutral in the sense that the net charge is zero. If we displace the electrons a distance δx less than half of the uniform ion spacing, $\delta x < \frac{1}{2}(x_{02} - x_{01})$, then each electron is attracted back toward its parent ion; the electrons oscillate about the ions. Is the motion simple harmonic? *No!* The detailed motion is to be obtained as a problem. (A simpler problem is to consider an isolated electron-ion pair; the force on either particle is independent of the separation in one dimension, so that while there is oscillation, it is not simple harmonic.)

The next step is to resolve this difficulty. One way is to put in more ions (say, five ions each of charge $q/5$ for each electron of charge $-q$) spaced uniformly; then, excite the electrons so that they pass several ions in their oscillatory swing. Between ions, the force is more or less constant (it is for an isolated group of ions and one electron), jumping up each time an ion is crossed; this begins to approximate the desired force (proportional to displacement from equilibrium). Indeed, the difficulty goes away entirely if the ions are distributed continuously, in a *uniform background.*

If we now add a spatial grid, smearing out the ions to make a uniform background is relatively easy. In a *periodic system,* where there is no net charge in a period, the simulation program can zero-out the net charge (in effect putting in a uniform background equal to the imbalance in charge) simply by zeroing the average term in the Fourier series for charge density, $\rho(k = 0) = 0$. Or, we may put in, say, one ion particle per cell and skip the ion motion, or make $m_i \gg m_e$ (or $\omega_{pi} \ll \omega_{pe}$); then

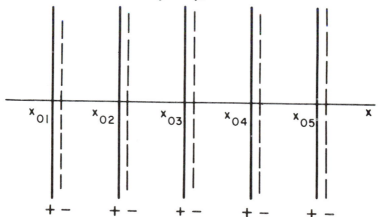

Figure 5-3a Electron-ion pairs in a nongridded space. The ions are massive and stay near the equilibrium positions x_{0i}. If the electrons are separated from their ions by less than the equilibrium sheet spacing, then the motion is oscillatory, but not simple harmonic. Electrons must cross many ions to approximate plasma oscillations.

correctly, ES1 sets each $\rho_j = \rho_{ion}$ before accumulating the charge density of active species.

At this point, it is helpful to work out the linear longitudinal ($\mathbf{E}$ along $\mathbf{k}$) dielectric function $\epsilon(\omega,k)$, in one dimension for a cold plasma with one mobile species (electrons only, $m_i / m_e \rightarrow \infty$); in addition the electrons may have a zero-order drift velocity $\mathbf{v}_0$, along $\mathbf{k}$. As the stream is cold, we use the fluid equations.

$$m\frac{d\mathbf{v}}{dt} = q\mathbf{E} = \mathbf{F} \tag{3}$$

$$\nabla \cdot \rho\mathbf{v} + \frac{\partial\rho}{\partial t} = 0 , \quad \mathbf{J} = \rho\mathbf{v} \tag{4}$$

$$\nabla \times \mathbf{H} = \mathbf{J} + \epsilon_0\frac{\partial\mathbf{E}}{\partial t} \tag{5}$$

For a stream with drift velocity $\mathbf{v}_0$,

$$\mathbf{v}(x,t) = \mathbf{v}_0 + \mathbf{v}_1(x,t) \tag{6}$$

The total derivative in (3) is

$$\frac{d}{dt} = \frac{\partial}{\partial t} + \mathbf{v} \cdot \nabla \tag{7}$$

Assuming x, t dependence as $\exp i(\mathbf{k} \cdot \mathbf{x} - \omega t)$, the *linearized* equation of motion is

$$-mi(\omega - \mathbf{k} \cdot \mathbf{v}_0)\mathbf{v}_1 = q\mathbf{E}_1 = \mathbf{F}_1 \tag{8}$$

The current is linearized to

$$\mathbf{J}(x,t) = \mathbf{J}_0 + \mathbf{J}_1(x,t) = \rho_0\mathbf{v}_0 + \rho_0\mathbf{v}_1 + \rho_1\mathbf{v}_0 \tag{9}$$

so that the continuity equation reads

$$i\mathbf{k} \cdot (\rho_0\mathbf{v}_1 + \rho_1\mathbf{v}_0) - i\omega\rho_1 = 0 \tag{10}$$

or

$$\rho_1 = \rho_0\frac{\mathbf{k} \cdot \mathbf{v}_1}{\omega - \mathbf{k} \cdot \mathbf{v}_0} \tag{11}$$

Taking $\mathbf{v}_1$ from (8) produces

$$n_1 = n_0\frac{i\mathbf{k} \cdot \mathbf{F}}{m(\omega - \mathbf{k} \cdot \mathbf{v}_0)^2} \tag{12}$$

Putting $\rho_1 = qn_1$ from (12) into (9) and using

$$i\mathbf{k} \times \mathbf{H}_1 = \mathbf{J}_1 - i\omega\epsilon_0\mathbf{E}_1 = -i\omega\epsilon\mathbf{E}_1 \tag{13}$$

produces

$$\frac{\epsilon}{\epsilon_0} = 1 - \frac{\omega_p^2}{(\omega - \mathbf{k} \cdot \mathbf{v}_0)^2} \tag{14}$$

The solutions for longitudinal waves are obtained from $\epsilon = 0$; the roots are

$$\omega = \mathbf{k} \cdot \mathbf{v}_0 \pm \omega_p \tag{15}$$

as shown in Figure 5-3b, either plasma (Langmuir) oscillations ($v_0 = 0$) or space-charge waves ($v_0 \neq 0$).

PROBLEMS

5-3a Obtain the motion of the electron sheets around the massive ion sheets ($m_i / m_e \rightarrow \infty$), for the infinite model shown in Figure 5-3a; there is no grid. Sketch the force F on e as a function of separation from its equilibrium position (the x_0's); obtain the force potential ψ where $\nabla \psi = -\mathbf{F}$ and sketch the potential well. Consider all pairs excited at a wavelength which is a multiple of the equilibrium spacing, $\lambda = m(x_{02} - x_{01})$. Your answer should show that the motion is *not simple harmonic* and that the frequency of oscillation is dependent on the initial excitation.

5-3b Re-do last problem, only this time let there be, say, 5 ion sheets (fixed, $m_i / m_e \rightarrow \infty$, $q_i = q_e / 5$) for each electron sheet. If the electron is excited so as to cross only the equilibrium sheet, then the result is like that of 5-3a. However, let the electron sheets be allowed to cross several sheets in one oscillation period; sketch the potential well, $\psi(x)$, to show that the well is approaching that for simple harmonic motion (parabola). Next, let the number of ion sheets per electron sheet become very large, so that the ions may be considered to form a *continuous background;* show that the motion is simple harmonic, with frequency

$$\omega^2 = \frac{1}{\epsilon_0} \left| \frac{q}{m} \right|_e (n_0 q)_i = \frac{1}{\epsilon_0} \left| \frac{q}{m} \right|_e \rho_{0i}$$

5-3c Let the model be electron sheets in a continuous immobile ion background. Let the initial excitation be uniform spacing of electrons but with *velocity modulation.*

$$v(x, t=0) = v_1 \sin k_0 x$$

Find what value of v_1 just produces crossing of the electron sheets and show which sheets cross

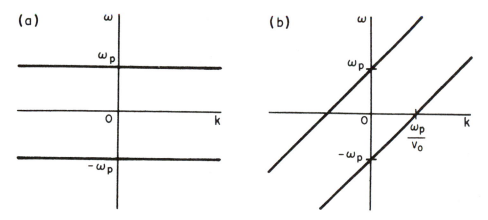

Figure 5-3b (a) Cold plasma dispersion, or ω-k, diagram showing plasma oscillations at $\omega = \omega_p$. (b) Similarly, for electrons drifting at velocity v_0 through heavy ions, showing the space-charge waves at $\omega = k v_0 \pm \omega_p$, essentially Doppler shifted oscillations.

first, and the time of crossing. There are N_e electrons uniformly spaced in length $L = 2\pi / k_0$. Carefully sketch the motion in a trajectory diagram, x versus t, for N_e large and N_e small. *Answer*: Crossing occurs for $v_1 / \omega_p =$ sheet spacing in equilibrium $= L / N_e$.

5-3d Let the model be periodic, period L, with N_e electron sheets in a uniform immobile ion background. What is the minimum number of sheets needed to obtain single harmonic (plasma) oscillations, $(N_e)_{min}$? What occurs if $N_e = 1$? For $N_e = 2$, what should be their velocity or displacement excitation phase to produce oscillations; sketch the motion (trajectories), x versus t. $E(k = 0) = 0$ here.

5-4 COLD PLASMA OSCILLATIONS; PROJECT

For this first project to be attempted, parameters to use on the initial run are: NSP $= 1$, L $= 2\pi$, DT $= 0.2$, NT $= 150$, NG $= 32$, IW $= 2$, EPSI $= 1$, plotting frequencies all 20, mode plots 1, 2,.... For species 1, N $= 128$, WP $= 1$, WC $= 0$, QM $= -1.0$, MODE 1 excited at X1 $= 0.001$ (all other values zero). Some of these values are the defaults in ES1 and need not be specified in the input.

Run the program with these parameters. Examine the plots most carefully. In the snapshots (at fixed times t) of $\rho(x)$, $\phi(x)$, $E(x)$ and phase-space plots, check to see that these quantities have the proper amplitudes and spatial phase, which exchange back and forth from PE to KE as is characteristic of simple harmonic motion (much larger than in a thermal plasma, where PE$/$KE$\approx 1 / N_D \ll 1$); the variation of total energy (if any) should be much smaller than that of PE (or else the validity of the program is open to question).

In the 150 steps suggested, with $\omega_p \Delta t = 0.2$, the program runs to $\omega_p T = 30$, or $30 / 2\pi \approx 5$ plasma periods. Using NG $= 32$, with 16 modes (0 to $k\Delta x = \pi$) makes the first mode occur at $k\Delta x = \pi / 16$; here we expect

$$\omega(k) = \omega_p \left[\frac{k \kappa(k) S^2(k)}{K^2(k)} \right]^{1/2} = 0.9952 \quad \text{for } k\Delta x = \frac{\pi}{16} \quad (1)$$

which is very close to ω_p. However, as suggested in Figure 4-6a, ω decreases with $k\Delta x$ so that the ω's for higher-numbered modes are less well-resolved in the 150 steps. The justification for use of $S^2(k)$ in (1) was given in Section 4-6 and for $\kappa(k)$ and $K^2(k)$ in Section 2-5. The result (1) is for the force averaged over all displacements of the grid and, as such, is approximate; the exact answer is

$$\omega(k) = \omega_p \cos\left(\frac{k\Delta x}{2}\right) \quad (2)$$

Both results are derived in Chapter 8.

From the plots of ESE$(k) \propto \frac{1}{2}\rho_k \phi_k^*$ for each value of k versus time, the frequency of each mode can be obtained from just one run. Remember that $\rho_k \phi_k^*$ is a product of $(\cos\omega t)(\cos\omega t)$ for a standing wave, producing 2ω.

Use a large enough excitation of mode 1 so that all modes are detectable, but small enough so that the motion is nearly linear, meaning that the modes are nearly independent. Plot the estimated ω from (1), the exact value from (2), and your measured values versus $k\Delta x$ (0 to π), with error bars on your measurement. The estimated ω from (1) does not include the effect of spatial aliases, so that the estimated and measured values should not agree; however, your results should agree very well with the exact calculation (2) which includes all aliases.

Try the NGP, CIC, and energy conserving weights (IW = 1, 2, 3). Do the differences in dispersion come from particle shape or from dynamics or the finite-differencing method (for $\nabla\phi$, $\nabla^2\phi$)? For IW = 3 how well is energy conserved as Δt is increased? [Energy-conserving ideas are developed in Part Two; for our use here, you need only know that energy (but not momentum) is conserved in the limit of $\Delta t \rightarrow 0$ using the weighting IW = 3.]

Try using 32 particles in 16 cells uniformly spaced (2 particles per cell), with particles *on* the grid points and in between (which is *not* the normal ES1 loading). Note the spikes in density with small initial sinusoidal displacement which was dubbed the *Kaiser-Wilhelm effect* by our good friend Dieter Fuss at LLNL. Try this initialization with no initial displacement. Try using NP/NG $\neq$ integer, .e.g. 33 particles and 16 cells with no initial excitation, for IW = 2. Account for the spike in charge density at $t = 0$ (*hint*: sketch the grid density for 3 particles in 2 cells). W. M. Nevins developed a theory (unpublished) which shows this *particle aliasing effect*. He obtained the charge density, complete with spike, from a sum over Bessel functions. *Denavit* (1974, Appen. B) presents such theory in detail.

Try increasing $\omega_p\Delta t$ above 2 to find the leap-frog instability of Section 4-2. Plot complex ω versus parameter $\omega_p\Delta t$ (for excitation of one mode at small $k\Delta x$). For $\omega_p\Delta t > 2.0$, which output would you choose to show the characteristic π phase shift each step, the so-called *odd-even separation*? The growth may be hard to observe directly without some special care, except that energy conservation should be lost.

Try letting the first species be electrons ($q_1 < 0$) and the second species be mobile ions ($|q_1| = q_2 > 0$) with the ratio $(\omega_{p1}/\omega_{p2})^2 = m_2/m_1$ on the order of 100 and with $N_1 = N_2$. Check the motion to see if peak values of density and velocity of the electrons are on the order of the mass ratio larger than values for the ions.

Add whatever additional diagnostics are needed.

Additional suggestions:

Excite with large amplitude, enough that particles touch and then cross. Describe wave breaking. Use $x_1 = 0$, large v_1 and observe the odd modes.

In the small-amplitude plasma oscillations, why do PE and KE have clear zeros, indicating exact exchange between PE and KE, but have different peak values?

Excite at short wavelength, beyond $k\Delta x = \pi$, and observe what occurs. Aliases should be excited. For example, NG = 16 has 8 modes (9 cosines, 7 sines); hence, exciting mode 10 produces mode 6 on the grid; exciting mode 12 produces mode 4.

Excite locally with small amplitude, using IW = 2 and then IW = 3, and look for propagation away from the region of excitation. Explain the differences. *Hint*: What are the group velocities for the two weightings? (See Chapters 8 and 10.)

5-5 HYBRID OSCILLATIONS; PROJECT

Plasma oscillations are changed into so-called *hybrid oscillations* by adding a uniform static magnetic field B_{0z} normal to the displacements in x, y. The equation of motion for any particle (*i.e.*, sheet) is

$$\delta \ddot{x} = -\omega_H^2 \delta x \tag{1}$$

where δx represents x or y displacements from the particle guiding center and the hybrid frequency is defined by

$$\omega_H^2 \equiv \omega_p^2 + \omega_c^2 \tag{2}$$

The motion of a particle is given by [for $v_{x1}(0) = 0$, $y_1(0) = 0$],

$$x = x_0 + x_1 \cos \omega_H t \tag{3}$$

$$y = y_0 + \frac{v_{y1}}{\omega_H} \sin \omega_H t \tag{4}$$

which are the equations for an ellipse centered at x_0, y_0 (the guiding center) in the x, y plane.

Suppose that we choose to find $\omega_H(k)$ at some k by putting in the same initial perturbation as in the $B_{0z} = 0$ plasma oscillation model, that is, a displacement $x_1 \propto \sin kx$, but with no initial velocity. The result (as you may see for yourself) is not a clear-cut hybrid oscillation in any of the grid quantities; however, each particle is oscillating at ω_H, simply not in the proper phase with other particles to produce a coherent spatial pattern (a mode). The desired hybrid oscillations are obtained by using a v_x perturbation with $x_1 = 0$, or by making the initial x and v_y perturbations consistent, such as

$$v_y = -\omega_c x_1 \tag{5}$$

which eliminates the drift in y and constant acceleration in x; *i.e.*, the guiding center remains stationary. Use, for example, $x_1 = A$ and $v_{y1} = -\omega_c A$; the hybrid oscillations should be just as clear as were the plasma oscillations measured in Section 5-4. In order to obtain ω's for all k's in one run, use an initial small-random-velocity excitation in v_x, using VT1 and X1 = 0; there is no need to put in the consistent displacement in y in this 1d model. Or, use a random-velocity excitation in y, but with consistent x displacement given by (5).

Plot ω versus k, as with plasma oscillations, and compare with the prediction

$$\omega(k) = \left[\omega_p^2 \cos^2 \frac{k\Delta x}{2} + \omega_c^2\right]^{1/2} \tag{6}$$

Add a program to trace out the trajectory of one particle in the x, y plane and plot the trajectories for several representative particles. Check the predicted size of the ellipses with those computed.

Do the hybrid simulation with excitation of one mode at short wavelength $\pi/2 < k\Delta x < \pi$. Use a small initial displacement x_{11}, with y velocity of $-\omega_c x_{11}$. Note that ESE oscillate at $2\omega_H$ with a flat envelope as expected, but that KE, also with $2\omega_H$ oscillations, has an envelope that slowly rises and falls at a beat frequency $\omega_H - \omega_c \approx \omega_p^2/2\omega_c$. An individual particle orbit slowly increases in radius then decreases. This result was observed by *Thomas and Birdsall* (1980) and shows that excitation at large $k\Delta x$ is to be avoided.

PROBLEMS

5-5a Verify (1)-(4). Obtain the guiding-center motion for arbitrary initial excitation.

5-5b (Due to John Cary.) There is a *vortex mode* of a cold magnetized plasma, $\mathbf{E}$ longitudinal, normal to $\mathbf{B}_0$, which has $\omega = 0$. Do you detect this mode? Is it the mode excited by $x_1 \neq 0$, $\dot{y}_1 = 0$, which produces a drift motion in y? See papers by *Taylor and McNamara* (1971), *Okuda and Dawson* (1973), *Montgomery and Joyce* (1974), and *Langdon* (1969).

5-5c (Due to Robert Littlejohn.) Use the fluid equation of continuity, the equation of motion in v_x and v_y, and $\nabla \cdot \mathbf{D} = \rho$ to produce

$$\frac{\partial^3 n_1}{\partial t^3} + \omega_H^2 \frac{\partial n_1}{\partial t} = 0$$

with solution

$$n_1(x,t) = n_1(x,0) + \frac{1}{\omega_H}\dot{n}_1(x,0)\sin\omega_H t + \frac{1}{\omega_H^2}\ddot{n}_1(x,0)(1 - \cos\omega_H t)$$

Find $\dot{n}_1(x,0)$ and $\ddot{n}_1(x,0)$ in terms of $n_1(x,0)$, $v_x(x,0)$, and $v_y(x,0)$. Show that with initial displacement only $v_x(x,0) = 0 = v_y(x,0)$ and $\omega_c = \omega_p$,

$$n_1(x,t) \approx 1 + \cos\sqrt{2}\omega_p t$$

which has but one zero in a hybrid cycle. Similarly

$$\rho\phi^* \propto E^2 \propto (1 + \cos\sqrt{2}\omega_p t)^2$$

has but one zero in $\sqrt{2}\omega_p t = 2\pi$, so that if we are counting two zeros (or two peaks) to a cycle (in ESE), then we would claim $\omega = \omega_H/2$, an incorrect result. Complete the solutions explicitly for v_x, v_y, and E; note the drift motion in y for all initial excitation except v_x.

5-6 TWO-STREAM INSTABILITY; LINEAR ANALYSIS

The model consists of two opposing streams of charged particles as sketched in Figure 5-6a. Models with relative motion between two sets or streams of charged particles have been studied in great detail since papers by *Haeff* (1949) and *Pierce* (1948). Detailed knowledge of the nonlinear behavior of opposing streams came much later, from the simulations done by *Dawson* (1962). The fluid analog was given much earlier, as by H. Hertz in the 1880's; see comprehensive books on hydrodynamics and acoustics, such as *Lamb* (1945) or *Rayleigh* (1945).

One can readily see that an opposing stream system is unstable. When two streams move through each other one wavelength in one cycle of the plasma frequency, a density perturbation on one stream is reinforced by the forces due to bunching of particles in the other stream and vice versa; hence

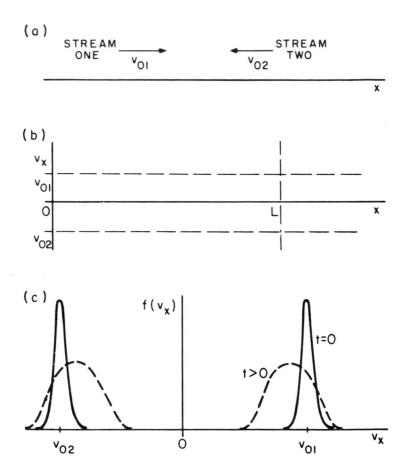

Figure 5-6a (a) Two opposing streams as seen in the laboratory. (b) The streams in phase space at the start of the problem, $t = 0$. (c) The streams in velocity space at $t = 0$ and $t > 0$.

$\Delta n_1 \propto n_1$, so that the perturbation *grows exponentially in time*. This simple relation was put forth in 1948 by Professor M. Chodorow of Stanford [and buried in Birdsall's dissertation (*Birdsall*, 1951)] for two streams moving in the same direction (*Chodorow and Susskind*, 1964). The phase relation for reinforcement is written as

$$(v_{\text{relative}})\left|\frac{2\pi}{\omega_p}\right| = \frac{2\pi}{k} \tag{1}$$

which for $v_{\text{relative}} = v_0 - (-v_0) = 2v_0$ is

$$k = \frac{\omega_p}{2v_0} \tag{2}$$

This k is very close to that found from analysis for maximum growth rate.

The longitudinal linear dielectric function for two independent cold streams may be obtained as was done in Section 5-3 by applying the equations of motion and continuity separately for each stream and adding the currents of each in the field equation. The result is

$$\frac{1}{\epsilon_0}\epsilon(\omega,k) = 1 - \frac{\omega_{p1}^2}{(\omega - \mathbf{k}\cdot\mathbf{v}_{01})^2} - \frac{\omega_{p2}^2}{(\omega - \mathbf{k}\cdot\mathbf{v}_{02})^2} \tag{3}$$

for two streams with drift velocities $\mathbf{v}_{01}$ and $\mathbf{v}_{02}$. This result is also obtainable directly from the usual Vlasov-Poisson set by letting the velocity distribution be two delta functions,

$$f_0(\mathbf{v}) = A\delta(\mathbf{v} - \mathbf{v}_{01}) + B\delta(\mathbf{v} - \mathbf{v}_{02}) \tag{4}$$

A system of N independent cold streams produces a sum over streams or species s:

$$\frac{1}{\epsilon_0}\epsilon(\omega,k) = 1 - \sum_{s=1}^{N}\frac{\omega_{ps}^2}{(\omega - \mathbf{k}\cdot\mathbf{v}_{0s})^2} \tag{5}$$

[Extension of the sum to an integral, for $N \to \infty$, must be done carefully, both analytically as shown by *Dawson* (1960), and also in simulation when a discrete set of beams is used to approximate a smooth distribution $f(v)$ as shown by *Byers* (1970), and *Gitomer and Adam* (1976), and discussed in Chapter 16.]

The solutions for complex ω, assuming real k (*i.e.*, an absolute instability, growth in time only, no convection in space), opposing streams of equal strength, $\omega_{p1} = \omega_{p2} \equiv \omega_p$, $v_{01} = -v_{02} \equiv v_0$, is found from $\epsilon(\omega,k) = 0$ which is quartic in ω with four independent solutions. These are

$$\omega = \pm[k^2v_0^2 + \omega_p^2 \pm \omega_p(4k^2v_0^2 + \omega_p^2)^{1/2}]^{1/2} \tag{6}$$

for which

$$0 < \frac{kv_0}{\omega_p} < \sqrt{2} \quad \begin{cases} \text{two roots are real} \\ \text{two roots are imaginary} \end{cases} \tag{7}$$

$$\sqrt{2} < \frac{kv_0}{\omega_p} \quad \text{all four roots are real} \tag{8}$$

$$\frac{kv_0}{\omega_p} = \frac{\sqrt{3}}{2}, \quad \omega_{\text{imaginary}} = \frac{\omega_p}{2}, \quad \text{maximum growth rate} \tag{9}$$

This behavior is sketched in Figure 5-6b; the growth ($\omega_{\text{imaginary}}$) is given in more detail in Figure 5-6c.

In this model, where there is growth ($\omega_{\text{imaginary}} > 0$), we find that $\omega_{\text{real}} = 0$; that is, there is no oscillatory part associated with the growth, a situation which is not generally true.

A point of Figure 5-6c is to make clear the existence of a *minimum unstable length L of the system;* in this model (normalized)

$$\frac{\omega_p L}{v_0} > \frac{2\pi}{\sqrt{2}} \quad \text{(unstable)} \tag{10}$$

in order to obtain growth. This is the same as (7) using $L = 2\pi / k_0$, where k_0 is the smallest wavenumber in the system.

Growth which begins at small amplitude continues until the streaming is destroyed; indeed, the distribution becomes nearly Maxwellian. Hence, we say that "the colliding streams have thermalized," although not by collisions.

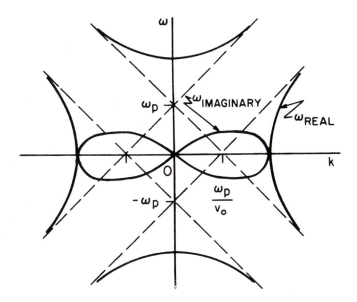

Figure 5-6b Dispersion, or ω-k, diagram for two equal opposing streams, real k, complex ω. The uncoupled space-charge waves are shown dashed. For each value of k, there are four values of ω that correspond to four linearly independent waves.

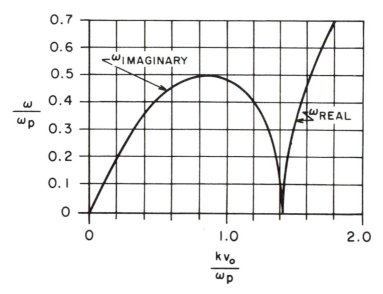

Figure 5-6c Growth rate $\omega_{imaginary}$ for two opposing streams.

Instead, collective effects build up large electric fields at long wavelengths ($\lambda \gg$ particle spacing) and these *scatter the particles* in phase space.

As the instability grows, two changes are readily observed in $f(v)$ as indicated for $t > 0$ in Figure 5-6a(c). The *width* of each beam increases [measured directly on an $f(v)$ plot or by $(\overline{v^2} - \overline{v}^2)$ of one stream], which is taken as an increase in the *temperature* of each beam (but perhaps carelessly so, for if the electric field were suddenly shut off—and you should try this—the spread might decrease). The drift or mean velocity $\overline{v}$ decreases.

We might expect, as $v_{thermal}$ increases and v_{drift} decreases, that the conditions for linear growth would cease to be met [see *Stringer* (1964), who shows the threshold for growth for electron-electron streams to be $v_{drift} \approx 1.3 v_{thermal}$] and that the exponential growth would stop. However, at this time, the conditions for linearity are largely violated, with perturbed charge densities comparable to the zero-order density; particles in one stream are about to pass their neighbors and wrap into *vortices in phase space,* that is, become *trapped*. Hence, the growth need not stop, although we might be tempted to look for a change in character of the growth (*e.g.*, away from exponential in time) at the time where v_t exceeds $\overline{v}/1.3$; keep this in mind in your project. Of course, ES1 can readily be run with warm beams; hence, look for growth with $v_0 = 2v_t$ (Section 5-9), but stability with $v_0 = v_t$.

PROBLEMS

5-6a Obtain the solution for the four waves given in (6). Show that Im $(\omega)_{max}$ occurs as given in (9).

5-6b At $t = 0$, let $x_{11}(x)$, $v_{11}(x)$, $x_{21}(x)$, and $v_{21}(x)$ be given (first-order perturbations in position and velocity), at one value of k as

$$x_{11}(x) = \text{Re} \, [x_{11} \exp (ikx)]$$

or

$$x_{11}(x) = x_{11} \cos kx$$

From these four values, obtain the excitation of the four waves, that is, the A_n's and B_n's of

$$\phi(x,t) = \sum_{n=1}^{4} A_n \exp i (\omega_n t - kx)$$

$$\rho(x,t) = \sum_{n=1}^{4} B_n \exp i (\omega_n t - kx)$$

You may use initial perturbation $\rho_{11}(x)$ rather than $x_{11}(x)$ if you like, making the translation shown in Section 5-2. You may choose to Laplace transform 5-3(3), 5-3(4), and 5-3(5) to facilitate obtaining the solution. Sketch $\phi(t)$ [or ln $\phi(t)$] at a fixed x to show the dominance of the growing wave at late time.

5-6c Show that growth of Im $(\omega) = 0.5\omega_p$ means growth of 27 dB in one plasma cycle. The definition of decibel is, state 1 to state 2,

$$1 \text{ dB} = 10 \log_{10} \left(\frac{\text{power}_2}{\text{power}_1} \right)$$

This is really a large growth, going from, say, noise at 10^{-12} W to power station output of 10^9 W (210 dB) in less than 10 plasma cycles! Obviously any laboratory double streaming lasts only a few plasma cycles.

5-7 TWO-STREAM INSTABILITY;
AN APPROXIMATE NONLINEAR ANALYSIS

A simulation project may be started with rather imperfect knowledge of what to expect. However, starting a project (especially one with instability growth) totally blind is not very wise. Most professional simulators can tell of at least one direct experience of incomplete planning that lead to considerable waste computation. The usual problems are with poor parameters or initial conditions (too noisy, wrong modes excited, not enough modes, etc.) and with poor or insufficient diagnostics [lack of resolution for small changes in $f(v)$, no temporal Fourier analysis, etc.]. A very common problem is with not knowing very well at what energy level an instability saturates and what t and x (ω and k) resolution is needed at that level. It is usually worthwhile to make some estimates of the expected saturation level and shift, if any, in ω and k. These estimates are not to be taken too seriously but they can be most helpful in choosing initial conditions, parameters, and diagnostics. It is also helpful to make some preliminary short runs to uncover unanticipated problems.

Two opposing streams, each of density n_0, start out with drift energy only. Let the streams have the same sign of charge ($q_1 q_2 > 0$) drifting through a background of immobile charges of opposite sign, of density

$$\rho_{background} = -2n_0 q \tag{1}$$

As the streams interact and form large bunches of charge, they also produce large electric forces which accelerate and decelerate the charges. The electric field energy is obtained from the initial kinetic energy of the drift motion. Hence, we expect (roughly)

$$(TE)_{t=0} = (KE)_{\substack{drift \\ t=0}} \rightarrow (KE)_{\substack{drift \\ t>0}} + (KE)_{thermal} + (PE)_{fields} \tag{2}$$

which implies that the drift energy must decrease once fields are generated by the instability; *i.e.*, both streams slow down.

If the parameter of linear analysis $\omega_p L / v_0$ is calculated from mean values

$$\frac{\omega_p^2 L^2}{v_0^2} \rightarrow \frac{q}{m} \frac{-\rho_{background}}{\epsilon_0} \frac{L^2}{<v>^2} \tag{3}$$

then (as $\rho_{background}$ is invariant) this ratio increases as $<v>$ decreases; this may increase or decrease ω_{imag} (see Figure 5-6c) but does not move the system to a region of no (linear) growth.

With streams of particles with charges of like sign ($q_1 q_2 > 0$), the interaction tends to produce charge bunches separated by $\lambda/2$. This bunching generates an electric field at the second spatial harmonic (*i.e.*, at twice the original k); hence, this wavelength should *not* be smoothed away in the simulation. [For unlike signs, ($q_1 q_2 < 0$, e.g., electrons and positrons), the bunching is different.] Indeed, *the nonlinear generation of spatial harmonics* ($k \rightarrow 2k, 3k, 4k$, etc.) *is a trademark of most instabilities,* in which the initial sinusoidal bunches "sharpen up" toward spikes. These harmonics can qualitatively alter the evolution of instabilities (*Ishihara et al.,* 1980, 1981)

As a first guess at large amplitude behavior, let the stream charges, at some stage, form two sinusoidal bunches separated by $L/2$, as shown in Figure 5-7a. (Any larger bunching would produce many harmonics). The average values, $<\rho>$, $<E>$, $<\phi>$, are all zero, as required in periodic models. Only one harmonic

$$k_2 = 2k_0, \quad k_0 \equiv \frac{2\pi}{L} \tag{4}$$

is present, associated with

$$\rho_{total}(x) = -|\rho_b| \cos 2k_0 x \tag{5}$$

$$E(x) = -\frac{|\rho_p|}{2k_0 \epsilon_0} \sin 2k_0 x \tag{6}$$

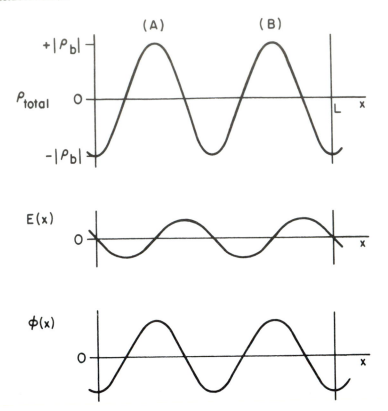

Figure 5-7a Large-amplitude (yet still sinusoidal) density, field, and potential for two opposing streams of *like charge sign* in an immobile background of opposite sign of charge, charge density ρ_b. Physically, the charges of one stream tend to be found in bunch A and, of the other, in bunch B. (The initial small-amplitude growth had one wavelength in this space, $\lambda = L$.)

$$\phi(x) = -\frac{|\rho_p|}{(2k_0)^2\epsilon_0} \cos 2k_0 x \tag{7}$$

We start the calculation by asking two questions: (a) Does the total electric field energy PE exceed the total initial kinetic energy KE_0, which is the total energy of the system TE? If so, then we must reduce the estimate of ρ_{total}. (b) Is the potential so large as to reflect particles? The answer to (a) comes from the ratio of $(PE)_{total}/TE$, which is

$$\frac{(PE)_{total}}{TE} = \frac{\frac{1}{2}\epsilon_0 \int_0^L E^2 dx}{2n_0(\frac{1}{2}mv_0^2)L} = \frac{\rho_b^2 L^2}{64\pi^2\epsilon_0 n_0 mv_0^2} = \frac{1}{32\pi^2}\left(\frac{\omega_p L}{v_0}\right)^2 \tag{8}$$

At the threshold for growth, $\omega_p L / v_0 = 2\pi/\sqrt{2}$, there is no problem, for there,

$$\frac{(PE)_{total}}{TE} = \frac{1}{16} \tag{9}$$

At $(\omega_{imaginary})_{max}$, $\omega_p L / v_0 = 4\pi / \sqrt{3}$, we find

$$\frac{(PE)_{total}}{TE} = \frac{1}{6} \tag{10}$$

which is still less than unity, hence physical; indeed, this value implies that we might expect tighter bunching than assumed. The answer to (b) comes from the ratio of single-particle energies,

$$\frac{|q\phi|_{max}}{\frac{1}{2}mv_0^2} = \frac{-q\rho_b}{(2k_0)^2\epsilon_0} \frac{1}{\frac{1}{2}mv_0^2} = \frac{1}{8\pi^2}\left[\frac{\omega_p L}{v_0}\right]^2 \tag{11}$$

which is four times larger than $[(PE)_{total}/TE]$. Hence, at threshold, this ratio is $1/4$; a particle at velocity v_0 is not be stopped at $|q\phi|_{max}$. At maximum rate of growth, however, $|q\phi|_{max} = 2/3(\frac{1}{2}mv_0^2)$, indicating that a particle initially at velocity v_0 would be considerably slowed by the potential hill. Just beyond $\omega_p L / v_0 = 4\pi / \sqrt{3}$, at $\omega_p L / v_0 = \sqrt{8\pi^2}$, we would expect a particle to be stopped and turned back. The fact that the particle energy ratio is four times larger than the total energy ratio may indicate that the mechanism limiting growth is that of reflecting particles rather than not allowing tight bunches. (Please ignore the mistake of equating $\frac{1}{2}mv_0^2$ and $|q\phi|$ in time varying fields.)

As a second guess, we look at tighter bunching. Let us try zero-thickness bunches (delta functions) as shown in Figure 5-7b. For bunch thickness τ essentially zero (meaning $\tau / L \ll 1$), we find the field and potential in region I to be (region I: $0 < x < L/4$):

$$E_I(x) = \frac{\rho_b}{\epsilon_0}x \quad (\rho_b < 0) \tag{12}$$

$$\phi_I(x) = A + Bx^2 \tag{13}$$

B is obtained from $\nabla\phi = -\mathbf{E}$ and A is obtained from $<\phi> = 0$; the result is

$$\phi_I(x) = \frac{\rho_b}{\epsilon_0}\left[\frac{L^2}{96} - \frac{x^2}{2}\right] \tag{14}$$

From this model

$$\frac{(PE)_{total}}{TE} = \frac{1}{48}\left[\frac{\omega_p L}{v_0}\right]^2 \tag{15}$$

and

$$\frac{|q\phi|_{max}}{\frac{1}{2}mv_0^2} = \frac{1}{24}\left[\frac{\omega_p L}{v_0}\right]^2 \tag{16}$$

These values are $2\pi^2/3$ and $\pi^2/3$ larger than the values obtained earlier,

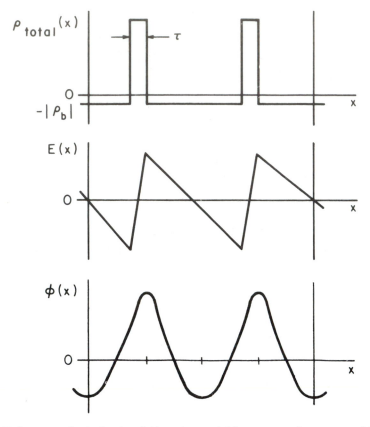

Figure 5-7b Large amplitude density, field, and potential for two opposing streams of like charge sign, with very tight bunches of thickness $\tau \ll L$.

and are now in the ratio of 1 to 2 (1 to 4 earlier). At $(\omega_{imaginary})_{max}$, $\omega_p L / v_0 = 4\pi / \sqrt{3}$, we find

$$\frac{(\text{PE})_{total}}{\text{TE}} = \frac{\pi^2}{9} \approx 1.09 \tag{17}$$

$$\frac{|q\phi|_{max}}{\frac{1}{2}mv_0^2} = \frac{2\pi^2}{9} \approx 2.19 \tag{18}$$

Both values exceed unity, with (17) indicating that there is not enough kinetic energy in the system to make zero-thickness bunches and (18), that particles (streaming at speed v_0) would be stopped and reflected by such bunches, never to be trapped. Thus *our third guess* should allow finite (but still small) thickness bunches. Indeed, using the above model, one finds

$$\frac{(PE)_{total}}{TE} = \frac{1}{48}\left[\frac{\omega_p L}{v_0}\right]^2\left[1 - \frac{2\tau}{L}\right]^2 \qquad (19)$$

At maximum growth rate, as in (17), making $(PE)_{total} = TE$ (meaning that all of the energy is now in the fields and the particles are completely stopped), requires that

$$\frac{\pi^2}{9} = \left[1 - \frac{2\tau}{L}\right]^{-2} \qquad (20)$$

or

$$\frac{\tau}{L} = \frac{1}{2}\left[1 - \frac{3}{\pi}\right] \approx 0.022 \qquad (21)$$

which is pretty thin. The answer for $|q\phi|_{max}/\frac{1}{2}mv_0^2 = 1$, left for a problem, probably indicates a somewhat thicker bunch.

These estimates, generally termed *back-of-the-envelope calculations,* are to be taken lightly; however they are advisable when exploring new areas. The real fun comes in comparing these with the simulation runs, which support some guesses and not others, and guide further theory and simulation.

PROBLEMS

5-7a Obtain $|q\phi|_{max}/\frac{1}{2}mv_0^2$ for a finite-thickness bunch, Figure 5-7b, in order to correct (16) for finite thickness. At what thickness τ/L does this ratio $=1$?

5-7b Let the two beams of velocities v_0, $-v_0$ "thermalize" (via the instability) into a Maxwellian distribution, $f(v) \propto \exp(-v^2/v_t^2)$, with a field level $(PE)_{total} \ll (KE)_{total}$. Find v_t in terms of v_0.

5-7c Read *Friedberg and Armstrong* (1968). Can you think up more estimates based on their work?

5-7d The fluid equations used in the linear analysis are valid until particles in a given stream overtake particles in that stream (*i.e.*, to the point when velocity v begins to be double-valued). Using the displacement as worked out in Section 5-2, and the x_1, E, ϕ, ρ of just the growing wave for two streams (Section 5-6), find the values of E and ϕ *at particle overtaking.* Put your results in terms of $(PE)_{total}/TE$ and $|q\phi|/\frac{1}{2}mv_0^2$ at overtaking, using parameter $(\omega_p L/v_0)^2$, as done in this section. These levels are additional nonlinear bench marks.

5-7e For opposing streams of *charges of unlike sign* ($q_1q_2 < 0$, say, electrons and positrons or D^+ and D^- ions), a first guess at the large amplitude behavior is as sketched in Figure 5-7c. Letting ρ, E, ϕ vary as $-\sin k_0 x$, $\cos k_0 x$ and $-\sin k_0 x$ respectively [as in (5), (6), and (7)] work out $(PE)_{total}/TE$ and $|q\phi|/\frac{1}{2}mv_0^2$. Show how these values differ from those derived for particles of like signs.

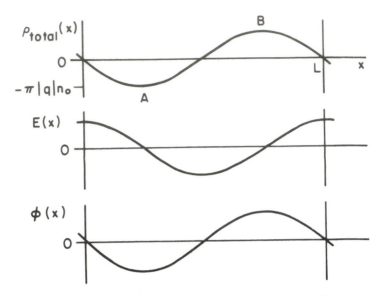

Figure 5-7c Like Figure 5-7a, but for particles of *unlike sign,* $q_1q_2 < 0$. The negative charges tend to collect in bunch A, and the positive, in bunch B.

5-8 TWO-STREAM INSTABILITY; PROJECT

The *linear behavior* is to be observed, such as:

(a) Growth rates and oscillation frequencies, to be checked with those given in Figure 5-6c;
(b) Phase of bunching, $\rho(x)$ relative to $\phi(x)$ and $E(x)$, for $q_1q_2 > 0$ and for $q_1q_2 < 0$

The *nonlinear behavior* is to be observed, such as:

(a) Peak values of charge densities;
(b) Tightness (thickness) of bunches;
(c) PE(at saturation)/KE(t = 0);
(d) Change of behavior (from oscillation to growth) of modes outside the linear growth region (*i.e.*, for $kv_0/\omega_p > \sqrt{2}$) due to nonlinear coupling;
(e) The decrease in average (streaming) velocities, $<v_1>$ and $<v_2>$ as PE becomes large;
(f) The increase in velocity spread as given by $[<v^2> - <v>^2]$, a measure of the change from cold to warm plasma;
(g) The difference in $(PE)_{max}$ between electron-electron $(q_1q_2 > 0)$ and electron-positron $(q_1q_2 < 0)$ two streaming.

A list of initial values follows to aid in getting you started: NSP = 2, L = 2π, DT = 0.2, NT = 300, NG = 32, IW = 2, EPSI = 1.0, A1, A2 = 0, plots every $50\Delta t$, mode plots 1, 2, 3. For species 1, N = 128, WP = 1.0, WC = 0.0,

QM = −1.0, V0 = 1.0, MODE = 1, X1 = 0.001 (the other values are zero). For species 2, use V0 = −1.0. Note that these values put the first mode at $k_0 v_0 / \omega_p = 1$, in the region of linear growth and the second mode at 2, outside the linear growth region, where ω is real. Hence when mode 1 (which is growing) becomes large enough to produce harmonics at $n k_0$, $n = 2, 3, \ldots$, then mode 2 begins to grow. The interesting part here is that these *nonlinearities* appear at amplitudes (for both mode 1 and 2) which are *far below their maximum values* (at saturation). This should give you some concern for the "boundary" between linearity and nonlinearity.

You might choose to use larger initial displacements, say, 10 or even 100 times larger, but run fewer steps (say, 50), just to see how the program runs and to obtain some idea of growth rates, saturation amplitudes, and whether you want more- or less-frequent diagnostics. Of course, such a run probably is nonlinear almost at $t = 0$, and is to be used in order to plan better, longer runs at smaller excitation. You might also consider moving k_0 to smaller values to allow for more growing modes; you may observe changes in saturation amplitude as the k's become more dense, pointing up some deficiencies of single-mode excitation or use of discrete k's inherent in simulation.

If you like further challenge, you may go back to the linear cold-beam theory and work out the initial values of displacements and velocities (perturbation amplitudes) needed to excite just the growing wave. See if there is a change in the nonlinear behavior from that for excitation of a single mode (all four waves) or for excitation of many modes (say, all modes excited, with small but random values of the amplitudes). Try to excite just the *decaying* mode. What terminates the decay?

In contrasting like-sign and opposite-sign results, return to the linear cold-beam theory and obtain the relative phases of ρ_1, x_1, ϕ_1, and E_1 of the growing wave. Apply this information to explain the differences in phase space (v_x, x) plots of the *e-e* and *e-p* runs.

5-9 TWO-STREAM INSTABILITY; SELECTED RESULTS

Simply as examples of results, we present a few results of two-stream runs made with ES1.

First, let us look at an *electron-electron* run. At $t = 0$, the streams have drift velocities ±1.0 and are perturbed slightly in position. Figure 5-9a shows the evolution of the instability. The grid charge density, the potential, and the electric field stay sinusoidal in space until about time $t = 15$, when the perturbation becomes quite large and the charge density has two peaks (not shown). Then, with a large electric field retarding the stream, some particles stop at about $t = 17$ and reverse as seen at about time $t = 18$; a phase-space vortex forms, and persists. Note that the large fields also accelerate some particles out to $|v| \approx 3$, *i.e.*, 9 times their original kinetic energy. At the end, the program plots the energies versus time. The

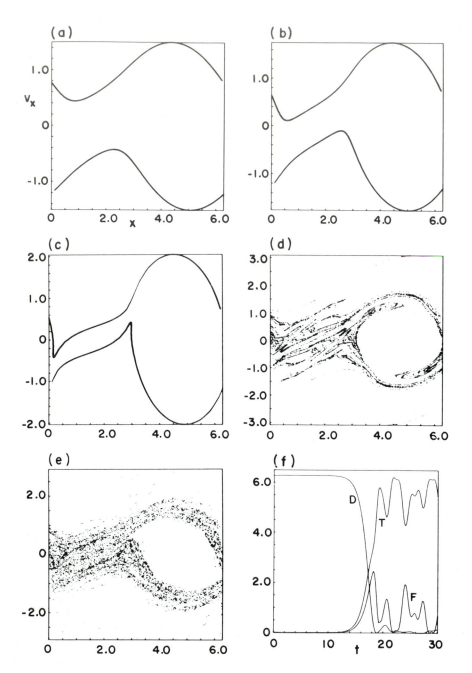

Figure 5-9a Evolution of electron-electron two stream-instability in v_x-x phase-space plots (a, b, c, d, and e at $t = 16, 17, 18, 34,$ and 60; note changes in v_x scale), and the energy histories, drift (D), thermal (T) and field(F), for $t = 0$ to 30 (in f). The initial velocities are ± 1.0. There are 4096 particles in each stream; the grid has 32 cells, $\omega_{p1} = 1 = \omega_{p2}$.

electrostatic or field energy reaches a peak of about 40 percent of total initial kinetic energy; this energy is almost wholly in mode 1 ($k_1 v_0 / \omega_0 = 1$), with exponential growth over many decades at very close to the theoretical rate as measured on a semilog plot (not shown). Not shown are modes 2 and 3 which are linearly stable but grow due to nonlinear coupling (at twice and three times $\omega_{imaginary}$ of mode 1), starting at about $t = 8$, $t = 10$ when mode 1 energy is still about 1000 and 100 times smaller than at saturation. This occurs in other models with a single growing mode; *e.g.*, *Crume et al.* (1972; see especially Figure 5).

Second, let us look at the *electron-positron* colliding stream model which has the same dispersion relation as the electron-electron model but has different phases among ρ, ϕ, and E for the growing wave. Figure 5-9b shows some of the results. We see an entirely different form of bunching than with like-sign particles at $t = 15.9$, when positive and negative bunches are formed nearly together. Beyond this time, there is a lot of scrambling rather than a persistent vortex. In the history plots, the field energy maximum is only 16 percent of total initial kinetic energy and has relatively smaller peak electric fields, a direct result of the form of bunching. Also, while the drift kinetic energy drops suddenly again in about one plasma cycle, it does not fall to zero.

Third, let us look at a warm electron-electron two-stream instability run, as shown in Figure 5-9c. The drift velocities are ± 1.0, with $v_{te} = v_0 / 2$; the initial velocity distribution is shown both in v_x-x space (a quiet start Maxwellian plus a small random thermal component) and integrated over x to give $f(v_x)$ *versus* v. Initially the instability grows exponentially in mode 3, with mode 2 about 10 dB behind, and forms three phase-space vortices which coalesce to two as shown and then finally to one vortex (not shown). The drift kinetic energy again does not go to zero but persists even out to $t = 300$. The reader may note that, since mode 2 and 3 grow at about the same rate, the coalescence observed may be due to the initial conditions—a challenge to pursue further.

These runs were done for fun, without the care and variation in parameters that should be used in scientific study. They serve as examples of initial runs that might be made in starting a thorough study and as examples of observations and diagnostics that are widely used.

PROBLEM

5-9a Since the equations of motion are reversible in time, if you insert statements into the code to reverse all particle velocities at some time during the growth of the instability, the system subsequently evolves back to its initial state! (What happens if you continue the problem further?) Work out how to do this reversal in ES1, remembering that v_x and x are at different times. Because of computer arithmetic errors, the code does not retrace accurately over a long time interval, especially in a physically unstable situation.

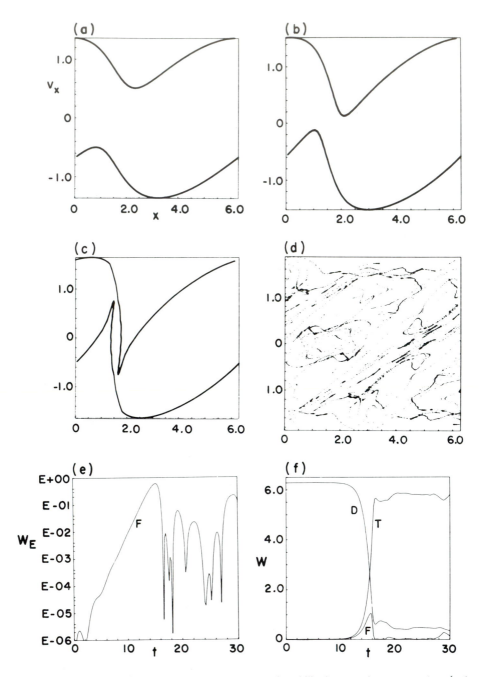

Figure 5-9b Evolution of electron-positron two-stream instability in v_x-x phase-space plots (a, b, c, and d at $t = 14, 15, 16$, and 33), mode 1 field energy history (e), and energy histories (f). The parameters are the same as in Figure 5-9a.

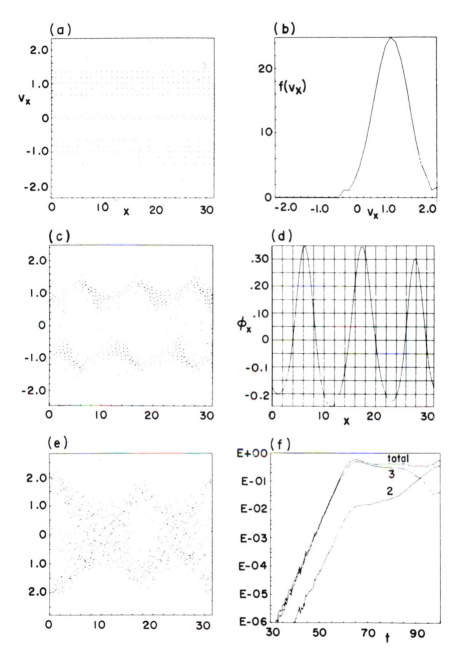

Figure 5-9c Evolution of warm electron-electron two-stream instability with v_x-x phase-space plots (a, c, and e at $t = 0, 60$, and 100), initial velocity distribution of one stream $f(v_x)$ *versus* v_x (in b), potential $\phi(x)$ *versus* x (in d at $t = 60$), and field energy history (in f; total, modes 2 and 3, $t = 0$ to 100).

5-10 BEAM-PLASMA INSTABILITY; LINEAR ANALYSIS

This model consists of a beam injected into a stationary plasma and has long been a favorite of theorists and simulators. There are many variations: a cold or warm stream and plasma; a weak or strong beam; growth in time or space; particle trapping and passing effects; linear, quasilinear, and nonlinear effects; infinite, periodic, or bounded models; etc. We cover only one or two aspects by following growth in time in our periodic model; the reader may take up variations as projects.

The beam-plasma system follows naturally on the two stream work. An example of early theory and experiments is the amplifier work of *Boyd et al.* (1958), who obtained microwave amplification at centimeter wavelengths using an electron beam injected into a long gas discharge column. They modulated the beam before it went into the plasma and stripped off the modulation after the beam left the plasma. The device aroused great interest for producing waves at millimeter wavelengths because the beam and plasma were intimately mixed and did not require a slow-wave circuit which was tiny and fragile to be grazed by an intense electron beam. More recent work by *Gentle and Lohr* (1973) emphasizes nonlinear behavior. The model is still fascinating.

The physical behavior is an example of double streaming as presented in Section 5-6. Those ideas can be lifted and used here by going to a new frame of reference. The largest change is that the beam density n_{b0} is generally much smaller than that of the plasma n_{p0} such that $\omega_{pb}^2 \ll \omega_{pp}^2$, the *weak-beam model*. The first interactions which we follow are electron-electron with an immobile neutralizing ion background (*i.e.*, $m_i / m_e \to \infty$). Hence, for uncoupled stream and plasma, the dispersion diagram is as shown in Figure 5-10a. The dominant frequency obviously is ω_{pp} and the wave numbers of interest are near $k = \omega_{pp} / v_0$.

For a cold plasma and a cold beam, the dispersion relation is

$$0 = \frac{\epsilon(\omega,k)}{\epsilon_0} = 1 - \frac{\omega_{pp}^2}{\omega^2} - \frac{\omega_{pb}^2}{(\omega - \mathbf{k}\cdot\mathbf{v}_0)^2} \tag{1}$$

Using reduced variables, defined by

$$W \equiv \frac{\omega}{\omega_{pp}}, \quad K \equiv \frac{\mathbf{k}\cdot\mathbf{v}_0}{\omega_{pp}}, \quad R = \left(\frac{\omega_{pb}}{\omega_{pp}}\right)^2 \tag{2}$$

the equation to be solved is

$$0 = F(W,K) = 1 - \frac{1}{W^2} - \frac{R}{(W-K)^2} \tag{3}$$

We choose to consider real k (or K) and seek complex values of ω (or W), with four ω's to each k. The solutions are shown in Figure 5-10b for R varying from 0.001 to 0.1. (Note the utility of having a root solver and a plotting routine.)

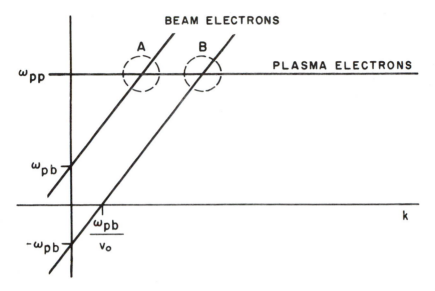

Figure 5-10a Sketch of dispersion diagram for the *weak-beam model*, with beam of density ω_{pb} injected at velocity v_0 into a plasma of density ω_{pp}, uncoupled. Coupling between the beam space-charge waves and plasma oscillations is expected at A and B.

The maximum growth rate $\omega_{\text{imaginary}}$ is plotted in Figure 5-10c along with ω_{real} and k at the maximum growth. It is seen that $(\omega_{\text{imag}})_{\text{max}}/\omega_{pp}$ increases first as about $2R^{1/3}/3$ (called *weak beam*) and then as about $R^{1/4}/2$ (called *strong beam*). The "boundary" is found graphically or by equating these two to be at

$$R = \left(\frac{3}{4}\right)^{12} = 0.032 \tag{4}$$

as noted on the figure. This *boundary* is not to be taken seriously, but regarded merely as some kind of transition region from *3-wave interaction* $\omega_i \propto (n_{b0})^{1/3}$, Figure 5-10b(a) and (b), and *4-wave interaction*, $\omega_i \propto (n_{b0})^{1/4}$, Figure 5-10b(c).

For *weak beams*, we see that maximum growth occurs very close to intersection B in Figure 5-10a, $\omega_{\text{real}} \approx \omega_{pp}$, $k \approx \omega_{pp}/v_0$. Note that even a very weak beam, say, with 10^{-4} the density of the plasma, sets up measurable growth; in this case the growth is 1.8 dB/plasma cycle, or a factor of 10 after 5.5 plasma cycles. For more exact results, see *Bers* (1972). For *strong beams,* the interactions at A and B in Figure 5-10a move apart widely to C and D in Figure 5-10d, and the three-wave weak-beam interaction becomes a four-wave interaction with $\omega_{\text{real}} \approx \omega_{pp}$, $k \approx \omega_{pb}/v_0$. This regime is puzzling physically with respect to neutralization and so deserves comment. Inequality 5-2(23) implies a beam electron density no longer trivial relative to background plasma electron density; the latter is usually taken as equal to the background ion density and is now insufficient for system neutrality.

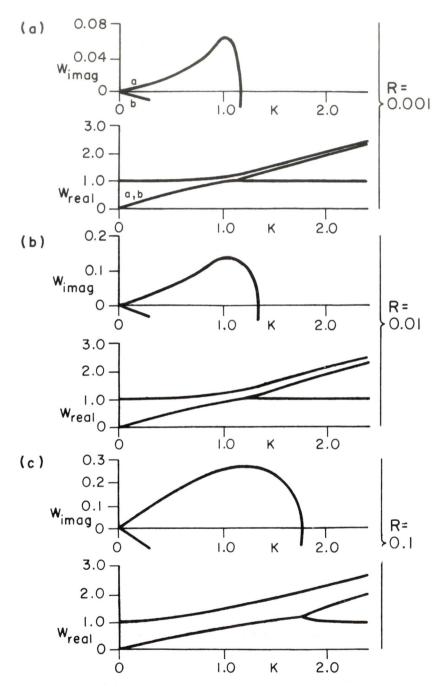

Figure 5-10b Real and imaginary roots of the dispersion relation (3). (a) For $R = 0.001$; (b) for $R = 0.01$; and (c) for $R = 0.1$. The complex roots are paired by the letters (a,b) in (b); the other roots are wholly real. The root at $W_{real} \approx -1.0$ is not shown.

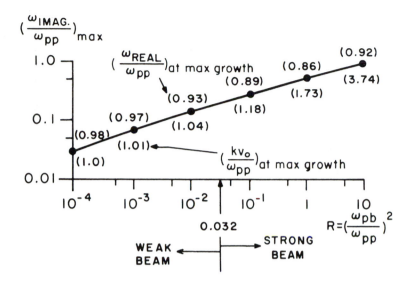

Figure 5-10c Maximum growth rate for cold-beam-plasma system as a function of beam to plasma density ratio. The density ratio $R = 0.032$ mark is taken as the boundary between weak and strong beams, the break point between growth at $R^{1/3}$ and $R^{1/4}$.

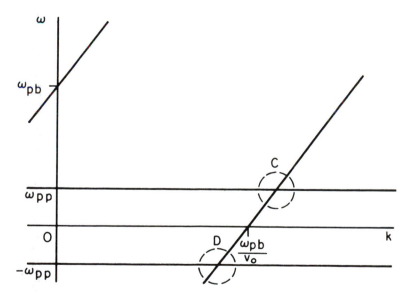

Figure 5-10d Sketch of dispersion-dispersion diagram for the *strong-beam model* uncoupled. Coupling between the beam and plasma is expected at C and D, roughly two two-wave couplings.

Hence, the plasma ion density must exceed the plasma electron density by

$$\frac{n_{pi}}{n_{pe}} = \frac{1 + R}{1 - \dfrac{m_e}{m_i}} \tag{5}$$

This is quite different from our initial discussion, as the beam is no longer just slightly perturbing a neutral plasma but is actually providing many of the electrons.

Of course, *the background could be all ions* such that $\omega_{pp}^2 = (q / m_i)(n_{b0}q / \epsilon_0)$, which forces $\omega_{pp}^2 \ll \omega_{pb}^2$. This model of electrons streaming through ions is the simulation model of *Buneman* (1959), one of the early milestones in plasma simulation. This is also the model used by *Pierce* (1948) in seeking to explain spurious signals, separated from the microwave signal by the ion plasma frequency (sidebands), observed in microwave tubes (triodes, tetrodes, etc.); Pierce solved the same dispersion relation as (1), but for real ω and complex k. This model is treated in non-linear theory and simulation by *Ishihara et al.* (1980, 1981), who show the key role in the saturation played by the appearance of spatial harmonics.

In your simulations, you might consider two distinct models:

(a) The *weak-beam model* $\omega_{pb}^2 \ll \omega_{pp}^2$ is that of a weak electron beam launched into cold but mobile electrons of much higher density. The plasma ions have $m_i \rightarrow \infty$ with no motion; neutralization is done simply by making the net charge density zero in the simulation program.

(b) The *strong-beam model* $\omega_{pb}^2 \gg \omega_{pp}^2$ is that of an electron beam launched into cold but mobile ions of equal density but with $m_i \gg m_e$.

Thus for (a), the model is electron-electron; for (b), it changes to electron-ion. Of course, one is free to choose other models as well.

As the beam temperature is increased, the nature of the roots changes (*O'Neil and Malmberg, 1968*).

5-11 BEAM-PLASMA INSTABILITY; AN APPROXIMATE NONLINEAR ANALYSIS

The nonlinear model of a cold weak beam launched into a plasma has been treated by theorists (*e.g.*, *Drummond et al.*, 1970), experimentalists (*e.g.*, *Gentle and Lohr*, 1973), and simulators (*e.g.*, *Kainer et al.*, 1972). The predictions and results fit together almost too well. We draw on these publications freely, especially on the interpretations and summary given by *Hasegawa* (1975).

We first use the linear analysis given earlier to obtain the relative density and velocity modulation needed to develop ideas on *trapping*. The dispersion equation 5-10(1), using

$$\omega \equiv \omega_{pp}(1 - \delta) \tag{1}$$

with $\delta \ll 1$, can be approximated as a cubic equation,

$$\delta^3 = -\frac{1}{2}\left[\frac{\omega_{pb}}{\omega_{pp}}\right]^2 = -\frac{1}{2}R \tag{2}$$

The three roots for δ are the three values of $(-1)^{1/3}$ times $(R/2)^{1/3}$. The complex frequency of the growing wave $(\text{Im}\,\omega > 0,\ \text{Im}\,\delta < 0)$ is

$$\omega_{\text{growing}} = \omega_{pp}\left[1 - \frac{1}{2}\left[\frac{R}{2}\right]^{1/3} + i\left[\frac{3}{2}\right]^{1/2}\left[\frac{R}{2}\right]^{1/3}\right] \tag{3}$$

The ordering [for $(q/m)_p = (q/m)_b$] of magnitudes of first-order variations in density and velocity is found to be

$$\left[\frac{n_1}{n_0}\right]_p \approx \left[\frac{v_1}{v_0}\right]_p \approx \delta\left[\frac{v_1}{v_0}\right]_b \approx \delta^2\left[\frac{n_1}{n_0}\right]_b \tag{4}$$

Even for 100 percent beam-density modulation, $n_{1b} \approx n_{0b}$, the velocity modulations are still small, especially for the plasma, as

$$(v_1)_b \approx \delta v_0 \quad \text{and} \quad (v_1)_p \approx \delta^2 v_0 \tag{5}$$

This ordering supports the idea that the beam may progress well into non-linearity while the plasma remains essentially linear. Indeed, in simulation of weak beams, taking the plasma background to be linear allows representing the plasma simply by a dielectric function, as developed by *O'Neil et al.* (1971), or as a linearized fluid, as done by many, for example *Lee and Birdsall* (1979a, b).

Even well into the nonlinear buildup, the potential can be assumed to remain essentially sinusoidal,

$$\phi_1\cos(kx - \omega t) = \phi_1\cos\omega_{pp}\left[\frac{x}{v_0} - t(1 - \delta_r)\right] \tag{6}$$

times the growth factor, $\exp(\omega_{pp}\delta_i t)$. Trapping of the beam is followed in this potential, in the frame moving at the phase velocity v_p. Hence, particle i sees a *stationary potential*

$$\phi_1 = \overline{\phi}\cos kx_i \tag{7}$$

and the particles behave as in a non-time-varying potential (*i.e.*, $\partial\phi/\partial t = 0$), obeying energy conservation

$$\frac{1}{2}mv_i^2 + q\overline{\phi}\cos kx_i = c_i \tag{8}$$

where c_i is the total energy of the ith particle. (Yes, we recall that in time varying potentials, $\frac{1}{2}mv^2 + q\phi = $ constant is not quite correct.) Beam particles have phase space behavior as shown in Figure 5-11a; low-energy particles, $c_i < q\bar{\phi}$, are trapped; high-energy particles, $c_i > q\bar{\phi}$, are passing particles; particles with energy $c_i = -q\bar{\phi}$ are at the bottom of the potential well and have $v = 0$. The critical velocity of escape (energy $c_i = q\bar{\phi}$) is

$$v_c^2 = \frac{4q\bar{\phi}}{m} \tag{9}$$

That is, when the potential $\bar{\phi}$ has increased sufficiently so that relative slip between beam and wave, $\Delta v = v_0 - v_{phase} = \delta_{real}v_0$ is v_c, then some of the beam begins to be trapped; this stage is defined by ϕ reaching ϕ_T, at time t_1

$$\phi_T \equiv \frac{m(\Delta v)^2}{4q} \tag{10}$$

which is related to the average field energy density of

$$\frac{1}{4}\epsilon_0 E^2(t_1) = \frac{1}{4}\epsilon_0 k^2 \phi_T^2 = 2^{-31/3}R^{1/3}\left|\frac{1}{2}n_{0b}m_b v_0^2\right| \tag{11}$$

that is, less than $1/1000$ of the beam kinetic energy density. *Drummond et al.* (1970) surmised that the trapping would proceed as shown in their sketches, Figure 5-11b. When the beam has been trapped, it falls to the far side of the well, in half a bounce time. At that time, t_2 roughly, the beam particles have lost

$$\Delta(KE) \approx \frac{1}{2}m_b n_{0b}[(v_0 + \Delta v)^2 - (v_0 - \Delta v)^2] \tag{12}$$

$$\approx 2m_b n_{0b} v_0 \Delta v \tag{13}$$

of which half goes into kinetic energy of oscillations, half into field energy (as in linear theory, here stretched a bit). This field energy density is, at

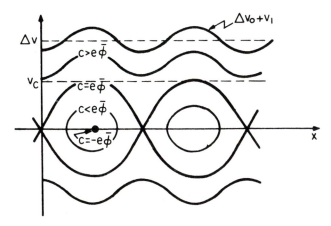

Figure 5-11a Phase space trajectories for beam particles, as viewed in the frame moving at the phase velocity of the wave. *(From Hasegawa, 1975, fig. 36.)*

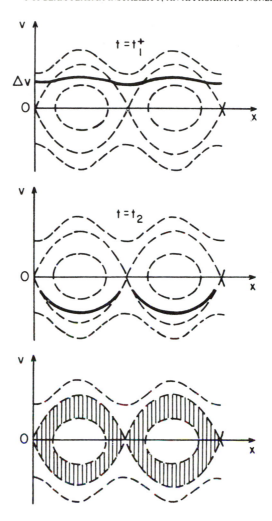

Figure 5-11b Phase space for the beam particles, just at the beginning of trapping ($t = t_1^+$), at a later time when they have made half a bounce ($t = t_2$) and, finally, smeared out after many bounces. *(From Drummond et al., 1970.)*

time t_2,

$$\frac{1}{4}\epsilon_0 E^2(t_2) = m_b n_{0b} v_0 \Delta v = \left[\frac{R}{2}\right]^{1/3}\left[\frac{1}{2}m_b n_{0b} v_0^2\right] \qquad (14)$$

This is $2^{10} = 1024$ times larger (30 dB more) than at $t = t_1$. [Note for the strongest weak beam allowed, $R = 0.032$, that $(R/2)^{1/3} = 0.251$, which says that 25 percent of the beam energy may appear as field energy, a rough estimate of the saturation or efficiency.] As the particle velocities bounce back up, the field energy goes back into kinetic energy; but, as the particles oscillate at different frequencies (the well is not parabolic), they mix in phase and the particle bouncing and field oscillations, Figure 5-11c, die out. The final field energy is half the difference between the initial beam energy and the beam energy of the smeared-out distribution. The latter is symmetric about the wave phase velocity, of value $\frac{1}{2}m_b n_{0b}v_p^2$. Hence, one-half the difference is the field energy density for $t \to \infty$,

$$\frac{1}{4}\epsilon_0 E^2(t \gg t_2) = \frac{1}{2}m_b n_{0b}(v_0^2 - v_p^2)\left(\frac{R}{2}\right)^{1/3}\left[\frac{1}{2}m_b n_{0b}v_0^2\right] \qquad (15)$$

which is just half the peak value (3 dB down). (See the expression of ϕ_{max} given by *Walsh and Hagelin, 1976*.)

Of course, the picture presented is quite approximate. *Hasegawa* (1975) notes that the beam velocity modulation v_1 is Δv just at trapping (see Figure 5-11a) and also that at this time $n_{1b} = n_{0b}$, *i.e.*, well-developed bunching. Hence, the slight velocity modulation, no-bunching sketch of Figure 5-11b is too simple. To be sure, the bunches are trapped and rotate in phase space and the gross behavior of electric energy, Figure 5-11c, does occur; however, the details are different, as are seen in simulations.

This gives us a pretty good idea of weak-beam nonlinear behavior. We refer the reader to the calculations of *Kainer et al.* (1972) which provide some tie between weak and strong beams in the nonlinear regime. With these insights and special values of $R = n_{pb}/n_{0p}$ in mind, we proceed to design useful beam-plasma simulations.

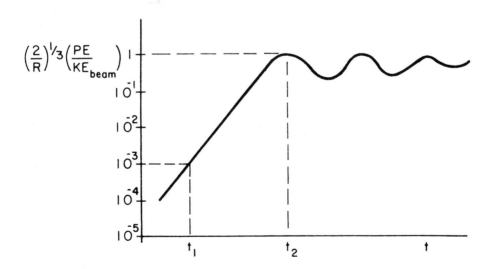

Figure 5-11c Electric field energy density as a function of time. At t_1, the first particles are trapped and at t_2, the beam has given up its maximum kinetic energy. The subsequent oscillations and particle bouncing die out as the particles do not all have the same frequency of oscillation in the wave potential well. *(From Drummond et al., 1970.)*

5-12 BEAM-PLASMA INSTABILITY; PROJECT

The object of the project is to observe the development of the weak $R = 0.001$ beam-plasma instability from initial exponential growth through to the end of growth at large amplitude. (The strong beam-plasma instability, especially $\omega_{pb} = \omega_{pp}$, $R = 1$, is very close to the two-stream instability already done.) The observations to be made are the checking of values of complex ω against those from linear theory, the saturation level ("efficiency" of conversion of kinetic to field energy), the slowing and broadening of the beam velocity distribution, and the bounce or trapping motion of the beam bunches following saturation.

For $R \ll 1$, the plasma (in contrast to the weak beam) remains linear by various tests (*e.g.*, n_1/n_0 remains $\ll 1$). Therefore, the plasma particles could be replaced by a linearized fluid or treated by a linear susceptibility. The particle-fluid model proved useful in a related model (magnetized ring-plasma) where the noise of the plasma had obscured the desired interaction (*Lee and Birdsall,* 1979a). We use another trick: Since only the plasma frequency of a linearly-responding species matters, we are free to choose a small value for q/m to ensure that the plasma remains linear, which permits us to use a small number of plasma particles (as few as one per cell!). The fixed background ions are put in by default by ES1.

Let the ratio $(\omega_{pb}/\omega_{pp})^2 \equiv R = 0.001$ in the weak-beam region ($R \leqslant 0.032$). The beam becomes highly distorted in v_x-x phase space, wrapping itself into a vortex. Therefore, for adequate accuracy, far more particles are required than for the plasma. Hence, try N = 512 for the beam and N = 64 for the plasma, with NG = 64 cells. Use $\omega_{pp} = 1.0$ and $\omega_{pb} = (1000)^{-1/2}$, with beam $v_0 = 1$. For a test run, place the first mode near the peak of the growth rate, $kv_0/\omega_{pp} = 1.01$, where $\omega/\omega_{pp} \approx 0.97 + i0.006$; hence, set $k_0 = 1$, $L = 2\pi$. For later runs, put the tenth mode at this peak, using $L = 20\pi$, and excite all modes initially using VT1. Excitation of the beam at X1 = 0.01 (to be decided on basis of best excitation of the growing wave) produces several decades of exponential growth, ending in saturation by time $t = 115$; using $\Delta t = 0.2$ requires NT = 600. Running for a longer time allows viewing of bounce, or trapping motion.

Check the nonlinear predictions of Section 5-11 and *Kainer et al.* (1972). Predict the potential when trapping is supposed to begin; does it do this according to your phase-space plots? Use phase-space plots made at saturation and several subsequent peaks and valleys (this takes a preliminary run to determine the times to make the plots) in order to verify (or not) the early trapping and subsequent mixing expected from Figure 5-11b. Is the peak of electric field energy that predicted in Section 5-11? Is the beam $f(v)$ spreading and slowing development as predicted by *Kainer et al.* (1972)? Are there some very hot particles, as are best seen in $\log f(v)$ versus v or v^2 plots? Follow the decrease in beam $<v>$ and increase in $<v^2>$ in time.

There are more variations worth noting, which you may try.

First, suppose only the beam is allowed to move, with the plasma held immobile. Then, only the fast and slow space-charge waves are expected, with no growth. However, there is a nonphysical instability produced by aliasing, as shown analytically by *Langdon* (1970a, b) and *Chen et al.* (1974) and verified by them and *Okuda* (1972); see Chapter 8. This instability causes heating of the beam to a level of $\lambda_D / \Delta x \approx 0.05$ (where $\lambda_D = v_t / \omega_p$ of the beam) and then stops, leaving a stable but slightly warm (and noisy) beam, as shown by *Birdsall and Maron* (1980). The maximum growth rate is $0.2\omega_{pb} = 0.2\omega_{pp} R^{1/2}$ which is roughly $\omega_{pp} / 150$ for $R = 0.001$, about a factor of 10 less than the beam-plasma growth rate expected. Hence, this self-heating effect (treated in detail in Part Two) may be viewed as transient here.

Second, examine what happens if the plasma is excited at relatively large amplitude, for example, by using X1 on the order of the unperturbed particle spacing. Does the beam-plasma interaction take place?

Third, ties may be made to linear and nonlinear experiments, where an electron gun launches a cold beam into a plasma. The simulation may be altered to fit the experiment more closely, where the beam electrons are continuously injected at $x = 0$ and collected at $x = L^-$ (not returned to $x = 0^+$, as in the usual periodic model) and the plasma electrons are reflected, in some careful manner, at $x = 0^+$, $x = L^-$. The excitation may be a velocity and/or density modulation of the beam by a sinusoidal (in time) source at $x = 0^+$. The level of modulation and length L should be chosen to produce saturation at $x < L$. These changes fit experiments more closely where spatial growth (real ω and complex k) is expected. See *Briggs* (1971, p. 76) on the distinction among absolute, convective, and "neither" instabilities.

Selected results for $R = 0.001$ are shown in Figure 5-12a. The v_x-x phase space plots were chosen to coincide with peaks of field energy, hence, valleys of beam kinetic energy of $t \approx 146$, 202, and 258 and valleys at $t \approx 173$, 230, and 286. Thus, the peak times should exhibit bunching at $v < v_0 = 1.0$ and valley times should exhibit bunching at $v > v_0$; the reader may see that this is roughly the case. Using ϕ_{max} in $\omega_{trap} = k(q\phi_{max}/m)^{1/2}$ produces $\tau_{trap} \approx 56$, very close to that observed. Note that the vortex takes on considerable structure so that simple bounce ideas fail.

PROBLEMS

5-12a Although we can choose q/m arbitrarily small for the plasma particles without altering the essential physics, show that there is a practical lower limit imposed by the finite precision of computer arithmetic.

5-12b What happens if the excitation X1 = 0.01 is applied to the plasma electrons instead of to the beam?

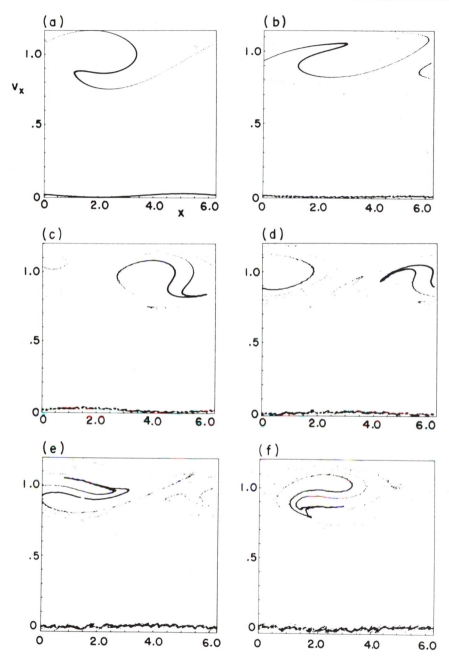

Figure 5-12a Evolution of weak-beam plasma instability in v_x-x phase space with $R = 0.001 = (\omega_{pb}/\omega_{pp})^2$ with beam velocity $v_0 = 1.0$. The left-hand set occurs very close to the first, second, and third peak of field energy and the right-hand set occurs very close to the subsequent valleys, showing the bounce motion and mixing. The sequence is $t = 146, 173, 202, 230, 258, 286$.

5-13 BEAM-CYCLOTRON INSTABILITY; LINEAR ANALYSIS

Another instability which is interesting is that due to relative drift of ions through electrons in an applied magnetic field $B_0\hat{z}$. The interaction is in the plane normal to B_0 (or nearly so). The particles have position variable x, and velocity variables v_x and v_y (and possibly v_z); the wave number is $k\hat{x}$.

The magnetic field strength is characterized by

$$\omega_{ci} \ll \omega_{pi} \ll \omega_{pe} \approx \omega_{ce} \tag{1}$$

For simplicity, both ions and electrons are treated as cold, *i.e.*, $v_{te} = 0 = v_{ti}$. We choose to work in the electron frame (electrons not drifting) with the ions having a steady drift v_0 in x. For such perturbations, the electrons are magnetized (*i.e.*, the electron $\mathbf{v}_1 \times \mathbf{B}_0$ term is kept); the ions are treated as unmagnetized as the waves have $|\omega - \mathbf{k} \cdot \mathbf{v}_0| \gg \omega_{ci}$. This model (and similar ones) has been used widely for theoretical studies of drift-excited electrostatic instabilities in magnetized plasmas; for more details and answers to threshold values of v_0 (*i.e.*, should $v_0 > v_{te}$ or $v_0 > v_{ti}$ for instability), see *Forslund et al.* (1972a).

Let us redo some parts of the theory, with results usable for comparison with results of simulation. The coordinates used are shown in Figure 5-13a. We look for hydrodynamic nonresonant instabilities, assuming

$$\frac{kv_{te}}{\omega_{ce}} < 1, \quad |\omega - kv_0| > kv_{te}, \quad \omega > k_\parallel v_{te} \tag{2}$$

The first assumption means

$$\frac{v_{te}}{v_0} < \frac{\omega_{ce}}{\omega_{uh}} \approx 1 \quad \text{for } \omega_{ce} \gtrsim \omega_{pe} \tag{3}$$

where $\omega_{uh}^2 \equiv \omega_{pe}^2 + \omega_{ce}^2$. The second assumption requires

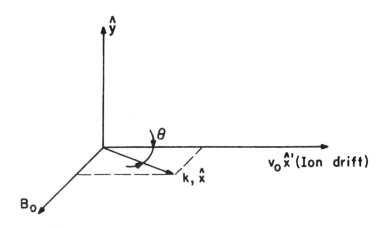

Figure 5-13a Coordinates for drift-excited instabilities.

$$\frac{v_{ti}}{v_0} < \left(\frac{m_e}{m_i}\right)^{1/3} \left(\frac{\omega_{pe}}{\omega_{uh}}\right)^{4/3} \tag{4}$$

For $T_e \approx T_i$ or $m_e v_{te}^2 \approx m_i v_{ti}^2$ and $\omega_{pe} \approx \omega_{ce}$, (4) becomes

$$\frac{v_{te}}{v_0} < \left(\frac{m_i}{m_e}\right)^{1/6} \left(\frac{\omega_{pe}}{\omega_{uh}}\right)^{4/3}$$

The cold dispersion relation, as investigated by *Buneman* (1962), is

$$1 - \cos^2\theta \frac{\omega_{pe}^2}{\omega^2 - \omega_{ce}^2} - \sin^2\theta \frac{\omega_{pe}^2}{\omega^2} - \frac{\omega_{pi}^2}{(\omega - kv_0\cos\theta)^2} = 0 \tag{5}$$

The $\cos^2\theta$ term is due to the electron motion across B_0; the $\sin^2\theta$ term comes from electron motion along B_0.

While our simulation project has $k_{\parallel} = 0$, $\theta = 0$, it is instructive to display the solutions of (4) for $\theta = 0$ and for small $\theta \approx \sqrt{m_e/m_i}$, showing both the *upper-hybrid-two-stream instability* ($\theta = 0$) predicted by Buneman and the *modified-two-stream instability* predicted by *Krall and Liewer* (1971, 1972). Figure 5-13b is a sketch of ω_{real} and $\omega_{imaginary}$ showing both instabilities. The physics of the two regimes are discussed in *Chen and Birdsall* (1973).

5-14 BEAM-CYCLOTRON INSTABILITY; PROJECT

This instability grows with Re $(\omega) \approx \omega_{uh}$ and Im $(\omega) \ll \omega_{uh}, \omega_{pe}, \omega_{ce}$. These dictate that Δt be chosen such that $\omega_{uh}\Delta t$ is small (say, 0.2) and that the program run several hundred steps (say, 600). The parameters recommended initially are those used in the linear analysis section:

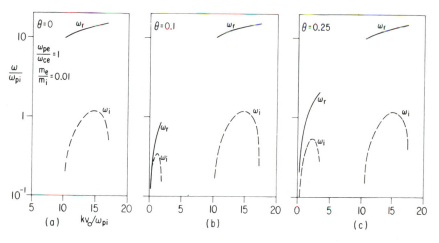

Figure 5-13b Plots of unstable roots for $\theta = 0$, $(m_e/m_i)^{1/2}$, and $2.5(m_e/m_i)^{1/2}$. Here $m_e/m_i = 0.01$ and $\omega_{pe}/\omega_{ce} = 1$. The scales are semilogarithmic. *(From Chen and Birdsall, 1973.)*

$$\omega_{pe} = |\omega_{ce}| = 10\omega_{pi}$$

The maximum growth rate $\gamma/\omega_{pe} \approx 0.12$ occurs at $kv_0/\omega_{pe} \approx 1.5$, with $\omega_{real}/\omega_{pe} \approx 1.38$. For example, choose $\omega_{pe} = 1.0$, $v_{0e} = 0$ (the interaction is viewed in the electron frame), $\omega_{ce} = -1.0$, $\omega_{pi} = 0.1$, $\omega_{ci} = 0$ (ions are taken to be unmagnetized), $v_{0i} = 1.0$. Let the first mode be placed at the k for the maximum growth rate, $k_0 \approx 1.5\omega_{pe}/v_0 = 1.5$, meaning that $L = 2\pi/k_0 = 4\pi/3$. Use $(q/m_e) = -1.0$ and $(q/m_i) = 0.01$. The ions are relatively massive and so move little away from their simple drift while the electrons wrap up in a phase space vortex. Hence, fewer ions may be used than electrons; try $N = 512$ for the electrons and $N = 128$ for the ions. Excitation of the first mode at electron X1 = 0.001 produces many orders of magnitude of growth (maybe too many, allowing use of larger X1) to saturation in the 600 steps already suggested.

One object is to verify the linear analysis. Part of the verification is to measure complex ω from the time histories and compare with the predicted values as a function, say, of k or other parameters. Why does ω_{real} appear to be zero in the mode 1 energy history? How would you find ω_{real}? What is the measured phase between, say, ϕ and ρ? Does this phase agree with linear analysis? Predict and observe the v_x, v_y plot for electrons.

Another object is to observe the large amplitude behavior. Note the value of electrostatic energy at saturation, say, as a fraction of the system initial kinetic energy. Does the ion drift energy drop? Is there growth in the electron drift energy? Observe the growth in ion and electron thermal energies. Validate the simulation by comparing the error in total energy with the variations in any of the components of the total; *e.g.*, is the maximum electrostatic energy much less than the error in the total?

5-15 LANDAU DAMPING

Electrostatic plasma waves are damped in a warm plasma, even without collisions. This surprising result was first found from consideration of analytic continuation in the complex ω plane of the Laplace-transformed ϕ. Detailed physical understanding and analysis may be found in *Jackson* (1960) and *Dawson* (1961).

With a Maxwellian velocity distribution, there are particles moving both faster and slower than the wave phase velocity ω/k. If we transform to a frame moving with the phase velocity, the potential is sinusoidal in x and decaying (but not oscillating) in time at a rate ω_i. Ignoring the decay for the moment, we see easily that the electrons in a velocity band $\omega/k \pm v_{tr}$ are trapped by the wave, oscillating about the velocity ω/k with frequency $\approx \omega_{tr}$, where $\frac{1}{2}mv_{tr}^2 = q\phi$ and $m\omega_{tr}^2 = qk^2\phi$, in which ϕ is the peak amplitude of the wave component at this phase velocity. Returning to the stationary frame, we see that the *resonant electrons* in the range $(\omega/k - v_{tr}, \omega/k)$ exchange with those in the range $(\omega/k, \omega/k + v_{tr})$ in a time $\approx \pi/\omega_{tr}$. If there are more electrons initially in the slow range, then there is a net gain in particle energy which must be supplied by the wave. This is the collision-less dissipation mechanism of Landau damping, and suggests why the damp-ing rate is proportional to $(\partial f_0/\partial v)_{v=\omega/k}$.

We begin with a straightforward example which demonstrates the effect but also points out some difficulties in performing a clean simulation. Our parameters are: DT = 0.1, NT = 200, NG = 256, A2 = 10^4, IPHI = 10, IXVX = 10, MPLOT = 1, 2, 3; and N = 2^{14}, VT2 = 0.5, NLG = 1, MODE = 1, X1 = 0.02.

The wave damps at a rate $\omega_i = -0.15\omega_p$ close to that predicted by the linear theory, but the wave energy drops only an order of magnitude, then oscillates slowly. The same results are reported by *Denavit and Walsh* (1981) using a different quiet-start technique.

With a random initialization (VT1 = 0.5, VT2 = 0) exponential decay is not seen. The distribution $f_0(v)$ must be well-represented by the particle loading, especially near ω/k. For this reason, we cannot decrease the very rapid rate of decay in this example by choosing smaller kv_t/ω_p to place the phase velocity out in the "tail" of the distribution, which is sparsely popu-lated. Instead, we use a trick, as follows.

First, we divide the electrons into two groups, one cold, the other Maxwellian. In ES1 these groups are conveniently handled as separate "species," c and h. By choosing $\omega_{pc}^2 \gg \omega_{ph}^2$ and $kv_h \approx \omega_{pc}$ we place the phase velocity at the steepest slope of f_{0h}, so that there are many particles in the trapping range, but the damping rate is $\ll \omega_{pc}$.

Second, an artifice enables us to decrease the number of particles needed. In the code we can choose $N_c \ll N_h$ which means that a "cold" particle carries a much larger charge than a hot particle. By choosing $q/m = 0.01$ for the cold particles, we can avoid nonlinearity in their response and use only N_c = NG. All this does not affect the behavior at the low amplitudes we use.

The parameters are: DT = 0.1, NT = 1000, NSP = 3, NG = 256, A2 = 10^4, IPHI = 20, IXVX = 20, MPLOT = 1, 2, 3, 4 for the main pro-gram; hot species is specified by N = 2^{14}, WP = 0.383, VT2 = 0.9; cold species by N = 256, QM = 0.01, WP = 0.924, VT2 = 0, MODE = 1, V1 = 2.5×10^{-4}; and marker species by N = 1020, NLG = 1020/3, QM = -1.0, WP = 10^{-10}, V0 = 0.9, with VT2 chosen to obtain velocities of 0.8, 0.9, and 1.0. In this example, the field energy drops by two orders of magni-tude, then rises as shown in Figure 5-15a. The initial damping rate is easily estimated from the linear theory by examining the imaginary part of the dispersion relation $\epsilon(k, \omega_r + i\omega_i) = 0$, with ω_i small:

$$0 = \text{Im } \epsilon \approx 2\frac{\omega_{pc}^2}{\omega_r^3}\omega_i - \pi\frac{\omega_{ph}^2}{k|k|}f_0'\left(\frac{\omega_r}{k}\right) \tag{1}$$

from which, for a Maxwellian,

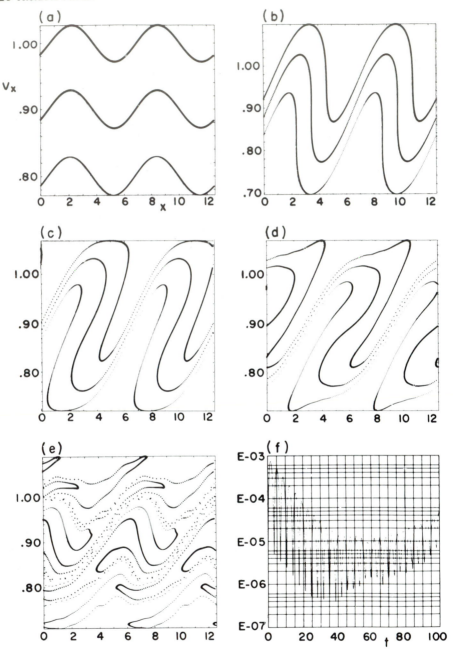

Figure 5-15a Landau damping as observed in v_x-x phase-space plots (a, b, c, d, and e at $t = 2, 16, 30, 40,$ and 90) and field energy history (f, for $t = 0$ to 100). The phase-space plots show marker particles (same q/m as the main particles, but much smaller q and m) originally at $v = 0.8, 0.9$ (the wave phase velocity) and 1.0, shown by $-$, $*$, and $+$. To clarify the spatial structure, the plotting interval is $0 \leqslant x \leqslant 2L$, and each particle is plotted twice, at (x, v) and $(x + L, v)$.

$$\frac{\omega_i}{\omega_r} = -\left(\frac{\pi}{8}\right)^{1/2} \frac{\omega_{ph}^2}{\omega_{pc}^2} \left(\frac{\omega_r}{kv_t}\right)^3 e^{-\omega_r^2/2k^2v_t^2} \tag{2}$$

In the simulation we find $\omega_r = 0.90$; this expression then gives $\omega_i = -0.058$ which agrees adequately with the rate $\omega_i = -0.062$ observed.

The rise in amplitude after $t = 45$ is due to oscillations of trapped electrons in the wave field. The rise can be made to occur sooner by raising the initial amplitude and hence the trapping frequency, but it cannot be delayed much without using more particles.

By changing a few statements in the code we can plot the phase space and $f(v)$ over the trapping range only. We do see $f(v)$ change, but the detailed information which should be available from phase space is not discernible in the plot because we do not know where a given particle was earlier. By plotting only the particles whose velocities were *initially* greater than 0.9 we can see the trapping. Another way is to plot a third species of marker particles, which have the same q/m but low charge density so they move with the hot particles but do not affect the evolution. We have loaded these particles in three beams at $v = 0.8$, 0.9, and 1.0. Phase space plots of these particles, Figure 5-15a, carry information on the *history* of the resonant particles; such information is not available from the distribution function $f(x, v, t)$ in a code which integrates the Vlasov equation.

PROBLEM

5-15a In a quiet-start in ES1, the velocities in each group are monotonically increasing while the positions are scrambled by the bit-reversal procedure. What difference would there be if instead the positions were monotonically increasing and the velocities were scrambled?

5-16 MAGNETIZED RING-VELOCITY DISTRIBUTION; DORY-GUEST-HARRIS INSTABILITY; LINEAR ANALYSIS

Plane waves propagating perpendicular to a uniform magnetic field, $\mathbf{B}_0$ ($k_\perp \neq 0$, $k_\parallel = 0$), in a uniform plasma, are stable or unstable depending on the velocity distribution (*Harris, 1959*). Consider warm ring-like velocity distributions with

$$dn = F_0^p(v_\perp) 2\pi v_\perp dv_\perp = \left[\int_{-\infty}^{\infty} dv_\parallel f_0(v_\perp, v_\parallel)\right] 2\pi v_\perp dv_\perp$$

$$= \frac{1}{\pi \alpha_\perp^2 p!} \left(-\frac{v_\perp}{\alpha_\perp}\right)^{2p} \exp\left(-\frac{v_\perp^2}{\alpha_\perp^2}\right) 2\pi v_\perp dv_\perp \tag{1}$$

For $p = 0$, F_0^0 is Maxwellian and the roots of the dispersion relation fall near the cyclotron harmonics (of both electrons and ions), moving asymptotically to $n\omega_c$ as $k_\perp v_\perp/\omega_c \to \infty$; these are the harmonics of *Bernstein* (1958).

See *Crawford and Tataronis* (1965) for dispersion curves, and *Krall and Trivelpiece* (1973, pp. 407-412) for a general discussion. For a Maxwellian distribution ($p = 0$), the dispersion relation predicts only real roots for any ratio ω_p/ω_c. Yet, if we hold ω_p constant and decrease ω_c to zero, we would expect to recover Landau damping; at least when ω_c is smaller than the damping rate. *Kamimura et al.* (1978) review the theoretical resolution of this paradox and do a sophisticated simulation study of this damping and its relations to autocorrelation and spectra of thermal field fluctuations and to particle diffusion across the magnetic field.

For $p \neq 0$, F_0^p is like a warm ring in $v_\perp$ space (say, v_y *versus* v_x) and is unstable for $p \geqslant 3$ for the so-called zero-frequency mode (meaning $\omega_{\text{real}} = 0$), as shown by *Dory et al.* (1965).

For $p \to \infty$, the ring has zero thickness (is cold), with F_0^∞ given by

$$F_0^\infty (v_\perp) = \frac{1}{2\pi v_{\perp 0}} \delta (v_\perp - v_{\perp 0}) \tag{2}$$

All particles have the same speed $v_{\perp 0}$. The cold ring is unstable for $(\omega_p/\omega_c)^2 > 6.62$ as shown by *Crawford and Tataronis* (1965) and *Tataronis and Crawford* (1970), with solutions of the dispersion relation given in Figure 5-16a. For use in the project, the growing complex roots for $(\omega_p/\omega_c)^2 = 10$ are $\omega/\omega_c = 1.37 + i0.265$ at $k_\perp v_{0\perp}/\omega_c = 4.5$ and $2.34 + i0.273$ at $k_\perp v_{0\perp}/\omega_c = 5.5$. Note the zero-frequency mode, $\omega_{\text{real}} = 0$, centered at $k_\perp v_{0\perp}/\omega_c \approx 3$. The $k_\perp = 0$ values are $n\omega_c$ plus the hybrid. The $k_\perp \neq 0$ values for $\omega = n\omega_c$ are at zeros of J_n and J'_n of argument $k_\perp v_0/\omega_c$.

The real parts of the frequency fall roughly midway between cyclotron harmonics and the maximum imaginary parts are comparable to ω_c [even for $(\omega_p/\omega_c)^2 \gg 1$]. This limit on growth may appear open to question. That is, why is the maximum $\gamma \leqslant \omega_c$ rather than $\propto \omega_p$? After all, the one-dimensional distribution function [$F_0^\infty (v_\perp)$ integrated over v_y, the projection onto the v_x axis] clearly has two sharp peaks, appearing like two streams, as shown in *Schmidt* (1966, p. 254, not in 2nd ed., 1979). However, when the Penrose criterion (for unmagnetized plasma) is applied to this distribution, the Penrose integral is zero, meaning no instability; this result is noted by Langdon in *Lindgren et al.* (1976). Hence, the ring instability requires the B_0 magnetic field. If γ exceeded ω_c, the magnetic field would not play a role; therefore the maximum growth rates for a single ring are of the order of ω_c.

The ring distribution is of interest for several reasons. One is that a neutral beam injected into a magnetic field (as in fusion experiments) produces a ring distribution. Another is that the ring is a model loss-cone distribution which is relatively easily handled in theory. Another is that a Maxwellian distribution in $F(v_\perp)$ can be constructed from a set of rings. Hence, the project to follow is a practical exercise.

A closely related instability occurs with both a warm ring and a warm core plasma. *Tataronis and Crawford* (1970) presented stability boundaries for a cold ring and warm core. *Mynick et al.* (1977) presented extensive

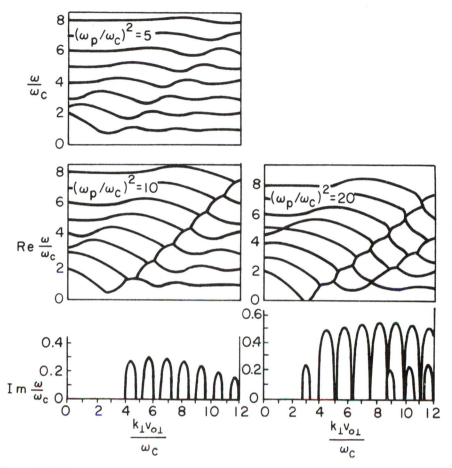

Figure 5-16a Complex ω *versus* $k_\perp$ for ring velocity distribution, speed $v_{0\perp}$, perpendicular to B_0 for $k_\parallel = 0$, below threshold [which is $(\omega_p/\omega_c)^2 = 6.62$ for $\omega_{real} \neq 0$ and 17.06 for $\omega_{real} = 0$] and above.

results for this two-ion component plasma, but unmagnetized. They find stability boundaries using the Penrose criterion, and growth rates for a wide range of ratios of plasma to core densities and temperatures; these results are helpful guidelines to cases where $(\omega_{pi}/\omega_{ci})^2 \gg 1$ and growth rate $\omega_{imag}/\omega_c \gtrsim 1/\pi$, meaning the ions behave as if not magnetized (with some exceptions). *Lee and Birdsall* (1979a, b) extended the magnetized model theory results considerably and provided verifying simulations. This two-component model is also a practical model for study; the reader may wish to simulate this model after doing the cold ring model.

PROBLEM

5-16a Relate $v_{\perp 0}$ in (2) to $\alpha_\perp$ in (1) with $p \gg 1$.

5-17 MAGNETIZED RING-VELOCITY DISTRIBUTION; PROJECT

The project is to observe the behavior of a uniform magnetized plasma, with a cold ring velocity distribution, speed v_0. We choose $(\omega_p / \omega_c)^2 = 10$, for which complex ω for real k (suiting the ES1 periodic model) are shown in Figure 5-16a.

Consider the particles to be ions. The electrons are neglected, put in by ES1 as a uniform neutralizing background.

One problem is to devise an initial particle-loading scheme which both puts particles in a ring of radius $v_{\perp 0}$ in velocity as uniformly as possible in phase space x, v_x, v_y. For example, let the particles be spaced uniformly in x, with the velocities given by

$$v_{xi} = v_{\perp 0} \cos \theta_i \tag{1}$$

$$v_{yi} = v_{\perp 0} \sin \theta_i \tag{2}$$

We do not want θ_i to be too obviously correlated with x_i (why?). A small change to subroutine INIT enables efficient loading of a ring. In the input specify WC, V0 $= v_{\perp 0}$ and NLG $= 1$. Alter the logic to skip over the velocity loading involving VT2 but retain the scrambling of positions. The coding between statements numbered 50 and 51 then rotates the velocity to fill in the ring. However, the angle THETA is intended for loading the distribution 5-16(1) (Problem 5-17b); we change this to THETA $=$ TWOPI $*$ $(I-1)/NGR$. The particles are now loaded with evenly spaced angles increasing monotonically with index i, while the positions x_i are scrambled. As phase space is now highly ordered, the electric field at the start is noiseless or quiet.

Using $\omega_p = 1$, and $\omega_c = (10)^{-1/2}$, choose $v_{\perp 0}$ to make $k v_{\perp 0} / \omega_c = 4.5$ (or 5.5), which places the fundamental mode at one of the maximum growth rates. For $\Delta t = 0.2$, NT $= 600$, X1 $= 10^{-3}$, and N $= 4096$, the instability grows over many decades (allowing accurate measurement of complex ω) and saturates.

Check the results early in time against the linear theory. Observe the rapid collapse of the cold ring toward a thermal distribution, as shown by *Byers and Grewal* (1970, figure 7). See if you duplicate their observation of a shift in the frequency spectrum of the electric field from roughly $3\omega_c / 2$ (or $5\omega_c / 2$) in the linear regime, to $n\omega_c$ in the postsaturation regime, where $f(v)$ is becoming more thermal (Gaussian), and the Bernstein harmonics are expected.

PROBLEMS

5-17a Relate NV2 and VT2 (in subroutine INIT in ES1) to exponent $2p$ and $\alpha_\perp$ in 5-16(1). NV2 and $2p$ may differ by 1 because of the difference in $v_\perp$ space volume elements.

5-17b Explain how the (unaltered) INIT sets up the distribution 5-16(1), using NLG = NG and N >> NG. What happens if NLG = 1, the default? Consider scrambling the angles using a radix-three digit-reversal (*Denavit and Walsh, 1981*) so that x, $v_\perp$, and θ are uncorrelated.

5-18 RESEARCH APPLICATIONS

From the earliest papers to the present, one dimensional electrostatic particle simulation continues to be a productive research tool. So much so that, in addition to papers already cited in this chapter, we mention here only one paper from each of a few of the research areas where these codes are used.

The presence of trapped particles can lead to unstable growth of side-bands at frequencies $\approx \omega_p \pm \omega_{tr}$ (*Kruer and Dawson, 1970*), A classic example of "parametric" instability in plasma (in which a driven wave is coupled nonlinearly to other modes to which its energy is transferred) is treated by *Kruer and Dawson* (1972), who demonstrate the resulting enhanced resistivity, and a relativistic version by *Lin and Tsintsadze* (1976). Another interesting consequence of nonlinear mode coupling is the "explosive" instability, so called because its growth is faster than exponential; examples are found in *Jones and Fukai* (1979). Plasma expansion into a vacuum is treated in *Denavit* (1979). With some modification, ES1 has been used for many projects. *Lee and Birdsall* (1979b) add a linearized fluid-like electron response. *Cohen and Maron* (1980) model linearized electron motion in the presence of a density gradient. These simulations are inexpensive with ES1 on present computers. Scaling of electron heating and density profile modification in a plasma driven by an oscillating electric field are treated by *Albritton and Langdon* (1980). Conditions for formation of electrostatic shocks ("double layers") in the auroral magnetosphere are studied by *Hudson and Potter* (1981). Many of these papers cite other applications of one dimensional electrostatic simulation.

SIX

A 1D ELECTROMAGNETIC PROGRAM EM1

Michael A. Mostrom and Bruce I. Cohen

6-1 INTRODUCTION

One dimensional codes, even with three velocity components, are almost always easier to set up and use, and more economical to run, than two and three dimensional codes. Also, there are 1d algorithms which have especially useful properties; this is true in the EM code given here, which uses a field solver that combines simplicity and stability.

The first fully electromagnetic algorithm was devised by J. M. Dawson in 1d (unpublished, *ca.* 1965). He separated the transverse fields into left- and right-going components. By choosing $\Delta x = c\Delta t$, these components are advanced in time simply by shifting the values over one cell and adding current contributions from particles whose paths intersect the light rays. We discuss a variation of this 1d algorithm as formulated and implemented by *Langdon and Dawson* (1967) and revived for laser-plasma studies by B. I. Cohen, M. A. Mostrom, D. R. Nicholson, and A. B. Langdon (*Cohen et al., 1975*).

Two and three dimensional fully electromagnetic programs are given in Chapter 15.

6-2 THE ONE-DIMENSIONAL MODEL

For linearly polarized electromagnetic waves, the variables used have components oriented as shown in Figure 6-2a. Electromagnetic and electrostatic waves propagate along x; there are no variations in y or z ($k_y =$

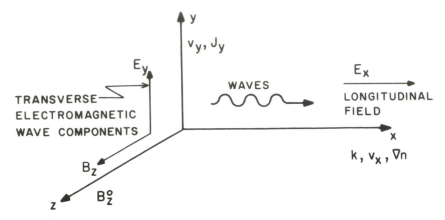

Figure 6-2a Location of the particle and field quantities for the linearly polarized 1½d EM1 code.

$0 = k_z)$. The plasma may be nonuniform along x. There may be magnetic fields; for convenience **B**, the sum of the wave and applied magnetic fields, is taken along z. The longitudinal electric field is E_x, the radiation or transverse field is E_y. The particles have three phase-space coordinates x, v_x, and v_y. EM1 for periodic systems and EM1BND for bounded systems were written by Cohen, Mostrom, Nicholson, and Langdon.

B. I. Cohen has proposed an arbitrarily-polarized wave model by adding in another static field B_x^0 component, plus wave fields B_y and E_z, velocity v_z, and current density J_z. As given at the end of the next section, these alter the Maxwell equation set somewhat.

6-3 ONE-DIMENSIONAL FIELD EQUATIONS AND INTEGRATION

The Maxwell equations for the transverse fields in the model are, in rationalized cgs (Heaviside-Lorentz) units,

$$\frac{\partial \mathbf{E}}{\partial t} = c \nabla \times \mathbf{B} - \mathbf{J} \tag{1}$$

$$\frac{\partial \mathbf{B}}{\partial t} = -c \nabla \times \mathbf{E} \tag{2}$$

By adding and subtracting these equations, for E_y, B_z, and J_y, we obtain

$$\left[\frac{\partial}{\partial t} \pm c \frac{\partial}{\partial x} \right] {}^{\pm}F = -\frac{1}{2} J_y \tag{3}$$

for the left- and right-going field quantities,

$$^{\pm}F \equiv \frac{1}{2}(E_y \pm B_z) \tag{4}$$

The transverse fields are recovered from $^{\pm}F$ by

$$E_y = {}^{+}F + {}^{-}F \tag{5a}$$

$$B_z = {}^{+}F - {}^{-}F \tag{5b}$$

The Poynting flux is

$$P = c({}^{+}F^2 - {}^{-}F^2) \tag{6}$$

and the energy density is $^{+}F^2 + {}^{-}F^2$. E_y, B_z, J_y, $^{\pm}F$, and P are grid quantities.

The fields are readily advanced, as the left-hand side of (3) is the *total derivative* dF/dt for an observer moving at velocity $\pm c$. That is, the convective derivative is taken along the vacuum characteristic so that (3) may be written as

$$\frac{^{\pm}F(t+\Delta t, x \pm c\Delta t) - {}^{\pm}F(t,x)}{\Delta t} = -\frac{1}{2}J_y^{\pm}\left[t + \frac{\Delta t}{2}, x \pm c\frac{\Delta t}{2}\right] \tag{7}$$

where $J_y^{\pm}$ is an appropriately-averaged current, space- and time-centered, as noted by the arguments. Hence, we may integrate exactly along the vacuum characteristics ($x = \pm ct + $ constant) using grid spacing $\Delta x = c\Delta t$.

The formal statement is that the field at grid j, time $n+1$, is

$$^{\pm}F_j^{n+1} = {}^{\pm}F_{j\mp 1}^{n} - \frac{\Delta t}{4}(J_{y,j\mp 1}^{-} + J_{y,j}^{+}) \tag{8}$$

J_y^{-} and J_y^{+} denote current densities computed from velocities $v_y^{n+\frac{1}{2}}$ assigned to the grid by linear weighting according to the positions x^n and x^{n+1}, respectively. This says that fields at $j \mp 1$ propagate to j, affected by the source terms at $j, j \mp 1$. The values are shown in Figure 6-3a.

The current J_y^{-} comes from the ith particle $[qv_y(t + \Delta t/2)]_i$, linearly weighted to the grid from the particle position $x_i(t)$; J_y^{+} comes from the ith particle $[qv_y(t + \Delta t/2)]_i$, linearly weighted to the particle position $x_i(t + \Delta t)$. These values are defined in this manner so that the current given in (8), an average of J_y^{-} and J_y^{+}, is centered halfway between the grid point x and the grid points $x \pm c\Delta t = x \pm \Delta x$, as required by (7).

Note that ^{+}F and ^{-}F represent right- and left-going electromagnetic quantities traveling with constant speed $c (= \Delta x/\Delta t)$, the speed of light in vacuum. However, in the plasma, electromagnetic waves travel with $v_{phase} > c$. Hence ^{+}F and ^{-}F are the usual right- and left-going electromagnetic fields *only* in vacuum, as in regions outside the plasma.

The longitudinal field E_x is obtained from $-\nabla\phi$ and ϕ from Poisson's equation just as in ES1.

This scheme (7) was designed to be time-centered (reversible and second-order accurate) and to render the radiation self-force accurately.

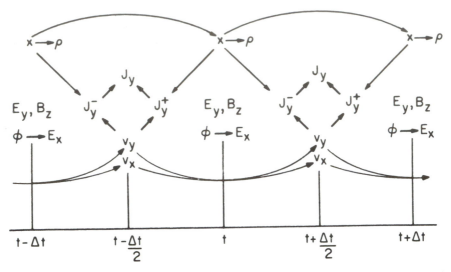

Figure 6-3a Time stepping used in the EM1 program. The particles are advanced by a leap-frog method, shown by the curved lines. ρ is the charge density needed for calculating ϕ, then E_x.

[The radiation drag $(\dot{v}_y)_{rad} = -v_y q^2/2mc$ is proportional to velocity in one dimension—a surprising result which can be reconciled with the point-particle three- dimensional case by considering the radiation from a moving spherical charge shell in the limits of large and small ratio of (radius/wavelength). The fact that a plasma slab does not quickly radiate away its thermal energy can be understood in terms of the radiative normal modes of the slab (*Langdon, 1969*).] The stability against nonphysical beam modes (*i.e.*, no numerical Cerenkov instability) noted by B. I. Cohen (private communication, 1975) and analyzed by *Godfrey and Langdon* (1976), is an unforeseen bonus, and is discussed further in the next section.

There are other advantages to this algorithm: (i) the fields are known at integral positions j and times n, for the convenience of particle movers and diagnostics, without the averaging and unaveraging procedures required in typical 2d, 3d electromagnetic programs; (ii) outgoing wave boundary conditions for $\pm F$ are trivial; the original application by *Langdon* (1969) involves a bounded plasma radiating into a vacuum.

An arbitrarily polarized wave model is mentioned at the end of the previous section, with variations along only the x coordinate, but with static magnetic field in the x-z plane and all three velocity components. Such a code is 1d3v, an example of which appears in Section 4-5. The additional field components, E_z and B_y, and new current density J_z require solving the additional transverse field equations,

$$\left[\frac{\partial}{\partial t} \pm c\frac{\partial}{\partial x}\right](E_z \mp B_y) = -\frac{1}{2}J_z \tag{9}$$

These equations are not included in the present version of EM1. The extension, however, would be very useful, and would allow simulation of a variety of electromagnetic and electrostatic waves which propagate at arbitrary angles with respect to an applied magnetic field. The computational simplicity and stability of the basic one dimensional algorithm are retained.

PROBLEM

6-3a Devise an explicit Maxwell's equation solver à la *Morse and Nielson* (1971) for the same fields as the Langdon-Dawson 1½ dimensional model of this section. Design a mesh; label where quantities are calculated in space and time; construct first-order, centered (if possible), finite-difference equations. Can you find a qualitative or quantitative condition on $\Delta x / c \Delta t$, *i.e.*, a Courant condition? Compare and contrast to the Langdon-Dawson electromagnetic scheme.

6-4 STABILITY OF THE METHOD

The numerical stability of one dimensional electromagnetic plasma simulation algorithms was investigated by *Godfrey* (1974), with emphasis on the numerical Cerenkov instability (*Boris and Lee,* 1973; *Godfrey,* 1975). This instability arises when the transverse wave vacuum dispersion (which should be $\omega = \pm kc$), obtained from the numerical methods, produces $|\omega / k| < c$, usually near $k\Delta x = \pi$; this phase velocity allows particle-wave interaction of a traveling-wave or Cerenkov type which is nonphysical. In Section 6 of *Godfrey* (1974), an advective differencing scheme was incorrectly attributed to A. B. Langdon; *Godfrey and Langdon* (1976) clarified that discussion and analyzed the actual Langdon-Dawson one dimensional algorithm. This section is taken from the 1976 article.

The basic approach of advective differencing is to integrate numerically Maxwell's equations along their vacuum characteristics. This is straightforward in one dimension, where right- and left-going ($\pm$) transverse waves explicitly decouple, and leads to

$$(E_y \pm B_z)_{j+1}^{n+1} = (E_y \pm B_z)_j^n - (J_y)_{j\pm\frac{1}{2}}^{n+\frac{1}{2}}\Delta t \tag{1}$$

with a similar equation for E_z and B_y. The integers n and j designate time and space, respectively. Units are chosen such that the speed of light and the plasma frequency each equal one. Note that (1) requires $\Delta x = \Delta t$.

It can be shown that the improved stability associated with advective differencing schemes is due not so much to the dispersionless vacuum transport of the fields, *per se*, as to the less conventional methods of determining the mesh current usually employed with advective differencing. Thus, for the case considered in *Godfrey* (1974), J^+ and J^- are equal, defined as

$$J_{j\pm\frac{1}{2}}^{n+\frac{1}{2}} = \sum_i q_i \, v_i^{n+\frac{1}{2}} \tfrac{1}{2}[S(X_{j\pm\frac{1}{2}} - x_i^{n+1}) + S(X_{j\pm\frac{1}{2}} - x_i^n)] \tag{2}$$

$S(x)$ is a spatial interpolation function, while J is the particle current; x_i and

v_i are the position and velocity of particle i. In words, currents are interpolated onto the mesh with particle positions at times t^{n+1} and t^n, but with velocities from $t^{n+½}$, and then averaged to give $J^{n+½}$. The principle effect of so defining J is to smooth the current term in the dispersion relation for (1) by the velocity dependent factor $\cos(kv\Delta t/2)$. For v large the factor suppresses nonphysical effects for $k\Delta x$ near $\pm\pi$. On the other hand, this definition also distorts physical phenomena in this region of wave-number space. This shortcoming is, however, overstated in *Godfrey* (1974). For any algorithm, and not for this one only, caution must be exercised in the interpretation of the behavior of large k modes. This definition of J is successfully employed in two and three dimensional electromagnetic codes by *Langdon and Lasinski* (1976) and *Boris* (1970b).

The differencing scheme actually developed by Langdon defines mesh current not as in (2), but as

$$J_{j\pm½}^{n+½} = \sum_i q_i\, v_i^{n+½} \tfrac{1}{2}[S(X_{j\pm1} - x_i^{n+1}) + S(X_j - x_i^n)] \qquad (3)$$

Current is averaged along vacuum characteristics rather than at fixed points in space.

For a single-species cold beam, with small particle velocities, $v \ll c$, the term due to current in the electromagnetic dispersion relation reduces to an expression characteristic of many differencing schemes, multiplied by $\cos(kc\Delta t/2)$, which is $\cos(k\Delta x/2)$ since $c = \Delta x/\Delta t$. This multiplier smooths the current contribution at short wavelengths independent of the beam velocity. Any numerical problems resulting from tangency (coupling) of the electromagnetic wave curve and a beam curve near $k\Delta x = \pm\pi$ are eliminated. For large particle velocities in the same problem, the curves move apart; even for beam velocity c, no numerical instability occurs.

6-5 THE EM1 CODE, FOR PERIODIC SYSTEMS

The code EM1 is very much like ES1. The main program calls subroutines CREATOR, FIELDS, ACCEL, MOVE, SETV, SETRHO, RPFT2, RPFTI2, CPFT, and the plot routines, much as in ES1. Of course, while FIELDS obtains the longitudinal field E_x from $-\nabla\phi$ and ϕ from ρ (through $\nabla^2\phi \sim \rho$), it also solves Maxwell's equations for the transverse fields E_y and B_z. Also, CREATOR inserts a relativistic Maxwellian distribution. ACCEL, also relativistic, follows the scheme laid out by Boris (Chapter 15) and uses Buneman's fast rotation algorithm (Section 4-4). The user with experience in using ES1 should be able to pick up EM1 quickly.

6-6 THE EM1BND CODE, FOR BOUNDED SYSTEMS; LOADING FOR $f(x,v)$

The version of EM1 for *bounded systems* is called EM1BND. The parts that are different are the loader, presented here, and the boundary conditions, given in the next section.

In the bounded code EM1BND, the particles are loaded in space according to a specified density profile FDENS(x) over a length $L_p \equiv L - $ DMOAT $< L$ and surrounded on both sides by a vacuum "moat" of width DMOAT/2 as indicated in Figure 6-6a. The function FDENS(x) is normalized so that its integral over the length L_p is equal to the total number N_s of particles of the particular species being loaded. The actual loading process then just involves integrating FDENS from XMIN=DMOAT/2 to x and placing a particle at each value of x which causes this integral to increase by 1 over its previous integer value. This use of the cumulative distribution is repeated for each species and a fixed nonuniform background charge density is added to the grid to ensure local and overall charge neutrality. Finally, the charge density is subjected to possible sinusoidal variations as in ES1. The code is presently set up to handle general quadratic density profiles ($f = a + bx + cx^2$) with N_s, DMOAT, and the linear and quadratic scale lengths (in units of L_p) introduced as input data.

Both codes EM1 and EM1BND are set up to load initial velocity distributions such as cold beams or drifting Maxwellians. The Maxwellian loader is similar in structure to the nonuniform density loader but with the density profile function FDENS(x) replaced by an appropriate velocity distribution function $f(v_x)$, $f(v_y)$, or $f(p = \gamma|\mathbf{v}|/c)$ for relativistically high temperatures.

For nonrelativistic temperatures, the nonrelativistic Maxwellian velocity distribution functions $f(v_x)$ and $f(v_y)$ are normalized so that their integral over v_x or v_y from $-\infty$ to $+\infty$ is equal to the total number N_s of particles of the species being loaded. The velocity initialization then involves integrating f from 0 to v and assigning the velocity v to a particle each time the integral increases by 1 over its previous integer value. This is carried out up to a v of four times the thermal velocity with the result that the first $N_s/2$ particles are assigned increasing velocities. The negatives of these velocities are then

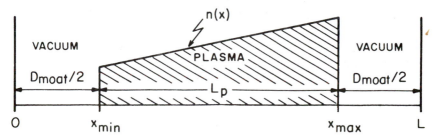

Figure 6-6a Typical density profile set up by the particle loader in EM1BND.

assigned to the remaining $N_s/2$ particles. If N_s is odd, the final odd particle is given zero velocity. The v_x and v_y distributions are constructed independently and are initially decorrelated from one another by means of velocity exchanges of randomly selected particle pairs.

Finally, to allow a possibly relativistic drift velocity and yet maintain particle velocities less than c, the drift velocity $v_0\hat{x}$ is added relativistically to each particle velocity $\mathbf{v}$ by [e.g., *Jackson*, 1975, eq. (11.33), p. 524]

$$v_x' = \frac{v_x + v_0}{1 + \dfrac{v_x v_0}{c^2}} \quad \text{and} \quad v_y' = \frac{\sqrt{1 - v_0^2/c^2}}{1 + \dfrac{v_x v_0}{c^2}} v_y \tag{1}$$

For relativistic temperatures, and isotropic velocity distributions, the relativistic Maxwellian distribution for $p \equiv \gamma|\mathbf{v}|/c$ is, in two dimensions (p_x, p_y),

$$f(p) = \frac{N_s}{2\pi} \frac{mc^2}{T} \frac{1}{(1 + T/mc^2)} \exp\left[-\frac{mc^2}{T}(\gamma - 1)\right] \tag{2}$$

where

$$\gamma = \sqrt{1 + p^2}$$

f is normalized by requiring that

$$\int_0^{2\pi} d\theta \int_0^{\infty} fp \, dp = N_s \tag{3}$$

The angle θ is defined by $p_x = p\cos\theta$ and $p_y = p\sin\theta$. If we perform the integration of $f(p)$ numerically over $\int_0^p p \, dp$ and multiply by $\pi/2$, we obtain

$$\frac{\pi}{2} \sum_{n=1}^{\text{IP}} (n\Delta p)\Delta p \, f(n\Delta p) = \tfrac{1}{4}N_p \quad (p \leqslant \text{IP}\Delta p) \tag{4}$$

which is the number of particles in the first quadrant, $0 \leqslant \theta < \pi/2$, with $p \leqslant \text{IP}\Delta p$, and N_p is the total number of particles assigned to all quadrants up to the value of p. Thus, every time running index IP reaches a value which increases this integral by 1 over its previous integer value, we assign to a particle the quantity $p = \text{IP}\,\Delta p$ and also a random angle θ between 0 and $\pi/2$. Actually, to ensure no drift or bias in v_x and v_y, four particles are simultaneously loaded with the angles θ, $\theta + \pi/2$, $\theta + \pi$, and $\theta + 3\pi/2$. Particle velocities are then determined from

$$v_x = \frac{p\cos\theta}{\sqrt{1 + p^2}} \quad v_y = \frac{p\sin\theta}{\sqrt{1 + p^2}} \tag{5}$$

The number of integration steps needed to load all N_s particles is set at a large number, NSTEPS = 100000 to ensure accuracy, and the maximum integration and loading range is set equal to PMAX = $4\sqrt{T/mc^2} + 4(T/mc^2)^2$ which gives $f(\text{PMAX})/f(p=0) = \exp(-8)$. The integration step size is then determined by $\Delta p =$ PMAX/NSTEPS.

As the final particle loading step, correlations between x and v_x or v_y are reduced by performing random pair exchanges of the particle positions. The density profile is left unchanged by this process.

The listing for EM1BND is available on request from C. K. Birdsall.

PROBLEM

6-6a (i) Using a nonrelativistic Maxwellian, one dimensional distribution function, discuss graphically and semiquantitatively [use table of $\mathrm{erf}(v/v_t)$] the effect of having a finite total number of particles, say 1000 and 4000, upon the population of the tail. Characterize velocity ranges by v_t, $2v_t$, $3v_t$, $4v_t$, etc. (ii) Discuss the impact this may have upon wave-particle resonance (like Landau damping) and other physics. (iii) Compare/contrast to a relativistic Maxwellian in the limit v_t^2/c^2 is appreciable (select a convenient value to keep arithmetic simple). (iv) (Optional) Suggest a relativistic quiet-start, weighted q and m scheme, to obtain a smoother Maxwellian tail.

6-7 EM1BND BOUNDARY CONDITIONS

The codes EM1 and ES1 are classified as periodic because they treat all spatially varying quantities as though they were periodic with a repetition length equal L of the system. Specifically, all field quantities $F(x)$ whether electromagnetic or electrostatic, are constrained to satisfy $F(x=L) = F(x=0)$. This implies that, whenever a left- or right-going electromagnetic field quantity reaches a boundary, its image must simultaneously appear at the opposite boundary. When applied to the electrostatic potential ϕ the above constraint implies that the spatial average of electrostatic field, $<E_x> = -<\partial\phi/\partial x>$, must vanish; the constraint applied to E_x implies a charge neutral system. To remain consistently periodic, the codes must also require that whenever a particle exits the system at one boundary it must simultaneously re-enter the system at the opposite boundary with unaltered velocity; relative to the new boundary, the particle is positioned the distance it otherwise would have traveled outside the system during the time step.

The code EM1BND is classified as bounded because, in general, $F(x=L) \neq F(x=0)$. Once radiation fields propagate outside the system, they do not return and are no longer considered. External radiation fields (*e.g.*, lasers) are allowed to enter the system and change their intensity before exiting. The system walls are transparent to radiation. The electrostatic field E_x is assumed to vanish at the boundaries and outside the system, in agreement with the charge neutrality constraint, $E_x(x) = 0$ for $0 \geqslant x \geqslant L$.

We would still like to have the capability of solving for the electrostatic potential ϕ by Fourier transforming Poisson's equation, but this requires that the charge density ρ at least appear to be periodic [*i.e.*, $\rho(x=L) = \rho(x=0)$]. The only choice consistent with a nonperiodic charge-neutral system is

$$\rho(x = L) = 0 = \rho(x = 0) \tag{1}$$

The particular solution ϕ_p of the inhomogeneous equation

$$\frac{\partial^2 \phi}{\partial x^2} = -\rho(x) \tag{2}$$

is then periodic. One must add to this the solution

$$\phi = a + bx \tag{3}$$

of the homogeneous equation

$$\frac{\partial^2 \phi}{\partial x^2} = 0 \tag{4}$$

with the constant b determined by the boundary condition

$$E_x(0) = -\frac{\partial \phi}{\partial x}\bigg|_0 = -\frac{\partial \phi_p}{\partial x}\bigg|_0 - b = -\frac{\partial \phi_p}{\partial x}\bigg|_L - b = E_x(L) = 0 \tag{5}$$

Using a two-cell centered-difference derivative and the periodicity $\phi_p(x + L) = \phi_p(x)$, we have

$$b = -\frac{\partial \phi_p}{\partial x}\bigg|_0 = \frac{\phi_p(-\Delta x) - \phi_p(+\Delta x)}{2\Delta x} = \frac{\phi_p(L - \Delta x) - \phi_p(+\Delta x)}{2\Delta x} \tag{6}$$

$$\phi(X_j) = \phi_p(X_j) + bX_j \qquad \text{for } 0 \leqslant X_j \leqslant L \tag{7}$$

$$E_x(X_j) = \frac{\phi(X_j - \Delta x) - \phi(X_j + \Delta x)}{2\Delta x} \qquad \text{for } \Delta x \leqslant X_j \leqslant L - \Delta x \tag{8}$$

Finally, particles are prevented from approaching within one grid-space of the boundaries (assuming linear or CIC weighting) in order to ensure $\rho(x = 0) = \rho(x = L) = 0$. The code EM1BND accomplishes this by placing elastically reflecting walls at positions $x = \Delta x$ and $x = L - \Delta x$. Whenever a particle collides with the walls (Figure 6-7a) during a time step, its velocity vector is reversed in direction and it is repositioned from the wall the distance it otherwise would have traveled into the wall.

6-8 EM1, EM1BND OUTPUT DIAGNOSTICS

As in ES1, the electromagnetic codes EM1 and EM1BND allow for the usual output plots of phase space, the various field quantities (electrostatic and electromagnetic) and sources as a function of position, and the time-history of the various energies. In addition, the codes plot the local temperature as a function of position for each species and the mode energies of the Fourier-transformed electrostatic and right- and left-going electromagnetic fields.

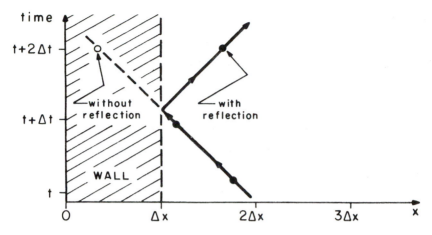

Figure 6-7a Treatment of particles at left-hand boundary, by reflection, in EM1BND.

The local temperature is obtained by first finding the average or fluid oscillation velocity $\mathbf{v}_{osc}(X_j)$ at each grid point; this is the average of the velocities of all particles whose positions lie within half a grid space of the grid point X_j. The random or thermal velocities of these particles are then obtained by subtracting out the oscillation velocity relativistically [as done in 6-6(1) (where a drift was added)] and used to determine a relativistic kinetic energy for each particle. Averaging these energies over all the particles (within $\Delta x/2$ of X_j) gives a local thermal energy per particle which is assigned to grid point X_j.

The electrostatic mode energies are obtained from the spatially Fourier-transformed electrostatic potential $\phi(k)$ which is already known from an intermediate step in the solution of Poisson's equation. The codes plot the time history of the energy in each mode separately.

As mentioned before, $^\pm F$ is the usual right- and left-going electromagnetic fields only in the vacuum moat surrounding the plasma in EM1BND. The code EM1BND, therefore, saves $^+F(x=L)$ and $^-F(x=0)$ each time step and temporally Fourier analyzes the two resulting arrays over NF values of time: $t = 0$ to $t = (NF - 1)\Delta t$, $t = NF\Delta t$ to $t = (2NF - 1)\Delta t$, etc. There are then $(NF/2 + 1)$ frequency modes (including $\omega = 0$) with the lowest nonzero or fundamental frequency, $\omega_0 = 2\pi/NF\Delta t$ and the maximum frequency, $\omega_{max} = \pi/\Delta t$. Since the frequency and wave number are related through the dispersion relation $\omega = ck$ (in vacuum), we also have a fundamental wave number $k_0 = 2\pi/NF\Delta x$ and a maximum wave number $k_{max} = \pi/\Delta x$. Thus, the diagnostic is capable of recognizing modes with a maximum wavelength $2\pi/k_0 = NF\Delta x$ which is greater than or less than the system length L, depending upon the ratio NF/NG. Our fast Fourier analysis requires that NF be a power of 2.

The complete list of output plots is as follows:

Phase Space

v_x vs. x, v_y vs. x, v_y vs. v_x

Distribution Functions

$f(x)$ vs. x, $f(v_x)$ vs. v_x, $f(v_y)$ vs. v_y

Electrostatic Fields

ϕ vs. x, E_x vs. x

Electromagnetic Fields

^+F vs. x, ^-F vs. x, E_y vs. x, B_z vs. x

Sources

ρ vs. x, J_y^+ vs. x

Temperature (each species)

T_s vs. x, T_s vs. t

Field Energies

E_x energy vs. t, E_y energy vs. t, ^+F energy vs. t, ^-F energy vs. t

Mode Energies

E_x mode energy vs. t (each mode) ^+F mode energy vs. mode number
^-F mode energy vs. mode number

Average Drift Momentum per Particle (each species)

P_x^s vs. t, P_y^s vs. t

SEVEN

PROJECTS FOR EM1

Bruce I. Cohen

7-1 INTRODUCTION

The code EM1 makes no physical approximations in solving Maxwell's equations and the particle equations of motion; hence, only finite differencing considerations, one-dimensionality, and polarization limit usage. The two versions of the code, EM1 and EM1BND, invoke periodic and finite boundary conditions respectively. In the periodic code, all spatially varying quantities, *i.e.*, transverse-wave amplitudes, potential, charge density, etc., are made periodic. In the bounded code, particles are elastically reflected at the system edges, while the electromagnetic waves are absorbed and the longitudinal electric field is required to vanish.

The codes have been used successfully to study linear instabilities, such as the relativistic two-stream and weak beam-plasma, and stimulated Raman and Brillouin scattering. Linear and nonlinear wave propagation of electromagnetic waves in unmagnetized plasma has also been studied.

The bounded code is well suited to studying laser-plasma interactions, since transverse waves can be launched quite trivially at the system walls. In this way Raman and Brillouin backscatter instabilities and laser beat heating of plasma have been successfully studied by *Cohen et al.* (1975). The code has been especially useful in studying the anomalous absorption of electromagnetic radiation in the presence of kinetic phenomena, *e.g.*, nonlinear

Landau damping, particle trapping in large (resonantly or nonresonantly excited) beat waves (beat heating or induced scattering), and wave breaking. Use of this code has extended our insight and theory into resonant excitation of nonlinear normal modes (*Cohen, 1975*), and the role of trapping in beat heating (*Cohen et al., 1975*).

Simulation represents the primary means of understanding the trapping of particles in potential wells whose time-dependent variation is far outside the reach of linear perturbation theory.

To trouble-shoot the periodic code, one can monitor the total (field and kinetic) energy, which should be conserved. In the bounded code, one must include the Poynting flux at the system boundaries as well. Examining linear phenomena (*e.g.*, linear dispersion relations for small-amplitude electromagnetic and electrostatic waves and/or instabilities, Landau or cyclotron damping rates) is encouraged as the best way to both check the code and gain confidence and experience in its use. We have verified that the code produces results in agreement with linear analytical theory for the propagation of transverse waves, electron plasma waves, and ion-acoustic waves in unmagnetized plasma, for linear Landau damping, for the growth rates of the electrostatic two-stream and weak beam-plasma instabilities and parametric Raman and Brillouin scattering, and for linear beat heating.

7-2 BEAT HEATING OF PLASMA

One application of EM1 by *Cohen et al.* (1975) is the heating of plasma by two lasers (of frequencies ω_0, ω_1) whose beat frequency ($\Omega \equiv \omega_0 - \omega_1$) is near the plasma frequency ω_p (Figure 7-2a). The nonlinear interaction may be considered as an induced decay ($\omega_0 \rightarrow \omega_1 + \Omega$), in which a fraction R of the incident power at ω_0 is converted to ω_1 and Ω, with the fraction $R \Omega / \omega_0$ appearing as a longitudinal plasma oscillation and, because of damping, ultimately as heat. It is the aim of the theory and simulations to determine the dependence of this efficiency parameter R on the parameters of the problem, such as laser intensity, density scale length, and temperature.

The process studied here involves three electron waves (two transverse and one longitudinal), with no externally-imposed magnetic field, and is illustrative of the more general three-wave process, possibly involving ions, and in a magnetic field. Thus, the principle of electron heating, by the damping of a resonant excitation from the beat of two high-frequency waves, can be extended to the analogous heating of ions in a magnetically-confined plasma.

Two linearly-polarized transverse waves are oppositely incident on a finite, inhomogeneous, underdense plasma (Figure 7-2b). There can be a resonant interaction with a longitudinal normal mode of the plasma if the electrostatic disturbance, driven by the ponderomotive force at the beat frequency and wave number ($\Omega \ll \omega_0$, ω_1; $\kappa \equiv k_0 - k_1$), approximate the

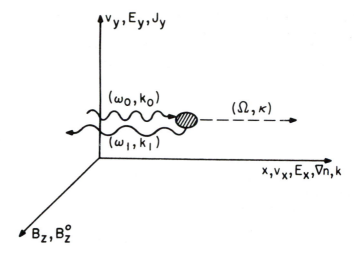

Figure 7-2a Sketch of the variables in the problem. Wave propagation and density variation occur along x. Transverse waves are linearly polarized, with E in the y direction. Magnetic fields are parallel to z. The three-wave interaction is diagrammed. *(From Cohen et al., 1975.)*

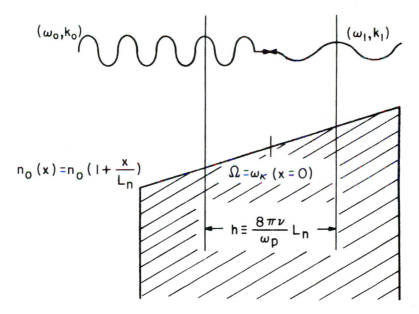

Figure 7-2b Beat heating in an inhomogeneous medium. Because of the resonance condition, there is a region h about the plane of exact resonance. The density gradient, described by the scalar length $L_n \equiv [d[\ln(n)]/dx]^{-1}$, is parallel to the propagation direction of waves. *(From Cohen et al., 1975.)*

Bohm-Gross dispersion relation somewhere in the plasma. Because of the plasma inhomogeneity, the three-wave interaction is resonant only in a finite region around the position of exact matching, shown in Figure 7-2b. The dissipation of the electron plasma oscillation introduces irreversibility into the three-wave process, and is the mechanism for the eventual thermalization of part of the energy provided by the electromagnetic waves. The dissipation may be due to collisions, Landau damping, convective loss, or non-linear mode-coupling processes.

The results, Figure 7-2c, show both the large amplitude electron plasma wave and beating. The initial thermal component of v_x/c was $v_{te}/c \approx 0.05$.

The project is to simulate the excitation of a resonant three-wave interaction in a warm, homogeneous electron plasma with neutralizing background:

$$\omega_0 = \omega_1 + \Omega$$

$$\mathbf{k}_0 = \mathbf{k}_1 + \kappa$$

(a) Discuss, itemize, and specify the parameters needed to guarantee the desired physics; *i.e.*, conditions on Δt, Δx, N, etc., such that accurate simulation is made for the behavior of a plasma in which two high-frequency waves and one low-frequency wave are present. Assume initial constraints of $N \leqslant 4000$, $NT \leqslant 1000$, $NG \leqslant 256$.

(b) Assume Landau damping to be the only dissipation mechanism damping an electron plasma oscillation driven *resonantly* at $\omega_0 - \omega_1$. Steady state is achieved after several inverse damping decrements $O(1/\nu)$. For two opposed light waves, suggest actual parameters for a simulation. Assume $\omega_{pe} = 1$, $10 \geqslant (\omega_0, \omega_1) \geqslant 1$, and $0.01 \leqslant \nu \leqslant 0.1$ and determine values of Δt, Δx, N, λ_{De}, etc. Calculate all frequencies and wave numbers, specify a moat, and check mesh corrections to ω_{pe}.

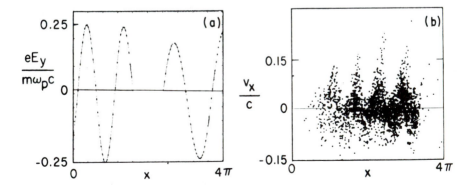

Figure 7-2c Beat heating in a finite, inhomogeneous medium. (a) The right- and left-going electromagnetic waves before onset of beating; (b) (x, v_x) phase space after a fairly large amplitude electron plasma wave has been established; $e/m = 1$, $\omega_p = 1$, $c \approx 2$. *(From Cohen et al., 1975.)*

(c) (Optional) Let the plasma be inhomogeneous, $n = n_0(1 + x/L_{n_e})$ and $\omega_{pe} \approx \omega_{pe}(0)(1 + x/2L_{n_e})$. Assume that the longitudinal wave is resonantly-excited at $x = 0$ with finite damping, and off-resonant at $x \neq 0$; then a simple Lorentzian model gives a plasma response to the ponderomotive driving force with shape factor

$$\left[\left(\frac{\nu}{\Omega} \right)^2 + \left(\frac{x}{2L_{n_e}} \right)^2 \right]^{-1}$$

Derive a criterion for an effective resonance length. Comment on how a particular choice of kL_{n_e} might affect your choice of parameters in (a) and (b).

7-3 OBSERVATION OF PRECURSOR

This project is to examine a traveling electromagnetic wave at frequency ω_0, propagating from vacuum into an unmagnetized, underdense plasma, $\omega_{pe} < \omega_0$ (Figure 7-3a). It is known that a linearly-polarized transverse wave enters a finite-length, uniform plasma from the left and propagates to the right with phase velocity $\omega_0/k = c/\sqrt{1 - \omega_{pe}^2/\omega_0^2}$ and group velocity kc^2/ω_0 once the main body of the wave train has arrived at the observer. However, one can observe precursors preceding the main body of the wave.

A simple model for describing precursors is given by *Sommerfeld* (1954, p. 118), who solved for the fields in the limit of small wave amplitude by Fourier-Laplace transforming the linear wave equation. In one dimension, this is

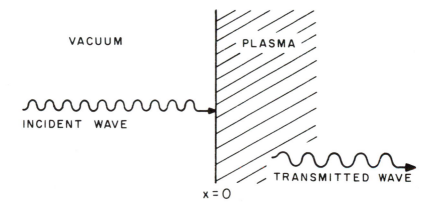

Figure 7-3a Right-going normally-incident electromagnetic waves.

$$f(x,t) = \frac{-\omega_0}{2\pi} \int_{-\infty+i\epsilon}^{\infty+i\epsilon} d\omega \frac{e^{i(kx - \omega t)}}{(\omega^2 - \omega_0^2)} \tag{1}$$

where $kc = \sqrt{\omega_0^2 - \omega_p^2}$. The branch points, cuts, and poles are diagrammed in Figure 7-3b(a). Closing the contour up, for $x > 0$ and $t > 0$, yields $f(x, t) = 0$. If we close the contour down, for $x > 0$ and $t > 0$, and select the contour pictured in Figure 7-3b(b), letting $R/\omega_0 \gg 1$, we obtain asymptotically

$$f(x, t) \approx \begin{cases} \dfrac{\omega_0}{\omega_p} 2^{1/2}(ct - x)^{1/2} J_1\left(\omega_p\left[\dfrac{2x}{c}\left(t - \dfrac{x}{c}\right)\right]^{1/2}\right) & x \leqslant ct \\ 0 & x > ct \end{cases} \tag{2}$$

From the zeroes of the Bessel functions one can predict the nodal positions as well as the amplitude itself of the transverse electric or magnetic field as a function of time. Typical results are shown in Figure 7-3c.

A suitable parameter set for a simulation of the precursor using the bounded code might consist of: N1=2000, NT=400, NG=128, DT=0.05, K0=0.5, IRHO=0, IRHOS=0, IPHI=0, IXVX=10, IVXVY=10, IEX=5, IEY=5, IBZ=5, IEYL=20, IEYR=20, IFT=10, WPMPR=9.75, W1=7.9, EPMPR=0.1, MOAT=40, ITHERM=25; other parameters are set by default. Simulation results using these parameters agree quite favorably with Sommerfeld's asymptotic solution.

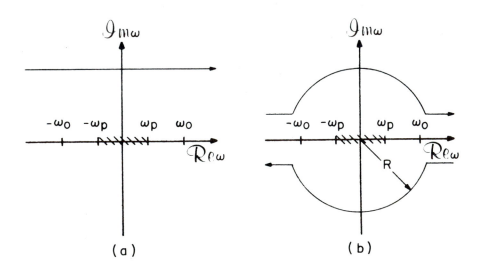

(a) (b)

Figure 7-3b (a) Topology of the poles, branch points and cut in the complex ω-plane; (b) Contour for derivation of asymptotic response.

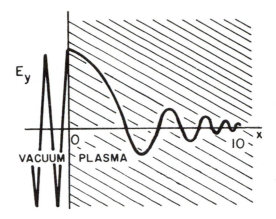

Figure 7-3c Transverse electric field, showing precursor, at $x = t = 18$.

THEORY

PLASMA SIMULATION USING PARTICLES IN SPATIAL GRIDS WITH FINITE TIME STEPS

WARM PLASMA

Simulations generally produce useful results, that is, meaningful physics. Just how meaningful depends on understanding the *approximations used in the physical modeling* and appreciation of the *additional effects incurred, wittingly or not, in the act of simulation,* as given in this part.

We have already seen some of the relatively straightforward effects of using finite difference (Δt and Δx) expressions in place of the true continuous partial differential equations. We saw that use of too large a time step leads to error and even to nonphysical instability. We saw that the spatial gridding could act as a filter, attenuating (smoothing) physical information. In fact, if you were adventurous in the projects in Part One, you probably found some strange results that were well beyond the expected inaccuracies.

We now take a closer look at physical and simulation modeling, to consider as complete a theory as can be mustered. Simulators are plagued with results that cannot be explained as bad physics or too large Δt or Δx more often than they like to admit. Just improving the accuracies of various Poisson solvers or particle movers, by themselves, may not always remove the problem. Hence, some idea of the *over-all system accuracy and stability* is needed and presented.

It is expected that most readers, both novice and experienced, will skip through Part Two initially with some quick reading. To be sure, one can do considerable simulation without understanding all or even part of simulation theory. But, we think (and come close to guaranteeing) that you will find need for the theory and its implications sooner than you may realize at first.

This part is due largely to Langdon and was done mostly in Berkeley. The rudimentary ideas on the use of (or implied use of) finite-size particles originated with many people (*e.g.*, Hockney, Buneman, Byers, Dawson, and ourselves). This formulation has been used widely and verified. We draw primarily on articles by Langdon and by our graduate students, our Quarterly Progress Report (QPR's) on our research, particularly for the years 1968, 1969, and 1970. Additional explanations or examples or modifications are given where deemed helpful. We are grateful to the *Journal of Computational Physics, Methods of Computational Physics*, and *Physics of Fluids* for permission to use these articles or chapters.

Chapter 8 presents the effects of using a spatial grid and the intimate ties to particle shape factor $S(x)$. The details of spatial aliasing, including numerical instabilities, are derived.

Chapter 9 adds the modifications to theory caused by use of a finite time step.

Chapter 10 discusses in detail "energy-conserving" models, including their Hamiltonian formulation.

Chapter 11 discusses multipole models, their advantages and disadvantages.

Chapter 12 is the kinetic theory of finite size particles in a spatial grid with finite time steps. Fluctuations, velocity drag, and diffusion are obtained.

Chapter 13 is drawn in part from the plasma simulation article by *Dawson* (1962) on the statistical mechanics of a sheet plasma. This work was valuable in promoting plasma simulation twenty years ago. Dawson used charge sheets with no grid, as the field may be obtained directly in 1d from Gauss' law. The reader may wish to repeat the experiments on Debye shielding, drag, relaxation, and thermalization using the gridded code ES1 and compare results with Dawson.

EIGHT

EFFECTS OF THE SPATIAL GRID

8-1 INTRODUCTION;
EARLY USE OF GRIDS AND CELLS WITH PLASMAS

This chapter uses the works of *Langdon* (1970a, b) and *Langdon and Birdsall* (1970), plus complementary work from our unpublished Berkeley reports for 1968-1970. Let us begin with an introduction and historical review.

Computer simulation has become a powerful tool for the study of plasmas. Much effort and computer time is being expended in applications to new and more difficult problems. In support of this work, we have performed theoretical analyses for a common class of many-particle simulation methods. As one does to learn basic properties of real plasmas, we examine oscillations, fluctuations, and collisions in the idealized case of uniform and infinite or periodic plasma. Even in this simple situation there are several instances in which the models fail (mildly to grossly) to reproduce plasma behavior. This chapter and the next four contain the theory and discuss such *nonphysical behavior caused by the finite-difference methods*. The plasma interacts coherently with the periodicity of the spatial grid on which the electromagnetic fields are defined and with the periodicity of the finite-difference time integration. Various parametric instabilities are sometimes induced, which may be either weak or strong and may be difficult to distinguish from real instabilities. There is also high-frequency noise associated with the rate at which particles cross the spatial-grid cells. If the time step is large enough that the frequency of this noise exceeds the time sampling frequency, then this noise degrades normal fluctuations and collisions and can become excessive. The theory has helped others examine experimentally the nature of such nonphysical effects.

In early studies of simulation plasmas, the difference schemes for each step of the calculation were analyzed, but their overall performance taken together with plasma behavior was not treated carefully. We begin with a rigorous treatment of the spatial grid, giving here a formulation which includes most codes now in use. This is done in such a way that the role of each step in the calculation is easily identified in the results, and also the expressions for plasma properties are easily compared with the corresponding "real" plasma properties. Details are given for the electrostatic case. The formulation is applied to linear wave dispersion and instability, and to the question of energy conservation in Chapter 10. The effect of the spatial grid is to smooth the interaction force somewhat and to couple plasma perturbations to perturbations at other wavelengths, called *aliases*. The strength of the coupling depends on the smoothness of the interpolation methods used. Its importance depends roughly on how well the plasma would respond, in the absence of the grid, to wave numbers $k \approx 2\pi / \Delta x$; e.g., if the Debye length is too small the coupling can destabilize plasma oscillations even in a thermal plasma.

The plasma models we discuss originated at Stanford University in 1963. In order to make simulations in *two dimensions* economically practical, Buneman (in *Yu et al.,* 1965) and *Hockney* (1965, 1966) developed a model which uses a *spatial grid* in which the charge density is found from the particle positions, Poisson's equation is solved in finite-difference form, and then the particle forces are interpolated from the grid. This is *much more efficient* than summing N^2 Coulomb interactions among N particles. They also realized that computational and physical problems associated with the divergent character of the Coulomb field are eliminated; the interactions at small separations are smoothed, reducing the large-angle binary collisions which are of little interest in hot plasmas and which had been exaggerated in simulation because of the small number of particles used (*Hockney,* 1966). Although simulation in one dimension was possible by other means, the new model offered simplicity and speed there, too (*Dunn and Ho,* 1963; *Burger et al.,* 1965; *Buneman and Dunn,* 1966). Part of the gain was in the method of integrating the system forward in time. The algorithms were fast and preserved certain physical properties (*Buneman,* 1967). The advantages of the grid approach were not immediately recognized elsewhere, and much simulation was done for which the gridded model would have been more efficient.

Other people then developed versions of the gridded model with more accurate interpolation and different methods of solving Poisson's equation (*Birdsall et al.,* 1968; *Birdsall and Fuss,* 1968, 1969; *Morse and Nielson,* 1968, 1969; *Boris and Roberts,* 1969; *Alder et al.,* 1970). Capping these efforts were the large-scale simulations done at Los Alamos Scientific Laboratory of several problems in the controlled thermonuclear fusion research program in 1968. Since then, simulation by these methods has been widely accepted as a plasma research tool.

We are less interested in the accuracy of individual particle orbits and more in the accuracy of *collective plasma phenomena.* Therefore, our analytical approach, terminology, and criteria are more that of the plasma physicist than of the numerical analyst studying, say, an initial-value problem for a small system of differential equations with no particular application in mind. In several places the collective properties of warm plasmas play a crucial role.

The models obviously do *not* accurately reproduce the microscopic dynamics of a plasma. One must consider if and how such errors modify the macroscopic behavior. There are usually far too few particles. This causes discrete-particle effects such as exaggeration of fluctuations and collisions. There may also be too few particles for adequate representation in phase space of plasma phenomena such as the Landau damping wave-particle resonance. There can be serious problems with initial and boundary conditions. Roundoff errors can usually be made negligible compared to other errors. Such sources of error are difficult to assess in practice. Some study of their nonphysical effects was needed. Much of this has been done through experiments with the models (*Birdsall et al.,* 1968; *Hockney,* 1968, 1970; *Montgomery and Nielson,* 1970; *Okuda,* 1970, 1972b).

In addition to empirical results, some theoretical analysis is desirable. This is true for the usual reasons that adding a good theory implies a better understanding than a stack of oscillograms or computer output alone. In Part Two, we develop a theoretical understanding of errors caused by the finite-difference representation in space and time of the field equations and particle dynamics. One hopes the theory also predicts unexpected interesting results that can be verified experimentally; we describe some results which we think are in this category.

There has been some approximate discussion of the models. Smoothing and noise aspects of the spatial grid were recognized early (*e.g., Hockney,* 1966; *Birdsall and Fuss,* 1968, 1969) and approximate theoretical descriptions made of each (*Hockney,* 1968; *Langdon and Birdsall,* 1970; *Okuda and Birdsall,* 1970). The time integration was considered heuristically (*Buneman,* 1967).

However, regarding the space-time grid simply as a source of smoothing and of noise fails to uncover some very important effects involving coherent interaction between the plasma and the sampling in space and time. Some theory has appeared which includes exactly the effects of the finite-differencing and is applied to linear wave dispersion and stability and to energy conservation (*Lindman,* 1970; *Langdon,* 1970a, b). We have also looked at fluctuations and collisions (*Langdon,* 1979, chap. 12). The theory is quite complete, and in about as tidy a form as is possible for such a system. Different parts of the simulation algorithms are easily identified and changes made. Where possible, we keep the results in a form permitting easy comparison with real plasma theory.

Although our techniques can be used for more general cases, we have confined our examples to models having only Coulomb interactions; in one instance an external magnetic field is imposed.

8-2 SPATIAL-GRID THEORY; INTRODUCTION

This chapter presents a mathematical framework with which one can apply conventional plasma theory to simulation plasma using a spatial grid on which charge and current densities, and electromagnetic fields, are defined. The use of fields is almost universal even in the electrostatic approximation with simple shapes, rather than summing the Coulomb interaction over all particle pairs (except in the one-dimensional sheet models, where the Coulomb force is a simple step function and the particles may be ordered). This chapter is directed toward understanding the physical properties of simulation plasma rather than to development of new algorithms for the codes.

In this chapter, *we treat time as continuous* to concentrate on the consequences of finite Δx, and defer finite Δt to the next chapter. The condition for treating time as continuous is $\omega_{max}\Delta t \ll 1$, with ω_{max} the largest frequency of concern when $\Delta t \rightarrow 0$, assuming the time integration is stable numerically. It is usually true that $\omega_p \Delta t$, $\omega_c \Delta t$, etc., are small, but when $\lambda_D / \Delta x$ is large, $(v_t / \Delta x)\Delta t = (\omega_p \Delta t)(\lambda_D / \Delta x)$ may not be small. The theory to follow helps delineate this domain of validity.

The spatial grid is a simple example of a periodic spatial nonuniformity, and we begin with some remarks on the more general case of a plasma with nonuniform interaction. Then each step in the calculation of interactions through use of a spatial grid is studied with enough generality to include most codes. Next we consider momentum conservation. Then we relate Fourier transforms of densities, forces, and fields. Finally, the formalism is applied to linear waves and instabilities.

8-3 SOME GENERAL REMARKS ON THE EFFECTS OF A PERIODIC SPATIAL NONUNIFORMITY

In this section, we make some general remarks about a plasma system whose interaction force has a spatial nonuniformity which is periodic and time-dependent (*e.g.*, a Fermi gas in a crystal, some electron-beam devices). In Section 8-4, we specialize to the grid problem; this general section is not a prerequisite for the rest of this chapter.

Let us consider the interaction force $F(x_1, x_2)$, defined as the force on a particle at x_2 due to a particle at x_1 in one dimension. In a normal physical system, which is invariant under displacement, F depends only on the separation $x \equiv x_2 - x_1$. However, in computer simulation using a spatial grid, invariance does not exist under all displacements (displacing particles but not the grid). Thus F *also depends on the location relative to the grid,* $\bar{x} \equiv \frac{1}{2}(x_1 + x_2)$ as well as x (Figure 8-3a). We have already seen some effects of the grid on the force in Section 4-8. In most simulations a grid with constant spacing Δx is used; in this case the *force* $F(\bar{x} - \frac{1}{2}x, \bar{x} + \frac{1}{2}x)$, considered as a function of displacement $\bar{x}$ with separation constant, *is periodic with period* Δx.

In order to study the effect of the nonuniformity on a plasma, we need the Fourier transform of $F(x_1, x_2)$. For an *infinite system* we use a Fourier integral transform in x and a Fourier series in $\bar{x}$;

$$F(\bar{x} - \tfrac{1}{2}x, \bar{x} + \tfrac{1}{2}x) = \int_{-\infty}^{\infty} \frac{dk}{2\pi} e^{ikx} \sum_{p=-\infty}^{\infty} e^{ipk_g\bar{x}} F_p(k) \tag{1}$$

where

$$k_g \equiv \frac{2\pi}{\Delta x}$$

is the *grid wave number,* and

$$F_p(k) = \int_{-\infty}^{\infty} dx F_p(x) e^{-ikx} \tag{2}$$

$$F_p(x) = \frac{1}{\Delta x} \int_{\Delta x} d\bar{x} \, e^{-ipk_g\bar{x}} F(\bar{x} - \tfrac{1}{2}x, \bar{x} + \tfrac{1}{2}x) \tag{3}$$

This sign and normalization for the Fourier integral are followed throughout.

Those properties of the plasma which are little affected by the lack of displacement invariance are expected to be similar to those of a plasma with two-particle force equal to the $p = 0$ or *average force* $F_0(x)$; such properties can be analyzed by the gridless theory (*Langdon and Birdsall,* 1970). The difference $\delta F = F - F_0$ is a *nonphysical grid force.* In some respects δF is like

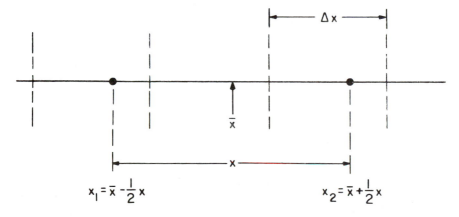

Figure 8-3a Notation in one dimension. Two particles are located at x_1 and x_2, with separation $x = x_2 - x_1$ and mean position $\bar{x} = (x_2 + x_1)/2$. A grid of spacing Δx is imposed on the space.

a "noise" force; however, it is coherent with the plasma perturbation. More is said about this later.

Let the particle density be $n(x, t)$ (actually n is a sum of δ functions when the number of particles is finite). Then the force $F(x)$ on a particle at x is (time dependence ignored for now),

$$F(x) = \int dx' F(x', x) n(x') \tag{4}$$

When transformed this becomes, using (1),

$$F(k) = \sum_{-\infty}^{\infty} F_p(k - \tfrac{1}{2} pk_g) n(k_p) \tag{5}$$

where

$$k_p \equiv k - pk_g \tag{6}$$

We see that the effect of δF (corresponding to the $p \neq 0$ terms) is to *couple density perturbations and forces at wave numbers which differ by integral multiples of the grid wave number k_g*. Such wave numbers are said to be *aliases* of one another (see *Blackman and Tukey, 1958*).

As an illustration, we derive a dispersion relation for small-amplitude plasma oscillations. Linearizing the Vlasov equation and adding time-dependence $\exp(-i\omega t)$ to n and F, we find the density response $n(k, \omega)$, of an unmagnetized uniform plasma, to the force field $F(k, \omega)$ to be

$$n(k, \omega) \equiv F(k, \omega) \psi(k, \omega) \tag{7}$$

where

$$\psi(k, \omega) = \frac{n_0}{im} \int dv \frac{f'_0}{\omega + i0 - kv} \qquad \mathrm{Im}\,\omega \geqslant 0 \tag{8}$$

For $\mathrm{Im}\,\omega < 0$, this must be analytically continued from the upper half ω plane. Combining (5) and (8) yields

$$n(k, \omega) = \psi(k, \omega) \sum_p F_p(k - \tfrac{1}{2} pk_g) n(k_p, \omega)$$

or, alternatively,

$$F(k, \omega) = \sum_p F_p(k - \tfrac{1}{2} pk_g) F(k_p, \omega) \psi(k_p, \omega)$$

If one replaces k by $k_q \equiv k - qk_g$ and p by $p - q$ ($q = 0, \pm 1, \pm 2, \ldots$), each set of equations may be written in infinite matrix form:

$$0 = \sum_p \{\delta_{q,p} - F_{p-q}(\tfrac{1}{2}[k_p + k_q]) \psi(k_q, \omega)\} n(k_p, \omega) \tag{9}$$

$$0 = \sum_p \{\delta_{q,p} - F_{p-q}(\tfrac{1}{2}[k_p + k_q]) \psi(k_p, \omega)\} F(k_p, \omega) \tag{10}$$

where $\delta_{q,p}$ is the Kronecker delta. We can now see several important features.

The possible free oscillations of the plasma are given by the zeros of the determinant of either matrix. The presence of off-diagonal terms, due to the coupling together of many wavelengths, shows that the normal coordinates (in the terminology of the small-oscillation problem in classical mechanics) for n and F are *not* the exponentials $\exp(ik_p x - i\omega t)$, but are some (as yet unknown) linear combinations of such exponentials, so that n or F varies as $\exp(ikx - i\omega t)$ times a periodic function of x with period Δx (Bloch function). Thus we have brought our problem into the classical form (see *Brillouin*, 1953).

If $|k| \ll |k_g|$, one may expect the $p = q = 0$ element to be much the largest in the matrix. Also, if $k_g v_t \gg \omega_p$, i.e., Debye length $\lambda_D = v_t/\omega_p \gg \Delta x$, we expect $n(k_p, \omega)$ to be largest when $p = 0$. Therefore we have an approximate dispersion relation $\epsilon_0 = 0$, where $\epsilon_0 \equiv 1 - F_0(k)\psi(k, \omega)$. This is exactly the same as we would get for a uniform plasma whose interaction force is F_0. The validity of this approximation is made clearer in Section 8-11 and following.

In Section 8-11, we show that the zeros of the determinant are the same in our application as the zeros of a much simpler series. In the next section, we begin the mathematical description specialized to plasma simulation.

8-4 NOTATION AND CONVENTIONS

Particle quantities are subscripted with particle name i, as

$$x_i, v_i, F_i = F(x_i, t) \tag{1}$$

and also include number density of cloud centers, number densities of clouds, and cloud charge densities,

$$n(x, t), \quad \rho_c(x, t) = qn_c(x, t) \tag{2}$$

Grid quantities carry the grid point name j, where they are known, as

$$\rho_j, \phi_j, E_j, X_j \tag{3}$$

where $$X_j = j\Delta x \tag{4}$$

In a one-dimensional periodic system, of length $L = N_g \Delta x$, particle functions satisfy $P(x + L) = P(x)$ and grid functions satisfy $G_{j+N_g} = G_j$. The Fourier transforms become sums of δ functions and the inverse transforms are therefore sums also. The coefficients of the δ functions are $2\pi/L$ times what one obtains by integrating the transforms over only one period:

$$P(k) = \int_0^L dx P(x) e^{-ikx} \tag{5}$$

$$G(k) = \Delta x \sum_{j=0}^{N-1} G_j e^{-ikX_j} \tag{6}$$

with $k = (2\pi n)/L$, $n = 0, \pm 1, \pm 2, \ldots$ (Some expressions should be interpreted as generalized functions, as in *Lighthill*, 1962. All results are well-behaved for finite systems.) In terms of (5) and (6), the inverse transforms are

$$P(x) = \frac{1}{L} \sum_{n=-\infty}^{\infty} P(k) e^{ikx} \tag{7}$$

$$G_j = \frac{1}{L} \sum_{n=0}^{N_g-1} G(k) \, e^{ikX_j} \tag{8}$$

The first is of course the conventional Fourier series and the finite discrete Fourier transform. The expressions for $P(k)$ and $G(k)$ differ from the infinite case only in the limits and in that they are evaluated only for $k = (2\pi n)/L$. The k integrals become sums according to the rule

$$\int_{-\infty}^{\infty} \frac{dk}{2\pi} \rightarrow \frac{1}{L} \sum_{n=-\infty}^{\infty} \tag{9}$$

$$\int_g \frac{dk}{2\pi} \rightarrow \frac{1}{L} \sum_{n=0}^{N_g-1} \tag{10}$$

where

$$\int_g dk \equiv \int_{-\frac{1}{2}k_g}^{\frac{1}{2}k_g} dk, \qquad k_g = \frac{2\pi}{\Delta x} \tag{11}$$

8-5 PARTICLE TO GRID WEIGHTING; SHAPE FACTORS

The *grid charge density* ρ_j is obtained from the charges q_i located at positions x_i from

$$\rho_j \equiv \rho(X_j) = \sum_i q_i S(X_j - x_i) \tag{1}$$

This can be interpreted as the charge density for *finite-size particles*, sampled on the grid through zero-, first-, or second-order interpolation (higher order is seldom used).

From ρ_j an *electric field is found on the grid*, usually the same grid. In electrostatic problems this is usually done by solving finite-difference forms of $\nabla^2 \phi = -\rho$ and $\mathbf{E} = -\nabla \phi$; there is nothing in the present analysis to require use of any particular form of smoothing or emphasis. We use particular forms only when we need numerical examples.

The *force on the particle* is interpolated from the grid electric field, as

$$F_i = q_i \Delta x \sum_j E_j S(X_j - x_i) \tag{2}$$

using the *same* weighing function S as in (1). Although using the same weight functions in (1) and (2) is not a necessary feature of this discussion, there are good reasons for doing so. Using different weight functions in (1) and (2) corresponds to using different cloud shapes, which can lead to a gravitation-like instability (Problem 4-6a). Also, if the difference equations relating ρ_j to E_j are symmetric in space, *use of the same weight function eliminates the self-force* and ensures conservation of momentum (see Section 8-6). Various interpolating functions, with which we already are familiar as shape factors, are shown in Figure 8-5a. The "assignment function shape" of *Hockney and Eastwood* (1981, section 5-3-4) corresponds to our S; their "cloud shape" does not arise in our analysis.

S is designed so that charge on the grid is the same as the total particle charge,

$$\Delta x \sum_j \rho_j = \sum_i q_i \tag{3}$$

The statement about a particle at x, which follows from (3),

$$\Delta x \sum_j S(X_j - x) = 1 \tag{4}$$

says that the contribution to the grid charge density ρ is the same and correct no matter where the particle is.

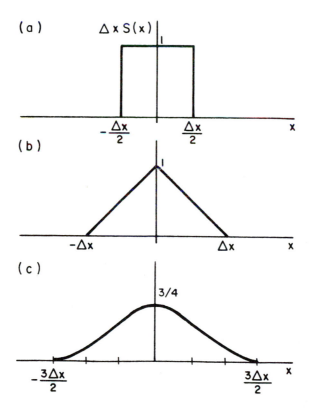

Figure 8-5a Various interpolating functions for charge and force: (a) zero-order (NGP), (b) first-order (CIC, PIC), (c) second-order (parabolic or quadratic) spline, consisting of three parabolic sections of length Δx, joined with no discontinuities in the slope; see Section 8-8.

The statement, in linear (CIC) and higher-order interpolation,

$$\Delta x \sum_j X_j S(X_j - x) = x \tag{5}$$

says that the charge at x makes the same, and correct, contribution to dipole moments independent of the particle location (Problem 8-5b).

Let us look at the force field $F(x)$ for *linear interpolation,* as shown in Figure 8-5b. It varies as a set of straight line segments; the segments give rise to *spatial harmonics* of period $\sim\!\Delta x$; stated another way, the particles "feel" the grid. In Section 8-7, we evaluate the amplitude of the Fourier components of F.

PROBLEMS

8-5a Show that (4) follows from (3).

8-5b Show that (5) is equivalent to

$$\Delta x \sum \rho_j X_j = \sum q_i x_i \tag{6}$$

i.e., the dipole moment on the grid is the same as the dipole moment of the particles.

8-6 MOMENTUM CONSERVATION FOR THE OVERALL SYSTEM

As an application of the notation, not requiring inquiry into any particular weighting (so long as it is the same for charge and force), we look at the *total momentum* of the system P. From Newton's equation of motion

$$\frac{dP}{dt} = \sum_i F_i \tag{1}$$

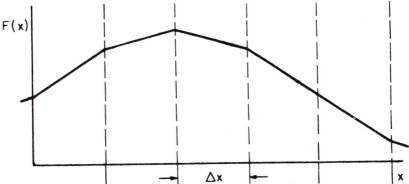

Figure 8-5b Force field $F(x)$ as a function of x is a set of straight-line segments, from grid point to grid point, in linear weighting.

which is

$$\frac{dP}{dt} = \sum_i q_i \Delta x \sum_j E_j S(X_j - x_i) \tag{2}$$

By changing the order of the sums, this becomes

$$\frac{dP}{dt} = \Delta x \sum_j E_j \sum_i q_i S(X_j - x_i) \tag{3}$$

which is recognized to be

$$\frac{dP}{dt} = \Delta x \sum_j \rho_j E_j \tag{4}$$

with no shape factor present. Hence, the question of momentum conservation, $dP/dt = 0$, is reduced to the properties of the calculation of E_j from ρ_j.

In an infinite or *periodic system,* if the algorithms treat all grid points in the same way (a limited form of *translation invariance*) and has left-right symmetry (*reflection invariance*), then

$$\Delta x \sum_j \rho_j E_j = 0 \tag{5}$$

Hence, *system momentum* is conserved. On the other hand, in the presence of metal boundaries, $\Delta x \sum_j \rho_j E_j \neq 0$ and the plasma momentum is not conserved, as is correct physically.

PROBLEM

8-6a Prove (5) under the stated conditions.

8-7 FOURIER TRANSFORMS FOR DEPENDENT VARIABLES; ALIASING DUE TO FINITE FOURIER SERIES

As we already know well, we find it most convenient to work in k-space by Fourier transforming charge, potential, field, and force. For ease of comprehension, we repeat some definitions and earlier steps.

The *transform pair* for ρ (and similarly for ϕ and E) is

$$\rho(k) \equiv \Delta x \sum_j \rho_j e^{-ikX_j} \tag{1}$$

$$\rho_j \equiv \int_{-\pi/\Delta x}^{\pi/\Delta x} \frac{dk}{2\pi} \rho(k) e^{ikX_j} \tag{2}$$

(infinite system, integration over the first Brillouin zone)

$$= \frac{1}{L} \sum_{n=-N_g/2}^{N_g/2-1} \rho(k) e^{ikX_j} \tag{3}$$

(finite system, $k = \dfrac{2\pi n}{L}$, length $L = N_g \Delta x$)

Using these definitions, the common Poisson finite-difference form, in rationalized cgs (Heavyside-Lorentz) units (*Panofsky and Phillips*, 1962, p.461; *Jackson*, 1975, pp. 817-818) is

$$\frac{\phi_{j+1} - 2\phi_j + \phi_{j-1}}{\Delta x^2} = -\rho_j \tag{4}$$

becomes

$$K^2(k)\phi(k) = \rho(k) \tag{5}$$

with

$$K^2(k) = k^2 \mathrm{dif}^2\left(\frac{k\Delta x}{2}\right) \tag{6}$$

where we have introduced the *diffraction function*

$$\mathrm{dif}\, \theta \equiv \frac{\sin\theta}{\theta} \tag{7}$$

(This is the same as the sinc θ function used by others.) The gradient finite-difference form

$$-\frac{\phi_{j+1} - \phi_{j-1}}{2\Delta x} = E_j \tag{8}$$

becomes

$$-i\kappa(k)\phi(k) = E(k) \tag{9}$$

with

$$\kappa(k) = k\,\mathrm{dif}\,(k\Delta x) \tag{10}$$

For the *force,* let us define

$$F(k) \equiv \int_{-\infty}^{\infty} dx\, F(x) e^{-ikx} \tag{11}$$

and use $F(x)$ from 8-5(2), so that

$$F(k) = \int_{-\infty}^{\infty} dx [q\Delta x \sum_j E_j S(X_j - x)] e^{ik(X_j-x)} e^{-ikX_j} \tag{12}$$

Reversing the order of the integral and sum leads to

$$F(k) = q[\Delta x \sum_j E_j e^{-ikX_j}]\left[\int dx\, S(X_j - x) e^{ik(X_j-x)}\right] \tag{13}$$

which is simply [with $S(-k)$ independent of j]

$$F(k) = qE(k)S(-k) \tag{14}$$

Now the grid quantities have the basic period Δx; hence, *the Fourier transforms of gridded quantities are periodic,* as

$$E(k - pk_g) = E(k) \tag{15}$$

Therefore, from (14), we see that how much of $E(k)$ at large k, $k\Delta x > \pi$ or $\lambda < 2\Delta x$, gets back into $F(k)$ depends on how fast $S(k)$ vanishes as $k\Delta x \to \infty$. This is because the smoothness of $F(x)$ depends on the smoothness and continuity of $S(x)$.

For the *charge density,* we first define *cloud density* ρ_c as in Section 4-6,

$$\rho_c(x) \equiv qn_c(x) \equiv q\int dx' \, S(x - x') \, n(x') \tag{16}$$

where n is the density of cloud centers, so that

$$\rho_c(k) = qS(k) \, n(k) \tag{17}$$

clearly showing the *filtering action of the shape factors.* The *grid charge density* ρ_j defined by 8-5(1), is

$$\rho_j \equiv \rho_c(X_j) = \int_{-\infty}^{\infty} \frac{dk}{2\pi} \rho_c(k) \, e^{ikX_j} \tag{18}$$

$$= \int_{-\pi/\Delta x}^{\pi/\Delta x} \frac{dk}{2\pi} e^{ikX_j} \left[\sum_{p=-\infty}^{\infty} \rho_c(k_p) \right] \tag{19}$$

by going from integrating over all k to integrating over one period and summing ρ_c over all spatial harmonics. We recognize the term $[\ ]$ as $\rho(k)$,

$$\rho(k) = \sum_p \rho_c(k_p) \tag{20}$$

$$= q\sum_p S(k_p) \, n(k_p) \tag{21}$$

It is here that the *aliases become coupled through the grid.* One can think of the infinite sum in this way: We are taking information defined on a continuum and trying to squeeze it into a discrete grid. The difficulty shows itself here in that different particle wavelengths (aliases) appear the same at the grid points.

The same phenomenon is familiar in the analysis of sampled time series. If one does not sample often enough, differing frequencies become indistinguishable. This can be improved by low-pass filtering the signal before sampling, and that is what one is doing here with a smooth S. In simulation models, the sampling effects are fed back into the system. Even a sinusoidal density perturbation produces forces with many wavelengths, which cause density perturbation at the new wavelengths, and all these perturbations act back on the original perturbations. The point to stress here is that density perturbations with $k\Delta x > \pi$ (beyond the fundamental Brillouin zone) contribute to $\rho(k)$ for $k\Delta x < \pi$. As with the force, how much depends on how fast $S(k_p)$ decays for large k. The reason for the coupling back is that k's differing by multiples of k_g look the same if we look only at the grid points X_j.

We have noted that *grid quantities* ρ, ϕ, E are periodic, so that $\phi(k) = \phi(k_p)$, etc., and can be pulled out of sums over p. However, the *normal modes for particle quantities F and n are not sinusoidal*, due to the alias coupling. The situation is like vibrations of atoms in a crystal rather than wave propagation in a continuum which has a periodic nonuniformity.

Where we use the fast Fourier transform to solve for the field, *we can choose K and κ freely* (over the fundamental period in k space) to best achieve the desired physics, even though there may not be any reasonable corresponding finite-difference relation involving a small number of grid points (see Appendix E). For instance, one can get more interaction smoothing, corresponding to widening the cloud, by truncating the k space here in some smooth manner (see Appendix B). This is computationally cheaper than using a complicated grid-particle interpolation involving many grid points. At least in one dimension, the most economical way to get a very smooth interaction without grid effects may be to use a fine mesh, low order weighting, and do the smoothing in k space as just described. Note that E in the model corresponds more closely to F rather than E in the grid-less cloud system. Unfortunately, NGP, even with a fine mesh, produces more self-heating than higher-order weighting, more than may be tolerable.

PROBLEMS

8-7a Show that 8-5(4) implies that

$$S(k) = 1 \quad \text{for } k = 0$$

$$= 0 \quad \text{for } k = pk_g \ (i.e., \ k\Delta x = 2\pi p) \ p \neq 0$$

and that 8-5(5) implies that

$$S'(pk_g) = 0$$

i.e., the *zeros* of $S(k)$ at $k = pk_g$ *are of order two* for any interpolation which is at least linearly exact. We then expect $S(k)$ to be smaller near multiples of k_g than when 8-5(5) is not true. Therefore $F(k)$ contains weaker alias contributions for small k, so $F(x)$ is smoother, as we would expect for linear as compared with NGP interpolation.

8-7b In order to make the filtering action of S clearer, first sketch an arbitrary density of cloud centers $n(x)$; then choose your favorite $S(x)$ and obtain the cloud density $n_c(x)$ from (16).

8-7c Instead of (8), define E at *half*-integer positions, $X_{j+\frac{1}{2}} = (j+\frac{1}{2})\Delta x$, say by $E_{j+\frac{1}{2}} = -(\phi_{j+1} - \phi_j)/\Delta x$. Show that (9) still holds, but the periodicity of κ (and E) is now $\kappa(k_p) = (-1)^p \kappa(k)$. Find the form of κ explicitly. Does (14) change? The discussion of Section 8-6 no longer applies; Show that momentum conservation is in fact no longer perfect.

8-8 MORE ACCURATE ALGORITHM USING SPLINES FOR $S(x)$

An "obvious" way to improve accuracy is to use *higher-order Lagrange interpolation*, say $S_L(x)$. However, the second-order $S_L(x)$ is then discontinuous and therefore unsuitable. The third order $S_L(x)$ can be differentiated but yields a discontinuous force. Because of these discontinuities, $S_L(k)$ drops off slowly for large k, implying large coupling to aliases.

If we arrange to have several continuous derivatives, then S drops off more rapidly and has small coupling to aliases. This suggests the use of *splines*, which we now examine in the present formalism (*Lewis et al., 1972; Buneman, 1973; Langdon, 1973; Brown et al., 1974; Denavit, 1974; Lewis and Nielson, 1975*).

Define $S_m(x)$ as the convolution of the square nearest-grid-point (NGP) weighting function [Figure 8-5a(a)] with itself m times. (For large m, S_m approaches a Gaussian in the sense of the central limit theorem; S_m is an analogue of the Gaussian for these systems.) Hence, S_1 is the linear interpolation case CIC [Figure 8-5a(b)]. The basis functions $S_m(X_j - x)$, and therefore $F(x)$, are piecewise polynomials of order m, as discussed in the caption of Figure 8-5a. Derivatives exist through to order m, which is discontinuous. Note also that $S_m(x) \geqslant 0$, which is not true for higher-order Lagrange interpolation. The relation to conventional spline interpolation is discussed in *Langdon* (1973, appendix D). See also *Hockney and Eastwood* (1981, section 5-3-2).

The transform is

$$S_m(k) = \left[\frac{\sin \frac{1}{2} k \Delta x}{\frac{1}{2} k \Delta x} \right]^{m+1} = \text{dif}^{m+1}\left[\frac{k \Delta x}{2} \right] \tag{1}$$

in one dimension. For large k, $S_m = O(k^{-m-1})$. At non-zero multiples of $k_g = 2\pi / \Delta x$, $S_m(k)$ has a zero of order $m+1$, and so is expected to be small nearby. This also true for Lagrange interpolation. However, for small k, $S(k \pm k_g)$ is about 5 times larger, and $S(k \pm 2k_g)$ about 21 times larger, for second order Lagrange than for S_2 [Figure 8-5a(c)]. Thus the spline reduces aliasing errors for long wavelengths yet takes the same amount of work to evaluate numerically. For small k, $S_m(k) \approx 1 - (m+1)(k\Delta x)^2 / 24$, whereas $S - 1 = O(k^{m+1})$ or $O(k^{m+2})$ for Lagrange interpolation. However this "error" in the spline can be compensated for by the Poisson algorithm, as noted earlier.

As an example, let us consider an algorithm for the quadratic spline (QS) S_2 in a one-dimensional periodic system [see Figure 8-5a(c)]. We have

$$S_2(X_j - x) = \frac{1}{\Delta x}\left[\frac{3}{4} - \left(\frac{x - X_j}{\Delta x} \right)^2 \right] \tag{2}$$

and

$$S_2(X_{j \pm 1} - x) = \frac{1}{2\Delta x}\left[\frac{1}{2} \pm \frac{x - X_j}{\Delta x} \right]^2 \tag{3}$$

with, for both (2) and (3),

$$|x - X_j| \leqslant \frac{\Delta x}{2} \qquad (4)$$

PROBLEM

8-8a Show that S_2 and S'_2 from the above two equations are continuous at $x = \pm \Delta x / 2 + X_j$.

8-9 GENERALIZATION TO TWO AND THREE DIMENSIONS

We now have a complete formulation of interactions in most one-dimensional models that we generalize to two or three dimensions in this section.

The grid label j becomes a vector $\mathbf{j} = (j_x, j_y, j_z)$ with integer components. The coordinate of grid point $\mathbf{j}$ is, in a three-dimensional oblique grid (for instance, triangular meshes have also been used),

$$\mathbf{X}_j = \mathbf{j} \cdot \Delta \mathbf{x} \qquad (1)$$

where the rows of the tensor $\Delta \mathbf{x}$ are the basis vectors for the grid (see *Brand*, 1957). This defines the edges of a grid cell whose volume is

$$V_c = \det \Delta \mathbf{x} \qquad (2)$$

In the usual rectangular grid, we have

$$\Delta \mathbf{x} = \begin{bmatrix} \Delta x & 0 & 0 \\ 0 & \Delta y & 0 \\ 0 & 0 & \Delta z \end{bmatrix} \qquad (3)$$

The transform becomes, for example,

$$\mathbf{E}(\mathbf{k}) = V_c \sum_j \mathbf{E}_j e^{-i\mathbf{k} \cdot \mathbf{X}_j} \qquad (4)$$

For a *point particle* at $\mathbf{x}$,

$$\mathbf{F}(\mathbf{x}) = q V_c \sum_j S(\mathbf{X}_j - \mathbf{x}) \mathbf{E}_j \qquad (5)$$

$$\rho_j = q V_c S(\mathbf{X}_j - \mathbf{x}) \qquad (6)$$

For *finite-size particles,* the transforms are

$$\mathbf{F}(\mathbf{k}) = q S(-\mathbf{k}) \mathbf{E}(\mathbf{k}) \qquad (7)$$

$$\rho(\mathbf{k}) = q \sum_p S(\mathbf{k}_p) n(\mathbf{k}_p) \qquad (8)$$

where

$$\mathbf{k}_p = \mathbf{k} - \mathbf{p} \cdot \mathbf{k}_g \qquad (9)$$

$$\mathbf{k}_g = 2\pi (\Delta \mathbf{x}^{-1})^T \qquad (10)$$

and $\mathbf{p}$ is a vector with integer components. The rows of tensor $\mathbf{k}_g$ are basis vectors, reciprocal to those given by Δx, times 2π. They define the periodicity of transforms of grid quantities, since

$$\exp(i\mathbf{k}_p \cdot \mathbf{X}_j) = \exp(i\mathbf{k} \cdot \mathbf{X}_j - 2\pi i\mathbf{p} \cdot \mathbf{j}) = \exp(i\mathbf{k} \cdot \mathbf{X}_j) \tag{11}$$

For a rectangular grid

$$\mathbf{k}_g = 2\pi \begin{bmatrix} \Delta x^{-1} & 0 & 0 \\ 0 & \Delta y^{-1} & 0 \\ 0 & 0 & \Delta z^{-1} \end{bmatrix} \tag{12}$$

The integral in the inverse transform is taken over one period in $\mathbf{k}$ space.

$$E_j = \int_g \frac{d\mathbf{k}}{(2\pi)^3} E(\mathbf{k}) e^{i\mathbf{k} \cdot \mathbf{X}_j} \tag{13}$$

The forms taken in one, two, or three dimensions are simply seen if one remembers that $d\mathbf{k}/(2\pi)^3 \rightarrow d\mathbf{k}/(2\pi)^d$, where d is the dimensionality. The relations between transformed grid quantities are

$$\rho(\mathbf{k}) = K^2(\mathbf{k})\phi(\mathbf{k}) \tag{14}$$

$$E(\mathbf{k}) = -i\kappa(\mathbf{k})\phi(\mathbf{k}) \tag{15}$$

Quantities ϕ_j, E_j, and ρ_j are defined only on the grid, while $n(\mathbf{x})$ and $F(\mathbf{x})$ are defined on a continuum of particle positions.

For a spline of order m in all three coordinates, the shape becomes, in a three-dimensional rectangular grid,

$$S_m(\mathbf{k}) = [\text{dif}(\tfrac{1}{2}k_x\Delta x)\,\text{dif}(\tfrac{1}{2}k_y\Delta y)\,\text{dif}(\tfrac{1}{2}k_z\Delta z)]^{m+1} \tag{16}$$

This includes NGP ($m=0$) and CIC-PIC (linear, $m=1$). The field finite-difference equations, in their simplest generalization, yield

$$K^2 = k_x^2 \text{dif}^2 \tfrac{1}{2}k_x\Delta x + k_y^2 \text{dif}^2 \tfrac{1}{2}k_y\Delta y + k_z^2 \text{dif}^2 \tfrac{1}{2}k_z\Delta z \tag{17}$$

and

$$\kappa_x = k_x \text{dif}\, k_x \Delta x, \quad \text{etc.} \tag{18}$$

PROBLEM

8-9a Write out the shape factor $S(\mathbf{k})$ for linear weighting CIC for both two- and three-dimensional rectangular grids.

8-10 LINEAR WAVE DISPERSION

If we use 8-7(14) in 8-3(9), we find for each row the same result,

$$\epsilon(k, \omega)\, qS(-k)\, E(k) = 0 \tag{1}$$

so that $\epsilon = 0$ is the *dispersion relation,* where

$$\epsilon(k,\omega) = 1 - S^{-1}(-k)\sum_p F_p(k - \tfrac{1}{2}pk_g)\,S(-k_p)\psi(k_p,\omega) \tag{2}$$

The solutions $\omega(k)$ are (multivalued) periodic functions of k. Equation (2) can be rewritten, using K^2 and κ, as

$$\epsilon(k,\omega) = 1 + \frac{iq^2}{K^2}\sum_p \kappa\,|S(k_p)|^2\psi(k_p,\omega), \qquad \kappa = \kappa(k_p) \tag{3}$$

where $n(k,\omega) = \psi(k,\omega)F(k,\omega)$. (This reduction suggests that the same result may be obtained more directly; this is done in the next section.) The only approximation is that the linearized plasma response is used; no approximation is made about the smallness of the grid effects. The function $\epsilon(k,\omega)$, for grid quantities, plays the usual role of dielectric function in kinetic theory results on fluctuations and collisions; see Chapter 12.

The *normal modes* are given by a scalar equation $\epsilon = 0$ rather than an infinite determinant because the aliases are equivalent for grid quantities. The normal modes are sinusoidal in space for grid quantities (ρ, ϕ, E), though *not* for particle quantities (F, n) due to the alias coupling.

8-11 APPLICATION TO COLD DRIFTING PLASMA: OSCILLATION FREQUENCIES

An easy-to-follow model with simple applications is one-dimensional with a beam of particles of velocity v_0 drifting through a background of immobile neutralizing particles. We assume that there are enough particles per cell so that the fluid equations of continuity and motion are valid; these produce the perturbed density from the force, as [see 5-3(12)],

$$n_1(k,\omega) = \frac{n_0}{m}\frac{ikF(k,\omega)}{(\omega - kv_0)^2} \tag{1}$$

Inserting this into 8-7(21) produces

$$\rho(k,\omega) = \left(\frac{n_0 q}{m}\right)\sum_p \frac{ik_p S(k_p)F(k_p,\omega)}{(\omega - k_p v_0)^2} \tag{2}$$

Using 8-7(9) and 8-7(14) produces

$$F(k_p,\omega) = -\,iqS(-k_p)\kappa(k_p)\phi(k_p,\omega) \tag{3}$$

Inserting F into (2) leaves

$$\rho(k,\omega) = \left(\frac{n_0 q^2}{m}\right)\sum_p \frac{k_p\kappa(k_p)S(k_p)S(-k_p)}{(\omega - k_p v_0)^2}\phi(k_p,\omega) \tag{4}$$

[We do *not* use $\kappa(k_p) = \kappa(k)$, as we wish to be free in choosing algorithms; as in Problem 8-7c, and Chapter 10 in which, for energy conserving programs, $\kappa(k_p)$ becomes k_p.] We take advantage of the periodicity of grid quantities by using

$$\phi(k_p, \omega) = \phi(k, \omega) \tag{5}$$

to remove ϕ from the sum. Also

$$S(k_p) S(-k_p) = |S(k_p)|^2 = S^2(k_p) \tag{6}$$

as $S(x)$ is real and even. The last step is to use Poisson's equation 8-7(5) to relate ϕ and ρ. We obtain the *dispersion equation,*

$$1 - \frac{\omega_p^2}{K^2(k)} \sum_p \frac{k_p \kappa(k_p) S^2(k_p)}{(\omega - k_p v_0)^2} = \epsilon(k, \omega) = 0 \tag{7}$$

where $\omega_p^2 = n_0 q^2 / m$.

First, consider no drift, $v_0 = 0$. Then the dispersion relation is simply

$$\omega^2 = \omega_p^2 \left[\frac{1}{K^2(k)} \sum_p k_p \kappa(k_p) S^2(k_p) \right] \tag{8}$$

The [ ] contains the modifications to Langmuir oscillations due to the imposition of the spatial grid. The sum on p converges fairly rapidly, and allows use of only, say, $p = 0, \pm 1, \pm 2$ to obtain fairly trustworthy results in the *fundamental Brillouin zone,* $|k \Delta x| < \pi$.

Or, the sum can sometimes be done analytically exactly, once K, κ, and S are specified. For example, if we use the *momentum-conserving algorithm,* meaning force is an interpolation of the differenced potential, then κ is given by 8-7(10), and using linear interpolation (CIC),

$$S(k) = \text{dif}^2 |\text{½} \, k \Delta x| \tag{9}$$

the sum becomes

$$\sum_p k_p \kappa(k_p) S^2(k_p) = \sum_p k_p^2 \, \text{dif} \, (k_p \Delta x) \, \text{dif}^4 \, (\text{½} \, k_p \Delta x) \tag{10}$$

$$= \frac{2^4 \sin(k \Delta x) \sin^4(\text{½} \, k \Delta x)}{(\Delta x)^5} \sum_p k_p^{-3} \tag{11}$$

Now we use the identity (*Abramowitz and Stegun,* 1964, eq. 4.3.92, p. 75)

$$\sum_p (k - pk_g)^{-2} = \left[\frac{2}{\Delta x} \sin \text{½} \, k \Delta x \right]^{-2} \tag{12}$$

which, when differentiated with respect to k, gives

$$\sum_p (k - pk_g)^{-3} = \left[\frac{2}{\Delta x} \sin \text{½} \, k \Delta x \right]^{-3} \cos \text{½} \, k \Delta x \tag{13}$$

Using the common three-point algorithm for Poisson's equation 8-7(4) completes the work. The result is simply

$$\omega^2 = \omega_p^2 \cos^2 \tfrac{1}{2} k\Delta x, \qquad \omega = \pm \omega_p \cos \tfrac{1}{2} k\Delta x \tag{14}$$

Note that this exact result is *not* the same as the approximate result obtained by keeping just the $p=0$ term corresponding to $\epsilon_0 = 0$ in Section 8-3 and as used in Chapter 4, where we ignored the spatial harmonics (aliases); using just $p=0$, the approximate result is

$$\omega^2 \approx \omega_p^2 \cos^2 \tfrac{1}{2} k\Delta x \left[\frac{\tan \tfrac{1}{2} k\Delta x}{\tfrac{1}{2} k\Delta x} \frac{\sin^2 \tfrac{1}{2} k\Delta x}{(\tfrac{1}{2} k\Delta x)^2} \right] \tag{15}$$

with error of less than 3 percent for $k\Delta x < \pi/2$, relative to (14).

On the other hand, if we use the *energy-conserving algorithm* (yet to come in Chapter 10), then we can show that $\kappa(k_p) = k_p$, and the result is simply *no dispersion,*

$$\omega^2 = \omega_p^2, \qquad \omega = \pm \omega_p \tag{16}$$

independent of k, as in Langmuir oscillations of a cold plasma. The error in using only $p=0$ terms here is much larger, as this approximation gives

$$\omega^2 \approx \omega_p^2 \left[\frac{\sin \tfrac{1}{2} k\Delta x}{\tfrac{1}{2} k\Delta x} \right]^2 \tag{17}$$

If the accuracy of dispersion, here for cold-plasma oscillations, were of prime importance, then we might modify the Poisson algorithm to compensate for the error in the momentum-conserving algorithm, or use an energy-conserving algorithm. However, we find in Sec. 8-13 that for a *warm* plasma keeping only $p=0$ is often a very good approximation, and both types of algorithm benefit by compensation, as discussed in Appendix B.

PROBLEM

8-11a For momentum-conserving codes, use spline weighting of order m with $S_m(k)$ given by 8-8(1). For $v_0 = 0$, show that (8) becomes

$$\omega^2 = \omega_p^2 \frac{1}{K^2(k)} \frac{\sin k\Delta x \sin^{2m+2} (\tfrac{1}{2} k\Delta x) \, 2^{2m+2}}{\Delta x^{3+2m}} \sum_p k_p^{-(1+2m)} \tag{18}$$

Then, with K^2 given by 8-7(8), show that

$$m = 0, \text{ NGP}, \quad \omega^2 = \omega_p^2 \cos^2 \tfrac{1}{2} k\Delta x \tag{19}$$

$$m = 1, \text{ CIC}, \quad \omega^2 = \omega_p^2 \cos^2 \tfrac{1}{2} k\Delta x \tag{20}$$

$$m = 2, \text{ QS}, \quad \omega^2 = \omega_p^2 \cos^2 \tfrac{1}{2} k\Delta x \frac{1}{3} [2 + \cos^2 \tfrac{1}{2} k\Delta x] \tag{21}$$

8-12 COLD BEAM NONPHYSICAL INSTABILITY

We now look at a cold (single-velocity) electron beam in a fixed ion background, moving relative to the grid. Physically, this model should be stable. The dispersion relation is 8-11(7). For any fixed k, the dispersion relation has a structure similar to that for many beams. One difference is that the effective plasma frequency squared for half of the "beams" is negative. The particles all react strongly when the Doppler-shifted frequency, $\omega - kv_0$, is near a harmonic of the grid-crossing frequency $k_g v_0$ that is, near the resonances defined by

$$\omega - kv_0 + pk_g v_0 = 0 \tag{1}$$

One can show that there are two roots corresponding to each alias term in the dispersion relation 8-11(7). The roots are either real or occur in conjugate pairs.

In Figure 8-12a, we sketch ϵ for small $k\Delta x$. One easily sees the pairs of real roots associated with $p = 0$ ($\omega - kv_0 = \pm\omega_p$), $p = 1$, and $p \leqslant -2$. For each of the other resonances there is a pair of complex roots, one unstable. For larger $|p|$, the modes are highly resonant and therefore hard to excite and also easily destroyed by any dissipation or spread in resonance, e.g., if $|\omega - kv_0| < |k_p v_t|$.

For each p term, for real k, there are two roots in ω, either both real or complex conjugates. For very small $k\Delta x$, $S^2 \ll 1$ for $p \neq 0$ (higher zones); for $p = 0$, the small $k\Delta x$ roots are $\omega \approx \pm\omega_p$; for $p \neq 0$, keeping the $p = 0$ term {calling $[k\kappa(k)S^2/K^2(k)] \approx 1$} and one $p \neq 0$ term, the solution near the resonance is

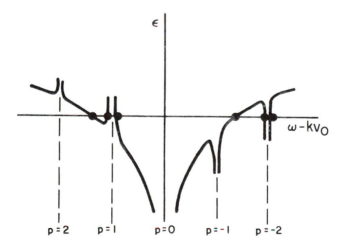

Figure 8-12a Sketch of $\epsilon(k, \omega)$ versus $\omega - kv_0$ for cold-beam, momentum-conserving code. The poles due to each p term are indicated. This figure is for $k\Delta x \ll 1$ and $v_0/\Delta x \omega_p < 1/2\pi$; for $v_0/\Delta x \omega_p > 1/2\pi$, the poles are outside $\pm\omega_p$. *(From Langdon, 1970b.)*

$$\omega \approx kv_0 \pm \omega_p \frac{S(k_p)}{K(k)} \left[\frac{k_p \kappa(k_p)}{1 - (\omega_p / p k_g v_0)^2} \right]^{\frac{1}{2}} \tag{2}$$

We see the possibility of instability (purely numerical) where $k_p \kappa(k_p) < 0$ (occurring only for momentum-conserving programs) or for $\omega_p > k_g v_0$ [for energy- or momentum-conserving programs; the inequality is questionable within a numerical factor, because of the approximation used to obtain (2)].

The growth rates from the $p = -1$ and 2 modes, and from $p = 1$ at larger $k\Delta x$, are substantial. In numerical solutions for $v_0 = 0.12 \Delta x \omega_p$, *Langdon* (1970b) found a growth rate of $0.017\omega_p$ at $k\Delta x = 1$ (due to the $p = 1$ alias), and a maximum growth rate greater than $0.22\omega_p$ at $k\Delta x = 2.2$. The cold-beam (and warm-plasma, next section) instabilities have been verified in simulations by *Okuda* (1972b) for both momentum- and energy-conserving algorithms, by *Langdon* (1973) for energy-conserving, and by *Chen et al.* (1974) for momentum-conserving models, with each author adding detail. In all simulations, the telltale mark(s) of numerical instability, loss of energy conservation (system momentum), was clear in momentum-conserving (energy-conserving) codes.

Birdsall et al. (1975, 1980) made extensive solutions of 8-11(7) for the linear growths, and verified these very closely using a momentum-conserving code. (For energy-conserving codes, the threshold was found to be $k_g v_0 = 2\omega_p$.) In addition, they pointed out various "cures," mostly means for reduction of the growth rates. However, by letting the cold beam nonphysical instability grow to saturation (in both electric field energy and thermal energy—the instability heats the beam), they found that growth stopped (and energy conservation returned) rather abruptly at a given level, roughly at

$$\frac{\lambda_D}{\Delta x} = \frac{v_t}{\omega_p \Delta x} = 0.046 \quad \text{for} \quad \frac{\lambda_B}{\Delta x} \equiv \frac{v_0}{\omega_p \Delta x} \geq \frac{1}{\pi} \tag{3}$$

The level of saturation, by particle trapping, was worked out analytically by Albritton and Nevins (in *Birdsall and Maron*, 1980). A stability diagram for all $\lambda_B / \Delta x$ is given by Figure 8-12b.

Initiating simulations with v_t larger than that given by (3) led to no growth. Hence, if one chooses to avoid the cold-beam nonphysical instability, then the prescription is to add a small thermal spread as prescribed by (3). (If not, then this v_t will be added automatically!) Hence, this nonphysical particle instability tends to be *self-quenching;* this behavior is in contrast with numerical instabilities in fluid codes, which tend to be self-destructing. It has been stated that the instability is benign because its effect is to move the initial state toward one which the algorithm represents more accurately. However, this change may be fatal to the purpose of the simulation! For example, if inappropriate parameters for ES1 had been chosen by *Ishihara et al.* (1981, 1982), the cold electron beam would have been warmed unacceptably by the grid instability in a time far shorter than the saturation time for the physical instability.

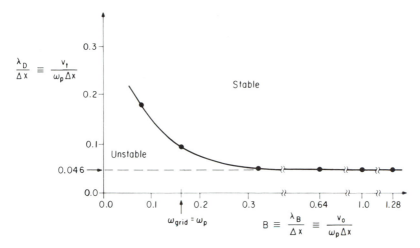

Figure 8-12b Experimental determination of the thermal spread needed for stability of a beam in a (mathematically) gridded periodic system, using a momentum-conserving program. For energy-conserving programs, stability is found analytically for $\lambda_B / \Delta x > 1 / \pi \approx 0.3$. *(From Birdsall et al., 1975; Birdsall and Maron, 1980.)*

Langdon (1973), using an energy-conserving program, and *Birdsall and Maron* (1980), using a momentum-conserving program, both obtained excellent phase space $(v_x - x)$ indication of short-wavelength activity (particles showing structure for $k\Delta x > \pi$), with proper alias relations; e.g., with $N_g = 32$ (16 distinct modes) one would observe mode 12 in ρ_{grid}, ϕ_{grid}, E_{grid}, and modes 12 and $32 - 12 = 20$ in phase space; see Section 10-9.

Birdsall and Maron (1980) also ran the purely thermal case, no drift, $(v_0 = 0)$, starting with $\lambda_D / \Delta x = 0.1$, CIC. The thermal (and total) energy increased with time, tending to behave asymptotically as

$$\frac{\lambda_D(t)}{\Delta x} \approx \frac{\lambda_D(\infty)}{\Delta x}[1 - \exp(-\alpha t)]$$

with $\lambda_D(\infty) / \Delta x \approx 0.26$ and α is a constant $\ll \omega_p$, where the linear growth rate has become very small. The theory for the thermal case is in the next section.

8-13 SOLUTION FOR THERMAL (MAXWELLIAN) PLASMA; NONPHYSICAL INSTABILITIES CAUSED BY THE GRID

The dielectric function for an unmagnetized electrostatic Vlasov plasma is (Problem 8-13a)

$$\epsilon(\mathbf{k}, \omega) = 1 + \frac{\omega_p^2}{K^2(\mathbf{k})} \sum_p |S(\mathbf{k}_p)|^2 \int \frac{d\mathbf{v}}{\omega + i0 - \mathbf{k}_p \cdot \mathbf{v}} \boldsymbol{\kappa} \cdot \frac{\partial f_0}{\partial \mathbf{v}}, \quad \text{Im} \, \omega \geq 0 \quad (1)$$

For f_0 a Maxwellian velocity distribution with no drift, (1) becomes

$$\epsilon(\mathbf{k}, \omega) = 1 - \frac{\omega_p^2}{2K^2 v_t^2} \sum_p |S(\mathbf{k_p})|^2 \frac{\boldsymbol{\kappa} \cdot \mathbf{k_p}}{k_p^2} Z'\left(\frac{\omega}{\sqrt{2}|\mathbf{k_p}|v_t}\right) \tag{2}$$

where Z' is the derivative of the plasma dispersion function (*Fried and Conte*, 1961).

When $\lambda_D \gtrsim \Delta x$ the principal term is the one whose $\mathbf{k_p}$ is nearest $\mathbf{k}$. The modes are heavily damped when $\mathbf{k}$ differs much from this $\mathbf{k_p}$. In this case we expect no important interaction among the different aliases. In the first zone let us take *only* the $\mathbf{p}=0$ term, obtaining the average force $\mathbf{F_0}$ dielectric function ϵ_0 discussed earlier in Sections 8-3 and 8-11. We then view the model as approximately a gridless cloud plasma with Coulomb interaction having

$$\epsilon_0 = 1 - \frac{\omega_p^2}{2k^2 v_t^2} |S_0|^2 Z'\left(\frac{\omega}{\sqrt{2}kv_t}\right) \tag{3a}$$

with $$|S_0(\mathbf{k})|^2 = |S(\mathbf{k})|^2 \frac{\mathbf{k} \cdot \boldsymbol{\kappa}}{K^2}, \qquad i.e., \quad \mathbf{p} = 0 \text{ only} \tag{3b}$$

where S_0 is the cloud shape factor to be used in the dispersion relation.

Solution of these two dispersion relations (2) and (3a) are given in Figure 8-13a for $\lambda_D = \Delta x$ for the Maxwellian in one dimension with $S(k) = \text{dif}\,(\frac{1}{2}k\Delta x)$ for NGP, for $\lambda_D/\Delta x = 1$. The difference between $\text{Im}\,\omega$ in the two cases is too small to be shown on the graph, and $\text{Re}\,\omega$ differs significantly only where the wave is heavily damped. Thus alias coupling is not very important for $\lambda_D/\Delta x = 1$ and the averaged force works very well. This conclusion is stronger for CIC-PIC.

There are now many wave phase velocities ω/k_p with which particles may resonantly interact. Unless the coupling is very strong there is no qualitative change in $\text{Re}\,\epsilon$ for real ω; in particular the sign of its derivative is unchanged. This need not be true for the sign of the imaginary part,

$$\text{Im}\,\epsilon = -\pi \frac{\omega_p^2}{K^2} \sum_p \kappa S^2(k_p) \frac{1}{|k_p|} f_0'\left(\frac{\omega}{k_p}\right) \tag{4}$$

so that the plasma stability may be changed. For half the aliases $\mathbf{k_p}$ has the opposite direction to $\boldsymbol{\kappa}$ so that the factor $\boldsymbol{\kappa} \cdot \mathbf{k_p}/k_p^2$ has the wrong sign. For small $k\Delta x$, this can make $\omega\,\text{Im}\,\omega$ negative, while $\omega\,\partial\text{Re}\,\epsilon/\partial\omega$ remains positive; this leads to an *instability* (*Lindman*, 1970; *Langdon*, 1970a, b). In the jargon of real plasmas, the wave has positive energy and experiences negative absorption. With $\lambda_D = \Delta x$, *Langdon* (1970a) found weak growth occurring only at wavelengths much *greater* than Δx (a surprise). Growth became significant when $\lambda_D/\Delta x$ was decreased, with appreciable growth over a wide range of $k\Delta x$, as in Figure 8-13b for $\lambda_D = \Delta x/10$. We now explain these results and come to additional conclusions.

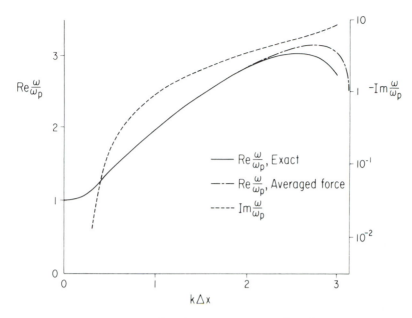

Figure 8-13a Solutions of the exact (all p) and average-force ($p=0$) dispersion relations for a Maxwellian velocity distribution with $\lambda_D/\Delta x = 1$, NGP interpolation. The Im ω is the expected Landau damping. *(From Langdon, 1970a.)*

The dispersion relation is periodic in k; we keep $k < k_g/2$, $k\Delta x < \pi$ (*i.e.*, in the fundamental Brillouin zone), which is the k one would think of physically. Then the physical phase velocity ω/k is larger than the alias velocities ω/k_p, $p \neq 0$. Thus ω/k may be much larger than v_t, leading to negligible Landau damping, at the same time that the slow waves are interacting strongly with the thermal particles, as shown in Figure 8-13c, if $k_g v_t \gtrsim \omega_p$. The contributions from waves with equal $|p|$ nearly cancel; the net effect turns out to be destabilizing. Although it is small for small $k\Delta x$ [being $\propto (k\Delta x)^2$ for NGP and $(k\Delta x)^4$ for linear weighting], the Landau damping of the principal wave goes to zero even faster as $k \to 0$. Thus the grid can destabilize oscillations even with long wavelengths $\lambda \gg \Delta x$.

For $\lambda_D \gtrsim \Delta x/2$ the alias wave velocities fall on the flat part of the particle velocity distribution, and Landau damping occurs unless $k\Delta x \lesssim 2k\lambda_D$ is small. Therefore, the instability is confined to long wavelengths and is very weak. A rough rule of thumb is that this nonphysical instability has ignorable growth for $\lambda_D/\Delta x \gtrsim 1/\pi \approx 0.3$ for linear weighting.

However, when $\lambda_D \sim 0.1\Delta x$, the lowest and strongest aliases interact with the steep sides of f_0 and there is little Landau damping even for $k\Delta x \sim 2$ where the coupling is strong. The result is strong instability; Im ω is as large as $0.1\omega_p$ for NGP; however, Im $\omega < 0.014\omega_p$ for linear weighting, which is about 1 dB per cycle, negligible in many applications.

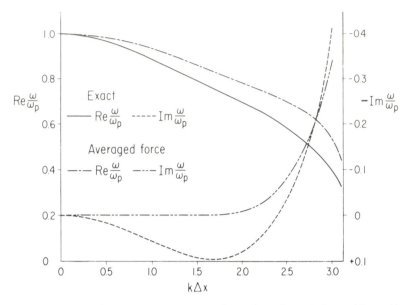

Figure 8-13b Solutions of the exact and average-force dispersion relations with $\lambda_D/\Delta x = 0.1$, NGP interpolation. The $\mathrm{Im}\,\omega > 0$ for $0 < k\Delta x \leqslant 2.5$ is nonphysical growth, due to aliasing, arising only if $|p| > 0$ terms are kept. In CIC-PIC, the maximum growth rate is about $0.014\,\omega_p$, about 10 times smaller. *(From Langdon, 1970a.)*

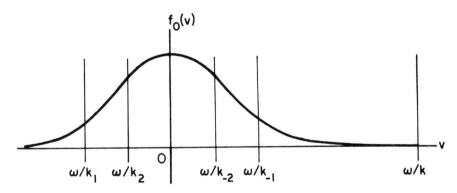

Figure 8-13c Wave phase velocity ω/k and alias wave phase velocities ω/k_p and $f_0(v)$ for $k_g v_t \sim \omega_p$. *(From Langdon, 1970b.)*

If $\lambda_D/\Delta x$ is decreased further, only the weaker, large p aliases contribute and the instability goes away, as it should since, of course, a cold stationary plasma is inactive (oscillatory only—stable). In many applications, such as that of Section 5-12, a cold plasma component provides accurate, noise-free collective behavior.

In studying this instability experimentally, there are difficulties without using a very large number of particles to ensure that the linear approximation is not violated by too-large fluctuations and grid noise forces or that the instability is not damped by collisions. One might puzzle what an instability looks like in a plasma which is already Maxwellian, guessing that it just gives enhanced fluctuations. These occur and cause a gradual *heating* of the plasma. This is not forbidden since the momentum conserving codes are not energy conserving. In *the energy-conserving* codes (Chapter 10), the destabilizing factor $\kappa \cdot \mathbf{k_p} / k_p^2$ is unity, so that there is no grid-induced instability unless the plasma is drifting through the grid. The verifying experiments with Maxwellian plasmas were mentioned in the previous section.

The feature of the oscillations which is central is that plasma perturbations of different wavelengths are coupled together. This can be thought of as a *parametric interaction* in which the *spatial grid* plays the role of the *pump*. The pump wave numbers are k_g and its harmonics pk_g; the frequency is zero. As a result plasma perturbations characterized by $(\omega, k_p = k - pk_g)$, p integral, are coupled and in the dispersion relation we had a sum over all these sub-modes. The coupling strength is given by $S^2(k_p)$ and is weaker as the order of interpolation increases.

PROBLEMS

8-13a Derive (1), in analogy to Section 8-11 but using results from Section 8-9 and the Vlasov kinetic equation instead of cold fluid results.

8-13b Keeping only $p = 0$, show that the Bohm-Gross dispersion takes the form

$$\omega = \omega_p \left(1 + \frac{3}{2} k^2 \lambda_D^2 - \alpha k^2 \Delta x^2 \right)$$

and find α for NGP and for CIC-PIC. Note that for $\lambda_D^2 / \Delta x^2 < 2\alpha / 3$ the wave becomes backward, *i.e.*, its group velocity is opposite to its phase velocity; this change in dispersion may affect many types of plasma instabilities and shows the need to modify the field solver to compensate for inaccuracies in the dispersion.

8-13c How does the instability of this Section change when E is defined as in Problem 8-7c?

NINE

EFFECTS OF THE FINITE TIME STEP

9-1 INTRODUCTION

In this chapter, we present the effects of using a finite time step Δt. The first part is for the unmagnetized plasma, the second part is for the magnetized plasma, and the last part is on other time-integration schemes and the problems of long time steps.

In particle models for computer simulation of plasmas, the algorithms for advancing the system one time step forward usually divide into two parts: calculation of electromagnetic fields and particle forces, and advancing the particle positions and velocities. Chapter 8 deals with the former, while assuming the latter is performed exactly by the differential equations of Newtonian dynamics. In this chapter, we discuss properties of finite-difference equations used to advance the particles in time. In most of the discussion, we revert for clarity to continuum x space for the fields and forces. However, the results of this and Chapter 8 are combined in Section 9-5 to yield results which include exactly the effects of discreteness in space and time. This work is a prerequisite for development of a quantitative kinetic theory of simulation plasmas (Chapter 12).

Other authors have studied aspects of the time integration by other methods (*Lindman,* 1970; *Godfrey,* 1974) and by extension of the methods of this chapter (*Byers et al.,* 1978); their work has included both fully electromagnetic fields and the Darwin magnetoinductive approximation. This chapter is a rather complete account of the electrostatic case drawn from *Langdon* (1970b, 1979b, and unpublished reports) and other sources.

We first find the dispersion function which describes the response of the plasma to perturbing fields and whose zeros give the dispersion and stability of free oscillations. This is derived by two methods which illustrate different aspects and parameter limits. The analysis of plasma oscillations involves the same physics as in the classic Landau problem (*Jackson*, 1960), and the results reduce to it simply and correctly in the limit $\Delta t \to 0$.

Finite Δt makes a simple change in the linear dispersion relations. With leap-frog integration in the unmagnetized model, we show that the usual resonant denominator, $1/(\omega - \mathbf{k} \cdot \mathbf{v})$, becomes

$$\frac{\Delta t}{2} \cot (\omega - \mathbf{k} \cdot \mathbf{v}) \frac{\Delta t}{2} = \sum_q \frac{1}{\omega - \mathbf{k} \cdot \mathbf{v} - q\omega_g}, \quad \text{where } \omega_g \equiv \frac{2\pi}{\Delta t} \quad (1)$$

The result in the roots for $\omega(\mathbf{k})$ for plasma oscillations, small $\omega_p \Delta t$, is a relative upward shift of $(\omega_p \Delta t)^2/24$, a correction observed earlier in Chapter 4.

The periodicity in ω reflects the fact that frequencies differing by harmonics of $\omega_g = 2\pi/\Delta t$ represent the same change in phase during a time step $\{\exp[-i(\omega - q\omega_g)\Delta t] = \exp[-i\omega\Delta t]\}$ and are therefore equivalent as seen by the difference equations. This is a sort of stroboscopic effect that fools not only the observer but the system dynamics. The frequencies $(\omega - q\omega_g)$ are called *aliases* because they are different designations for the same thing.

For simple harmonic oscillations, as in the small $k\lambda_D$ limit, we know from Chapter 4 that the usual leapfrog scheme becomes unstable when $\omega_p \Delta t \geqslant 2$. Collective effects reduce this to $\omega_p \Delta t \geqslant 1.62$ for a Maxwellian velocity distribution. This is because Bohm-Gross dispersion increases ω above ω_p toward $\pi/\Delta t$, not because particles traverse a large part of a wavelength in one time step. However, no other nonphysical instabilities are found.

These results and those of Chapter 8 are then combined to describe both the spatial and temporal difference algorithm.

In the magnetized plasma, Section 9-6, the aliasing of cyclotron harmonics allows a nonphysical instability which is a finite Larmor radius effect, that is not seen in single particle analysis or in a cold plasma, and is possible even with a Maxwellian distribution.

Three approaches to efficient simulation of slowly-varying (compared, e.g., to ω_{pe}) phenomena are outlined in Section 9-7. The theory is generalized to a class of integration schemes in Section 9-8. Some examples are analyzed, and new algorithms are synthesized.

9-2 WARM UNMAGNETIZED PLASMA DISPERSION FUNCTION; LEAPFROG ALGORITHM

We consider perturbations of a uniform, infinite or periodic, unmagnetized, one-species plasma with fixed neutralizing background. We analyze *collective* motion of the plasma, a different emphasis than in the usual numerical analysis literature, in which different considerations of accuracy and speed are involved. For instance, the initial-value problem for ordinary differential equations has been extensively studied and a variety of very accurate algorithms are available. Such methods have been used, for example, for single-particle motion in the study of magnetic field configurations in fusion experimental devices. The methods used in many-particle simulation may seem comparatively simple. However, what is important is not that individual particle orbits be accurate, but that the *collective motions of many particles reflect real plasma behavior.* When computer capacity is limited, it has usually been better to use simple and fast difference algorithms rather than, say, having to use fewer particles. Even some errors in collective motion, for example, oscillation frequencies, may be acceptable if one understands quantitatively their origin and consequences. On the other hand, it is desirable that the codes retain certain physical properties. For instance, while many successful codes are not exactly time reversible, experience has shown often that unacceptable types of errors are avoided when one builds exact *reversibility* into the difference equations (*Buneman, 1967*).

It seems more informative to work with the deflections from zero-order orbits caused by the fields, rather than to seek a finite-difference analog to the Vlasov equation as done in *Lindman* (1970) and *Godfrey* (1974). Relevant features, such as the limitations of linearized analysis, are seen directly. With no complication, the theory applies to time-integration schemes which correspond to third-, or higher-, order differential equations of motion (so that the dimensionality of phase space necessary to describe the state of the system is $9N$ or higher instead of $6N$, for N particles in three dimensions), and to algorithms which are not measure-preserving, *i.e.*, particle motion does not preserve phase-space volume (see Section 9-8, especially Problems 9-8b and 9-8c). We retain the physical content of the Vlasov approximation but perform its bookkeeping function by another means.

In the Vlasov limit the particles travel along straight lines, not appreciably affected by collisions. We compute the deflection of a particle from this straight path due to an electric field of the form

$$\mathbf{E}^{(1)}(\mathbf{x}, t) = \mathbf{E}e^{i\mathbf{k}\cdot\mathbf{x} - i\omega t} \tag{1}$$

with Im $\omega > 0$, starting from $t = -\infty$ when the plasma was undisturbed. The extension to multiple species and to Im $\omega \leqslant 0$ becomes evident later.

The particle difference equations are, from Section 2-4:

$$
\begin{aligned}
\mathbf{v}_{n+\frac{1}{2}} - \mathbf{v}_{n-\frac{1}{2}} &= \mathbf{a}_n \Delta t \\
\mathbf{x}_{n+1} - \mathbf{x}_n &= \mathbf{v}_{n+\frac{1}{2}} \Delta t
\end{aligned}
\tag{2}
$$

It is clear that this algorithm is measure-preserving because the velocity advance and position advance are each only a shear in phase space. Splitting $\mathbf{x}$ and $\mathbf{v}$ into unperturbed and perturbed parts, *i.e.*, $\mathbf{x} = \mathbf{x}^{(0)} + \mathbf{x}^{(1)}$, eliminating $\mathbf{v}$ and linearizing, we have

$$\mathbf{x}_{n+1}^{(1)} - 2\mathbf{x}_n^{(1)} + \mathbf{x}_{n-1}^{(1)} = \frac{q}{m}\Delta t^2 \, \mathbf{E}e^{i\mathbf{k}\cdot\mathbf{x}_n^{(0)} - i\omega t_n} \tag{3}$$

where $t_n \equiv n\Delta t$. On the left-hand side, $\mathbf{x}^{(0)}$ drops out. On the right-hand side, the field is evaluated at the unperturbed orbit position $\mathbf{x}_n^{(0)} = \mathbf{x}_0^{(0)} + \mathbf{v}^{(0)}t_n$; thus the linearization condition is

$$\mathbf{k}\cdot\mathbf{x}^{(1)} \ll 1 \tag{4}$$

This condition reappears later in the calculation and *replaces* the more stringent condition

$$\mathbf{E}^{(1)}\cdot\frac{\partial f^{(1)}}{\partial\mathbf{v}} \ll \mathbf{E}^{(1)}\cdot\frac{\partial f^{(0)}}{\partial\mathbf{v}} \tag{5}$$

usually quoted (see *Jackson, 1960*), which is a *sufficient* condition for linearization of the Vlasov equation. Condition (4) suffices for the linearized calculation of the charge density perturbation $\rho^{(1)}$; (5) applies to calculation of the phase-space density $f^{(1)}(\mathbf{x}, \mathbf{v})$. The success of linear theory at amplitudes for which (5) is violated may be understood via this Lagrangian analysis. For example, growth rates for instability of counterstreaming cold beams are accurately predicted when (4) is satisfied, yet (5) is violated at *any* amplitude for cold beams.

The right-hand side of (3) varies in time as

$$e^{-i(\omega - \mathbf{k}\cdot\mathbf{v}^{(0)})t_n} \equiv e^{-i\omega_d t_n} \tag{6}$$

(defining ω_d); $\mathbf{x}_n^{(1)}$ also varies in this way. No linearly varying terms appear in $\mathbf{x}^{(1)}$ since $\mathbf{x}^{(1)} \to 0$ as $t_n \to -\infty$. Substituting this dependence into the left-hand side of (3), we find the solution

$$\mathbf{x}^{(1)}(\mathbf{x}_n^{(0)}, \mathbf{v}^{(0)}, t_n) = \frac{\Delta t^2}{e^{-i\omega_d\Delta t} - 2 + e^{i\omega_d\Delta t}}\frac{q}{m}\mathbf{E}e^{i(\mathbf{k}\cdot\mathbf{x}_n^{(0)} - i\omega t_n)}$$

$$= -\left[\frac{2}{\Delta t}\sin(\omega - \mathbf{k}\cdot\mathbf{v}^{(0)})\frac{\Delta t}{2}\right]^{-2}\frac{q}{m}\mathbf{E}^{(1)} \tag{7}$$

We have explicitly recognized the dependence on $\mathbf{x}^{(0)}$, $\mathbf{v}^{(0)}$, and t.

From this orbit deflection, we must calculate the resulting charge density perturbation. This density could be imagined as resulting from displacement of particles from their unperturbed positions $\mathbf{x}^{(0)}$ by amounts $\mathbf{x}^{(1)}$ and regarding the result as a superposition of monopoles q at $\mathbf{x}^{(0)}$ and dipoles consisting of $+q$ at $\mathbf{x}^{(0)} + \mathbf{x}^{(1)}$ and $-q$ at $\mathbf{x}^{(0)}$, *i.e.*, dipole moment $q\mathbf{x}^{(1)}$ located at $\mathbf{x}^{(0)}$. The monopole density is canceled by the neutralizing background. The dipole density $\mathbf{P}$ is obtained in the Vlasov approximation by an average of $q\mathbf{x}^{(1)}$ weighted by the velocity distribution:

$$\mathbf{P}(\mathbf{x}, t) = n_0 q \int d\mathbf{v} \, f_0(\mathbf{v})\mathbf{x}^{(1)}(\mathbf{x}, \mathbf{v}, t) \tag{8}$$

in which n_0 and f_0 are the particle density and velocity distribution in the absence of the perturbation. To obtain the change in charge density $\rho^{(1)}$, consider an arbitrary volume V and imagine moving particles from positions $\mathbf{x}^{(0)}$ to $\mathbf{x}^{(0)} + \mathbf{x}^{(1)}$; the change δQ in enclosed charge is due to particles which cross the bounding surface S:

$$\delta Q = -\int_S \mathbf{ds} \cdot \mathbf{P}$$

$$= -\int_V d\mathbf{x} \, \nabla \cdot \mathbf{P} \tag{9}$$

Thus the charge density is

$$\rho^{(1)} = -\nabla \cdot \mathbf{P} = -i\mathbf{k} \cdot \mathbf{P} \tag{10}$$

which varies as in (1). This result relies again on $\mathbf{k} \cdot \mathbf{x}^{(1)} \ll 1$ and also on $n_0 \gg k^3$.

Using (7) and (8), we express the polarization in terms of the *susceptibility* χ as

$$\mathbf{P} = \chi \mathbf{E} \tag{11}$$

Taken with (10) and Gauss's law, one finds the dispersion relation

$$\epsilon(\mathbf{k}, \omega) \equiv 1 + \chi(\mathbf{k}, \omega) = 0 \tag{12}$$

where the dispersion function ϵ is

$$\epsilon(\mathbf{k}, \omega) \equiv 1 - \omega_p^2 \int \frac{d\mathbf{v} \, f_0(\mathbf{v})}{[(2/\Delta t)\sin \frac{1}{2}(\omega + i0 - \mathbf{k} \cdot \mathbf{v})\Delta t]^2} \tag{13a}$$

$$= 1 + \frac{\omega_p^2}{k^2} \int d\mathbf{v} \, \mathbf{k} \cdot \frac{\partial f_0}{\partial \mathbf{v}} \frac{\Delta t}{2} \cot(\omega + i0 - \mathbf{k} \cdot \mathbf{v}) \frac{\Delta t}{2} \tag{13b}$$

and $\omega_p = (n_0 q^2/m)^{1/2}$ is the plasma frequency in rationalized cgs (Heaviside-Lorentz) units (*Panofsky and Phillips*, 1962, p. 461; *Jackson*, 1975, pp. 817-818). The second form is obtained by an integration by parts; in the limit $\Delta t \to 0$ it reduces to the familiar plasma result (*Jackson*, 1960). The dispersion function ϵ plays the usual role in other results such as for shielding and fluctuations in Chapter 12. In a multispecies plasma, each species contributes its χ to (12).

The term "$i0$" in (13) reminds us that it was derived assuming $\text{Im } \omega > 0$, but may be used for real ω or even damped oscillations by analytic continuation to or below the real ω axis, just as in the usual Landau damping analysis (*Jackson*, 1960). This can be shown formally by doing an initial-value problem using the z transform (*Jury*, 1964) analogously to Landau's treatment using the Laplace transform. In general, $g(\omega + i0)$ means the limit of $g(\omega + i\delta)$ as δ approaches zero through real positive values. The imaginary part of ϵ, a result needed frequently in later sections, is found in Problem 9-2b.

For a cold, drifting plasma, (13a) yields

$$\omega = \mathbf{k} \cdot \mathbf{v} \pm \frac{2}{\Delta t} \arcsin \frac{\omega_p \Delta t}{2}$$

$$= \mathbf{k} \cdot \mathbf{v} \pm \omega_p [1 + \frac{1}{24}(\omega_p \Delta t)^2 + \cdots] \tag{14}$$

This suggests that to order Δt^2 the finite-difference error might be accounted for simply by an adjustment to ω_p. This is true also for a warm plasma. Expanding the cotangent in (13b),

$$\epsilon = 1 + \frac{\omega_p^2}{k^2} \int d\mathbf{v} \, \mathbf{k} \cdot \frac{\partial f_0}{\partial \mathbf{v}} \left[\frac{1}{\omega - \mathbf{k} \cdot \mathbf{v}} - \frac{1}{3}(\omega - \mathbf{k} \cdot \mathbf{v}) \left(\frac{\Delta t}{2} \right)^2 + O(\Delta t^4) \right]$$

$$= \epsilon_0 - \frac{1}{12}(\omega_p \Delta t)^2 + O(\Delta t^4) \tag{15}$$

where ϵ_0 is the standard dispersion function for continuous time. The effect of the Δt^2 term on solutions $\omega(\mathbf{k})$ of the dispersion relation $\epsilon = 0$ is the same slight increase in frequency as in (14). The absence of terms $\propto \Delta t^3$ is due to time reversibility of the equations of motion (*Buneman*, 1967).

A useful form for ϵ is obtained by using an expansion for cotangent which is valid everywhere in the complex ω plane (*Abramowitz and Stegun*, 1964, p. 75). This form is

$$\epsilon = 1 + \frac{\omega_p^2}{k^2} \int d\mathbf{v} \, \mathbf{k} \cdot \frac{\partial f_0}{\partial \mathbf{v}} \sum_{q=-\infty}^{\infty} \frac{1}{\omega - \mathbf{k} \cdot \mathbf{v} - q\omega_g}$$

i.e.,
$$\chi(\mathbf{k}, \omega) = \sum \chi_0(\mathbf{k}, \omega - q\omega_g) \tag{16}$$

where $\chi_0 = \epsilon_0 - 1$ is the susceptibility for continuous time, and $\omega_g = 2\pi/\Delta t$ is the frequency characteristic of the time "grid." Each term in the sum is analogous to the continuum result, with ω replaced by $\omega - q\omega_g$. Thus if we can compute ϵ_0 for some f_0, then we can also compute ϵ using this series, whose convergence can be accelerated (see *Langdon*, 1979b, appendix). For example, the result for a Maxwellian with drift $\bar{\mathbf{v}}$ and thermal velocity $v_t = (T/m)^{1/2}$ is

$$\epsilon = 1 - \frac{\omega_p^2}{2k^2 v_t^2} \sum_{q=-\infty}^{\infty} Z' \left[\frac{\omega_q - \mathbf{k} \cdot \bar{\mathbf{v}}}{\sqrt{2}|k| v_t} \right] \tag{17}$$

where $\omega_q \equiv \omega - q\omega_g$ is the "aliased" frequency (*Hamming*, 1962; *Blackman and Tukey*, 1958), and Z' is the derivative of the dispersion function of *Fried and Conte* (1961). This form is easy to calculate since computer programs to evaluate Z are common.

Before discussing solutions of this dispersion relation, we digress to discuss aliasing and its implications. The aliases ω_q of ω satisfy the relation $\exp(-i\omega_q t_n) = \exp(-i\omega t_n + 2\pi i q n) = \exp(-i\omega t_n)$ for all integers n and q. Thus the aliases are different frequencies which produce identical variations in quantities defined only at times t_n on a temporal grid. This equivalence is reflected in our theory by the periodicity of all quantities as functions of ω, the periodicity being $\omega_q = 2\pi / \Delta t$. Thus the infinity of poles of the integrand of (13b) does not imply more than one resonance, nor does the periodicity of ϵ imply the existence of new modes of oscillation.

It is often helpful to replace the variable ω by $z = \exp(-i\omega\Delta t)$, the rotation per time step, thus eliminating the periodicities and multiplicities of dispersion roots. The least-damped solutions for z of (17) are shown in Figure 9-2a for the case $\omega_p \Delta t = 2\sin(0.5) = 0.95885$, a rather large time step whose exact value is chosen for later comparison with the continuous-time solutions. At $k = 0$ the solutions $z = \exp(\pm i)$ correspond to the plasma oscillations at $\omega = \pm \omega_p$. As $k\lambda_D$ is increased, the roots arc to the left, then toward the negative real axis, where they rapidly meet and move apart along the real axis. One of the roots moves left a little, then follows the other root toward the origin.

This interaction of the two plasma oscillation branches is a nonphysical aspect of the periodicity induced by the numerical methods. During and after the meeting of the roots, however, $|z|$ is well under unity, so that the modes are strongly damped and their interaction is harmless.

We now examine the preceding dispersion solutions plotted in the ω plane. We chose $\omega_p \Delta t$ so that at $k = 0$ the solutions are $\omega \Delta t = \pm 1$, plus aliases. Shown in Figure 9-2b is the branch passing through $\omega \Delta t = 1$, plus an alias of its negative which passes through $\omega \Delta t = 2\pi - 1$. One period in $\mathrm{Re}\,\omega\Delta t$ is shown. The $\mathrm{Im}\,\omega$ are all equal. To see the accuracy of the correction in (15) we also solve the continuous time (exact) dispersion relation with ω_p^{-1} chosen to be the time step in the simulation case. The solutions agree within 1 percent for $|k\lambda_D| \lesssim 1$. For larger $k\lambda_D$, $\mathrm{Re}\,\omega\Delta t$ is nearing π, aliasing effects become important, and the simple correction fails. At $k\lambda_D \approx 1.45$ the solution meets the alias of its negative (also a solution); this interaction is already described above and is of little consequence since the modes are strongly damped.

For $\omega_p \Delta t > 1.62$, one mode becomes unstable, as discussed in Section 9-4.

PROBLEMS

9-2a Show that the exact charge density may be obtained from the Jacobian

$$\rho(\mathbf{x}, t) = \rho_0 \int d\mathbf{v}_0 [\partial(\mathbf{x}(\mathbf{x}_0, \mathbf{v}_0, t)) / \partial(\mathbf{x}_0)]^{-1} f_0(\mathbf{v}_0) \tag{18}$$

Show that (10) is then obtained by linearization of the Jacobian.

9-2b Show that $\mathrm{Im}\,(\Delta t / 2)\cot \tfrac{1}{2}(\omega + i0 - \Omega)\Delta t$ can be replaced in an integral by $-\pi \sum_q \delta(\omega - \Omega - q\omega_g)$ over all q, and $\mathrm{Im}\,[(2/\Delta t)\sin \tfrac{1}{2}(\omega + i0 - \Omega)\Delta t]^{-2}$ is likewise equivalent to $+\pi \sum_q \delta'(\omega - \Omega - q\omega_g)$. *Hint:* Consider a contour for Ω lying along the real axis except that it passes under the poles at $\omega - q\omega_g$ along semicircles centered on the poles; let the radii of the semicircles approach zero.

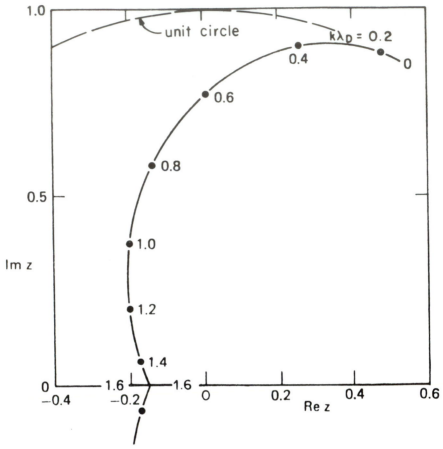

Figure 9-2a Locus of numerical solutions for $z = \exp(-i\omega\Delta t)$, varying $k\lambda_D$, for a Maxwellian velocity distribution, with arcsin $(\tfrac{1}{2}\omega_p\Delta t) = 0.5$ and neglecting spatial grid effects. At $k\lambda_D = 0$ we recover the simple harmonic oscillator solution $z = \exp(\pm i)$. As $k\lambda_D$ increases, the solutions move to the left, remaining just inside the unit circle, then toward the real z axis, meeting when $k\lambda_D = 1.45$ and then separating. (The root moving temporarily to the left along the real z axis is the one which becomes unstable for $\omega_p\Delta t > 1.62$; see Section 9-4.) For $k\lambda_D \geqslant 1.6$ both roots move toward $z = 0$ (increasingly damped) for the parameters of this example. *(From Langdon, 1979b.)*

Using either of these results in (13a) or (13b), show that

$$\text{Im } \epsilon = -\pi \frac{\omega_p^2}{k^2} \int d\mathbf{v}\, \mathbf{k} \cdot \frac{\partial f}{\partial \mathbf{v}} \sum_q \delta(\omega - q\omega_g - \mathbf{k} \cdot \mathbf{v}) \tag{19}$$

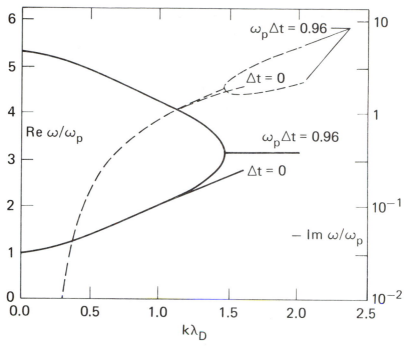

Figure 9-2b Real and imaginary parts of $\omega = (i/\Delta t)\ln z$ for the solution shown in Figure 9-2a. One period in $\mathrm{Re}\,\omega$ is shown (ω is multiple valued). The lower and upper curves for $\mathrm{Re}\,\omega$ correspond to $\omega \sim +\omega_p$ and an alias of $-\omega_p$. For comparison, solutions of the continuous-time limit are shown for corresponding parameters. The approximation of (15) is seen to be quite accurate until $\mathrm{Re}\,\omega$ approaches $\pi/\Delta t$ (see text), despite the large Δt which exceeds normal practice. *(From Langdon, 1979b.)*

9-3 ALTERNATIVE ANALYSIS BY SUMMATION OVER PARTICLE ORBITS

It is instructive to reconsider the problem by means of a summation, analogous to integration, over past accelerations, evaluated along the zero-order orbits. This approach leads to an understanding of the large $kv_t\Delta t$ limit, the "memory" of the system, and the role of phase mixing, and uncovers a point of distinction between leapfrog and other schemes which facilitate evaluation of the dispersion relation and motivates the classification in Section 9-8a.

To evaluate $\mathbf{x}^{(1)}$ in terms of $\mathbf{a}^{(1)}$, we use the *impulse response:* the deflection $\mathbf{x}_n^{(1)}$ due to impulse $\mathbf{a}_{n-s}^{(1)}\Delta t$ is just $(\mathbf{a}_{n-s}^{(1)}\Delta t)(s\Delta t)$. Summing over all past impulses,

$$\mathbf{x}_n^{(1)} = \Delta t \sum_{s=1}^{\infty} \mathbf{v}_{n-s+\frac{1}{2}}^{(1)}$$

$$= \Delta t^2 \sum_{s=1}^{\infty} s \mathbf{a}_{n-s}^{(1)} \tag{1}$$

Assuming the force field variations of 9-2 (1) we find

$$\mathbf{x}^{(1)}(\mathbf{x}^{(0)}, \mathbf{v}^{(0)}, t_n) = \frac{q}{m} \mathbf{E} \Delta t^2 e^{(i\mathbf{k} \cdot \mathbf{x}^{(0)} - i\omega t_n)} \sum_{s=1}^{\infty} s e^{i(\omega - \mathbf{k} \cdot \mathbf{v})s\Delta t}$$

in which $\mathbf{v} \equiv \mathbf{v}^{(0)}$ and the sum converges for Im $\omega > 0$. Proceeding as before to evaluate the charge density and then the dielectric function,

$$\epsilon = 1 + \omega_p^2 \Delta t^2 \sum_{s=1}^{\infty} \int d\mathbf{v} \, f_0(\mathbf{v}) \, s \, e^{i(\omega - \mathbf{k} \cdot \mathbf{v})s\Delta t} \tag{2a}$$

$$= 1 - i\omega_p^2 \Delta t \frac{\partial}{\partial \omega} \sum_{s=1}^{\infty} \int d\mathbf{v} \, f_0(\mathbf{v}) \, e^{i(\omega - \mathbf{k} \cdot \mathbf{v})s\Delta t} \tag{2b}$$

By performing the $\mathbf{v}$ integral next we Fourier transform f_0 in velocity space

$$\epsilon = 1 - i\omega_p^2 \Delta t \frac{\partial}{\partial \omega} \sum_{s=1}^{\infty} \tilde{f}_0(\mathbf{k}s\Delta t) e^{i\omega s\Delta t} \tag{3}$$

where

$$\tilde{f}_0(\tilde{\mathbf{v}}) \equiv \int d\mathbf{v} \, f_0(\mathbf{v}) e^{-i\mathbf{v} \cdot \tilde{\mathbf{v}}} \tag{4}$$

For a Maxwellian, $\tilde{f}_0(\tilde{\mathbf{v}}) = \exp(-\frac{1}{2}\tilde{v}^2 v_t^2)$ and the series converges for *any* complex ω; this expression requires no analytic continuation. As $kv_t\Delta t \to 0$ the sum becomes an integral and (3) reduces correctly to the continuum result,

$$\epsilon = 1 - i\omega_p^2 \frac{\partial}{\partial \omega} \int_0^{\infty} d\tau \tilde{f}_0(\mathbf{k}\tau) e^{i\omega\tau} \tag{5}$$

$$= 1 - i\omega_p^2 \frac{\partial}{\partial \omega} \int_0^{\infty} d\tau e^{i\omega\tau - \frac{1}{2}k^2 v_t^2 \tau^2} \tag{6}$$

for a Maxwellian; (6) can be manipulated by completing the square in the integrand to be expressed in terms of the complex error function.

If we first perform the summation (a simple geometric series) in (2b), we can recover the result of the preceding section:

$$\sum_{s=1}^{\infty} e^{i\omega_d s\Delta t} = (e^{-i\omega_d\Delta t} - 1)^{-1} \tag{7}$$

$$= -\frac{1}{2} + \frac{i}{2} \cot(\frac{1}{2}\omega_d\Delta t) \tag{8}$$

Substitution into (2b) yields 9-2(13b) (Problem 9-3b).

The rate of convergence of the series in (2b) indicates how long the force field at a given time continues to affect the charge density, *i.e.*, the memory time. For given ω, the memory is shorter (convergence faster) for large $kv_t\Delta t$. Of course, the deflections $\mathbf{x}^{(1)}$ do not decay, and may grow; only their net contribution to $\rho^{(1)}$ decreases when *averaged* over a *smooth* distribution f_0 in an oscillatory field. This process is familiar in plasma physics where it is called "phase mixing" (Problem 9-3c).

Equation (2b) is a power series in $z^{-1} = \exp(+i\omega\Delta t)$. Truncating the sum yields a polynomial in z. When enough terms are kept in the Maxwellian case, adding more terms adds z roots of small magnitude (heavily damped) without greatly changing the larger z roots. In this way we recover the infinity of roots expected in analogy to the continuum case (*Jackson*, 1960).

It is instructive to consider how 9-2(13b) and (3) change for another difference scheme. In a preview of Section 9-8a, we consider the scheme of *Feix* (1969, p. 157),

$$\mathbf{x}_{n+1} = \mathbf{x}_n + \mathbf{v}_n\Delta t + \tfrac{1}{2}\mathbf{a}_n\Delta t^2 \tag{9}$$

$$\mathbf{v}_{n+1} = \mathbf{v}_n + \mathbf{a}_n\Delta t + \tfrac{1}{2}\left[\frac{\mathbf{a}_n - \mathbf{a}_{n-1}}{\Delta t}\right]\Delta t^2 \tag{10}$$

We find (Problem 9-3d)

$$\mathbf{x}_n^{(1)} = -\tfrac{1}{2}\mathbf{a}_{n-1}^{(1)}\Delta t^2 + \Delta t^2\sum_{s=1}^{\infty} s\,\mathbf{a}_{n-s}^{(1)} \tag{11}$$

The second term is the leapfrog result. Thus the schemes differ in the response $\mathbf{x}^{(1)}$ to $\mathbf{a}^{(1)}$ at the preceding time *only*. Thus ϵ differs from 9-2(13b) and (3) only by

$$\epsilon - \epsilon_{\text{leapfrog}} = -\tfrac{1}{2}\omega_p^2\Delta t^2\tilde{f}_0(\mathbf{k}\Delta t)e^{i\omega\Delta t} \tag{12}$$

The occurrence of this sort of relation between the leapfrog and several other schemes with second-order accuracy motivates the classification of Section 9-8a.

PROBLEMS

9-3a Prove (1), then fill in the steps leading to (2a) (*Langdon*, 1979b, p. 209).

9-3b Use (8) in (2b) to recover 9-2(13b). *Hint*: the effect of $\partial/\partial\omega$ on the integrand is the same as $-k^{-2}\mathbf{k}\cdot(\partial/\partial\mathbf{v})$. Why does the term $-\tfrac{1}{2}$ in (8) drop out in the integral over velocity?

9-3c For another viewpoint on phase mixing, consider a system of noninteracting particles described by the Vlasov equation, uniform in space and having a Maxwellian distribution for $t < 0$, subjected to an impulsive force field $F_0\,\delta(t)\cos kx$. Find $f(x,v,t)$ for $t > 0$. Although perturbations in f grow in time (examine $\partial f/\partial v$), show that the density (a velocity moment) decays as $\exp(-\tfrac{1}{2}k^2v_t^2t^2)$.

9-3d Obtain (11) from (9) and (10) by considering the impulse response, or by appropriate sums of (9) and (10) over past times n, $n-1,\dots$. (Another method is to identify the powers of z^{-1} in the right side of 9-8(9) with the time levels in (11)).

9-4 NUMERICAL INSTABILITY

In the analysis of spatial grid effects in Chapter 8, an instability was found that was associated with the resonance of particles at the low phase velocities of alias waves; *i.e.*, $\omega / k_p \lesssim v_t$ for $p \neq 0$ while $\omega / k \gg v_t$. The finite time step introduces new resonances and we should consider whether these have inconvenient consequences. Considering only temporal aliasing, no analogous instabilities have been found. Assuming without loss of generality that $\omega < \pi / \Delta t$, the phase velocities of the aliases are larger in magnitude and contribute less to Im ϵ than the $q = 0$ term which contributes Landau damping. In fact, only when $kv_t\Delta t \gtrsim \pi$ can more than one q term in 9-2(16) resonate appreciably with thermal particles. Even for large values of $kv_t\Delta t$, we have found no instabilities of this sort.

Here we consider an instability which does arise in the linear theory, and an instability in a nonlinear oscillation.

First we discuss the *linear instability*. For $\omega_p\Delta t > 2$, we can see from 9-2(14) that instability occurs for a cold plasma. We now show that the Bohm-Gross dispersion reduces the instability threshold in a Maxwellian plasma. At the onset of instability, $\omega\Delta t = \pi + \mathbf{k}\cdot\bar{\mathbf{v}}$ or an alias, where Im $\epsilon = 0$. Making this substitution in 9-3(2a) the dispersion relation becomes

$$(\omega_p\Delta t)^{-2} = -\sum_{s=1}^{\infty} s\,(-1)^s\, e^{-\frac{1}{2}(kv_t\Delta t)^2 s^2} \tag{1}$$

The value of $kv_t\Delta t$ which maximizes the right-hand side shows where the plasma first becomes unstable as $\omega_p\Delta t$ is increased. We find instability at $kv_t\Delta t = 1.14$ when $\omega\Delta t > 1.62$ (Problem 9-4a).

One might expect this instability to be attributable to particles moving more than about half a wavelength per time step, thus seeing a very distorted impression of the field variation. This situation can cause errors, as discussed in Section 9-7b. However, at the onset of this instability a thermal particle travels less than one-fifth of a wavelength per time step. What occurs is not the destabilization of *one* mode; rather it is the unphysical interaction due to aliasing of the two plasma oscillation frequencies. Instability *may* occur whenever one branch of the dispersion curve closely approaches an alias of another branch, as in Figs. 9-2a and 9-2b.

As $\omega_p\Delta t$ is increased, the instability becomes stronger and occurs also at longer wavelengths, extending to $k = 0$ when $\omega_p\Delta t > 2$. (Note that $\omega_p\Delta t$ is not to be interpreted as modulo 2π). In actual simulations in this regime, the threshold values of $\omega_p t$ are affected by the field-solution algorithms used on the spatial grid.

Several examples have been tried using the one-dimensional code ES1, with these parameters: periodicity length 4π, 32 cells, $v_t = 1.14$, $\Delta t = 1$, and

4096 particles. To achieve a "quiet" nonthermal starting condition, 128 particles were loaded in the first cell distributed in velocity to approximate a
Maxwellian. These particles were replicated in the other cells. A small random velocity (rms 10^{-5}) is added to each particle; from this perturbation the
instability may grow. With $\omega_p = 1.8$, the second Fourier mode grew rapidly,
as predicted by the theory, with $\mathrm{Re}\,\omega\Delta t = \pi$. Trapping vortices form around
velocities $\pm\pi$ which resonate with the oscillation, since $(\omega - \mathbf{k}\cdot\mathbf{v})\Delta t = 0$
modulo 2π. Saturation of mode 2 occurs at time 60; by time 80 the kinetic
energy is 1.8 times the initial value and rising rapidly, with a superthermal
"tail" on the velocity distribution. As $\omega_p\Delta t$ is decreased to 1.6, saturation
levels for mode 2 and the total field energy drop rapidly, as does the increase
in kinetic energy. For $\omega_p\Delta t = 1.6$ and 1.5, *i.e.*, below threshold, mode energies saturate at values close to those of runs with noisy, random starting conditions, *i.e.*, velocities uncorrelated. The observations in this paragraph are
consistent with the predictions of the linear collisionless theory.

It is still mode 2 which reaches saturation most quickly when $\omega_p\Delta t = 1.5$.
This may be related to the theoretical result that the fluctuation spectrum is
proportional to $|\epsilon|^{-2}$, which is enhanced even for parameters below the colli-
sionless threshold. Such results are beyond the scope of this chapter, but we
confirm the empirical observation (*Hockney*, 1971) that large $\omega_p\Delta t$ and
$v_t\Delta t/\Delta x$ result in high noise levels and rapid nonphysical heating.

The predictions of linear theory are more readily confirmed for a
"square" distribution [$f_0 = (2a)^{-1}$ for $|v| < a$ and zero otherwise] because
nonlinear effects (particle trapping) begin at higher amplitudes than for the
Maxwellian case. Instability occurs for

$$(\tfrac{1}{2}\omega_p\Delta t)^2 \frac{\tan(\tfrac{1}{2}ka\Delta t)}{\tfrac{1}{2}ka\Delta t} > 1 \qquad (2)$$

We have tried two examples in ES1: $\omega_p = 1.77$ with $a = 2.33$, and $\omega_p = 1$ with
$a = 2.86$. Other parameters remain the same as for the Maxwellian cases. In
both, mode 1 was stable and mode 2 clearly showed exponential growth until
saturation by particle trapping began. Saturation occurred at a lower amplitude in the second case since the phase velocity $\omega/k = 1.1a$ is closer to the
particle velocities. It is clear that large Bohm-Gross frequency shifts, made
possible by this sharply-cutoff distribution, cause the instability by producing
large oscillation frequencies approaching $\pi/\Delta t$.

In our present view, destructive instabilities are found only for conditions so extreme that trouble should be expected even without having a formal theory.

Next, we consider a *nonlinear instability*. In a nonlinear oscillator the
acceleration is not sinusoidal but contains many harmonics of the fundamental oscillation frequency. If an alias of a harmonic is near the fundamental
frequency, then the oscillation may deviate from the correct result. For
example, we consider an oscillator of the form $d^2x/dt^2 + \omega_0^2 x = \delta T_6(x)$,
where T_6 is a Chebyshev polynomial with the property that $T_n(\cos\theta) =$

$\cos n\theta$ (*Abramowitz and Stegun*, 1964, eq. 22.3.15) so that the nonlinear term contains only the sixth harmonic when x varies sinusoidally with unit amplitude. Starting with $x_n = \cos \omega t_n$ and $\omega_0 \Delta t = 2 \sin (\pi / M)$ so that one period takes M time steps (with $\delta = 0$), we find empirically that the amplitude initially increases when $M = 7$ and *decreases* when $M = 5$, as predicted by an analysis in which one sets $x_n = A_n \cos (\omega t_n + \theta_n)$, where A_n and θ_n vary slowly. With small δ, these cases show *recurrent* behavior (limit cycles). For example, with $M = 5$, the energy decreases by about 30 percent and then returns almost to the initial value; the recurrence time is $1.36 / (\delta \omega_0)$. The energy ratio is *independent* of δ, so that even a very small nonlinearity can produce a significant error. Instability is also observed with $M = 6$. One usually associates instability with an increase in amplitude, but a decrease can be equally as damaging a divergence from correct behavior.

PROBLEMS

9-4a Find the instability boundary for (1). *Hint*: the series converges rapidly enough (say, two terms) to allow use of a pocket calculator.

9-4b Derive (2).

9-5 THE DISPERSION FUNCTION INCLUDING BOTH FINITE Δx AND Δt

The results in this chapter and in Chapter 8, 10, and 11 on the spatial grid are merged easily. For the dispersion relation $\epsilon = 0$ we find (Problem 9-5a) the result used in *Langdon* (1970b, 1979a), *Chen, Langdon, and Birdsall* (1974), and Chapter 12,

$$\epsilon = 1 + K^{-2} \sum_{\mathbf{p}} \mathbf{k_p} \cdot \boldsymbol{\kappa}(\mathbf{k_p}) S^2(\mathbf{k_p}) \chi(\mathbf{k_p}, \omega) \tag{1a}$$

$$= 1 + \frac{\omega_p^2}{K^2} \sum_{\mathbf{p}} S^2(\mathbf{k_p}) \int d\mathbf{v}\, \boldsymbol{\kappa}(\mathbf{k_p}) \cdot \frac{\partial f_0}{\partial \mathbf{v}} \frac{\Delta t}{2} \cot (\omega + i0 - \mathbf{k_p} \cdot \mathbf{v}) \frac{\Delta t}{2} \tag{1b}$$

This result serves as a dielectric function for quantities defined on the space-time grid, and, like them, is a periodic function of $\mathbf{k}$ and ω.

For the "momentum conserving" field algorithms of Chapter 8, $\boldsymbol{\kappa}(\mathbf{k_p}) = \boldsymbol{\kappa}(\mathbf{k})$ and so $\boldsymbol{\kappa}(\mathbf{k_p})$ may be removed from the sum. With $\boldsymbol{\kappa}(\mathbf{k_p}) = \mathbf{k_p}$, we find in Chapter 10 that this same expression holds for the "energy conserving" field algorithm. With suitable definitions of $\boldsymbol{\kappa}$ and S, (1a) holds for multipole algorithms as well (Chapter 11).

The alternate form for the Δt analysis gives

$$\epsilon = 1 + \omega_p^2 \Delta t^2 \sum_{\mathbf{p}} S^2 \frac{\mathbf{k_p} \cdot \boldsymbol{\kappa}}{K^2} \sum_{s=1}^{\infty} \tilde{f}_0(\mathbf{k_p} s \Delta t) s e^{i \omega s \Delta t} \tag{2}$$

One observation from this expression is that phase mixing makes the sum over **p** converge more rapidly than when time is continuous ($\Delta t = 0$). For a Maxwellian, the contribution of the terms for which $\mathbf{k_p} v_t \Delta t \geq 1$ in the sum over **p** is approximately

$$\omega_p^2 \Delta t^2 S^2 \frac{\mathbf{k_p} \cdot \boldsymbol{\kappa}}{K^2} e^{-\frac{1}{2} k_\mathbf{p}^2 v_t^2 \Delta t^2 + i\omega \Delta t} \tag{3}$$

since the $s \geq 2$ terms are much smaller. It is clear that, for $|\mathbf{p}| \geq (k_g v_t \Delta t)^{-1}$ in one dimension, the contribution is reduced by phase mixing. Thus if $v_t \Delta t \geq \Delta x$, the sum over **p** converges in only a few terms.

A more trivial remark is that a drift velocity equal to a multiple of $\Delta x / \Delta t$ is the same as no drift in infinite or periodic systems. This is not surprising in electrostatic codes, since field grid points are at the same positions relative to the plasma at every time step. Thus a limited form of Galilean invariance is restored.

More complicated examples of combined Δx and Δt analysis appear in *Chen, Langdon, and Birdsall* (1974), and Chapter 12.

There is a difference worth emphasizing between the finite spatial gridding and finite time stepping. As the particle dynamics places the particles essentially at all **x**, and interpolation within a cell is used to obtain charge density and force, spatial information exists for all **x** and for all **k**. However, in time, information is generated only at 0, Δt, $2\Delta t$, $3\Delta t$, ... with no interpolation within a step.

PROBLEMS

9-5a Derive (1a). Use 9-2(10) and 9-2(11) to relate *particle* charge density and force, $qn = -i\mathbf{k}\chi \cdot (\mathbf{F}/q)$. Use results from Section 8-9 to relate qn and **F** to *grid* quantities ρ and ϕ, as

$$\rho(\mathbf{k}, \omega) = -\sum S^2(\mathbf{k_p}) \mathbf{k_p} \cdot \boldsymbol{\kappa}\chi(\mathbf{k_p}, \omega)\phi(\mathbf{k_p}, \omega).$$

Finally, use Poisson's equation 8-9(14) and the periodicity of $\phi(\mathbf{k}, \omega)$ (but do not assume periodicity of $\boldsymbol{\kappa}$ in 8-9(15)).

9-5b In (1a), use the form of χ from 9-2(13a) and integrate by parts to obtain (1b).

9-6 MAGNETIZED WARM PLASMA DISPERSION AND NONPHYSICAL INSTABILITY

With a magnetic field, new collective modes appear. For propagation across the magnetic field at wavelengths approximating the Larmor radius, there are waves near *harmonics* of the cyclotron frequency. Both $\omega_p \Delta t$ and $\omega_c \Delta t$ may be small, but the cyclotron *harmonic* frequency may be comparable to $\pi / \Delta t$. We show how this leads to a nonphysical cyclotron harmonic instability, involving in its simplest form an interaction between $n\omega_c$, $-n\omega_c$, and ω_g when $\omega_c \approx \omega_g / 2n$.

(a) Derivation of the Dispersion Function

We now derive the collective behavior of magnetized plasma particles whose equations of motion are replaced by difference equations. As in Section 9-2 on the unmagnetized case, we ignore the effect of the spatial grid used for the fields and concentrate on the time integration. We consider small perturbations of a uniform plasma in a uniform magnetic field parallel to the z axis.

An external magnetic field can be incorporated into the particle equations in such a way that the zero-order orbits in constant fields are the exact helices plus $\mathbf{E} \times \mathbf{B}$ drift with the correct gyrofrequency ω_c. We shall use Hockney's algorithm, as in Problem 4-3b with $\mathbf{B}$ parallel to the z axis. Then for $\mathbf{x}_\perp = (x, y)$

$$\frac{\mathbf{x}_{\perp,s+1} - 2\mathbf{x}_{\perp,s} + \mathbf{x}_{\perp,s-1}}{\Delta t^2}$$

$$= \lambda \frac{q}{m} \left[\mathbf{E}_{\perp,s}(\mathbf{x}_s, t_s) + (\mathbf{x}_{\perp,s+1} - \mathbf{x}_{\perp,s-1}) \times \hat{\mathbf{z}} \frac{B}{2\Delta t} \right] \tag{1}$$

where
$$t_s = s\Delta t$$

$$\lambda = \frac{2}{\omega_c \Delta t} \tan \frac{\omega_c \Delta t}{2} \tag{2}$$

and $\omega_c \equiv qB / m$. The difference equation for z_s is the same as used in Section 9-2 for unmagnetized plasmas.

Setting $\mathbf{E} = 0$ we find the zero-order orbits

$$x_s^{(0)} = \frac{v_\perp}{\omega_c} \sin(\omega_c t_s - \psi) + x_c \tag{3}$$

$$y_s^{(0)} = \frac{v_\perp}{\omega_c} \cos(\omega_c t_s - \psi) + y_c \tag{4}$$

$$z_s^{(0)} = z_0^{(0)} + v_z t_s \tag{5}$$

in which we see that the positions lie on a helix of constant radius $v_\perp / \omega_c$ (which should be regarded as defining $v_\perp$) and rotate about the axis at x_c, y_c by an angle $-\omega_c \Delta t$ at each time step. The existence of the constants of motion $v_\perp$ and v_z is important in applications and in the following analysis. Apart from this, choosing other difference equations alters our results only in uninteresting ways (Problem 9-6e).

We consider a wave propagating in the x-z plane. The field along the zero-order orbit is

$$\mathbf{E}^{(1)}(\mathbf{x}_s^{(0)}, t_s) = \mathbf{E} \exp{(ik_x x_s^{(0)} + ik_z z_s^{(0)} - i\omega t_s)} \tag{6}$$

$$= \mathbf{E} \exp{(ik_x x_c + ik_z z_0^{(0)})} \sum_{n=-\infty}^{\infty} J_n(\chi)$$

$$\times \exp{(-in\psi - i[\omega - k_z v_z - n\omega_c]t_s)} \tag{7}$$

where $\chi \equiv k_x \times$ (actual Larmor radius) $= k_x v_\perp / \omega_c$, and we have used the Bessel function identity $\exp{(i\chi \sin\theta)} = \sum J_n(\chi)\exp{(in\theta)}$. This $\mathbf{E}^{(1)}$ is substituted into (1) written for perturbed quantities. The contribution of each term in the sum over cyclotron harmonics may be obtained separately by substituting $(\mathbf{x}_s^{(1)}, \mathbf{E}^{(1)}) = (\mathbf{X}, \mathbf{E})\exp{(-i\omega't_s)}$, where $\omega' = \omega - k_z v_z - n\omega_c$. We obtain

$$-4\mathbf{X}_\perp \sin^2{\tfrac{1}{2}\omega't} = \lambda\frac{q}{m}\mathbf{E}_\perp \Delta t^2 - 2i\sin\omega'\Delta t\tan{\tfrac{1}{2}\omega_c\Delta t}\ \mathbf{X}_\perp \times \hat{\mathbf{z}} \tag{8}$$

We solve (8) using $E_y^{(1)} = 0$ (from $k_y = 0$) and sum over harmonics using (7). With boundary conditions $\mathbf{x}_s^{(1)} = 0 = \mathbf{x}_s^{(1)} - \mathbf{x}_{s-1}^{(1)}$ at $t_s \rightarrow -\infty$, and $\text{Im}\,\omega > 0$, no solution of the homogeneous equation is to be added to get the full solution (Problem 9-6a)

$$\begin{bmatrix} x_s^{(1)}(\mathbf{x}_s^{(0)}, v_\perp, v_z, \psi) \\[1mm] z_s^{(1)}(\mathbf{x}_s^{(0)}, v_\perp, v_z, \psi) \end{bmatrix} = -\frac{q}{m}\exp{[i\mathbf{k}\cdot\mathbf{x}_s^{(0)} - i\chi\sin{(\omega_c t_s - \psi)} - i\omega t_s]}$$

$$\times \sum_n J_n(\chi)e^{in(\omega_c t_s - \psi)}\frac{\Delta t^2}{4}\begin{bmatrix} E_x\dfrac{\sin\omega_c\Delta t}{\omega_c\Delta t} \\[4mm] \dfrac{\sin^2{\tfrac{1}{2}\omega'\Delta t} - \sin^2{\tfrac{1}{2}\omega_c\Delta t}}{E_z} \\[4mm] \dfrac{E_z}{\sin^2{\tfrac{1}{2}\omega'\Delta t}} \end{bmatrix} \tag{9}$$

These are the deflections at time t_s from the zero-order orbit, as a function of the zero-order parameters. We assume that the unperturbed particles are distributed with uniform density n_0, uniformly in angle ψ, and with velocity distribution $f_0(v_\perp, v_z)$. With the time-reversible difference schemes used here, such a distribution of particles is constant in the absence of perturbing fields. Replacing $\mathbf{x}_s^{(0)}$ by $\mathbf{x}$ in (9), the perturbations in orbit, electric field and potential are related by

$$-\nabla^2\phi^{(1)} = -n_0 q \int d\psi\ v_\perp\ dv_\perp\ dv_z\ f_0(v_\perp, v_z)\ \nabla \cdot \mathbf{x}_s^{(1)}(\mathbf{x}, v_\perp, v_z, \psi)$$

$$\mathbf{E}^{(1)} = -\nabla\phi^{(1)} \tag{10}$$

Eliminating $E^{(1)}$ and $\phi^{(1)}$ and performing the integral over ψ yields the dispersion relation

$$0 = \epsilon(\mathbf{k}, \omega) = 1 - \frac{1}{k^2}\left(\frac{\omega_p \Delta t}{2}\right)^2 \int 2\pi v_\perp \, dv_\perp \, dv_z \, f_0(v_\perp, v_z)$$

$$\times \sum_n J_n^2(\chi)\left[\frac{k_x^2(\sin \omega_c \Delta t / \omega_c \Delta t)}{\sin^2 \frac{1}{2}(\omega - k_z v_z - n\omega_c)\Delta t - \sin^2 \frac{1}{2}\omega_c \Delta t}\right.$$

$$\left. + \frac{k_z^2}{\sin^2 \frac{1}{2}(\omega - k_z v_z - n\omega_c)\Delta t}\right] \qquad (11)$$

This result can be manipulated into a form which is closely analogous to the dispersion relation of *Harris* (1959) (Problem 9-6b):

$$0 = \epsilon = 1 + \frac{\omega_p^2}{k^2}\int d\mathbf{v}\sum_n J_n^2(\chi)\frac{\Delta t}{2}\cot\left[\frac{\Delta t}{2}(\omega - k_z v_z - n\omega_c)\right]$$

$$\times \left[\frac{n\omega_c}{v_\perp}\frac{\partial f_0}{\partial v_\perp} + k_z \frac{\partial f_0}{\partial v_z}\right] \qquad (12)$$

In the limit $\Delta t \to 0$ we recover the *Harris* (1959) result trivially. For a Maxwellian f_0 the integral can be done analytically (Problem 9-6c).

When $\omega - k_z v_z$ is near a harmonic $n\omega_c$, the resonance in (12) is accurate within $O(\Delta t^2)$ for *any* Δt. In constructing the difference scheme the design criteria were to make the particle respond accurately to forces at frequencies low compared to ω_c (*Hockney* 1966; *Buneman*, 1967; *Hockney and Eastwood*, 1981, Eq. 4-111; and Problem 4-3b). In a hot plasma with finite χ, even low frequency fields are felt by the particles as having frequency components near $\pm\omega_c$. Then it is surprising to find that the behavior of ϵ near the nth harmonic is very accurate even if $n\omega_c \Delta t$ is not small. The reason is that the same measures Hockney and Buneman used to give accuracy at low frequencies also happen to give accurate longitudinal response near the resonances at $\omega' = \pm\omega_c$ (see also Problems 9-6d and 9-6e).

(b) Properties of the Dispersion Relation

Let us examine in detail the simple case of perpendicular propagation, $k_z = 0$. The v_z integral may now be performed trivially.

We first note that, as in the real plasma, the frequencies ω for real k are either real or occur in conjugate pairs [ϵ is real for real ω and is analytic, therefore $\epsilon(k, \omega^*) = \epsilon^*(k, \omega) = 0$ if $\epsilon(k, \omega) = 0$].

It is helpful to change variables from ω to $z = \exp(-i\omega\Delta t)$ (Problem 9-6f):

$$\epsilon = 1 + \frac{\omega_p^2}{\omega_c^2}\int_0^\infty 2\pi v_\perp \, dv_\perp f_0(v_\perp)\frac{1}{\chi}\sum_n n(J_n^2)'\frac{i\Delta t}{1 - z^{-1}z_c^n} \qquad (13)$$

where $z_c = \exp(-i\omega_c \Delta t)$. This form is very convenient for root-finding because it is algebraic in z, in addition to the advantages found in the unmagnetized analysis (Section 9-2), such as clarifying the nature of aliasing.

In order to enumerate the roots we consider three cases: (a) $\omega_g / \omega_c = 2\pi / \omega_c \Delta t$ is an integer, (b) ω_g / ω_c is a noninteger rational number, and (c) ω_g / ω_c is irrational. In cases (a) and (b) there is a finite number, at most $N-1$, of distinct poles of $\epsilon(k,z)$, where N is the smallest positive integer such that $z_c^N = 1$. Then the poles may be removed by multiplying ϵ by $\prod_{n=1}^{N-1} (z - z_c^n)$, omitting the $n = N/2$ factor when N is even, without adding any new roots. The result is a polynomial of degree $N-1$ or $N-2$, whose roots are easy to calculate (we used a variant of Muller's method in order to avoid the awkward calculation of the polynomial coefficients and attendant loss of accuracy). Thus the finite-difference scheme introduces *no new modes,* and combines harmonic modes when their corresponding cyclotron harmonics are *aliases* of one another.

In case (c) the powers $\{z_c^n\}$ of z_c are dense on the unit circle and (13) is well-behaved only for $|z| > 1$. However, the terms in the series with $|n| \gg \chi$ contribute little to ϵ outside the unit circle, so the system is very close to a system whose ϵ is (13) with the sum truncated at n_{max} terms. The additional terms and modes, introduced by increasing n_{max} excessively, only slightly affect the roots corresponding to smaller n terms, especially if one thinks of the effect of collisions, finite k_z, slightly nonuniform magnetic field, etc., which wash out the higher harmonic resonances. Alternatively, one could say the system is similar to another system whose cyclotron frequency ω_c' is sufficiently close to ω_c and is such that ω_g / ω_c' is rational.

(c) Numerical Instability

We first consider the "cold" limit ($\chi \ll 1$) and waves perpendicular to **B**. The dispersion relation is (Problem 9-6g)

$$\sin^2 \tfrac{1}{2}\omega \Delta t = (\tfrac{1}{2}\omega_p \Delta t)^2 \frac{\sin \omega_c \Delta t}{\omega_c \Delta t} + \sin^2 \tfrac{1}{2}\omega_c \Delta t \tag{14}$$

$$= 1 - [1 - \lambda(\tfrac{1}{2}\omega_p \Delta t)^2] \cos^2 \tfrac{1}{2}\omega_c \Delta t \tag{15}$$

The condition for instability is $\lambda(\tfrac{1}{2}\omega_p \Delta t)^2 > 1$.

In a warm plasma, the aliasing of harmonics opens the possibility of nonphysical instability arising from interaction of harmonic modes which are aliases of each other. If $\Delta\omega = l\omega_c + m\omega_c + n\omega_g$ is small enough for some integers l, m and n, then the lth and the mth harmonics have artificially been brought close together. This numerical instability is a finite Larmor radius effect which cannot be seen in a single-particle analysis and does not occur in a cold plasma. Even a *Maxwellian* distribution can be unstable.

A simple case is interaction between terms $\pm n$, where $n\omega_c \approx \omega_g / 2$. If the contribution of the other terms may be neglected near this frequency, the dispersion relation may be written as

$$(\omega - \tfrac{1}{2}\omega_g)^2 = (n\omega_c - \tfrac{1}{2}\omega_g)^2 + 2(n\omega_c - \tfrac{1}{2}\omega_g)\frac{\omega_p^2}{\omega_c}\int dv \, f_0 \frac{n}{\chi}(J_n^2)' \quad (16)$$

For fixed ω_p, ω_c, k, and f_0, the most unstable choice of time step yields (Problem 9-6h)

$$\pm \mathrm{Im}\,\omega = n\omega_c - \tfrac{1}{2}\omega_g = -\omega_c \left(\frac{\omega_p}{\omega_c}\right)^2 \int dv \, f_0 \frac{n}{\chi}(J_n^2)' \quad (17)$$

with $\mathrm{Re}\,\omega = \omega_g / 2$. Thus the instability is of odd-even type. The growth rate is half the difference between the $2n^{th}$ harmonic and the sampling frequency ω_g; this is the rate at which a wheel with $2n$ spokes, rotating at frequency ω_c, appears to rotate when viewed in the light of a stroboscope flashing at intervals Δt.

The full dispersion relation has been solved numerically for the case $\omega_c \Delta t = 6\pi / 25$ (about 8 steps per period), $\omega_p^2/\omega_c^2 = 2$ with $k_{\parallel} = 0$, and a monoenergetic velocity distribution, Equation 5-16(2), a ring in $v_\perp$ space. (This is done to get a more viable harmonic wave, not to create a "negative energy" wave. Unlike harmonic mode instabilities in real plasma, the interacting modes do *not* need to have opposite "energies." The pair may have opposite signs of harmonic number, giving the same effect; only the required sign of $\Delta\omega$ is affected.) For these parameters the plasma should be stable (Section 5-16). There are harmonic resonances at every multiple of $\omega_c / 3$, 24 in all. Harmonics $4\omega_c$ and $-4\omega_c$ represent nearly the same phase change (π) per time step. The resulting odd-even instability is strongest at $\chi \approx 3.6$ (wavelength $\sim$ Larmor diameter) with $\mathrm{Im}\,\omega / \omega_c \approx 0.15$. Others are found with growth rates as large as $0.06\omega_c$ (at $\chi \approx 5.8$ from interaction of harmonics $\pm 2\omega_c$ and $\mp 6\omega_c$). The instability is absent when $\omega_c \Delta t / 2\pi = 1/8$, a small change which reduces the number of resonances to as few as possible (6 separated by ω_c). Thus the nonphysical instability is simply avoided by making the cyclotron period an *integral* number of time steps, when ω_c is a constant.

With ES1 configured as for Section 5-17, good agreement with the theory is found. With the parameters of this example, the v_x-v_y phase space plots develop eight lobes around the ring. The wave grows to amplitudes comparable to those for real cyclotron instabilities. A change of $\omega_c \Delta t$ to $6\pi / 24$ (exactly 8 steps per period) eliminates the instability, also as predicted by the theory.

This instability is weakened when the harmonic waves are weakened, *e.g.*, when the distribution function is smoothed (*e.g.*, Maxwellian, Problem 9-6i), the Larmor radii are $\leq \Delta x$ (suppressing the required large χ forces), or $\omega_c \Delta t$ is decreased (forcing any unstable interaction to be between higher harmonics for which the tuning is delicate and growth rates are small). It might be important, say, when one has warm electrons which are supposed to be stable, and one uses large $\omega_{ce} \Delta t$ because the interesting time scales are much larger than ω_{ce}^{-1}.

The instability mechanism is little affected by the choice of difference equations (Problem 9-6e), but $\omega_c \Delta t$ in (12) or (13) must be the gyrorotation actually produced by the *difference* equations.

Even if the electron cyclotron harmonic waves are not the object of study, if they exist physically then they should be taken into account in choosing Δt. As we have seen, it may not be enough that $\omega_{pe} \Delta t$ and $\omega_{ce} \Delta t$ be small.

PROBLEMS

9-6a In order to derive (9), first show that the solution of (8) can be written

$$X = -\frac{q}{m} E_x \frac{\Delta t^2}{4} \frac{(\sin \omega_c \Delta t)/(\omega_c \Delta t)}{\sin^2 \tfrac{1}{2}\omega' \Delta t - \sin^2 \tfrac{1}{2}\omega_c \Delta t}$$

Use (7) and adapt 9-2(7) to find the E_z term of (9).

9-6b Manipulate (11) into (12). *Hints*: In the perpendicular part use the trigonometric identity

$$\frac{\sin \omega_c \Delta t}{\sin^2 \tfrac{1}{2}\omega' \Delta t - \sin^2 \tfrac{1}{2}\omega_c \Delta t} = \cot \tfrac{1}{2}(\omega' - \omega_c)\Delta t - \cot \tfrac{1}{2}(\omega' + \omega_c)\Delta t \qquad (18)$$

and rearrange the sum making use of the Bessel identity

$$J_{n-1}^2 - J_{n+1}^2 = \frac{2n}{\chi}(J_n^2)' \qquad (19)$$

then integrate by parts with respect to $v_\perp$. In the parallel part integrate by parts with respect to v_z.

9-6c When f_0 is Maxwellian and $k_z = 0$ use the identity

$$\int dv\, f_0 J_n^2 = e^{-\Lambda} I_n(\Lambda)$$

where $\Lambda = (k_x v_t / \omega_c)^2$ and I_n is a modified Bessel function (*Abramowitz and Stegun*, 1964), to derive from (12) the dispersion relation

$$\epsilon = 1 - \frac{\omega_p^2}{\omega_c} \frac{e^{-\Lambda}}{\Lambda} \sum_{-\infty}^{\infty} n I_n(\Lambda) \frac{\Delta t}{2} \cot \tfrac{1}{2}(\omega - n\omega_c)\Delta t$$

9-6d Show from solution of (8) for Y that the *transverse* response, of the particle orbit to a perturbing field with frequency near ω_c, is *also* made more accurate by the Hockney-Buneman multiplier λ in the equation of motion (1).

9-6e Show that the dispersion relation holds for the difference equations used in ES1, or those of Problem 4-5b, if factors (≈ 1) are included in the k_x^2 term of (11) and the $\partial f_0 / \partial v_\perp$ term in (12).

9-6f Derive (13).

9-6g Derive (14) from (2) and (11), or set $qE_x/m = -\omega_p^2 X$ in the result of Problem 9-6a.

9-6h Derive (16) and show how to adjust ω_g to find the maximum growth rate (17).

9-6i When f_0 is a Maxwellian, show from (17) and Problem 9-6c that the maximum growth rate for the odd-even instability is given by

$$\operatorname{Im} \omega = \omega_c \left(\frac{\omega_p}{\omega_c}\right)^2 \frac{n}{\Lambda} e^{-\Lambda} I_n(\Lambda)$$

How does this compare in magnitude to the growth rates with a monoenergetic ring distribution?

9-7 SIMULATION OF SLOWLY-EVOLVING PHENOMENA; SUBCYCLING, ORBIT-AVERAGING AND IMPLICIT METHODS

In many applications one wishes to study plasma phenomena whose characteristic time-scale is much longer than ω_{pe}^{-1}, and perhaps even ω_{pi}^{-1}, while retaining a kinetic description. Examples are ion-acoustic turbulence and electromagnetic Weibel instabilities. Particle simulation as described so far in this book requires a very large number of time steps and may be too expensive. In this section we outline three approaches to the problem of efficient simulation of slowly evolving kinetic plasma phenomena, which is perhaps the most rapidly developing new reseach area in particle simulation of plasmas.

(a) Subcycling

A simple approach, which provides an appreciable saving of computer time, is *electron subcycling* (*Adam, Gourdin, and Langdon, 1982*). The standard leapfrog scheme is used for both electrons and ions, but the electron time step is a fraction of the ion time step. For each complete cycle of time integration, there is one cycle for the ions and several sub-cycles for the electrons. The cost of integrating the ions becomes negligible. The electrons are integrated on their time-scale while the slower, massive ions can be integrated with a larger time-step, using their own field plus the low-pass filtered field of the electrons. This coupling of the species can be made time-centered. Stability and design of the low-pass filter are analyzed by extension of the methods of this chapter and examined empirically in an implementation of the algorithm. Unlike the implicit codes, there is no limitation on wavelength or field gradient, and high-frequency electron waves are retained. Although the potential gain in computer time is much smaller than for the implicit codes, subcycling is more widely applicable. To obtain advantages of both, an amalgam is proposed which has much in common with implicit orbit-averaged schemes.

(b) Implicit Time Integration

Another approach is to seek an integration scheme that remains stable even when $\omega_{pe}\Delta t \gg 1$. This usually requires an *implicit* scheme, in which the calculation of the positions x_n requires knowledge of the fields at the *same* time (Problem 9-7a), rather than the preceding time, as in the *explicit* methods discussed so far. Since the fields at time t_n depend on the unknown positions $\{x_n\}$, the field and particle equations represent a very large system of coupled nonlinear equations. An approximate solution must be quite accurate if stability is to be retained.

Recently there has been experimentation with schemes for accurate prediction of the fields at the next time step (*Mason, 1981; Friedman, Langdon, and Cohen, 1981; Denavit, 1981; Brackbill and Forslund, 1982; Langdon, Cohen, and Friedman, 1983; Barnes et al., 1983*). Once found, the particles can be advanced one at a time. If their new charge density is not consistent with the predicted electric field, then convergent iteration is possible. In this way, the number of coupled equations to be solved is the order of the number of cells, rather than the number of particles. We will describe these developments in Chapters 14 and 15. Parallel work on design of implicit time differencing schemes applies the analysis of this chapter (next section, and *Cohen, Langdon, and Friedman, 1982b*).

There is a fundamental limitation in using particle electrons (*Langdon, 1979b; Langdon, Cohen, and Friedman, 1983*). For $kv_t\Delta t \gg 1$ the dispersion function becomes of the form $\epsilon \approx 1 + \beta\omega_p^2\Delta t^2$ (Problem 9-8d), which is large for $\omega_p\Delta t \gg 1$, whereas in fact what we want is $\epsilon = 1 + \omega_p^2/(kv_t)^2 + \ldots$. Thus when $kv_t\Delta t \geq 1$ we are unable to reproduce even Debye shielding correctly! To see how restrictive this condition is, consider that $k\lambda_D = kv_t\Delta t/\omega_p\Delta t$ must be much less than unity when $\omega_p\Delta t \gg 1$. A Vlasov equation model for the electrons may be more practical than a particle model when $kv_t\Delta t$ cannot be kept small.

A limitation on electric field gradient is noted by *Denavit* (1981) and elaborated upon by *Langdon, Cohen, and Friedman* (1983). At a potential energy minimum electrons oscillate at the "trapping" frequency ω_{tr}, given by $\omega_{tr}^2 = (q/m)\partial E/\partial x$ in one dimension. When $(\omega_{tr}\Delta t)^2 > 1$, the linearization used in the field prediction is unreliable.

There has been considerable experience accumulated in the implicit time integration of the equations of fluid flow, diffusion, chemical kinetics, magnetohydrodynamics, and many other fields. Now particle simulation is also beginning to profit from the use of implicit methods.

(c) Orbit Averaging

Another class of methods found to improve the efficiency of particle codes takes direct advantage of the widely separated time-scales typically present in plasma physics. In an orbit-averaged magneto-inductive algorithm (*Cohen et al., 1980; Cohen and Freis, 1982*), particles are advanced with a small time-step that accurately resolves their cyclotron orbits. An explicit solution for the electromagnetic fields, dropping radiation and electrostatics, is obtained using a current density that is accumulated from the particle data at each small step and temporally averaged over the fast, orbital time-scale. The calculations required per particle are not reduced; the real gain is the great reduction in the number of particles needed. The averaged contributions from each particle can substitute for those from many particles in a conventional code. This increased efficiency allows use of realistic parameters, such as the ratio of the ion cyclotron frequency to the rate of ion slowing due to collisions with electrons, in simulations of "magnetic mirror" experiments.

In order to apply orbit-averaging to a model including electrostatic fields and extend the simulation to large $\omega_p \Delta t$, an implicit field solution must be performed (*Cohen, Freis, and Thomas, 1982*).

PROBLEMS

9-7a In great generality, a time-integration algorithm is a linear combination of x_n, $a_n \Delta t^2$, and some earlier positions and accelerations. Show that such a scheme becomes unstable, as $\Delta t \to \infty$, when applied to a single particle simple harmonic oscillator, if the a_n term is omitted. Including a_n makes the scheme *implicit* (*Cohen, Langdon, and Friedman, 1982*).

9-8 OTHER ALGORITHMS FOR UNMAGNETIZED PLASMA

Alternatives to leap frog integration include, for example, the implicit schemes (last section), and schemes with damping. In this section we see how to derive properties of an integration scheme in a direct and reliable manner.

Generalization of the derivation of 9-2(13a) leads to

$$\epsilon = 1 + \omega_p^2 \int d\mathbf{v}\, f_0(\mathbf{v}) \left. \left(\frac{\mathbf{X}}{\mathbf{A}} \right) \right|_{\omega + i0 - \mathbf{k} \cdot \mathbf{v}} \tag{1}$$

where $\mathbf{X}/\mathbf{A}$ refers to the ratio of Fourier amplitudes of $x^{(1)}$ and $a^{(1)}$. This expression applies very generally to isotropic schemes in which $\mathbf{v}$ remains constant when $\mathbf{a} = 0$ (*i.e.*, zero-order orbits do not decay), and is easy to apply.

A different approach often seen uses a finite-difference analog of the Vlasov equation (*Lindman, 1970; Godfrey, 1974; Hockney and Eastwood, 1981*). Based on the "measure-preserving" property of the equations of motion (they map a volume of $\mathbf{x}, \mathbf{v}$ phase space onto an equal volume at another time), the Vlasov equation states that the phase space density $f(\mathbf{x}, \mathbf{v}, t)$ is constant along any particle trajectory. As applied to finite-Δt integration that takes $\{\mathbf{x}_n, \mathbf{v}_n\}$ to $\{\mathbf{x}_{n+1}, \mathbf{v}_{n+1}\}$, one uses $f(\mathbf{x}_{n+1}, \mathbf{v}_{n+1}, t_{n+1})$ $= f(\mathbf{x}_n, \mathbf{v}_n, t_n)$. When $\mathbf{v}$ is defined at half-steps, as in the leap frog scheme, one can define a velocity $\mathbf{v}_n$ (see, *e.g.*, *Hockney and Eastwood, 1981*, sec. 7-3-2). It is also possible to define f at time t_n as the density of $\{\mathbf{x}_n, \mathbf{v}_{n-\frac{1}{2}}\}$ and use $f(\mathbf{x}_{n+1}, \mathbf{v}_{n+\frac{1}{2}}, t_{n+1}) = f(\mathbf{x}_n, \mathbf{v}_{n-\frac{1}{2}}, t_n)$ (Problem 9-8a).

This method fails when applied as stated to *other* equations of motion which are not "measure-preserving" (Problem 9-8b), such as the damped schemes described later in this section. For example, it is clear in the case of damped oscillations in a potential well that measure decreases and f increases. While it is possible to repair this derivation (Problem 9-8c), we find it far simpler to use (1).

In the rest of this section, we consider two classes of time integration schemes with error of the same (or better) order as the leap-frog method. Included are *implicit* schemes which were designed and analyzed using the methods of this chapter.

(a) Class C Algorithms

We now consider a class of algorithms for which the impulse response differs from that of the leapfrog result for only a few time steps, recalling the discussion of 9-3(11), *i.e.*,

$$\mathbf{x}_n = \Delta t^2 (c_0 \mathbf{a}_n + c_1 \mathbf{a}_{n-1} + \cdots + c_{k-2} \mathbf{a}_{n-k+2} + \sum_{s=1}^{\infty} s \mathbf{a}_{n-s}) \tag{2}$$

Assuming exponential time dependence, $(\mathbf{x}_n^{(1)}, \mathbf{a}_n^{(1)}) = (\mathbf{X}, \mathbf{A}) z^n$,

$$\frac{\mathbf{X}}{\mathbf{A} \, \Delta t^2} = c_0 + \frac{c_1}{z} + \frac{c_2}{z^2} + \cdots + \frac{c_{k-2}}{z^{k-2}} + \frac{z}{(z-1)^2} \tag{3}$$

where again the last term is the same as for the leapfrog algorithm, which is therefore the simple special case in which $k = 2$ and $c_0 = 0$. The corresponding difference equation can be written in the form

$$\frac{\mathbf{x}_n - 2\mathbf{x}_{n-1} + \mathbf{x}_{n-2}}{\Delta t^2} = \mathbf{a}_{n-1} + c_0(\mathbf{a}_n - 2\mathbf{a}_{n-1} + \mathbf{a}_{n-2})$$

$$+ c_1(\mathbf{a}_{n-1} - 2\mathbf{a}_{n-2} + \mathbf{a}_{n-3}) + \cdots + c_{k-2}(\mathbf{a}_{n-k+2} - 2\mathbf{a}_{n-k+1} + \mathbf{a}_{n-k}) \tag{4}$$

It is evident that the unperturbed $(\mathbf{a} = 0)$ motion is rectilinear. The order of this equation is k (because there is a span of $k + 1$ times involved), and it is *implicit* if c_0 is nonzero.

The warm plasma dispersion function, found by substitution of (3) into (1), is

$$\epsilon = \epsilon_{\text{leapfrog}} + \omega_p^2 \Delta t^2 \sum_{s=0}^{k-2} c_s \int d\mathbf{v} \, f_0(\mathbf{v}) e^{i(\omega - \mathbf{k} \cdot \mathbf{v}) s \Delta t} \tag{5}$$

Evaluation of $\epsilon_{\text{leapfrog}}$ is discussed in Section 9-2 and *Langdon* (1979b, Appendix), and the new terms are easily expressed in terms of the velocity transform $\tilde{f}_0(\tilde{\mathbf{v}})$. Equation (5) can be used to check the collective response as modelled by implicit ($c_0 > 0$) time integration schemes (Problem 9-8d).

At long wavelengths the particles undergo simple harmonic oscillations at the plasma frequency. For an oscillator with frequency ω_0 there are two roots of (3) corresponding to frequencies near $\pm\omega_0$, and $k-2$ strongly damped roots, for small $\omega_0 \Delta t$. As shown later, the two roots near $\pm\omega_0$ have an error in their real part which is second or higher order in Δt, and an error in their imaginary part which is third or higher order in Δt.

To analyze the harmonic oscillator, we put $A = -\omega_0^2 X$ in (3). For small $\omega_0 \Delta t$ it is evident from inspection of (3) that there are $k-2$ roots near $z=0$ (strong damping), and two near $z=1$, corresponding to the oscillation. We find, in Problem 9-8e

$$\text{Re} \frac{\delta\omega}{\omega_0} = \frac{(\omega_0 \Delta t)^2}{2} \left[\frac{1}{12} - c_0 - \cdots - c_{k-2} \right] + O(\Delta t^3) \tag{6}$$

$$\text{Im} \frac{\delta\omega}{\omega_0} = -\frac{(\omega_0 \Delta t)^3}{2} \left[c_1 + 2c_2 + \cdots + (k-2)c_{k-2} \right] + O(\Delta t^4) \tag{7}$$

Thus this class has second-order error in $\text{Re}\,\omega$, but the more damaging error in $\text{Im}\,\omega$ is third order, as claimed above. For the leapfrog algorithm, we find $\delta\omega / \omega_0 = (\omega_0 \Delta t)^2 / 24$, as in 4-2(9) and 9-2(14); $\delta\omega$ is in fact real to all orders for this time-reversible, second-order scheme.

Example: (*Feix*, 1969) This scheme, 9-3(9) and 9-3(10), can be motivated by a Taylor series expansion about time t_n, with a first-difference estimate for $d\mathbf{a}/dt$. The dispersion function has already been given, 9-3(12). Setting $(\mathbf{x}_n, \mathbf{v}_n, \mathbf{a}_n) = (\mathbf{X}, \mathbf{V}, \mathbf{A})z^n$ and eliminating $\mathbf{V}$, we find

$$\frac{\mathbf{X}}{\mathbf{A} \Delta t^2} = \frac{\left[\dfrac{3}{2} - \dfrac{z}{2} \right]}{(z-1)^2} + \frac{1}{2(z-1)} \tag{8}$$

$$= -\frac{1}{2z} + \frac{z}{(z-1)^2} \tag{9}$$

Comparing with (3), (6), and (7), we find $k=3$, $c_0 = 0$, and $c_1 = -\frac{1}{2}$. Thus

$$\text{Re}\,\frac{\delta\omega}{\omega_0} = \frac{7}{24}(\omega_0\,\Delta t)^2 \tag{10}$$

$$\text{Im}\,\frac{\delta\omega}{\omega_0} = \frac{1}{4}(\omega_0\,\Delta t)^3 \tag{11}$$

The error in oscillation period is seven times that for the leapfrog method, and there is also a weak instability. [This is a hint that the measure-preserving properties of the leapfrog algorithm have been lost in this scheme, even if this system could be described in a phase space with only the coordinates $(\mathbf{x}_n, \mathbf{v}_n)$.] Another disadvantage is that a past acceleration must be retained, as well as $\mathbf{x}$ and $\mathbf{v}$. However, for the parameters used in *Feix* (1969), $\omega_p\Delta t = 0.1$ typically, the growth is weak and was probably suppressed by collisional damping. Also it may sometimes be an advantage that $\mathbf{x}$ and $\mathbf{v}$ are given at the same times.

Example: Another scheme with this advantage is

$$\mathbf{v}_{n+1} = \mathbf{v}_n + \mathbf{a}_n\Delta t + \frac{1}{2}\left[\frac{\mathbf{a}_n - \mathbf{a}_{n-1}}{\Delta t}\right]\Delta t^2 \tag{12}$$

$$\mathbf{x}_{n+1} = \mathbf{x}_n + \frac{1}{2}(\mathbf{v}_{n+1} + \mathbf{v}_n)\Delta t \tag{13}$$

The velocity is advanced in the same way as in the first Example, but the position is advanced by the trapezoidal rule. We find

$$\frac{\mathbf{X}}{\mathbf{A}\Delta t^2} = \frac{(3 - z^{-1})(z + 1)}{4(z - 1)^2} \tag{14}$$

$$= -\frac{1}{4z} + \frac{z}{(z - 1)^2} \tag{15}$$

Thus $k = 3$, $c_0 = 0$, $c_1 = -1/4$, and

$$\text{Re}\,\frac{\delta\omega}{\omega_0} = \frac{1}{6}(\omega_0\,\Delta t)^2 \tag{16}$$

$$\text{Im}\,\frac{\delta\omega}{\omega_0} = \frac{1}{8}(\omega_0\,\Delta t)^3 \tag{17}$$

This scheme is more accurate than the preceding, but less accurate than the leap-frog method.

Example: This is an example of algorithm *synthesis*, as opposed to *analysis*, in which we eliminate the $O(\Delta t^2)$ error in Re $\delta\omega$ in the leapfrog scheme. If one could use an implicit scheme, one would simply set $c_0 = 1/12$; the resulting time-centered scheme has been very successful in other applications. Instead, we remain explicit at the expense of introducing an $O(\Delta t^3)$ damping. From (6) and (7) we are lead to choose $k = 3$, $c_0 = 0$, and $c_1 = 1/12$. This gives

$$\mathrm{Re}\,\frac{\delta\omega}{\omega_0} = O(\Delta t^3) \tag{18}$$

$$\mathrm{Im}\,\frac{\delta\omega}{\omega_0} = -\frac{1}{24}O(\omega_0\Delta t)^3 \tag{19}$$

which becomes

$$\frac{\mathbf{X}}{\mathbf{A}\,\Delta t^2} = \frac{1}{12z} + \frac{z}{(z-1)^2} \tag{20}$$

and can be factored as

$$\mathbf{V}z^{-1/2}(z-1) = \mathbf{A}\Delta t \tag{21}$$

$$\mathbf{X}(z-1) = \mathbf{V}z^{1/2}\Delta t + \frac{\mathbf{A}\Delta t^2}{12z}(z-1) \tag{22}$$

in order to introduce a velocity. The difference scheme may now be written by identifying powers of z wth time levels (e. g., $\mathbf{V}z^{-1/2}$ corresponds to $\mathbf{v}_{n-1/2}$),

$$\mathbf{v}_{n+1/2} = \mathbf{v}_{n-1/2} + \mathbf{a}_n\Delta t \tag{23}$$

$$\mathbf{x}_{n+1} = \mathbf{x}_n + \mathbf{v}_{n+1/2}\Delta t + \frac{\Delta t^2}{12}(\mathbf{a}_n - \mathbf{a}_{n-1}) \tag{24}$$

This is the same as the leapfrog method except for a term $\sim(d\mathbf{a}/dt)(\Delta t^3/12)$ which corrects $\mathrm{Re}\,\omega$. The damping arises because time centering is lost in this term. This scheme seems preferable to the preceding two, if one is willing to save $\mathbf{a}$ for use in the next time step. By going to fourth order with $c_0 = 0$, $c_1 = 1/6$, and $c_2 = -1/12$, we could eliminate $\mathrm{Im}\,\delta\omega$ as well, to the accuracy of (6) and (7).

(b) Class D Algorithms

One might wonder if all schemes with the accuracy properties of C schemes fall into the form of (2) and (3). A variant is

$$\frac{\mathbf{X}}{\mathbf{A}\Delta t^2} = \frac{1}{2-z^{-1}} + \frac{z}{(z-1)^2} \tag{25}$$

which corresponds to (3) with $c_s = \frac{1}{2}(2z)^{-s}$ and $k \to \infty$. The difference $\boldsymbol{\xi}_n$ between $\mathbf{x}_n$ and the leap-frog result is given by the *recursive* filter $\boldsymbol{\xi}_n = \frac{1}{2}(\boldsymbol{\xi}_{n-1} + \mathbf{a}_n\Delta t^2)$. Its impulse response decays rather than vanishing after several steps. The example (25) is the first member of a class dubbed "implicit D schemes" by *Cohen, Langdon, and Friedman* (1982), which are of the form

$$\frac{\mathbf{X}}{\mathbf{A}\Delta t^2}z^{-1}D(z^{-1}) = \frac{z}{(z-1)^2} \tag{26}$$

where $D(z^{-1}) = d_0 + d_1/z + d_2/z^2 + \cdots$ is a polynomial in z^{-1}.

Accuracy and stability constrain the choice of coefficients $\{d_i\}$ (Problem 9-8g). Difference equations can be written in several ways, such as

$$\mathbf{x}_n - 2\mathbf{x}_{n-1} - \mathbf{x}_{n-2} = \bar{\mathbf{a}}_{n-1}\Delta t^2$$

where
$$d_0\bar{\mathbf{a}}_{n-1} = \mathbf{a}_n - d_1\bar{\mathbf{a}}_{n-2} - d_2\bar{\mathbf{a}}_{n-2} - \cdots \tag{27}$$

i.e., leap-frog integration using recursively filtered accelerations (Problem 9-8i).

PROBLEMS

9-8a Use the phase space density method to derive the dispersion function for the leap-frog scheme. *Hint:*

Write
$$f(\mathbf{x}_n, \mathbf{v}_{n-\frac{1}{2}}, t_n) = f(\mathbf{x}_{n+1}, \mathbf{v}_{n+\frac{1}{2}}, t_{n+1}) \tag{28}$$

$$= f(\mathbf{x}_n + \mathbf{v}_{n-\frac{1}{2}}\Delta t, \mathbf{v}_{n-\frac{1}{2}}, t_{n+1}) + \mathbf{a}_n \Delta t \cdot \partial f_0(\mathbf{v}_{n-\frac{1}{2}})/\partial \mathbf{v} \tag{29}$$

to linear order. Then write

$$f(\mathbf{x}_n, \mathbf{v}_{n-\frac{1}{2}}, t_n) = f_1(\mathbf{v}_{n-\frac{1}{2}}) e^{i\mathbf{k}\cdot\mathbf{x}_n - i\omega t_n} + f_0(\mathbf{v}_{n-\frac{1}{2}}) \tag{30}$$

and
$$\mathbf{a}_n = \mathbf{A} e^{i\mathbf{k}\cdot\mathbf{x}_n - i\omega t_n} \tag{31}$$

and find
$$f_1 = \mathbf{A}\Delta t \cdot \frac{\partial f_0}{\partial \mathbf{v}}(1 - e^{i(\mathbf{k}\cdot\mathbf{v} - \omega)\Delta t})^{-1} \tag{32}$$

With 9-3(8), find $\epsilon(\mathbf{k}, \omega)$ in the form of 9-2(13b).

9-8b Apply the phase space density method to the Euler scheme

$$\mathbf{x}_{n+1} = \mathbf{x}_n + \mathbf{v}_n \Delta t, \qquad \mathbf{v}_{n+1} = \mathbf{v}_n + \mathbf{a}_n(\mathbf{x}_n) \Delta t \tag{33}$$

Assuming f is constant as in Problem 9-8a, show that f_1 thus derived is the *same* as for the leap-frog method! By Fourier analyzing the equations of motion, find

$$(\mathbf{X}/\mathbf{A})_{\omega - \mathbf{k}\cdot\mathbf{v}} = (e^{i(\mathbf{k}\cdot\mathbf{v} - \omega)\Delta t} - 1)^{-2} \tag{34}$$

and substitute into (1) to obtain the correct dispersion function.

9-8c To apply correctly the phase space density method to the Euler scheme (and many others), use $J^{-1}f(\mathbf{x}_n, \mathbf{v}_n, t_n) = f(\mathbf{x}_{n+1}, \mathbf{v}_{n+1}, t_{n+1})$ where the Jacobian J, given in one dimension by

$$J = \begin{vmatrix} \dfrac{\partial x_{n+1}}{\partial x_n} & \dfrac{\partial v_{n+1}}{\partial x_n} \\ \dfrac{\partial x_{n+1}}{\partial v_n} & \dfrac{\partial v_{n+1}}{\partial v_n} \end{vmatrix} = \begin{vmatrix} 1 & \Delta t \dfrac{\partial a_n}{\partial x_n} \\ \Delta t & 1 \end{vmatrix} \tag{35}$$

expresses the change in phase space volume. For the Euler scheme, show that

$$f_1 = \left[-ikA\Delta t^2 f_0(v) + A\Delta t \frac{\partial f_0}{\partial v} \right](1 - e^{i(kv - \omega)\Delta t})^{-1} \tag{36}$$

in which the f_0 term arises because $J \neq 1$. Derive the dispersion function and show that it agrees with the final result in Problem 9-8b.

9-8d Use (5) and 9-3(2a) to show that ϵ goes to $1 + c_0\,\omega_p^2\Delta t^2$ as $kv_t\Delta t$ is increased above one.

9-8e To derive (6) and (7), set $z = \exp\left[-i(\omega_0 + \delta\omega)\Delta t\right]$ in (3) and expand, keeping terms linear in $\delta\omega$.

9-8f With $k = 2$, show that $c_0 \geqslant 1/4$ is necessary and sufficient for stability of an oscillator as $\omega_0\Delta t \to \infty$. The case $c_0 = 1/4$ corresponds to the trapezoidal rule. With $k = 3$, show using (3) that $c_0 \geqslant c_1 + 1/4$ and $c_1 \geqslant 0$ are necessary for stability as $\omega_0\Delta t \to \infty$ (in fact, this is also sufficient; *Cohen, Langdon, and Friedman*, 1982).

9-8g Derive two conditions on the coefficients $\{d_i\}$ to ensure second-order accuracy (*Cohen, Langdon, and Friedman*, 1982). From these show that $d_0 = 2$, $d_1 = -1$ for the simple scheme, D_1. Show that D_1 is equivalent to (25). What would happen if any of the roots of $D(z^{-1}) = 0$ lie outside the unit circle $|z| = 1$?

9-8h Apply the D schemes to a simple harmonic oscillator. Show that, as $\omega_0\Delta t \to 0$, the roots are $z = 1$ and the zeroes of $D(z^{-1}) = 0$. Show that all the roots approach $z = 0$ as $\omega_0\Delta t \to \infty$; what feature of the difference equations causes this?

9-8i Factor (26) into $\mathbf{X}/\overline{\mathbf{A}}\,\Delta t^2 = z/(z-1)^2$ and $z^{-1}D(z^{-1})\overline{\mathbf{A}} = \mathbf{A}$ and identify powers of z with time levels to obtain the difference equations (27) (*Cohen, Langdon, and Friedman*, 1982; *Langdon, Cohen, and Friedman*, 1983; *Barnes et al.*, 1983).

ENERGY-CONSERVING SIMULATION MODELS

10-1 INTRODUCTION

It was inevitable that a variational formulation would be developed for plasma simulation; this was done by *Lewis* (1970a, b) and has been extended by many. Improved energy conservation is an ostensible benefit. It was equally inevitable that mathematical elegance would obscure the practical properties. In this chapter, we derive the algorithms and explore their properties using the methods of Chapter 8. This chapter draws heavily on *Langdon* (1973).

We begin by showing that the momentum-conserving algorithm cannot conserve energy, then show how it can be adjusted so that it does. The algorithm derived from the variational principle follows this prescription and in addition provides the Poisson algorithm. The loss of momentum conservation and the overall accuracy of the variational procedure are discussed in the remaining sections. Although the energy-conserving algorithms have not demonstrated superiority in practice, they have many interesting properties.

ES1 has an energy-conserving option (IW = 3). Although we do not go beyond electrostatic fields, the electromagnetic case is developed by *Lewis* (1970b, 1972) and applied by *Denavit* (1974).

10-2 NONEXISTENCE OF A CONSERVED ENERGY IN MOMENTUM CONSERVING CODES

We desire some combination of grid quantities which function as a field energy. Two commonly-used candidates are

$$W_E = \frac{\Delta x}{2} \sum E_j^2 \quad \text{and} \quad W_E = \frac{\Delta x}{2} \sum \rho_j \phi_j \tag{1}$$

We quickly discover that either of these (they are normally unequal) when added to the kinetic energy does not give a constant, no matter how accurate the time integration may be. To see why the sum of energies is not exactly constant, but is often very nearly so, we express the rates of change in terms of the particle current density J and particle force F, in one dimension:

$$\frac{d}{dt} \frac{\Delta x}{2} \sum_j E_j^2 = \frac{d}{dt} \int_g \frac{dk}{2\pi} \frac{|E(k)|^2}{2} = -\int_{-\infty}^{\infty} \frac{dk}{2\pi} \frac{F(-k)}{q} J(k) \frac{k\kappa}{K^2} \tag{2}$$

$$\frac{d}{dt} \frac{\Delta x}{2} \sum_j \rho_j \phi_j = \frac{d}{dt} \int_g \frac{dk}{2\pi} \tfrac{1}{2}\rho(k)\phi^*(k) = -\int_{-\infty}^{\infty} \frac{dk}{2\pi} \frac{F(-k)}{q} J(k) \frac{k}{\kappa} \tag{3}$$

whereas

$$\frac{d}{dt} \text{KE} = \int_{-\infty}^{\infty} \frac{dk}{2\pi} \frac{F(-k)}{q} J(k) \tag{4}$$

which is $\int d\mathbf{x}\, \mathbf{E} \cdot \mathbf{J}$ for a real plasma. The integrands are equal only for $k = 0$. This *cannot be corrected* by redefining κ and K, because they must be periodic; we can define $\kappa = K = k$ only in the first zone, while the integrals are over *all* k.

For example, suppose we use the usual three-point differencing for ∇^2 and two-point for ∇, with $K = k \operatorname{dif} (\tfrac{1}{2} k\Delta x)$ and $\kappa = k \operatorname{dif} (k\Delta x)$; then, for $\sum E_j^2$ in (2), we have

$$\frac{k\kappa}{K^2} = \frac{k\Delta x/2}{\tan(k\Delta x/2)} \leqslant 1 \quad \to 0 \quad \text{as } k\Delta x \to \pi \tag{5}$$

and for $\sum \rho_j \phi_j$ in (3), we have

$$\frac{k}{\kappa} = \frac{k\Delta x}{\sin(k\Delta x)} \geqslant 1 \quad \to \infty \quad \text{as } k\Delta x \to \pi \tag{6}$$

Depending on which form is chosen, we either under- or overemphasize the electric energy at short wavelengths.

Although we see that energy is not conserved microscopically, in many momentum conserving simulations, the observed macroscopic "total energy" changes by amounts small compared to other energies of importance, *e.g.,* the field energy, which is much less than the kinetic energy in a warm plasma. When this is so, our results suggest that most of the exchange of energy between fields and particles has taken place at long wavelengths. Since this is where the model most accurately simulates the plasma, a good energy check gives credibility to the simulation.

PROBLEM

10-2a In (2) and (3), show how to go from the integrals of grid quantities (E, ρ, ϕ) to the integrals of particle quantities F and J.

10-3 AN ENERGY-CONSERVING ALGORITHM

Let us propose an algorithm which conserves the sum of particle kinetic energy plus a field energy defined on the spatial grid. We decree that the total field (or potential) energy is given by

$$W_E = \frac{V_c}{2} \sum_j \rho_j \phi_j \tag{1}$$

where V_c is the cell volume, which is valid for any interaction force, not just Coulomb's (*Reitz and Milford,* 1960, Section 6-2). In particular, (1) applies when the Coulomb force is smoothed at short range. The charge density is defined at the grid points as usual:

$$\rho_j = \sum_i q_i S(X_j - x_i) \tag{2}$$

If we obtain the force on the ith particle from

$$F_i = -\frac{\partial W_E}{\partial x_i} \tag{3}$$

and the electric potential ϕ is obtained from ρ by some procedure yet to determined, then, assuming accurate time integration, total energy is conserved trivially.

From (1) and (3),

$$F_i = -\frac{V_c}{2} \sum_j \left[\frac{\partial \rho_j}{\partial x_i} \phi_j + \rho_j \frac{\partial \phi_j}{\partial x_i} \right] \tag{4}$$

From (2), $\partial \rho_j / \partial x_i$ in the first term is $q_i \partial S(X_j - x_i) / \partial x_i$ and is easily evaluated as only a few ϕ_j's are involved. However, $\partial \phi_j / \partial x_i$ is nonzero for *all* j. Since this term therefore contains contributions from all cells, its evaluation would be far too expensive. It is helpful to rewrite (4) as

$$F_i = -V_c \sum_j \frac{\partial \rho_j}{\partial x_i} \phi_j + \frac{V_c}{2} \sum_j \left[\frac{\partial \rho_j}{\partial x_i} \phi_j - \rho_j \frac{\partial \phi_j}{\partial x_i} \right] \tag{5}$$

We show later in this section that the second sum is generally zero.

The particle force is then obtained from the first sum in (5), as

$$F_i = -q_i V_c \sum_j \phi_j \frac{\partial}{\partial x_i} S(X_j - x_i) = -q_i \frac{\partial V}{\partial x_i} \bigg|_{\phi_{j,\text{fixed}}} \tag{6}$$

The gradient of S is performed analytically and is therefore exact. *It is this step which differs crucially* from the momentum-conserving algorithms, in which the potential is differentiated numerically to obtain $\mathbf{E}$ and then $\mathbf{E}$ is interpolated to the particle.

In a code, ϕ_j is calculated from ρ_j in one step, and $\mathbf{F}_i$ is calculated in a later step using that ϕ_j, which is now *fixed*. We can define a potential field by interpolation from ϕ_j,

$$V_i(\mathbf{x}_i) = q_i V_c \sum_j \phi_j S(\mathbf{X}_j - \mathbf{x}_i) \tag{7}$$

The particle force is then obtained from the gradient of this potential field, as in the last equality in (6), in which we remember that ϕ_j is regarded as constant in differentiating V_i. The prescription given by *Lewis* (1970a) in his Eq. (50a) is identical to our (6); see Section 10-5.

Note that the same interpolation function S is used in (2) and (6). Lewis gave examples using first-order (linear) interpolation in one and two dimensions. However, in *Lewis* (1970a, b) and *Langdon* (1970b, 1973), there was no restriction to these weights. Zero-order interpolation (NGP) in which S is a discontinuous function is not suitable because the gradient in (6) does not exist.

We said the second sum in (5) vanishes; this is true if (Problem 10-3a)

$$V_c \sum_j \rho_j^{(1)} \phi_j^{(2)} = V_c \sum_j \rho_j^{(2)} \phi_j^{(1)} \tag{8}$$

where (1) and (2) refer to two different density distributions and their corresponding potentials. In the limit $\Delta x \rightarrow 0$, this statement approaches Green's reciprocation theorem of "real" electrostatics (*Jackson,* 1975, problem 1.12, p. 51). The reciprocity result holds when ϕ is the solution of a difference equation of the form

$$\rho_j = -V_c \sum_m \Delta_{jm} \phi_m \tag{9}$$

with $$\Delta_{jm} = \Delta_{mj} \tag{10}$$

(Problem 10-3b). The symmetry of Δ_{jm} usually arises naturally in the formulation of Poisson's equation in a general curvilinear coordinate system. In a neutral plasma with periodic boundary conditions and a Poisson equation which is symmetric to reflection in the lattice planes, this symmetry of Δ_{jm} is ensured. Formally, $\Delta_{j,m} = \Delta_{-j,-m} = \Delta_{m,j}$. The first equality follows from reflection; the second, from translation by the amount $\mathbf{j} + \mathbf{m}$. Δ is symmetric in Lewis' prescription for the Poisson difference equation 10-5(6) since the integral is invariant under interchange of subscripts $\mathbf{j}$ and $\mathbf{j}'$. This latter property is very much less restrictive than 10-5(6), and therefore, the energy-conserving property is shared by a much wider class of algorithms than that derived by Lewis.

In curvilinear coordinates it should also be true that $\Delta_{jm} \geqslant 0$ for $\mathbf{j} \neq \mathbf{m}$, and that

$$0 = \sum_{m} \Delta_{jm} \tag{11}$$

since $\rho_j = 0$ if ϕ is uniform in space. These conditions affect the sign of the field energy (Problem 10-3d).

Another case is when the Poisson equation can be solved by a discrete Fourier transform, as in a periodic model, or with a rectangular boundary at a fixed potential, as in *Lewis* (1970a, b); one requires only that the ratio $\phi(\mathbf{k})/\rho(\mathbf{k})$ be real, as in *Langdon* (1970a). If this ratio is positive, then the self-potential energy is nonnegative.

The conclusions are: when there is reciprocity, as per (8), the force used in an energy-conserving code is identical to the negative gradient of the total field energy. The discussion of 10-4(3) shows also that reciprocity is required.

Using spline weighting of order m in one dimension (Section 8-8), from (6), the force is

$$F_i = - q_i \Delta x \sum_{j} \phi_j \frac{\partial}{\partial x_i} S(X_j - x_i) = + q_i \Delta x \sum_{j} \phi_j S'_m (X_j - x_i) \tag{12}$$

Using the identity for the derivative of S_m,

$$S'_m(x) = \frac{1}{\Delta x} [S_{m-1}(x + \tfrac{1}{2}\Delta x) - S_{m-1}(x - \tfrac{1}{2}\Delta x)]$$

which follows from the definition of S_m as a convolution of S_{m-1} with the nearest-grid-point weighting S_0, we can write the force as

$$F_i = q_i \Delta x \sum_{j} E_{j+\frac{1}{2}} S_{m-1}(X_{j+\frac{1}{2}} - x_i) \tag{13}$$

where $X_{j\pm\frac{1}{2}} = (j \pm \frac{1}{2})\Delta x$ and (Problem 10-3e)

$$E_{j+\frac{1}{2}} = - \frac{\phi_{j+1} - \phi_j}{\Delta x} \tag{14}$$

For linear weighting, $m = 1$, the force F_i is piecewise-constant as shown in Figure 10-3a, *i.e.*, the same as for zero-order (NGP) force weighting. This means that there are jumps as a particle moves through a cell boundary, leading to enhanced noise and self-heating, just as with NGP in momentum-conserving programs. For quadratic splines, $m = 2$, the force is continuous and piecewise-linear.

PROBLEMS

10-3a Show that the second sum in (5) vanishes, given (8). *Hint*: Let $\rho^{(1)} = \rho$, $\rho^{(2)} = \delta\rho = (\partial\rho/\partial x_i) \cdot d\mathbf{x}_i$.

10-3b Prove the reciprocity result by substituting (9) into (8) and using the symmetry of Δ_{jm}.

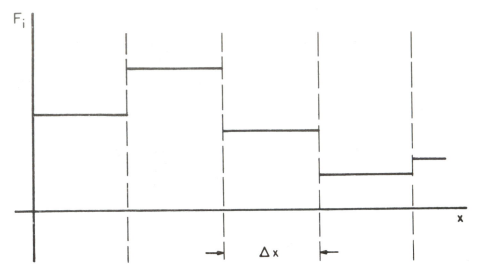

Figure 10-3a The force for an energy-conserving algorithm, using linearly weighted charge, is piecewise constant.

10-3c If the field algorithm is (14) with $E_{j+\frac{1}{2}} - E_{j-\frac{1}{2}} = \rho_j \Delta x$, show that both sides of (8) are equal to

$$\Delta x \sum_j E_{j+\frac{1}{2}}^{(1)} \, E_{j+\frac{1}{2}}^{(2)}$$

proving the reciprocity result, and also showing that the field energy is nonnegative. Generalize this to two dimensions.

10-3d Use (9), (10), and (11) to show that the field energy can be written

$$\tfrac{1}{2} V_c \sum_j \rho_j \phi_j = (\tfrac{1}{2} V_c)^2 \sum_{j \neq m} \Delta_{jm} (\phi_j - \phi_m)^2$$

and therefore that the field energy is nonnegative if $\Delta_{jm} \geq 0$ for $j \neq m$. (This result applies to the energy in most field solution methods, while Problem 10-5a is specific to Lagrangian formulations).

10-3e Provide the steps connecting (12) and (14).

10-3f Show that the undesirable jump in force (13) may be avoided by going to one-order-higher weighting in charge (to the quadratic spline, Section 8-8), which also increases the order in weighting the force (to linear).

10-4 ENERGY CONSERVATION

In this section we demonstrate the conservation of energy under rather general conditions on the field equations. Since the energy-conserving property applies exactly only when the time integration is exact for the particle equations of motions, we assume time is continuous.

The rate of change of kinetic energy is

$$\frac{d}{dt}(KE) = \frac{d}{dt}\sum_i \tfrac{1}{2}\, m_i \dot{x}_i^2 = -\sum_i \dot{\mathbf{x}}_i \cdot \frac{\partial}{\partial \mathbf{x}_i} q_i V_c \sum_j \phi_j S(\mathbf{X}_j - \mathbf{x}_i)$$

$$= -V_c \sum_j \phi_j \frac{d}{dt} \sum_i q_i S(\mathbf{X}_j - \mathbf{x}_i) \tag{1}$$

$$= -V_c \sum_j \dot{\rho}_j \phi_j$$

The electric potential is the solution of a discrete analogue to Poisson's equation, and is a linear combination of the $\{\rho_j\}$ and the boundary conditions if the latter are inhomogeneous. We therefore write the potential as

$$\phi_j = V_c \sum_m g_{j,m} \rho_m + \phi_{j,\text{ext}} \tag{2}$$

where $g_{j,m}$ is the Green's function for the difference Poisson's equation. Any fixed charge density can be either included in ρ and regarded as due to infinitely massive particles or regarded as a contributor to $\phi_{j,\text{ext}}$.

By analogy with real electrostatic field theory, we expect that the potential energy of the system due to the fields of the particles is (see *Jackson, 1975, p. 21*)

$$\tfrac{1}{2} \sum_i V_{i,\text{self}}(\mathbf{x}_i)$$

where $V_{i,\text{self}}$ is interpolated from the first term of (2), and the potential energy due to the external potential is $\sum V_{i,\text{ext}}$. Let us see when this is true. From the identity

$$\sum_i V_i(\mathbf{x}_i) \equiv V_c \sum_j \rho_j \phi_j \tag{3}$$

we find that the time rate of change of the prospective total energy is

$$\frac{d}{dt}\left(KE + \tfrac{1}{2}\sum_i V_{i,\text{self}} + \sum_i V_{i,\text{ext}}\right)$$

$$= \frac{d}{dt}\left[KE + \frac{V_c}{2}\sum_j \rho_j \phi_{j,\text{self}} + V_c \sum_j \rho_j \phi_{j,\text{self}}\right] \tag{4}$$

$$= V_c \sum_j \rho_j \dot{\phi}_{j,\text{ext}} + \tfrac{1}{2} V_c \sum_j (\rho_j \dot{\phi}_{j,\text{self}} - \dot{\rho}_j \phi_{j,\text{self}})$$

The first term on the right-hand side is the rate of change of total energy due to its explicit time dependence; it corresponds to $\partial H/\partial t$ and, therefore, its appearance is justified. To obtain an energy-conserving system, therefore, we want the second sum on the right-hand side to vanish. It does so if the potential solution satisfies the Green's reciprocation theorem 10-3(8) (just set $\rho^{(1)} = \rho$, $\rho^{(2)} = \dot{\rho}\, dt$). An alternate proof of the sufficiency of reciprocity for energy conservation was given in Section 10-3.

Reciprocity can be shown to be satisfied if ϕ is given by (2) and the Green's function $g_{j,m}$ is known to be symmetric, as may be expected when Δ_{jm} is symmetric. [There is an arbitrariness in specifying g because the total charge is zero in a periodic system, so that a transformation of the g obtained straightforwardly from a Poisson solver may be required before the symmetries are explicit (*Lewis et al.*, 1972). The relevance of symmetry is indicated by Problem 10-4b.]

In any case, it is clear that the energy-conserving property is easily obtained.

PROBLEMS

10-4a Derive (3) using 10-3(2) and 10-3(7).

10-4b Consider a single particle in an infinite system in which the Green's function is of the form $g_{i,j} \sim i - j$ (antisymmetric). Show that the particle accelerates in its own field, gaining kinetic energy while the field energy is constant (zero).

10-5 ALGORITHMS DERIVED VIA VARIATIONAL PRINCIPLES

The basic idea is to substitute into the exact Lagrangian an approximate representation of the fields and particles. This representation has a finite set of variables, and the usual variational principle provides equations governing these variables.

To avoid notational complexity, we retain vector coordinates instead of "generalized" coordinates and specialize to electrostatic fields. (*Lewis*, 1972, treats formally the case of generalized coordinates and the full electromagnetic field). In rationalized cgs (Heaviside-Lorentz) units (*Panofsky and Phillips*, 1962, p. 461; *Jackson*, 1975, p. 817-818), the Lagrangian is [*Goldstein*, 1950, eq. (11-73), p. 369]

$$L = \sum_i \tfrac{1}{2} m_i \dot{x}_i^2 - \sum_i q_i \phi(\mathbf{x}_i, t) + \int d\mathbf{x} \tfrac{1}{2} [\nabla \phi(\mathbf{x}, t)]^2 \tag{1}$$

We now replace ϕ by an interpolated potential

$$\Phi(\mathbf{x}, t) = V_c \sum_j \phi_j(t) S(\mathbf{X_j} - \mathbf{x}) \tag{2}$$

Applying the variational principle to a Lagrangian of the form $L(\{\mathbf{x}_i\}, \{\dot{\mathbf{x}}_i\}, \{\phi_j\})$ yields the Euler-Lagrange equations

$$0 = \frac{\partial L}{\partial \mathbf{x}_i} - \frac{d}{dt} \left(\frac{\partial L}{\partial \dot{\mathbf{x}}_i} \right) \tag{3}$$

$$0 = \frac{\partial L}{\partial \phi_j} \tag{4}$$

The second equation takes this form because $\dot{\phi}_j$ does not appear in L (this equation is the same as could be obtained using "finite-element" Galerkin methods). With the representation (2), we find

$$m_i \ddot{\mathbf{x}}_i = -q_i \frac{\partial}{\partial \mathbf{x}_i} V_c \sum_j \phi_j(t) S(\mathbf{X}_j - \mathbf{x}_i) \tag{5}$$

$$\rho_j = V_c \sum_{j'} \phi_{j'} \int d\mathbf{x} \left[\frac{\partial}{\partial \mathbf{x}} S(\mathbf{X}_j - \mathbf{x}) \right] \cdot \left[\frac{\partial}{\partial \mathbf{x}} S(\mathbf{X}_{j'} - \mathbf{x}) \right] \tag{6}$$

Equation (5) is the usual equation of motion with the same force term as in 10-3(6). It is the gradient of the interpolated potential, rather than the inter-polated first difference of the potential. The connection of this feature to the existence of a conserved energy is shown in Section 10-3. Equation (6) is the same, in our notation, as eq. (60) in *Lewis* (1970a).

In one dimension and with linear S, we recover from (6) the simplest difference approximation to the Poisson equation. However, in two or three dimensions the resulting Poisson difference equation is not familiar. We find in Section 10-8 that (6) produces the correct cold-plasma oscillation fre-quency at any wavelength! It automatically compensates for the increase in smoothing as one goes to higher-order splines. It is tempting to think that these simulation algorithms *ought* to be optimal in some sense. To answer that, one must decide what properties of the simulation are to be accurate, and then go outside the variational principle to the methods of Chapter 8 to analyze and adjust the algorithm. For example, we see in Section 10-10 that for warm plasmas, (6) does not give the most accurate oscillation frequen-cies. This is because a warm plasma responds less to short-wavelength noise than to errors at long and medium wavelengths. The variational principle cannot "know" this. In systems where such analysis is difficult or the better algorithms are difficult to implement (*e.g.*, where curvilinear coordinates are used), the variational principle may be useful.

PROBLEM

10-5a Show that the field energy is *nonnegative* when ϕ is obtained from (6). *Hint:* First show that the field energy can be written as

$$\frac{1}{2} \sum \rho_j \phi_j = \frac{1}{2} \int d\mathbf{x} (\nabla \Phi)^2$$

where Φ is the interpolated potential (2). This proof can be adapted to general coordinate systems. The nonnegative property and symmetry of the Poisson operator facilitate numerical solution of (6).

10-6 SPATIAL FOURIER TRANSFORMS OF DEPENDENT VARIABLES

As in Section 8-9, we relate the spatial Fourier transforms of the equations of the force calculation, and note differences between the momentum- and energy-conserving models.

The transform of 10-3(2) is, of course, the same as in Section 8-9, but the transform of 10-3(6) is

$$F(\mathbf{k}) = -iqS(-\mathbf{k})\kappa\phi(\mathbf{k}) \tag{1}$$

with $\kappa = \mathbf{k}$. (Compare 10-2(3) and 10-2(4)). Comparing to 8-9(7) and 8-9(15), we see that (1) is formally the same, differing only in the definition of κ. The transform of the Poisson equation 10-5(6) is

$$K^2(\mathbf{k})\phi(\mathbf{k}) = \rho(\mathbf{k}) \tag{2a}$$

which is formally the same as 8-9(14), but here K^2 is determined by the Lagrangian to be

$$K^2(\mathbf{k}) = \sum_{\mathbf{p}} k_{\mathbf{p}}^2 S^2(\mathbf{k_p}) \tag{2b}$$

In one dimension, $k_p = k - 2\pi p / \Delta x$. (Note $K^2 \geqslant 0$, so the field energy is nonnegative.)

10-7 LEWIS'S POISSON DIFFERENCE EQUATION AND THE COULOMB FIELDS

While Lewis's prescription of the form of Poisson's equation is not related to energy conservation, it does attempt to reproduce accurately the Coulomb interaction implicit in his Lagrangian and even compensates partially for errors in interpolation. To see this, we use the results of the last section to relate the transforms of the particle density and force field

$$F(\mathbf{k}) = \frac{-iq^2\mathbf{k}S(\mathbf{k})\sum_{\mathbf{p}}S(\mathbf{k_p})n(\mathbf{k_p})}{\sum_{\mathbf{p}}k_{\mathbf{p}}^2 S^2(\mathbf{k_p})} \tag{1}$$

Suppose an interpolation is used which is free of aliasing. This requires that $S(\mathbf{k}) = 0$ outside the first Brillouin zone, where the first zone is defined by $\max |k_x\Delta x, k_y\Delta y, k_z\Delta z| < \pi$ in a rectangular lattice. This is called *band-limited interpolation*. It is not necessary that $S(\mathbf{k})$ be constant ($=1$) within the first zone. In this case only the $p=0$ terms contribute, leaving

$$\mathbf{F}(\mathbf{k}) = \begin{cases} -iq^2 n(\mathbf{k})\mathbf{k}/k^2 & \text{in the first zone} \\ 0 & \text{elsewhere} \end{cases} \tag{2}$$

Thus the long wavelength fields are exactly Coulomb in the alias-free limit. If $S(k)$ is not constant in the first zone, so that there are errors in the interpolation, Lewis' Poisson algorithm makes compensating errors in calculating ϕ to yield good overall accuracy. This would be important to the practical realization of high-accuracy algorithms, since if $S(\mathbf{k})$ is constant in the first zone, then $S(\mathbf{x})$ drops off very slowly with increasing x and also does not remain positive. However, band-limited interpolation is not very practical in plasma simulation, and if it were, the momentum-conserving algorithm would also conserve energy and could be made as accurate.

E. L. Lindman (private communication) observed that small oscillations of a cold, nondrifting plasma in a linear-weighting Lewis model occur at exactly the correct frequency (except for time-integration errors). This interesting observation is true for any weighting function S, and in one, two, or three dimensions, as shown in Section 10-8. Here is an instance in which the variational principle does as well as can be done. However, this turns out to be an exceptional case, as seen in Section 10-10.

As a measure of accuracy in realistic cases, we examine the "averaged force" $\mathbf{F}_0(\mathbf{k})$, defined in Section 8-3. Imagine holding the particles fixed while displacing (not rotating) the grid. Then $\mathbf{F}_0(\mathbf{x})$ is the average of $\mathbf{F}(\mathbf{x})$ over all such displacements. One can show that $\mathbf{F}_0(\mathbf{k})$ is obtained from (1) by keeping only the $p=0$ term in the numerator,

$$\mathbf{F}_0(\mathbf{k}) = \frac{-iq^2 n(\mathbf{k})\,\mathbf{k}}{\sum_{\mathbf{p}} k_{\mathbf{p}}^2 S^2(\mathbf{k_p})} \tag{3}$$

The applicability of $\mathbf{F}_0$ is discussed in Sections 8-13 and 10-10. We make use of it in Sections 10-11 and 10-12 in discussing two examples.

10-8 SMALL-AMPLITUDE OSCILLATIONS OF A COLD PLASMA

In this section we show that small-amplitude oscillations of a cold plasma with no drift velocity have the correct frequency and spatial properties using a Lagrangian algorithm. This is done both with and without use of Fourier transforms.

The linear response of a cold plasma is

$$n(\mathbf{k},\omega) = \frac{in_0\mathbf{k}\cdot\mathbf{F}(\mathbf{k},\omega)}{m\omega^2} \tag{1}$$

We multiply this by $S(\mathbf{k})$, replace $\mathbf{F}$ using 10-7(1), replace $\mathbf{k}$ by $\mathbf{k_{p'}}$, sum over $\mathbf{p}'$ making use of the periodicity of the sums in 10-7(1), then cancel the sums $\sum Sn$ and $\sum k_p^2 S^2$. We are left with simply

$$\omega^2 = \frac{n_0 q^2}{m} \equiv \omega_p^2 \tag{2}$$

independent of k. This is the correct result, having no error due to finite Δx.

Although this derivation is very short [once 10-7(1) has been derived] it is instructive to repeat the derivation *ab initio* without using Fourier transforms. The meaning of linearization and the fluid limit is clarified, as is the nature of the oscillations. For brevity the discussion is kept to one dimension; the generalization is trivial.

We assume that the unperturbed particle positions x_{i0} are equally spaced, and there is an integer number of particles per cell. They are neutralized by a fixed background. After perturbing the particle positions by x_{i1},

$$\rho_j = q \sum_i x_{i1} \frac{\partial}{\partial x_{i0}} S(X_j - x_{i0}) \tag{3}$$

This Taylor expansion of 10-3(2) is where linearization first enters. We now differentiate this twice in time. The acceleration $\ddot{x}_{i1}$ is given by 10-3(6), but evaluated at x_{i0} (linearization again). Rearranging the sum, we have

$$\ddot{\rho}_j = -\frac{q^2}{m} \Delta x \sum_{j'} \phi_{j'} \sum_i \left[\frac{\partial}{\partial x_{i0}} S(X_{j'} - x_{i0}) \right] \left[\frac{\partial}{\partial x_{i0}} S(X_j - x_{i0}) \right] \tag{4}$$

Assuming that the number of particles per cell, $n_0 \Delta x$, is large, the particle sum may be replaced by an integral. Then, comparing with 10-5(6), we have simply

$$\ddot{\rho}_j = -\frac{n_0 q^2}{m} \rho_j \tag{5}$$

Each ρ_j oscillates at the plasma frequency and independently of the others. Alternatively the ϕ_j can oscillate independently. This can be understood by working through the linear case with only one ρ_j or ϕ_j oscillating. The oscillations of particles in the same cell are not independent.

If the kinetic energy is evaluated as in ES1 (Section 3-11) in a linear-weighting model, placing the particles so that grid points fall between them, and keeping oscillation amplitudes low enough that particles do not cross grid points, then total energy is conserved to within roundoff error (Problem 4-10e). However, this is a very special situation!

The derivation breaks down if the particles have any drift motion. In this case the dispersion relation 10-10(1) may be used. Some features of this case are discussed in Section 10-9 and in *Langdon* (1970b).

10-9 LACK OF MOMENTUM CONSERVATION

We have implied that these models do not conserve momentum. We now show how this failure is associated with aliasing. Both are manifestations of the nonuniformity of the system dynamics. When the Lagrangian is not invariant under displacement, momentum is not conserved in general. Consider the total force on the system of particles when band-limited interpolation is used:

$$\int d\mathbf{x}\, n(\mathbf{x})\, F(\mathbf{x}) = \int \frac{d\mathbf{k}}{(2\pi)^3}\, n(\mathbf{k})\, F(-\mathbf{k})$$

$$= \int \frac{d\mathbf{k}}{(2\pi)^3}\, n(\mathbf{k})\, i\mathbf{k}S(\mathbf{k})\, q\phi(-\mathbf{k})$$

$$= \int \frac{d\mathbf{k}}{(2\pi)^3}\, i\mathbf{k}\rho(\mathbf{k})\, \phi(-\mathbf{k}) = 0 \tag{1}$$

where $\rho(\mathbf{k}) = qS(\mathbf{k})\, n(\mathbf{k})$ in the absence of aliasing, and the integrand is odd. Thus the total particle force is zero and momentum is conserved. This is essentially because the particles can no longer sense the positions of the grid points; the nonuniformity of the grid is removed from the dynamics. [It should be noted that, in the absence of aliasing, the usual models, which conserve momentum, can be made to conserve energy also. See 10-2(2) with K^2 defined as $k\kappa$, or 10-2(3) with $\kappa = k$, in the first zone.]

A simple instance of the failure of momentum conservation is the force exerted on a particle by its own field. We examine this self-force later, but let us assume for the present that we are not interested in such a force error on the microscopic level (perhaps because it averages to zero) unless there is some macroscopic manifestation. We now give two examples in which a large change may take place in the total momentum.

A dramatic failure of momentum conservation is illustrated in Figure 10-9a and Figure 10-9b. Instability is predicted and observed in a cold plasma drifting through the grid with a fixed neutralizing background; see Section 8-12 and Problem 10-9a. There need not be two or more plasma components drifting relative to each other. Clearly this instability is not physically valid; its origin is in aliasing errors. Let us divide the energy into three nonnegative parts: kinetic energy associated with the mean motion, kinetic energy of motion relative to the mean, and field energy. The sum of these can be made to remain as nearly constant as desired, by decreasing the time step. As the instability develops, the latter two contributions to the total energy both increase, so the first contribution decreases. Therefore, the mean velocity and momentum must be decreasing. The force errors produce a drag on the mean motion. [Note that the existence of the energy constant means the instability amplitude is limited by the available energy, which is not the case for such instabilities in the usual models in which the total energy has been observed to increase several fold (*Okuda,* 1970, 1972).] This description is supported by the simulation results shown in Figure 10-9a

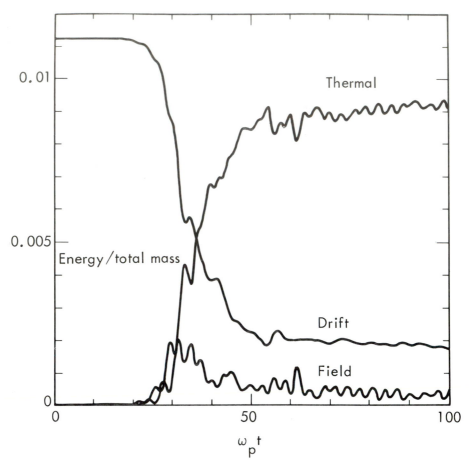

Figure 10-9a An example of macroscopic failure of momentum conservation. A cold beam passing through a fixed uniform neutralizing background is made unstable by the grid (via aliasing). Up to about $\omega_p t = 30$ the beam behaves as a linearized cold fluid. Drift kinetic energy ($\propto$ momentum^2) is converted to field energy and kinetic energy relative to the mean ("thermal") energy. Later behavior is more affected by the small number of particles used (960). The range of variation in the total energy is 0.6%; this variation is due solely to time integration errors ($\omega_p \Delta t = 0.1$). *(From Langdon, 1973.)*

and Figure 10-9b.

Collisions also produce a drag leading to decreasing momentum, as predicted for a warm, uniform, stable plasma drifting through the grid (Section 12-6). The loss of energy of mean motion is compensated for by an increase in temperature.

We do not claim to have shown that the lack of momentum conservation is necessarily damaging in practice, but only that it can have macroscopically visible consequences.

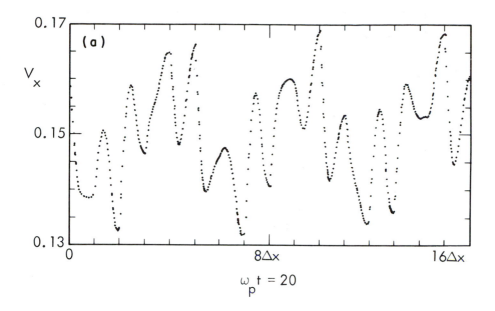

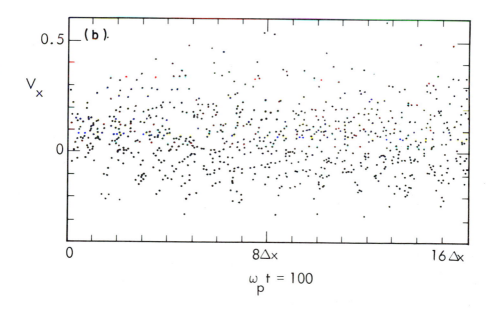

Figure 10-9b This shows phase space at times $\omega_p t = 20$ and 100. The initial drift velocity was 0.15 $\omega_p \Delta x$ and the third mode was excited. Soon mode 13 [= number of grid points (16) minus fundamental mode number] appears. The two are coupled by the grid and grow quickly together (a). [Note the expanded velocity scale in (a)]. The distribution after saturation shows little structure (b). *(From Langdon, 1973.)*

PROBLEM

10-9a Adapt the reasoning in Section 8-12 to show that a cold beam is unstable if its drift speed is *less* than about $\omega_p \Delta x / 2\pi$.

10-10 ALIASING AND THE DISPERSION RELATION FOR WARM PLASMA OSCILLATIONS

We stated that the dispersion of waves in a warm plasma is described very well by a dispersion relation based on the averaged force F_0. This has been observed in numerical solutions of the exact (including aliasing) dispersion relations (Section 8-13). In this section we show that the aliasing errors can be fourth or fifth order in Δx, in the linear-weighting case. Therefore, the dispersion errors at long wavelengths can be dominated by second-order errors in F_0 if the variational principle is used, as discussed at the end of Section 10-11.

From Section 9-5, the dispersion relation is

$$\epsilon \equiv 1 + \frac{\omega_p^2}{K^2} \sum_p S^2(\mathbf{k_p}) \int \mathbf{k_p} \cdot \frac{\partial f_0}{\partial \mathbf{v}} \frac{d\mathbf{v}}{\omega - \mathbf{k_p} \cdot \mathbf{v}} \tag{1}$$

in which the particles are treated as a linearized Vlasov plasma, and $\mathbf{x}$, $\mathbf{v}$, and t are continuous. The effects of finite grid spacing are treated exactly. For simplicity, we work in one dimension.

If one keeps only the $p=0$ term in the above, one has an approximate relation which we write as $\epsilon_0 = 0$ and which would be obtained if one started with F_0 as the interaction force. We now examine the difference between ϵ and ϵ_0, due to the aliasing terms. In the linear-weighting case, we have

$$S^2(k_p) = \left(\frac{2}{k_p \Delta x} \sin \frac{k_p \Delta x}{2} \right)^4 = \left(\frac{k \Delta x}{2\pi p} \right)^4 \left[1 + \frac{2k \Delta x}{\pi p} \right] + O(k \Delta x)^6 \tag{2}$$

when $p \neq 0$ and $k \Delta x \ll \pi$. From this alone one suspects that $\epsilon - \epsilon_0$ is fourth order in Δx. To be sure we must consider the response of the plasma to short wavelengths by evaluating the velocity integral. We use a Maxwellian with root-mean-square velocity spread v_t superimposed on a drift velocity v_0. Then

$$\epsilon = 1 - \frac{1}{2K^2 \lambda_D^2} \sum_p S^2(k_p) Z' \left(\frac{\omega - k_p v_0}{\sqrt{2}|k_p| v_t} \right) \tag{3}$$

Z' is the derivative of the plasma dispersion function of *Fried and Conte* (1961). We can proceed further analytically in the interesting case $\lambda_D \gtrsim \Delta x$, $v_0 \lesssim v_t$. Then the small-argument expansion of Z' is appropriate:

$$\frac{1}{2}Z'\left(\frac{\xi}{\sqrt{2}}\right) = -i\left(\frac{\pi}{2}\right)^{1/2} e^{-\xi^2/2} - 1 + \xi^2 - \frac{1}{3}\xi^4 + \frac{1}{15}\xi^6 \cdots \qquad (4)$$

Substituting this into (3) and using small-k expansions like (2), we have, to lowest nonvanishing order in Δx,

$$\epsilon - \epsilon_0 \approx \frac{1}{(k\lambda_D)^2}\left[\frac{(k\Delta x)^4}{720} + i\,2.65 \times 10^{-4}\frac{(k\Delta x)^5}{v_t}\left(\frac{\omega}{k} + 4v_0\right)\right] \qquad (5)$$

The corresponding error in $\omega(k)$ is nearly proportional to $\epsilon - \epsilon_0$, so the errors in $\mathrm{Re}\,\omega$ and $\mathrm{Im}\,\omega$ are fourth and fifth order in Δx, respectively, due to aliasing. However, using the variational principle, the error in F_0 is $O(\Delta x)^2$. An overall error of $O(\Delta x)^4$ in $\omega(k)$ requires a different Poisson operator, as discussed in Section 10-11.

As an aside, the imaginary part of (5) shows the damping influence of the aliasing terms when $v_0 = 0$. On the other hand, with $v_0 \neq 0$ these terms can become destabilizing.

10-11 THE LINEAR-INTERPOLATION-MODEL EXAMPLE

We now examine the nature of the particle force **F** in the linear interpolation examples of *Lewis* (1970). First, we note that **F** is *discontinuous*. For given, fixed $\{\phi_j\}$ in two dimensions, F_x is continuous and piecewise linear as a function of y alone, and is a step function when x alone is varied. In one dimension, F is a step function (Figure 10-3a). Thus this case may be expected to be *noisier* than the common linear-weighting algorithms, but the overall computational time is shorter because the expression for F is much simpler. It is also difficult to integrate in time accurately enough to realize an improvement in energy conservation. In empirical studies (*Lewis et al., 1972; Brown et al., 1974; Lewis and Nielson, 1975*), the variational algorithms have not demonstrated superiority.

(a) Momentum Conservation and Self-Forces

As mentioned above, a simple example of the failure of momentum conservation is the force exerted on a particle by its own fields. Since a particle has many neighbors, it is more significant that the *two*-particle interaction force is nonconservative. However, the single particle case is simple and of some interest. Consider a single particle in a large one-dimensional system, using linear interpolation. Place the particle between adjacent grid points located at $x = 0$ and Δx. Then the self-force is

$$F = -q\frac{\phi_1 - \phi_0}{\Delta x} = \frac{1}{2}q\Delta x(\rho_0 - \rho_1) \qquad (1)$$

$$= -q^2 \left[\frac{x}{\Delta x} - \frac{1}{2} \right] \qquad 0 \leqslant x \leqslant \Delta x$$

This is a simple harmonic oscillator potential well. Let us attempt to assess its importance in a single-species plasma. It yields an oscillation frequency $\omega_{\text{self}} = \omega_p / \sqrt{n\Delta x}$, much smaller than the plasma frequency ω_p if the number of particles per cell $n\Delta x$ is large. The well-depth energy W_{self} may be compared to the thermal energy:

$$\frac{W_{\text{self}}}{(\frac{1}{2} mv_t^2)} = \frac{1}{4} \frac{\left(\frac{\Delta x}{\lambda_D} \right)^2}{n\Delta x} = \frac{1}{4} \frac{\left(\frac{\Delta x}{\lambda_D} \right)}{n\lambda_D} \tag{2}$$

where m is the particle mass, v_t is the rms thermal velocity, and λ_D is the Debye length. Similar results are anticipated in two and three dimensions (Problem 10-11a).

Note that all these ratios are desirably small when $n\Delta x$ is large, given $\Delta x / \lambda_D$. Furthermore, since other particles in the cell contribute forces comparable to a particle's self-force, the latter becomes relatively small compared to normal many-body interactions.

We have so far ignored time-integration errors, in effect, assuming the time step Δt is kept negligibly small. If Δt is held constant while Δx is decreased then, when $\omega_{\text{self}} \Delta t \geqslant 2$, the self-force oscillation becomes unstable. This requires $n\Delta x < 1$. M. A. Lieberman (private communication) has shown that, even before the instability threshold, particle velocities can diffuse without limit—gross nonconservation of energy permitted by the time-integration errors. While such unphysical behavior in even the "unperturbed" particle orbits is undesirable, one should compare it to velocity diffusion due to normal collisions. To estimate the self-force diffusion rate, assume Δx is so small that the particle position within a cell is randomly and independently distributed from time step to step. The mean-square self-force in one dimension is $<F^2> = q^4 / 12$, leading to a diffusion given by

$$<\Delta v^2> = \frac{<F^2 \Delta t^2>}{m^2} \frac{t}{\Delta t} \tag{3}$$

Defining a diffusion time τ_s by $<\Delta v^2> = v_t^2$, we obtain

$$\omega_p \tau_s = 12 \frac{(n\lambda_D)^2}{\omega_p \Delta t} \tag{4}$$

whereas the normal collision times for a one dimensional plasma are $\omega_p \tau_c (n\lambda_D)$ or $(n\lambda_D)^2$. There seems to be no difficulty in making $\tau_s \gg \tau_c$, as desired. A similar argument may be made in two and three dimensions. Thus the time-integration errors appear to increase the significance of the self-force, in this limit, but not disastrously.

One might try to restore momentum conservation by adding a new force to each particle which cancels its self-force. However, consider the case in which the particles are evenly spaced at integral submultiples of Δx apart in a periodic system. The unmodified algorithm correctly calculates no forces.

To see that no such approach succeeds, note that in the Vlasov limit, the self-acceleration vanishes and the system is the same as without the self-force cancellation. The discussion of Section 10-9 shows that momentum is not conserved in the Vlasov limit. Nor does any smoothing of ρ or ϕ restore momentum conservation. The point is that the lack of conservation of momentum is *not primarily a question of the single particle self-force.*

(b) Macroscopic Field Accuracy

If a plasma phenomenon is not affected by displacing the grid relative to it, then it would not be affected by replacing the interaction force by the averaged force F_0, defined in Section 10-7. As an important example, we now consider oscillations of a warm plasma, using the accuracy of the period and rate of decay (or growth) as a measure of the accuracy of the fields.

In Section 10-10, we showed that the contribution of the $p \neq 0$ (aliasing) terms in the dispersion function $\epsilon(k, \omega)$ is fourth, or higher, order in Δx. Therefore, a lower-order error in F_0 reduces the order of overall accuracy. For linear weighting we find, using 10-5(6)

$$\frac{K^2}{k^2} = S^2(k) + \frac{1}{12}(k\Delta x)^2 + O(\Delta x)^4 \tag{5}$$

Thus F_0 has a relative error, $-(k\Delta x)^2/12$, causing a similar error in the oscillation frequency ω. Further, the error in $\operatorname{Re} \omega$ can cause an error in $\operatorname{Im} \omega$ by changing the phase velocity; for a Maxwellian, this contribution to relative error in $\operatorname{Im} \omega$ is $-\Delta x^2/24\lambda_D^2$, independent of k.

This $O(\Delta x)^2$ error may be removed by changing the Poisson algorithm. One way to do this is to solve the same Poisson equation but with

$$\rho'_j = \frac{1}{12}(-\rho_{j-1} + 14\rho_j - \rho_{j+1}) \tag{6}$$

as the source density. One is then left with an $O(\Delta x^4)$ error in F_0 and from aliasing terms, resulting in a fourth-order error in $\omega(k)$.

These remarks hold also in two or three dimensions. The Poisson algorithms obtained from the variational principle are not optimal from the present point of view. Whether they are optimal in some other situation (apart from the singular case of cold-plasma oscillations) remains to be shown.

There should be no surprise that problems arise with the variational principle when the basis functions are an incomplete set. Many other such examples are known, *e.g.*, Gibb's phenomenon in least-squares fitting of trigonometric sums (yielding a truncated Fourier series), in which, as here, better algorithms can be obtained after taking into account the nature of the result one is trying to compute.

PROBLEM

10-11a Use dimensional arguments to predict that, in two and three dimensions, (1) and (2) become

$$\frac{\omega_{self}^2}{\omega_p^2} \sim \frac{1}{N_c}, \qquad \frac{W_{self}}{\frac{1}{2} mv_t^2} \sim \frac{1}{N_c} \left(\frac{\Delta x}{\lambda_D}\right)^2 \tag{7}$$

where $N_c = nV_c$ is the number of particles per cell. The frequency ω_{self} is only an approximate indication of the average force gradient, since a single-particle oscillation is no longer simple harmonic.

10-12 THE QUADRATIC SPLINE MODEL

The reduction of aliasing provided by a higher-order spline enables the variational principle to achieve better accuracy. The Poisson algorithm may be found from 10-5(6) or from 10-6(2b) and 8-8(1) for $S_m(k)$, as

$$K^2(k) = \left(\frac{2}{\Delta x} \sin \frac{1}{2} k\Delta x\right)^6 \sum_p (k - pk_g)^{-4} \tag{1}$$

$$= \left(\frac{2}{\Delta x} \sin \frac{1}{2} k\Delta x\right)^2 \frac{1}{3}(2 + \cos k\Delta x)$$

The sum is evaluated with help from *Abramowitz and Stegun* [1964; take the second derivative of their Eq. (4.3.92)]. This is all that is needed, if Fourier transform methods are to be used in the simulation. The new factor in (1) is added by the variational derivation in order to compensate for the low-pass filtering (smoothing) effect of S_2 as compared to S_1.

For small $k\Delta x$, the relative "error" in both K^2 and S_2^2 is $-(k\Delta x)^2/4 + O(k^4)$, so that the second-order errors in $F_0(k)$ cancel (leaving a fourth-order relative error), showing how the long-wavelength errors due to S_1 have been reduced.

The difference-equation coefficients are the coefficients for an expansion of (1) in powers in $\exp(ik\Delta x)$:

$$-\rho_j \Delta x^2 = \frac{1}{6}(\phi_{j+2} + \phi_{j-2}) + \frac{1}{3}(\phi_{j+1} + \phi_{j-1}) - \phi_j \tag{2}$$

This is a fourth-order difference equation, so that in a nonperiodic finite system, two more boundary conditions are needed in addition to the usual two. These emerge naturally from Eq. (60) in *Lewis* (1970a) and depend on how the interpolation is modified at the ends of the system.

The self-force in this example scales as in 10-11(1) and 10-11(2), but with smaller coefficients. In general, the deficiencies due to aliasing (the instability, momentum nonconservation, grid noise) are reduced.

Returning to the question of oscillations of a warm plasma, we find that aliasing terms make a contribution to $\epsilon - \epsilon_0$ which is $O(\Delta x^6)$, while the error in F_0 is only fourth order. Again one achieves best accuracy in the complex frequency $\omega(k)$ with a Poisson algorithm different than that specified by the variational principle.

PROBLEM

10-12a Equation (2) leads to a matrix equation in which the nonzero elements are all in a band, five elements across, down the diagonal, plus a few in the other corners. However, (1) shows how to solve two tridiagonal systems instead, each corresponding to factors of K^2. Show that the corresponding difference equations can be written as

$$\frac{1}{6}\rho'_{j-1} + \frac{2}{3}\rho'_j + \frac{1}{6}\rho'_{j+1} = \rho_j \tag{3}$$

$$\phi_{j-1} - 2\phi_j + \phi_{j+1} = -\rho'_j \Delta x^2 \tag{4}$$

Is such a factorization possible in two dimensions?

ELEVEN

MULTIPOLE MODELS

W. M. Nevins and A. B. Langdon

11-1 INTRODUCTION

In this chapter we consider an alternative approach to deriving plasma simulation algorithms, relate them to those of Chapters 8 and 14, and offer our perspective. We begin with a little history and an outline of this chapter.

The first method considered by Dawson and co-workers for two-dimensional simulation employed a truncated Fourier series expansion of the field, evaluated at each particle position. In applications requiring much spatial detail (*i.e.*, many Fourier modes), evaluation of the Fourier sums is too expensive. Instead, the field and its derivatives are evaluated at a number of spatial grid points (many fewer than the number of particles), and the field at a particle is evaluated as a truncated Taylor series expansion about the nearest grid point. They interpret this expansion in terms of multipole moments.

In order to reduce storage requirements, the "subtracted multipole" method evaluates only the field from the Fourier sum; its derivatives are evaluated using finite differences. This algorithm is easily expressed in the formulations of Chapter 8; doing so facilitates comparison with other methods.

The multipole method is derived in Section 11-2 in its original form, and in Section 11-3 in the "subtracted" form. In Section 11-4 we show how to derive the standard one-dimensional linear weighting as a dipole scheme, and derive a new two-dimensional dipole scheme. Adding the quadrupole xy moment yields the familiar bilinear or area weighting! Compared to published dipole algorithms, those derived here give smoother spatial variation for comparable resolution. Section 11-5 develops the Fourier space relationships between particle and grid quantities, adapting results from Chapter 8. These are used in Section 11-6 to examine overall accuracy.

11-2 THE MULTIPOLE EXPANSION METHOD

Let us look at the over-all multipole simulation method, then consider the accuracy and explicit forms (monopole, dipole, quadrupole). The original papers are *Kruer, Dawson, and Rosen* (1973), *Chen and Okuda* (1975), and *Okuda* (1977).

The predecessor of the multipole method is a field algorithm in which the particle is considered to be a finite-size cloud and the field is represented by a truncated Fourier series (*Dawson*, 1970, p. 16 ff). The force on particle i is therefore

$$F_i = q_i \int dx' \hat{S} (x' - x_i) \left[\frac{-i}{L} \sum_k k \phi(k) e^{ikx'} \right] \tag{1}$$

$$= \frac{-i}{L} q_i \sum_k k \hat{S}(-k) \phi(k) e^{ikx_i}$$

where L is the length of the 1d system, and the smoothing factor $\hat{S}_k$ is chosen to be $\exp(-k^2 a^2/2)$. This sum must be evaluated for each particle. The charge density for each Fourier mode is

$$\rho(k) = \int dx' e^{ikx'} \left[\sum_i q_i \hat{S} (x' - x_i) \right] \tag{2}$$

$$= \hat{S}(k) \sum_i q_i e^{-ikx_i}$$

from which $\phi(k) = \rho(k)/k^2$. This method provides smooth variation of the fields but the computational effort per particle increases as spatial resolution (and therefore the number of Fourier modes) increases.

To speed up the force calculation, a spatial grid is introduced. The particle force is evaluated by Taylor expansion of the exponential in (1):

$$F_i = -\frac{i}{L} q_i \sum_k k \hat{S}(-k) \phi(k) e^{ikX_j} [1 + ik(x_i - X_j) - \frac{1}{2}k^2(x_i - X_j)^2 \cdots] \tag{3}$$

where X_j is the location of the grid point nearest the particle x_i. This can be written as

$$F_i = F_{0,j} + (x_i - X_j) F_{1,j} + \frac{1}{2}(x_i - X_j)^2 F_{2,j} + \cdots \tag{4}$$

$$= \sum_{l=0} \frac{1}{l!} F_{l,j} (x_i - X_j)^l$$

The force derivatives are given by

$$F_{l,j} = \frac{1}{L} \sum_k F_l(k) \, e^{ikX_j} \tag{5}$$

where

$$F_l(k) = (ik)^l \, F_0(k),$$

$$F_0(k) = -ik \, q_i \, \hat{S}(-k) \, \phi(k) \tag{6}$$

The $F_{l,j}$ are evaluated from $F_l(k)$ by Fast Fourier Transforms (FFT); F_i is then evaluated for each particle.

The charge density is similarly evaluated by Taylor expanding the exponential in (2):

$$\rho(k) = \hat{S}(k) \sum_j e^{-ikX_j} \sum_{i \in j} q_i \, [1 - ik(x_i - X_j) - \tfrac{1}{2}k^2(x_i - X_j)^2 \cdots] \tag{7}$$

where the outer sum is over grid points and the inner sum is over those particles nearest grid point j and can be written as a sum over multipole moments,

$$\rho(k) = \hat{S}(k)\Delta x \sum_j e^{-ikX_j} [\rho_{0,j} - ik\rho_{1,j} - \tfrac{1}{2}k^2\rho_{2,j} \cdots] \tag{8}$$

where

$$\rho_{l,j} = \sum_{i \in j} \frac{q_i}{\Delta x} (x_i - X_j)^l \tag{9}$$

is the multipole density. We identify the first two of these quantities as the monopole ($l = 0$) density

$$\rho_{0,j} = \sum_{i \in j} \frac{q_i}{\Delta x}$$

(which is the same as the Nearest-Grid-Point density), and the dipole ($l = 1$) density

$$\rho_{1,j} = \sum_{i \in j} \frac{q_i}{\Delta x} (x_i - X_j)$$

Finally, (8) becomes

$$\rho(k) = \hat{S}(k)[\rho_0(k) - ik\rho_1(k) - \tfrac{1}{2}k^2\rho_2(k) \cdots] \tag{10}$$

in which

$$\rho_l(k) = \Delta x \sum \rho_{l,j} \, e^{-ikX_j} \tag{11}$$

is evaluated by Fast Fourier Transform.

To summarize the multipole force calculation in a code, the multipole densities are collected from the particles, (9), transformed as in (11) and combined in (10) to form $\rho(k)$, from which $\phi(k) = \rho(k)/k^2$. The $F_l(k)$ are formed and transformed to the force derivatives $F_{l,j}$, using (6) and (5). The force on each particle is evaluated as the sum (4). The whole scheme is illustrated in Figure 11-2a.

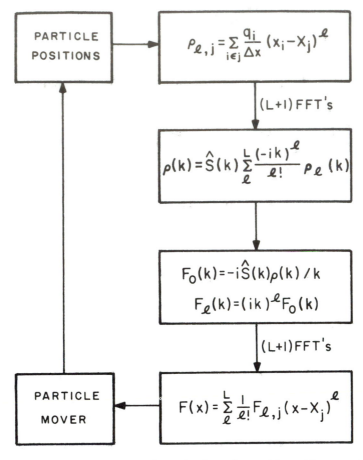

Figure 11-2a Multipole expansion method, starting from the particle positions x_i, with $(L + 1)$ Fourier transforms on the grid charge densities, $(L + 1)$ more on the force moments, through to the force at the particle. Dipole is $L = 1$, quadrupole is $L = 2$, etc.

Consider the storage requirements and the number of FFT's required for a one-dimensional force calculation retaining moments 0 through M. For each grid point, there are $M + 1$ densities $\rho_{l,j}$ and force derivatives $F_{l,j}$. These require a total of $2(M + 1)$ real transforms, normally done as pairs of complex transforms (Appendix A). When the particle coordinates reside in secondary storage, such as a rotating magnetic disk, one prefers to collect the new densities as the particles are advanced to their new positions, rather than going through the particle list twice per time step. In this case, both the old force and the new density must be in fast memory, so we must store $2(M + 1)$ quantities per grid point. In one dimension this is much easier than in two, where F and k become vectors. In a two-dimensional quadrupole ($M = 2$) code, there are 18 quantities to store per grid point, and 18 2d real Fourier transforms to perform (Problem 11-2a). The requirements for an electromagnetic code are more startling yet (Problem 11-2b).

In practice, multipole codes have often been used in the monopole (i.e., NGP!) mode, and rarely have more than dipoles been included. Hence, in our discussion and evaluation, we also go no further than $M = 2$.

Let us now examine the spatial variation of the force field using the dipole approximation, as compared to a standard linear-interpolation method. In the latter case, the force appears as in Figure 11-2b, continuous and piece-wise linear. In the multipole expansion method, the force is given by a truncated Taylor series (4), expanding about the nearest grid point. Accuracy is good near the grid point and degrades rapidly toward the midpoint between cells. Furthermore, after the particle crosses the midpoint, the expansion is about a *new* grid point. Hence the force jumps *discontinuously* at the midpoint, as in Figure 11-2c. A deleterious effect of this discontinuity is increased aliasing errors, as we see in Sections 11-6 and 11-7. In practice, the magnitude of this jump (Problem 11-2c) is decreased by choosing the parameter a in $\hat{S}$ to be Δx or larger. For a given Fourier mode, this does not decrease the size of the jump relative to the force itself. Rather, both are suppressed at short wavelenghths, at some cost in resolution.

PROBLEMS

11-2a Modify Figure 11-2a for 2d and 3d. Indicate: the number of FFT's required; which quantities are scalar, vector, tensor (with rank); the number of quantities stored per grid point.

11-2b Consider the multipole expansion method for both charge and current sources for a two-dimensional electromagnetic code.

11-2c For a sinusoidal force field $F_{0,j} = A \cos kX_j$ in a dipole code, show that the magnitude of the jump at the cell midpoints is as large as $2A \sin^2(k\Delta x / 4)$. At the shortest resolved wavelength, $\pi / \Delta x$, the jump is as large as $2A$!

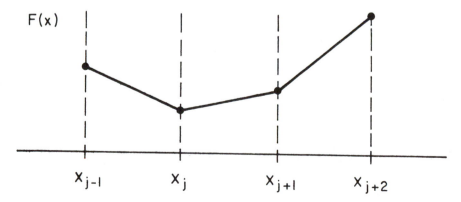

Figure 11-2b Conventional particle-grid force weighting, with linear interpolation between grid points. The force is continuous for all x, but $\partial F / \partial x$ is discontinuous at the X_j.

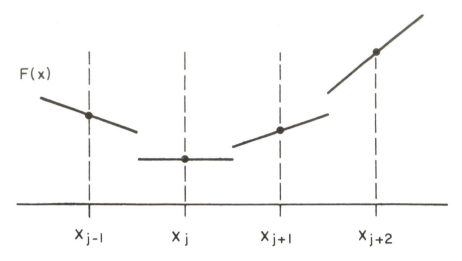

Figure 11-2c Dipole method force, $L = 1$, showing discontinuities at cell edges in both F and $\partial F / \partial x$.

11-3 THE "SUBTRACTED" MULTIPOLE EXPANSION

A method proposed by *Kruer et al.* (1973) to reduce the storage required by the multipole method is called the *subtracted multipole scheme.* In this approach, the derivatives of the force at the grid points are formed by using a finite-difference operator on the grid. Hence, one need only calculate and store the force at each grid point. The force may be obtained by finite-difference operation on the potential. The multipole densities are combined into a charge density by difference operators symmetric to those used for the force. FFT's are retained for the Poisson solution and smoothing, and perhaps also in in differentiating the potential.

Of the multipole schemes, a subtracted dipole expansion (SDPE) has been most used. In the force we replace $F_{1,k} = ikF_{0,k}$ by a centered difference,

$$F_{1,j} = \frac{F_{0,j+1} - F_{0,j-1}}{2\Delta x} \tag{1}$$

11-2(4) becomes

$$F_i = F_{0,j} + (x_i - X_j)\frac{F_{0,j+1} - F_{0,j-1}}{2\Delta x} \tag{2}$$

Similarly, $\rho_k = \hat{S}(\rho_{0,k} - ik\rho_{1,k})$ from 11-2(10) becomes

$$\rho_j = \rho_{0,j} - \frac{\rho_{1,j+1} - \rho_{1,j-1}}{2\Delta x} \tag{3}$$

The smoothing factor $\hat{S}$ (which is the factor SM in ES1) is applied during the Poisson solution which becomes $k^2\phi = \hat{S}^2\rho$.

This algorithm fits the formalism of Chapter 8, in which the force 8-5(2) is

$$F_i = q\,\Delta x\,[\,\cdots\,E_{j-1}S\,(X_{j-1}-x_i) + E_j\,S\,(X_j - x_i) + E_{j+1}S\,(X_{j+1}-x_i)\,\cdots\,]\quad(4)$$

The SDPE force (2) will be written as (4) if E is the smoothed field, i.e., $F_{0,j} = qE_j$, and the weighting function S is given by

$$\Delta x\,S\,(X_j - x_i) = 1$$

$$\Delta x\,S\,(X_{j\pm1} - x_i) = \pm\frac{x_i - X_j}{2\Delta x}\quad(5)$$

for $|X_j - x_i| < \Delta x / 2$ and zero elsewhere (Figure 11-3a(a)).

The charge density 8-5(1), $\rho_j = q\,S\,(X_j - x_i)$, with this same S gives the SDPE result (3) (Problem 8-3a). Therefore the results of Chapter 8 can be applied to the SDPE scheme.

It is worthwhile to make a simple model when trying to understand a new method. In the spirit of classical multipole expansions, let the origin be the grid point nearest x_i be called X_j. First, place all of the charge q there, as in the NGP weighting. Next add a rudimentary dipole; let the fraction of charge that is in the next cell be Δq, and place it at X_{j+1}; complete the

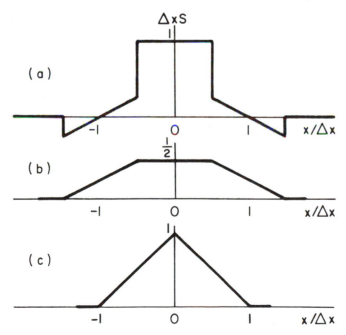

Figure 11-3a Shape functions for particle-grid interaction in (a) the "subtracted" dipole approximation, (b) the improved dipole approximation of *Kruer et al.* (1973), (c) the common linear weighting S_1.

dipole, with $-\Delta q$ placed at X_{j-1}. The dipole moment is now $(\Delta q)(2\Delta x)$, and must equal $q(x_i - X_j)$. The resulting contribution to charge *density* is $[\rho_{j-1}, \rho_j, \rho_{j+1}] = [-\Delta q, q, \Delta q]/\Delta x$, with $\Delta q = (q/2\Delta x)(x_i - X_j)$, while $|x_i - X_j| < \Delta x/2$. Each particle causes charges to be placed at three grid points (compared with two in linear weighting), implying an effective charge width of at least two cells.

The original multipole schemes and the subtracted schemes can be regarded in a unified viewpoint (Problem 11-3b) which we use in Section 11-5.

The SDPE weighting function is shown in Figure 11-3a. Its severe discontinuities result in poor aliasing properties reminiscent of NGP weighting (*Kruer et al.,* 1973, Section V). In Sections 11-6 and 11-7 we compare $S(k)$ for this and other schemes.

Kruer et al. (1973) propose an improved weighting function, shown in Figure 11-3a(b), which removes the discontinuities by distributing half of the monopole charge from the nearest grid point to the two next nearest grid points. They note that this improved S is equal to the average of the two linear S_1 functions shifted by Δx. As a result, with this improved subtracted dipole weighting, the forces are identical to those calculated using the standard linear S_1 function, with the addition of a simple spatial smoothing of ρ or ϕ (Problem 11-3d).

This observation suggests that the standard linear weighting scheme may be a multipole scheme. This and other possibilities are developed in the next Section.

PROBLEMS

11-3a Find $S(X_j - x_i)$, for $|X_j - x_i| < \Delta x/2$, by comparing the charge density 8-5(1), $\rho_j = q_i S(X_j - x_i)$, to (3) with $\rho_{0,j} = q/\Delta x$. Find $S(X_{j\pm 1} - x_i)$ by comparing to (3) rewritten as $\rho_{j+1} = \rho_{0,j+1} \pm (\rho_{1,j} - \rho_{1,j\pm 2})$ with $\rho_{1,j} = q(x_i - X_j)/\Delta x$. The answers agree with (5).

11-3b Show that the SDPE forces are identical to those from the DPE with $F_1(k)$ replaced by $iF_0(k)(\sin k\Delta x)/\Delta x$, and a similar change to 11-2(10).

11-3c Find $S(k)$ for the subtracted dipole scheme, e.g., by direct Fourier transform of $S(x)$.

11-3d Show that the forces calculated using the improved weighting of Figure 11-3a(b) are the same as obtained using the standard linear weighting of Figure 11-3a(c) with the addition of a spatial smoothing ($\frac{1}{4}$, $\frac{1}{2}$, $\frac{1}{4}$) applied to ρ or ϕ on the grid, separate from the particle move. Which is more efficient?

11-4 MULTIPOLE INTERPRETATIONS OF OTHER ALGORITHMS

Here we consider other particle-grid weighting algorithms which can be interpreted as multipole methods. First we show that the familiar one-dimensional linear-weighting algorithm can be derived as a dipole expansion about the *midpoint* between grid points. Next we derive an improved two-dimensional dipole algorithm by making an expansion about the nearest cell center. Then we show that area weighting is a dipole scheme which also includes the xy quadrupole moment. If the multipole interpretation is valued, one may use standard linear and area weighting with a clear conscience. We conclude with some opinions on the design of multipole algorithms which optimize accuracy and computational demands.

Let us construct a monopole plus dipole with charges at two grid points instead of three. For a charge q at position x between grid points j and $j + 1$, the monopole is constructed by placing half the charge at each grid point. The dipole is constructed by adding and subtracting Δq:

$$\rho_j \, \Delta x = \tfrac{1}{2}q - \Delta q$$

$$\rho_{j+1} \, \Delta x = \tfrac{1}{2}q + \Delta q \tag{1}$$

The dipole moment $(\Delta q)\, \Delta x$, centered at $X_{j+\frac{1}{2}}$, must equal the particle moment $q\,(x - X_{j+\frac{1}{2}})$. The result,

$$\rho_j \, \Delta x = q \left[\frac{1}{2} - \frac{x - X_{j+\frac{1}{2}}}{\Delta x} \right] = q \, \frac{X_{j+1} - x}{\Delta x}$$

$$\rho_{j+1} \, \Delta x = q \left[\frac{1}{2} + \frac{x - X_{j+\frac{1}{2}}}{\Delta x} \right] = q \, \frac{x - X_j}{\Delta x} \tag{2}$$

is just the standard linear weighting used in ES1! This is as smooth as the improved subtracted dipole scheme and can be evaluated just as quickly (Problem 11-4c).

In two dimensions, form the monopole and two dipole moments as indicated in Figure 11-4a, centered in the middle of the cell. At the grid point (j, k), for example, we add

$$\frac{q}{4} + \frac{q}{2\Delta x} (x - X_{j+\frac{1}{2}}) + \frac{q}{2\Delta y} (y - Y_{k+\frac{1}{2}}) \tag{3}$$

to $\rho_{j,k} \, \Delta x \, \Delta y$. The contribution of a charge in any of the four cells around (j, k) is given by

$$\rho_{j,k} \, \Delta x \, \Delta y \equiv q \, \Delta x \, \Delta y \, S\,(X_j - x, \; Y_k - y)$$

$$= \tfrac{3}{4}q - \tfrac{q}{2}|x - X_j| - \tfrac{q}{2}|y - Y_k| \quad \text{for } |x - X_j| < \Delta x$$

$$\text{and } |y - Y_k| < \Delta y \tag{4}$$

$$= 0 \qquad\qquad\qquad \text{otherwise}$$

(see Figure 11-6b). The smaller discontinuities in this weighting function, compared to that of *Kruer et al.* (1973), permit improved accuracy. It

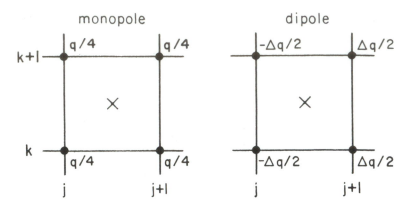

Figure 11-4a Construction of monopole and dipole moments, centered at the cell centers (x), by adding appropriate charges to the four cell corners.

happens that the discontinuities are removed by adding the *xy quadrupole* term (Problem 11-4a), which leads to the common bilinear (area) weighting (see Figure 11-6c).

Underlying these observations are the relations given in 8-5(3) and Problem 8-5b, which follow from 8-5(4) and 8-5(5). *Any* linear weighting can be regarded as a dipole method, in one, two, or three dimensions. Bilinear weighting includes, in addition, one quadrupole moment (Problem 11-4b).

PROBLEMS

11-4a Show that the *xy* quadrupole term to be added to (3) is

$$q \, \frac{(x - X_{j+\frac{1}{2}})(y - Y_{k+\frac{1}{2}})}{\Delta x \, \Delta y} \tag{5}$$

with which (4) becomes

$$\rho_{j,k} \, \Delta x \, \Delta y = q \left(1 - \frac{|x - X_j|}{\Delta x} \right)\left(1 - \frac{|y - Y_k|}{\Delta y} \right) \tag{6}$$

This is the same as *area weighting* (Chapter 14).

11-4b Show that the relations in 8-5(3) and Problem 8-5b generalize for bilinear weighting in two dimensions to

$$\Delta x \, \Delta y \sum_j \rho_j \, (1, X_j, Y_j, X_j Y_j) = \sum_i q_i \, (1, x_i, y_i, x_i y_i) \tag{7}$$

Show that this means bilinear weighting is a dipole scheme with the addition of *xy* quadrupole terms.

11-4c Show that, by accumulating the sum over particles in each cell of q and qx, that either the SDPE or the standard linear weighting cell charge density can be formed. Similarly, show that the particle force is a linear combination of q and qx in each cell, in both schemes. Therefore, the computational expense of the SDPE and the standard linear scheme are the *same*, when both are optimally calculated and use equal storage.

11-4d Make QS a quadrupole method by adding a grid filter. There are many possibilities; one is to apply the filter

$$\rho_j \leftarrow -\frac{1}{8}\rho_{j-1} + \frac{5}{4}\rho_j - \frac{1}{8}\rho_{j+1} \tag{8}$$

to ρ before, and similarly to ϕ after, the Poisson solution.

11-5 RELATIONS BETWEEN FOURIER TRANSFORMS OF PARTICLE AND GRID QUANTITIES

In preparation for deriving a dispersion relation and for comparing multipole and other algorithms, we derive in this section relations between Fourier transforms of particle density $n(x)$ and force field $F(x)$, and between grid quantities $\rho_{l,j}$, etc. We save effort by borrowing some results from Chapter 8, and facilitate comparisons by manipulating the relations into the same form as Chapter 8.

In Section 11-2, k is understood (without saying so) to be confined to the first Brillouin zone, i.e., $|k| < \pi/\Delta x$. This suffices for transforms of grid quantities, but for $n(k)$ and $F(k)$ we need *all* k. As in Chapter 8, it is convenient to use the fact that transforms of grid quantities have periodicity $2\pi/\Delta x$, e.g., $\rho(k + 2\pi/\Delta x) = \rho(k)$. To avoid confusion, we introduce the periodic function

$$\kappa = k \text{ modulo } \frac{2\pi}{\Delta x}, \quad |\kappa| \leqslant \frac{\pi}{\Delta x} \tag{1}$$

On substituting κ for k wherever the distinction matters, the results from Section 11-2 become valid for all k:

$$\rho(k) = \hat{S}(\kappa) \sum_l \frac{(-i\kappa)^l}{l!} \rho_l(k) \tag{2}$$

$$\phi(k) = \frac{\rho(k)}{\kappa^2} \tag{3}$$

$$F_l(k) = (i\kappa)^l F_0(k), \tag{4}$$

$$F_0(k) = -i\kappa q \hat{S}(-\kappa) \phi(k) \tag{5}$$

We begin by rewriting the multipole density, 11-2(9), using the nearest grid-point weighting function S_0 (Figure 8-5a) to select those particles closest to grid point j:

$$\rho_{l,j} = \sum_i q_i S_0 (X_j - x_i)(x_i - X_j)^l \tag{6}$$

Note the similarity to Equation 8-5(1), $\sum q_i S(X_j - x_i)$, whose Fourier transform is, from 8-7(21),

$$\sum q \, n(k_p) S(k_p) = \sum q \, n(k_p) \int dx \, e^{-ik_p x} S(x) \tag{7}$$

where $k_p = k - 2\pi p / \Delta x$. Comparing these, we see by inspection that

$$\rho_l(k) = \sum_p q \, n \, (k_p) \int dx \, e^{-ik_p x} \, S_0(x) \, (-x)^l \tag{8}$$

Substitution into (2) yields the desired relation in a form like that of Equation 8-7(21):

$$\rho(k) = \hat{S}(\kappa) \sum_p q \, n \, (k_p) \, S \, (k_p) \tag{9}$$

where

$$S(k) = \sum_l \int dx \, e^{-ikx} \, S_0(x) \, \frac{(i\kappa x)^l}{l!} \tag{10}$$

$$= \frac{1}{\Delta x} \int_{-\Delta x/2}^{\Delta x/2} dx \, e^{-ikx} \sum_l \frac{(i\kappa x)^l}{l!} \tag{11}$$

is the (transform of the) effective weighting function for particle-grid interaction. (The factor $\hat{S}$, not present in 8-7(21), describes spatial smoothing here included in the evaluation of $\rho(k)$ and $F(k)$, but included in K^2 in Chapter 8.)

Turning to the force, we rewrite 11-2(4), again using S_0 to select the correct terms in a sum over j:

$$F_i = \sum_l \left[\Delta x \sum_j F_{l,j} \, S_0(X_j - x_i) \frac{(x_i - X_j)^l}{l!} \right] \tag{12}$$

For each l, the term $[\,\cdots\,]$ has the same form as 8-5(2), whose transform is 8-7(14),

$$E(k) \, S(-k) = E(k) \int dx \, e^{ikx} \, S(x)$$

Again, by inspection we see that the transform of the multipole force field is

$$F(k) = \sum_l F_l(k) \int dx \, e^{ikx} \, S_0(x) \frac{(-x)^l}{l!} \tag{13}$$

Using $F_l(k) = (i\kappa)^l \, F_0(k)$,

$$F(k) = S(-k) \, F_0(k) = \hat{S}(-\kappa) \, S(-k) \, [-i\kappa q \phi(k)] \tag{14}$$

in which we encounter again the effective particle-grid weighting function (10). This result is in the same form as 8-7(14) with 8-7(9), with κ given by (1) and the addition of the spatial smoothing $\hat{S}$. Hence, it is possible to draw on other results in Chapter 8, such as the discussion of momentum conservation (Problem 11-5a).

If we could keep all moments, then (Problem 11-5b)

$$S(k) = 1 \quad \text{if } |k| < \pi / \Delta x \qquad (\text{i.e., } k = \kappa)$$

$$= 0 \quad \text{otherwise} \qquad (\text{i.e., } k - \kappa \text{ is a multiple of } 2\pi / \Delta x) \tag{15}$$

i.e., we recover the "band limited" interpolation of 11-2(1) and 11-2(2) which has no aliasing errors.

To make connection with the different representation used by *Chen and Okuda* (1975), we rewrite (10) as (Problem 11-5c)

$$S(k) = \sum_l \frac{1}{l!} \left[-\kappa \frac{d}{dk} \right]^l S_0(k) \tag{16}$$

where

$$S_0(k) = \frac{\sin\theta}{\theta} \equiv \text{dif } \theta, \quad \theta = \frac{k\Delta x}{2} \tag{17}$$

For $|k| < \pi/\Delta x$, $S(k_p)$ is the same as the function $I(k, k_p)$ used by Chen and Okuda.

For the dipole approximation we keep $l = 0$ and 1,

$$S(k) = S_0(k) - \kappa \frac{d}{dk} S_0(k)$$

$$= \frac{\sin\theta}{\theta} - \frac{\kappa}{k} \left[\cos\theta - \frac{\sin\theta}{\theta} \right] \tag{18}$$

Using results of Problem 11-3b, we adapt this result to the subtracted dipole (SDPE) scheme of Section 11-3 to find

$$S(k) = \frac{\sin\theta}{\theta} - \frac{\sin 2\theta}{2\theta} \left[\cos\theta - \frac{\sin\theta}{\theta} \right] \tag{19}$$

which was also found in Problem 11-3c.

The results of this section readily generalize to two and three dimensions (Problem 11-5e).

We can identify three sources of error in the force calculation: errors in the *magnitude* and *direction* of **F**, and coupling of different wavelengths due to *aliasing*. For **k** in the first Brillouin zone, there is no error in direction when $\mathbf{F}_0$ (or **E**) is obtained by Fourier transform of (5), and errors in magnitude, due to $\hat{S}S$ differing from unity, can be compensated for by adjusting $\hat{S}$. Aliasing errors, given by the $\mathbf{p} \neq 0$ terms, are determined at a given **k** by the magnitude of $S(\mathbf{k_p})$ and cannot be corrected or reduced without loss of spatial resolution.

We have created relations (9) and (14), analogous to 8-7(17) and 8-7(14). Section 8-7 and this Section can be made congruent if we move the factors $\hat{S}(\kappa)$, $\hat{S}(-\kappa)$ from (9) and (14) to (3), which becomes $K^2\phi = |\hat{S}|^2\rho$. This is the same as 8-7(5) with $K^{-2} = |\hat{S}|^2\kappa^{-2}$, and is also the way to implement this method efficiently.

In the following sections we use these results to construct a dispersion relation and consider the overall accuracy of the force calculation in the multipole approximation. The grid-particle weighting function plays the central role in limiting accuracy.

PROBLEMS

11-5a Show that the multipole force calculation conserves momentum if the same number of moments is used in the density and force calculations. (*Hint:* Argue that the multipole calculation can be written in the form of 8-5(1) and 8-5(2), so that Section 8-6 is applicable.)

11-5b When all moments are kept in (11), replace the series by $\exp(i\kappa x)$ and use (1) to obtain (15). Alternatively, the sum in (16) is recognized as a complete Taylor expansion of S_0 about k_p; that is

$$S(k_p) = S_0(k_p - k) \tag{20}$$

$$= S_0(-pk_g) \tag{21}$$

$$= \begin{cases} 1 & \text{if } p = 0 \\ 0 & \text{if } p \neq 0 \end{cases} \tag{22}$$

11-5c Evaluate each term in (10) as the l'th derivative of $S_0(k)$ to obtain (16).

11-5d Derive (19).

11-5e Generalize (9) $-$ (11) and (14) $-$ (19) to two and three dimensions. For example, in 2d, (19) becomes

$$S(\mathbf{k}) = \text{dif} \, \tfrac{1}{2}\theta_x \, \text{dif} \, \tfrac{1}{2}\theta_y - \text{dif} \, \theta_y \, \text{dif} \, 2\theta_x \, (\cos\theta_x - \text{dif} \, \theta_x)$$

$$- \text{dif} \, \theta_x \, \text{dif} \, 2\theta_y \, (\cos\theta_y - \text{dif} \, \theta_y)$$

with $\theta_x = k_x \Delta x / 2$, $\theta_y = k_y \Delta y / 2$.

11-6 OVERALL ACCURACY OF THE FORCE CALCULATION; DISPERSION RELATION

By showing the effect of the numerical methods on plasma waves, the dispersion relation clarifies the role of each of the steps in the force calculation in determining the overall accuracy. Using results of the last section, we proceed as in Section 9-5 to derive a dispersion relation for unmagnetized plasma,

$$\epsilon = 1 + \frac{\hat{S}^2}{\kappa^2} \, \boldsymbol{\kappa} \cdot \sum \mathbf{k_p} \, S^2(\mathbf{k_p}) \, \chi(\mathbf{k_p}, \omega) \tag{1}$$

which is identical to 9-5(1a) when we choose $K^{-2} = \hat{S}^2/\kappa^2$ and remove $\boldsymbol{\kappa}(\mathbf{k_p}) = \boldsymbol{\kappa}(\mathbf{k})$ from the sum.

Because the time integration does not interact in any unusual way with the multipole force calculation, we ignore the complications of finite time-step. From 9-5(1b),

$$\epsilon = 1 + \frac{\hat{S}^2}{\kappa^2} \, \boldsymbol{\kappa} \cdot \sum S^2(\mathbf{k_p}) \int \frac{\partial f_0}{\partial \mathbf{v}} \, \frac{d\mathbf{v}}{\omega + i0 - \mathbf{k_p} \cdot \mathbf{v}} \tag{2}$$

Also, our quantitative comparisons are limited to one-dimension. This dispersion relation and some of our discussion are equivalent to that of *Chen and Okuda* (1975).

The particle shape $\hat{S}$ referred to in the literature ("Gaussian-shaped particles") is *not* to be confused with the weighting function S. In the dispersion relation (1) we see clearly the different roles played by $\hat{S}$ and S. $\hat{S}^2(\kappa)$, appearing outside the sum on p, is simply *any* smoothing factor (in the fundamental Brillouin zone $|k\Delta x| < \pi$), usable in *any* method (it is the factor SM^2 in ES1). Hence, the relative accuracy of the multipole expansion will appear through $S^2(k)$.

If all moments could be kept, we know from 11-5(15) that there are no aliasing errors. Therefore only one p term is nonzero in the dispersion function, which is the same as in the *gridless* force calculation 11-2(1) and 11-2(2).

$$\epsilon_\infty(k,\omega) = 1 + \omega_p^2 \frac{\hat{S}(\kappa)}{\kappa} \int dv \, \frac{\partial f_0(v)/\partial v}{\omega - \kappa v} \tag{3}$$

This is the gridless dielectric function; hence for $L \to \infty$, indeed, the multipole expansion is independent of the grid and no nonphysical results due to aliasing arise. However, in practice at most the octopole moments are kept, so the relevant comparisons are at dipole and octopole order.

In the fundamental Brillouin zone $(k\Delta x < \pi)$, the subtracted dipole $S^2(k)$ is nearly flat, while the linear weighting drops off more rapidly, as $[(\sin\theta)/\theta]^2$. Of course, in the fundamental Brillouin zone we are free to compensate by adjusting $\hat{S}$ in any way we like, and so the flatness or drop-off can be altered to suit.

However, we cannot separately control $\hat{S}^2(\kappa)S^2(k)$ in the higher Brillouin zones, (the unhatched zones with $k\Delta x > \pi$ in Figure 11-6a. For SDPE with $k\Delta x \leq 1$,

$$S^2(k_p) = \left[\frac{k\Delta x}{2\pi p}\right]^4 \left[1 - \frac{k\Delta x}{2}\left[p\pi + \frac{2}{p\pi}\right]\right]^2 \qquad \text{for } p \neq 0 \tag{4}$$

while for standard linear (CIC) and quadratic spline (QS) we have

$$S^2(k_p) = \left(\frac{\sin k_p \Delta x/2}{k_p \Delta x/2}\right)^{2M} \qquad \text{where } M = 2 \text{ for CIC, 3 for QS} \tag{5}$$

$$\approx \left(\frac{k\Delta x}{2\pi p}\right)^{2M} \qquad \text{for small } k\Delta x \text{ and } p \neq 0 \tag{6}$$

For $k\Delta x/2 \ll (p\pi + 2/p\pi)^{-1}$, *i.e.,* wavelengths longer than about $10p$ cells, the alias coupling from SDPE and CIC are the same. In the opposite limit, $S^2(k_p)$ for CIC goes as (6) while the larger alias coupling of SDPE goes as $S^2(k_p) \approx (k\Delta x/2)^6(\pi p)^2$ which indicates that for $(k\Delta x/2) \sim 1$, there is the same order of alias coupling for dipole as for NGP, for which $S^2(k_p) = (k\Delta x/2\pi p)^2$. The slow $(1/p^2)$ drop-off is due to the discontinuity in force at the mid-cell planes.

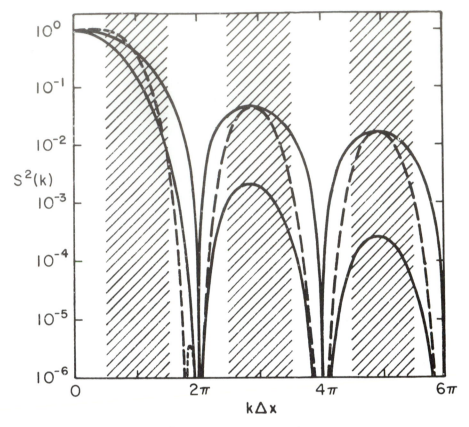

Figure 11-6a Weight functions $S^2(k)$ plotted against $k\Delta x / 2\pi$ for NGP and linear weighting (upper and lower solid curves) and the subtracted dipole weighting (dashed curve). Shading indicates intervals of $k\Delta x$ which are affected by smoothing of wavelengths $> 4\Delta x$ *on the grid.* Therefore the important comparison is in the unshaded intervals, where filtering of grid quantities cannot selectively suppress errors due to large values of S^2 for $k\Delta x$ near $2\pi, 4\pi, \ldots,$ without also suppressing the physical force (losing resolution).

The quadrupole scheme, QPE, stays below 1% coupling over almost all of the range shown in Figure 11-6a. This achievement requires substantial additional computation and memory. Hence, the QPE comparison should be made, for example, to higher-order particle weighting schemes such as quadratic spline QS, which has $M = 3$ in (5), (6). The QS $p = 1$ coupling is larger, rising to about 0.067 at $k\Delta x = \pi$; this is about 0.4 of CIC, a little less than DPE (for $p = -1$) and about 3 times QPE; for other values of p, QS is less than 10^{-4} with smaller maxima than all other methods shown.

Okuda and Cheng (1978) show that even quadratic splines and higher-order multipoles are numerically unstable for small $\lambda_D / \Delta x$ (just as we found for lower-order weighting). For $\lambda_D / \Delta x < 0.1$, they find better stability with quadratic and cubic splines than with the octupole expansion. This they attribute to the greater smoothness provided by splines, consistent with our viewpoint. With $\lambda_D / \Delta x = 0.01$, only cubic splines provided acceptable numerical stability.

In Figure 11-6b and Figure 11-6c we compare our improved subtracted dipole, 11-4(4), and the 2d bilinear (area) weighting. Standard SDPE is not as good. In bilinear weighting, alias coupling comes dominantly near the k_x and k_y axes where S^2 is largest. There, S^2 for improved SDPE is at least comparable, though larger.

11-7 SUMMARY AND A PERSPECTIVE

Much valuable research has been done using multipole codes. We have described the multipole method, as originally conceived and in its "subtracted" form. We saw how other algorithms can be interpreted as multipole algorithms. Dipole schemes with smoother spatial variation were derived;

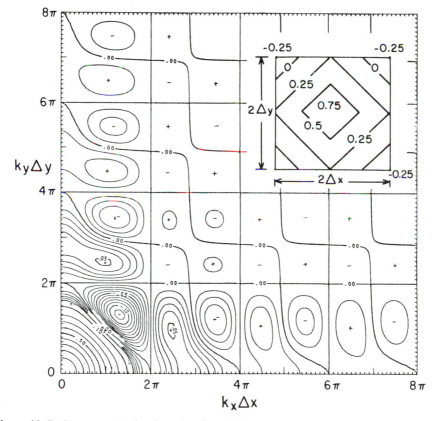

Figure 11-6b Contours of $S(x,y)$ and $S(k_x, k_y)$ for the improved subtracted dipole, equation 11-4(4). $S(x,y) = 0$ outside the region shown. $S(x,y)$ is discontinuous at the sides of the square; however, compared to the historic SDPE scheme, $S(k_x, k_y)$ outside the fundamental Brillouin zone is much improved.

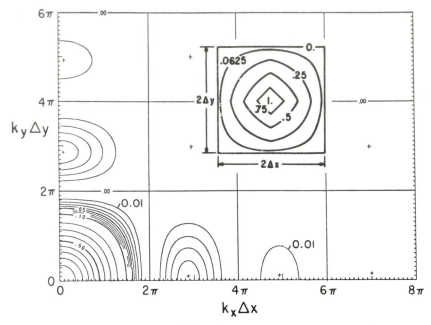

Figure 11-6c Contours of $S(x,y)$ and $S(k_x, k_y)$ for the common 2d bilinear ("area weighting", "charge sharing") scheme, which is a dipole plus xy quadrupole scheme (see Problem 11-4a). $S(x,y)$ is continuous everywhere, and is zero outside the square shown.

two of these improvements are the familiar 1d linear weighting and 2d bilinear (area) weighting. The overall force calculation was analyzed along the lines of Chapter 8. It is clear that aliasing errors in the particle-grid weighting, once made, *cannot* be removed by a shape factor (usually $\hat{S}(k) = \exp(-k^2 a^2 / 2)$ with $a \approx \Delta x$) applied to the grid quantities ρ or ϕ. With these analytic tools, we compared accuracy among the published subtracted dipole scheme, a suggested improvement, and the standard linear or bilinear (CIC, area-weighting) scheme.

Practical algorithms are a compromise between spatial resolution and freedom from aliasing errors, versus storage and computational effort required, and adaptability to more general boundary conditions, non-Cartesian coordinate systems, etc. In making comparisons or interpreting the literature, one or more factors must be held constant. For example, in an application where a Fourier-transform field solver is convenient, one might compare the spatial resolution possible on a given spatial grid, with a given tolerance for aliasing errors, using SDPE, bilinear, or other particle-grid weightings. For each weighting, the factor $\hat{S}(k)$ must be adjusted to trade between resolution and aliasing errors.

Whatever its motivation, the multipole method, as described in the literature, is no more than the use of truncated Taylor series expansions about the nearest grid points to represent the force field, with an analogous

procedure for inferring a charge density from particle positions. Taylor expansions (multipole method) or Lagrange interpolation (subtracted method) concentrate their accuracy near the grid points, at the expense of accuracy elsewhere. When smooth and accurate representation is desired *over an interval*, splines are preferred for similar computational effort and spatial resolution. Splines provide a systematic progression to higher order, and still may be interpreted just as well in terms of multipoles. Accuracy can be improved, by altering the mathematics, without affecting the physical interpretation. More optimal weightings, probably application-dependent, may yet be found. The tools derived in these chapters may be helpful.

TWELVE

KINETIC THEORY FOR FLUCTUATIONS AND NOISE; COLLISIONS

12-1 INTRODUCTION

In this chapter, an accurate theory of fluctuations, noise, and collisions in computer simulation of plasma is developed. The analytic method describes the space and time discretization exactly; the results reduce simply and correctly to the standard results of plasma kinetic theory in the limit of small space and time steps. If the particles are imagined to be a Monte-Carlo sampling of the phase-space, then the fluctuations in this sampling, as modified by collective effects, are a concern of this chapter. This theory is of interest in theoretical and empirical studies to understand the character of representation by simulation methods of plasma processes such as transport.

Fluctuations have been of interest in computer simulation of plasmas because they interfere with modeling of collisionless phenomena. However, one man's noise is another man's signal, and computer simulation has been used as a tool to study fluctuations and other processes involving discrete-particle effects in plasma, such as transport. Measurements of the fluctuation spectrum have also been used to check new simulation programs. Thermal fluctuations have been measured and analyzed theoretically in gridless "sheet" plasma models and in models using a spatial grid to mediate the particle interaction, the class of model considered here.

The particles can be regarded as Lagrangian markers embedded randomly in the Vlasov fluid and moving with it through phase-space (*Morse and Nielson,* 1969). This Monte-Carlo viewpoint explicitly recognizes the random sampling aspects of the particle-in-cell models, and this chapter provides information on the sampling statistics, *e.g.*, variance and correlations in the density which are evaluated as a Monte-Carlo integral over phase space. The statistics are modified by collective effects, however, since the particles influence each other through the self-consistent fields, rather than remaining independent markers in the phase-space fluid. Therefore, we use analytical methods analogous to those of normal plasma kinetic theory.

The results are cast in forms as similar as possible to the standard results of plasma kinetic theory, in order to facilitate comparison. As expected, qualitative differences result when the space or time differencing is too coarse, as often happens when computer time or memory are restrictive.

In deriving our results, we find it simpler and more general to take an approach different than the usual, which begins with moments of the Liouville equation or uses Klimontovich's formalism (*Rostoker,* 1961; *Rostoker and Rosenbluth,* 1960; *Klimontovich,* 1967; *Dawson and Nakayama,* 1966). There are several reasons for this. Some numerical time integration schemes correspond to third- or higher-order differential equations of motion, in that the dimensionality of phase space necessary to describe the state of the system is $9N$ or higher (for N particles in three dimensions) instead of $6N$. Furthermore, some time integration schemes are not measure-preserving; *i.e.*, particle motion does not preserve phase-space volume. Such features encumber the usual developments of kinetic theory. By contrast, an approach similar in spirit to that of *Hubbard* (1961) yields the desired results in a manner both simpler and more informative to physical intuition, and is readily adapted to plasma systems with altered dynamics, such as simulation models.

Only electrostatic forces are included in this chapter; even in electromagnetic models it will be mainly the longitudinal fields that govern density fluctuations and Debye shielding. The divisions of this chapter are as follows. Section 12-2 applies the results of Chapters 8 to 10 to the simple example of Debye shielding. The fluctuation spectrum is derived in Section 12-3, and its physical and nonphysical components are examined in several limits. Section 12-4 discusses the limitations on the validity of this theory and some other observations. These sections are from *Langdon* (1979a). In Sections 12-5a and 12-5b, velocity diffusion and drag due to density fluctuations are found; the other contribution to drag is due to the polarization of the plasma as sensed by a specified particle. In Section 12-5c, these results are used in the derivation of a collision operator which includes the familiar Balescu-Lenard operator in the Δx, $\Delta t \rightarrow 0$ limit. Finally, conservation properties and the H-theorem are studied in Section 12-6.

12-2 TEST CHARGE AND DEBYE SHIELDING

We begin by deriving the linear response of the simulation plasma to a perturbing charge and apply it to Debye shielding.

The total perturbed charge density resulting when a stable plasma with its uniform neutralizing background is perturbed by an imposed external charge ρ_e on the grid is

$$\rho(\mathbf{k}, \omega) = \rho_e(\mathbf{k}, \omega) + \sum_{\mathbf{p}} S(\mathbf{k_p}) qn(\mathbf{k_p}, \omega) \tag{1}$$

$$= \rho_e(\mathbf{k}, \omega) - \sum_{\mathbf{p}} S(\mathbf{k_p})\chi(\mathbf{k_p}, \omega) i\mathbf{k_p} \cdot [-i\kappa(\mathbf{k_p}) S(-\mathbf{k_p})\phi(\mathbf{k_p}, \omega)]$$

using 8-9(8), 9-2(10), and 9-2(11a), similarly to Problem 9-5a. Using the periodicity $\phi(\mathbf{k_p}, \omega) = \phi(\mathbf{k}, \omega)$ to remove ϕ from the sum, and Poisson's equation 8-9(14), we obtain

$$\epsilon(\mathbf{k}, \omega)\rho(\mathbf{k}, \omega) = \rho_e(\mathbf{k}, \omega) \tag{2}$$

$$\phi(\mathbf{k}, \omega) = \frac{\rho_e(\mathbf{k}, \omega)}{K^2(\mathbf{k})\epsilon(\mathbf{k}, \omega)} \tag{3}$$

where the dielectric function is

$$\epsilon(\mathbf{k}, \omega) = 1 + K^{-2}(\mathbf{k})\sum_{\mathbf{p}}\kappa(\mathbf{k_p}) \cdot \mathbf{k_p} S^2(\mathbf{k_p})\chi(\mathbf{k_p}, \omega) \tag{4}$$

$$= 1 + \frac{\omega_p^2}{K^2}\sum_{\mathbf{p}} S^2 \int d\mathbf{v}\, \kappa \cdot \frac{\partial f_0}{\partial \mathbf{v}} \frac{\Delta t}{2} \cot(\omega + i0 - \mathbf{k_p} \cdot \mathbf{v})\frac{\Delta t}{2}$$

as in Section 9-5.

In some cases it is possible to evaluate $K^2\epsilon$ in closed form. For a stationary test charge and a Maxwellian velocity distribution we find from 9-2(15) that

$$\chi(\mathbf{k_p}, \omega) = \omega_p^2\left[\frac{1}{k_\mathbf{p}^2 v_t^2} - \frac{\Delta t^2}{12}\right] \tag{5}$$

where we have assumed $\omega \ll k_\mathbf{p} v_t \lesssim \Delta t^{-1}$; this assumption is addressed below. The sum over $\mathbf{p}$ in (4) can be evaluated using *Abramowitz and Stegun* (1964, eq. 4.3.92, p. 75, and its derivatives). For linear weighting in a one-dimensional Hamiltonian model, for example,

$$K^2\epsilon = K^2 + \omega_p^2\sum_p k_p^2 S^2\left[\frac{1}{k_p^2 v_t^2} - \frac{\Delta t^2}{12}\right] \tag{6}$$

$$= \left\{1 - \frac{1}{12}\left[(\omega_p \Delta t)^2 + \frac{2\Delta x^2}{\lambda_D^2}\right]\right\}K^2 + \frac{1}{\lambda_D^2}$$

where $\lambda_D \equiv v_t/\omega_p$ is the Debye length and we have used 10-6(2b) and the derivative of 8-11(13). (We see now we should drop the Δt^2 term because, when it is significant compared with the Δx^2 term, the approximation $k_p v_t \Delta t \leqslant 1$ is violated for $|p|$ values too small for convergence to be good.) In three dimensions the triple sum $\sum S^2$ factors, and becomes a product of sums over the components of $\mathbf{p}$ separately.

The spatial decay of the Debye potential is given by the zeroes $\pm ik$ of the denominator of (3),

$$\left(\frac{2}{\Delta x} \sinh \frac{ik\Delta x}{2}\right)^{-2} = \lambda_D^2 - \frac{1}{6}\Delta x^2 \tag{7}$$

For small Δx, Δt, the correct shielding is recovered. However, as λ_D is decreased below Δx, the shielding length becomes comparable to Δx. Furthermore, we show below that it is possible for the potential to alternate sign as it decays, when $\lambda_D^2 < \Delta x^2 / 6$.

We can perform the inverse Fourier transform for the simple case of an external charge q at $X_j = 0$, for which $\rho_e(k) = q$. Changing variables to $\zeta = \exp(ik\Delta x)$, the Fourier inverse integral becomes a contour integral around the circle $|\zeta| = 1$, with one pole inside and one outside. The result is

$$\phi_j = q \frac{\lambda_D}{2} \zeta_0^{|j|} \left[1 + \frac{\Delta x^2}{12\lambda_D^2}\right]^{-\frac{1}{2}} \tag{8}$$

where

$$\zeta_0 = \frac{\lambda_D^2 - \Delta x^2 / 6}{\lambda_D^2 + \Delta x^2 / 3 + \Delta x (\lambda_D^2 + \Delta x^2 / 12)^{\frac{1}{2}}} \tag{9}$$

is the location of the pole inside the unit circle. As $\Delta x / \lambda_D \to 0$, $\zeta_0^{|j|} \to \exp(-|X_j| / \lambda_D)$ and therefore

$$\phi_j = q \frac{\lambda_D}{2} e^{-|X_j|/\lambda_D} \tag{10}$$

as expected. However, if we reduce $6\lambda_D^2 / \Delta x^2$ toward unity, $\zeta_0 \to 0$ and $\phi_j = q\lambda_D / \sqrt{6}$ at $j = 0$ and zero elsewhere. One could say the shielding length is Δx. For $6\lambda_D^2 / \Delta x^2 < 1$, the potential alternates sign from cell to cell as it decays. When $\lambda_D / \Delta x$ is very small,

$$\phi_j = \sqrt{3} q \frac{\lambda_D^2}{\Delta x} \zeta_0^{|j|} \tag{11}$$

with $\zeta_0 = -(2 + \sqrt{3})^{-1}$, showing the oscillating decay and the reduction in magnitude of $\phi_0 / (q\lambda_D)$ by the factor $2\sqrt{3}\lambda_D / \Delta x$; both features have been noted for a cloud plasma with no grid (*Langdon and Birdsall, 1970; Okuda and Birdsall, 1970*). Here, the effective cloud "radius" is approximately equal to Δx.

In a finite system, the Fourier inversion becomes a sum; if the $k = 0$ mode is absent, then ϕ_j differs from (8) or (10) by a constant and approaches a negative limit, rather than zero, for large $|X_j|$, as observed by *Hockney* (1971).

12-3 FLUCTUATIONS

In this section we derive the fluctuation spectrum, first for uncorrelated particles, then including collective effects, and examine the results in various limits.

(a) The Spectrum

To the lowest approximation, the particles move independently along straight-line orbits. Hence, the zero-order position of particle i at time $t_n \equiv n\Delta t$ is

$$\mathbf{x}_{in}^{(0)} = \mathbf{x}_{i0} + \mathbf{v}_i t_n \qquad (1)$$

and the Fourier transformed number density is

$$n^{(0)}(\mathbf{k}, \omega) = 2\pi \sum_i \exp(-i\mathbf{k} \cdot \mathbf{x}_{i0}) \delta(\omega - \mathbf{k} \cdot \mathbf{v}_i, \omega_g) \qquad (2)$$

where $\omega_g \equiv 2\pi / \Delta t$ and we have introduced the periodic delta-function comb

$$\delta(\omega, \omega_g) \equiv \sum_{q=-\infty}^{\infty} \delta(\omega - q\omega_g) \qquad (3)$$

which replaces the ordinary delta function of the continuum transform.

We now consider an example of systems such that its averages are independent of where and when they are taken, i.e., a uniform and stationary ensemble. This means that the ensemble average of the net charge density, say, will be zero, but the average of products need not vanish. We find the ensemble average

$$\langle \rho(\mathbf{k}, \omega) \rho(\mathbf{k}', \omega') \rangle$$

from which the fluctuations of other quantities may easily be found.

First, considering noninteracting particles, we use (2) substituted into 8-9(8). In performing the ensemble average we use the following information: the zero-order particle positions and velocities are independent, n_0 is the average particle density, and the velocity distribution is $f_0(\mathbf{v})$ normalized to unity. In the double sum over particles, terms corresponding to pairs of differing particles cancel terms due to the mean neutralizing charge density of other species. In the remaining terms,

$$\langle g(\mathbf{x}_{i0}, \mathbf{v}_i) \rangle \rightarrow n_0 \int d\mathbf{x} \, d\mathbf{v} \, f_0(\mathbf{v}) g(\mathbf{x}, \mathbf{v})$$

One is left with a result of the form (Problem 12-3a)

$$\langle \rho^{(0)}(\mathbf{k}, \omega) \rho^{(0)}(\mathbf{k}', \omega') \rangle = (\rho^2)_{\mathbf{k},\omega}^{(0)} (2\pi)^4 \delta(\mathbf{k} + \mathbf{k}', \mathbf{k}_g) \delta(\omega + \omega', \omega_g) \qquad (4)$$

where $\delta(\mathbf{k}, \mathbf{k}_g)$ is defined in analogy to (3). Then (4) defines the fluctuation spectrum

$$(\rho^2)^{(0)}_{\mathbf{k}, \omega} = \sum_{\mathbf{p}, q} q^2 S^2(\mathbf{k_p}) (n^2)^{(0)}_{\mathbf{k_p}, \omega_q} \tag{5}$$

where $\omega_q \equiv \omega - q\omega_g$ and

$$(n^2)^{(0)}_{\mathbf{k}, \omega} = 2\pi n_0 \int d\mathbf{v} \, f_0(\mathbf{v})\delta(\omega - \mathbf{k} \cdot \mathbf{v}) \tag{6}$$

is the number density spectrum for noninteracting particles, with $\Delta t = 0$. Note that the spectrum of ρ is periodic in $\mathbf{k}$ and ω, as are the transforms of other grid quantities.

The fluctuating density produces fields which deflect the particles slightly, altering the density. We take the density perturbation response to be the time-asymptotic response in the linearized Vlasov approximation, as in 12-2(3). The charge density is now given by 12-2(2) with ρ_e replaced by $\rho^{(0)}$; the difference between this and the actual density is higher-order than will be kept in our final expressions (*Hubbard*, 1961). We form the average

$$\epsilon(\mathbf{k}, \omega)\epsilon(\mathbf{k'}, \omega')\langle \rho(\mathbf{k}, \omega) \, \rho(\mathbf{k'}, \omega')\rangle = \langle \rho^{(0)}(\mathbf{k}, \omega)\rho^{(0)}(\mathbf{k'}, \omega')\rangle$$

use (4), and replace $\epsilon(\mathbf{k'}, \omega')$ by $\epsilon(-\mathbf{k_p}, -\omega_q) = \epsilon^*(\mathbf{k}, \omega)$ (Problem 12-3b). We find

$$(\rho^2)_{\mathbf{k}, \omega} = \frac{2\pi\rho_0 q}{|\epsilon(\mathbf{k}, \omega)|^2} \sum_{\mathbf{p}} S^2(\mathbf{k_p}) \int d\mathbf{v} \, f_0(\mathbf{v})\delta(\omega - \mathbf{k_p} \cdot \mathbf{v}, \omega_g) \tag{7}$$

which is the principal result of this section. Since the source term involves only straight-line motion, this expression holds for most time integration algorithms, although the form of ϵ will change. The generalization to more than one particle species is trivial.

The fluctuations of other grid quantities are related in the obvious way. For instance, the energy-density spectrum in a Hamiltonian model is

$$(\tfrac{1}{2}\rho\phi)_{\mathbf{k}, \omega} = \frac{(\rho^2)_{\mathbf{k}, \omega}}{2K^2} \tag{8}$$

This expression holds even for force laws other than Coulomb's (see Chapter 10). The sign and normalization conventions give (Problem 12-3c)

$$
\begin{aligned}
\langle \tfrac{1}{2}\rho_{\mathbf{j}, n}\phi_{\mathbf{j'}, n'}\rangle & \\
= \int_g \frac{d\mathbf{k}}{(2\pi)^3} \frac{d\omega}{2\pi} (\tfrac{1}{2}\rho\phi)_{\mathbf{k}, \omega} & \exp\left[i\mathbf{k} \cdot (\mathbf{X_j} - \mathbf{X_{j'}}) - i\omega(t_n - t_{n'})\right]
\end{aligned} \tag{9}
$$

The dependence on only the differences $\mathbf{X_j} - \mathbf{X_{j'}}$ and $t_n - t_{n'}$ is due to the form of (4), which is therefore a reflection of the constancy of the ensemble. For one and two space dimensional cases one makes the obvious choice for the power of 2π in the normalization of the $\mathbf{k}$ integral.

(b) Limiting Cases

In many-parameter regimes it is possible to obtain additional information analytically. Efficient numerical evaluation of the multiple sums in the general case has been discussed in *Langdon* (1979b).

(1) Fluctuation-dissipation theorem For the Hamiltonian models of Chapter 10 it is sometimes possible to write the spectrum (8) in the form of the fluctuation-dissipation theorem, *viz.*,

$$(\tfrac{1}{2}\rho\phi)_{\mathbf{k},\omega} = -\frac{T}{\omega}\operatorname{Im}\frac{1}{\epsilon(\mathbf{k},\,\omega)} \tag{10}$$

where T is the temperature in energy units. To this end, we use the Plemelj formula in 12-2(4) to generalize the result of Problem 9-2b

$$\operatorname{Im}\epsilon(\mathbf{k},\,\omega + i0) = -\pi\frac{\omega_p^2}{K^2}\sum_{\mathbf{p},q}S^2\int d\mathbf{v}\,\mathbf{k}_{\mathbf{p}}\cdot\frac{\partial f_0}{\partial\mathbf{v}}\delta(\omega - q\omega_g - \mathbf{k}_{\mathbf{p}}\cdot\mathbf{v}) \tag{11}$$

In order to progress to (10), set $\Delta t = 0$ (eliminating the sum over q), assume f_0 is Maxwellian with no drift relative to the grid and that all mobile species have the same temperature. Then, we can write

$$\operatorname{Im}\epsilon = \frac{\omega}{K^2 T}2\pi\rho_0\,q\sum_{\mathbf{p}}S^2\int d\mathbf{v}\,f_0(\mathbf{v})\,\delta(\omega - \mathbf{k}_{\mathbf{p}}\cdot\mathbf{v}) \tag{12}$$

in which we recognize the numerator of (7). The fluctuation-dissipation result, (10), follows immediately. Perhaps with $\Delta t \neq 0$ there exists a result similar to (10), but with ω replaced by an expression with the necessary periodicity ω_g.

(2) Spatial spectrum The spatial spectrum is more commonly measured than the full $\mathbf{k}$, ω spectrum. We can integrate (10) analytically to obtain the spatial spectrum

$$(\tfrac{1}{2}\rho\phi)_{\mathbf{k}} \equiv \int\frac{d\omega}{2\pi}(\tfrac{1}{2}\rho\phi)_{\mathbf{k},\omega} = -T\int\frac{d\omega}{2\pi\omega}\operatorname{Im}\frac{1}{\epsilon} \tag{13}$$

by taking the imaginary part of the following integral over the closed contour shown in Figure 12-3a,

$$-T\oint\frac{d\omega}{2\pi\omega}\left[\frac{1}{\epsilon} - 1\right] = 0$$

whose integrand is analytic in the upper half-plane (since ϵ has no roots there under the assumptions made here; see Chapters 8 and 10) and vanishes there as $|\omega| \to \infty$. The integral along C_1 (the real axis excepting the origin) is the right hand side of (13). From the integral over the vanishingly-small semicircle C_0 over the origin, we obtain the right-hand side of

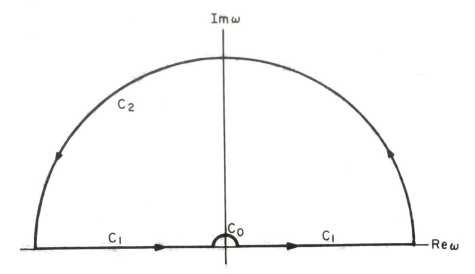

Figure 12-3a Contour used in evaluating (13) to obtain the spatial energy density spectrum. The radius of the large semicircle is taken to infinity.

$$(\tfrac{1}{2}\rho\phi)_{\mathbf{k}} = \frac{T}{2}\left[1 - \frac{1}{\epsilon(\mathbf{k},0)}\right] \tag{14a}$$

$$= \frac{T}{2}\left[\frac{\sum S^2}{\sum S^2 + K^2\lambda_D^2}\right] \tag{14b}$$

where 12-2(6) was used to express $\epsilon(\mathbf{k},0)$. In a finite system, this is also the energy corresponding to one Fourier mode (remembering that components of $\mathbf{k}$ take on both negative and positive values; see (9) and Sections 8-4 and 8-7). Equation (14b) reduces to the familiar result when $\Delta x \to 0$ with $\mathbf{k}$ fixed.

In the point-particle case ("sheet" in one dimension) the Debye shielded potential and the spatial spectrum have the same form. Since the latter is easier to measure in a computer "experiment," it may be preferable, as well as sufficient, to measure only the spatial spectrum in a study of the kinetic properties of simulation plasmas. Comparing 12-2(6) and (14b), we see that the Debye potential and spatial spectrum have the same denominator (collective modifications) but differ somewhat in the numerator (source). Thus, the two are no longer synonymous when $\lambda_D \lesssim \Delta x$.

As in Section 12-2 the sum over $\mathbf{p}$ can be performed analytically to evaluate $\epsilon(\mathbf{k},0)$, and the Fourier transform can be inverted analytically to find the spatial correlation function and total thermal field energy density. The limit $\lambda_D/\Delta x \to 0$ is trivial: different cells are uncorrelated and the field energy per cell is $T/2$ (Problem 12-3d).

(3) $\Delta x \neq 0$, $\Delta t = 0$ **high-frequency noise** Here, the main result is that the spectrum falls off slowly at very high frequencies, as a negative power of ω, instead of being proportional to $f_0(\omega/k)$ which decreases more rapidly. The high frequencies are associated with particles crossing the grid at frequencies $\approx \mathbf{p} \cdot \mathbf{k_g} \cdot \mathbf{v}$.

Assuming $\lambda_D \gtrsim \Delta x / 2\pi$, we have $k_g v_t \gtrsim \omega_p$ so that $\epsilon \approx 1$, leaving

$$(\rho^2)_{k,\omega} \approx 2\pi\rho_0 q \sum_p \frac{S^2}{|k_p|} f_0\left(\frac{\omega}{k_p}\right) \tag{15}$$

in one dimension. In the absence of the spatial grid, a particle contributes to fluctuations at all frequencies $\leqslant |\mathbf{k}||\mathbf{v}|$, but not higher, in two or three dimensions. At a frequency $\omega > k_g v_t$ there are contributions from all alias numbers $|p| \geq p_0 \equiv \omega/(k_g v_t)$. Assuming $f_0(v) \approx 1/2v_t$ for $|v| < v_t$ and $f_0 \approx 0$ otherwise, we can estimate the fluctuation level in the one-dimensional linear-weighting case as

$$(\rho^2)_{k,\omega} \approx 2\pi\rho_0 q \left(\frac{2}{\Delta x} \sin\frac{k\Delta x}{2}\right)^4 \sum_{|p|>p_0} \frac{1}{2v_t |k_p|^5} \tag{16}$$

$$\approx \frac{\pi}{2} \frac{\rho_0 q}{k_g v_t} \left(\frac{\omega}{kv_t}\right)^{-4}$$

approximating the sum by an integral. For nearest-grid-point weighting the noise falls off as ω^{-2}. Graphs of the spectrum and supporting comparisons with measurements on nearest-grid-point and linear-weighting simulations may be found in *Okuda* (1972).

This high-frequency noise need not be very harmful in electrostatic simulations; it occurs at frequencies above those of physical interest, and particles do not respond strongly to such nonresonant fields. (High-frequency grid noise *is* troublesome in electromagnetic simulations; see Chapter 15.) However, with finite Δt, the picture changes for the worse, as we show next.

(4) $\Delta x = 0$, $\Delta t \neq 0$ Finite time-step makes high-frequency noise appear the same as low-frequency noise. This is the meaning of the sum over q in (7). We will see that large Δt also tends to make the spectrum "flatter," *i.e*, vary less with ω, upsetting the balance between velocity diffusion and drag which preserves energy in thermal equilibrium.

With $\Delta x = 0$, and a Maxwellian velocity distribution, the spectrum (7) becomes

$$(\rho^2)_{k,\omega} = \frac{\sqrt{2\pi}\rho_0 q}{|\epsilon|^2 |k| v_t} \sum_q \exp\left(\frac{-\omega_q^2}{2k^2 v_t^2}\right) \tag{17}$$

For $\omega_p \Delta t \ll 1$ the spectrum is unchanged by finite Δt when $kv_t\Delta t \leq 1$, but for $kv_t\Delta t \geq 1$ several terms in the sum over q contribute simultaneously. It is then convenient to use an alternate expression (*Langdon, 1979b*)

$$(\rho^2)_{k,\omega} \approx \frac{\rho_0 q}{|\epsilon|^2} \Delta t \sum_{n=-\infty}^{\infty} \exp\left[i\omega n\Delta t - \frac{1}{2}(kv_t\Delta t)^2 n^2\right] \tag{18}$$

which can be obtained from (17) through application of the Poisson summation formula (*Lighthill, 1962*). For $kv_t\Delta t \gtrsim 1$, Problem 9-8d shows that $\epsilon \approx 1$, with which (18) shows that the spectrum is nearly constant (white noise)

$$(\rho^2)_{k,\omega} \approx \rho_0 q\Delta t \left[1 + 2\exp\left(-\frac{1}{2}k^2 v_t^2\Delta t^2\right)\cos\omega\Delta t + \cdots\right] \tag{19}$$

In this limit, thermal particles move a substantial fraction of a wavelength in one time step; their net contributions to ρ then show no correlation from one time step to the next, hence the white spectrum.

When Δx and Δt are both finite, a white-noise component can appear even when $kv_t\Delta t$ is small.

(5) Δx, Δt both nonzero The main effect discussed here is the appearance at low frequencies of the grid noise, (16). This effect occurs most strongly when $v_t\Delta t \gtrsim \Delta x$; we will consider the case $v_t\Delta t/\Delta x = 0.5$. Then $\omega_g/kv_t = (k_g/k)(\Delta x/v_t\Delta t) > 4$, so (17) is appropriate for the $p = 0$ source term, for a Maxwellian and $k < (v_t\Delta t)^{-1} < k_g/2$. On the other hand, $k_g v_t\Delta t = \pi$ so (18) is appropriate for $p \neq 0$ terms. This suggests rewriting the source as

$$(\rho^2)_{k,\omega} = \frac{\rho_0 q}{|\epsilon|^2}\left[\sum_p S^2\left(\sqrt{2\pi}\sum_q \frac{\exp\left(-\omega_q^2/2k_p^2 v_t^2\right)}{|k_p| v_t} - \Delta t\right) + \Delta t \sum_p S^2\right] \tag{20}$$

The first sum over p converges rapidly once $p \gtrsim (k_g v_t\Delta t)^{-1}$, while the second sum over p can be evaluated analytically as in Section 12-2. For $v_t\Delta t/\Delta x = 0.5$, it is sufficient to keep only the $q = 0$ term. At frequencies $\omega \lesssim kv_t$ the first term is dominant, as desired. For frequencies higher than kv_t and ω_p, this physical term is dominated by white noise,

$$(\rho^2)_{k,\omega} \approx \rho_0 q\Delta t \sum_{p\neq 0} S^2 \tag{21}$$

For linear weighting in one dimension (21) is proportional to $(k\Delta x)^4$ at long wavelengths. For shorter wavelengths, $(v_t\Delta t)^{-1} < k < k_g/2$, we use (18) for all p terms and obtain in one dimension

$$(\rho^2)_{k,\omega} \approx \rho_0 q\Delta t \left[1 - \frac{2}{3}\sin^2\frac{k\Delta x}{2}\right] \tag{22}$$

at all frequencies. This implies a lack of correlation between time steps which was first mentioned by *Hockney* (1966). Our results are in reasonable agreement with the heuristic discussion by *Abe et al.* (1975) of the effects of varying $v_t\Delta t/\Delta x$ on correlation times and other properties.

For nearest-grid-point weighting, $\sum S^2 = 1$ and the spectrum is constant both in $\mathbf{k}$ and ω. For subtracted-dipole weighting (Chapter 11) in one dimension, we find

$$\sum_p S^2 = 1 + \frac{1}{12} \sin^2 k\Delta x$$

which is worse than for nearest-grid-point, and more than three times larger at short wavelengths than for linear weighting. The causes are the discontinuity in $S(x)$, which makes $S^2(k) \propto k^{-2}$ (versus k^{-4} for linear weighting), and the negative values taken on by S, which increases

$$\int_g \frac{dk}{2\pi} \sum_p S^2(k_p) = \int dx\, S^2(x)$$

In practice, users of subtracted-dipole weighting filter out the short wavelengths where the differences are large. Of course, if one is willing to suffer the same loss of resolution in the linear weighting case, then the noise level again is lower than for subtracted-dipole weighting.

PROBLEMS

12-3a The ensemble-averaged fluctuation results (4)-(9) were derived for an infinite system. Reconsider the derivation for a *finite* periodic system taking care of the special cases in which $\mathbf{k}$ or $\mathbf{k}'$ is zero, and of terms $\propto$ (number of particles)$^{-1}$ which do not appear in an infinite system. Assume net charge neutrality.

12-3b Show that $\epsilon(-\mathbf{k}_p, -\omega_q) = \epsilon^*(\mathbf{k}, \omega)$ by extending the results of Probem 9-2b and using the periodicity of ϵ.

12-3c Verify (9).

12-3d Using (14b) in (9) with $\lambda_D \to 0$ and $t_n = t_{n'}$, show that $\langle \frac{1}{2}\rho_j \phi_{j'} \rangle = 0$ if $j \neq j'$, and the field energy per cell is $\Delta x \langle \frac{1}{2}\rho_j \phi_j \rangle = T/2$.

12-4 REMARKS ON THE SHIELDING AND FLUCTUATION RESULTS

Expressions for Debye shielding and the fluctuation spectrum have been derived using an exact mathematical description of the numerical algorithms. The results for a simulation plasma and a real plasma are compared in the case of a Maxwellian velocity distribution. As expected, exact agreement is found when the grid spacing and time step are small. Qualitative differences in Debye shielding and spatial spectrum arise when $\lambda_D \leqslant \Delta x/2$; for example, the Debye potential oscillates as it decays. In the fluctuation spectrum, noise is found at high frequencies on the order of $v_t/\Delta x$. A large time step ($\Delta t \geqslant \Delta x/v_t$) redistributes this noise to all frequencies, producing a flatter spectrum and contributing to velocity diffusion (Section 12-5).

The theoretical results have the same limitations on validity as do the conventional kinetic theory results to which they reduce, except that no short-range divergences arise in simulation plasmas. The theory is valid only for stable plasmas. Yet, we have shown in Chapters 8 and 10 that even a Maxwellian velocity distribution can be destabilized at long wavelengths by the spatial grid. Even if stable, the long wavelengths may not establish equilibrium fluctuations before f_0 has evolved collisionally, violating the adiabatic hypothesis implicit in the derivation. In these cases the expressions are meaningless at such wavelengths. However, it may be argued that the expressions will apply at other wavelengths which are Landau damped, providing amplitudes stay low enough so that linearization is valid.

When quiet starts are used (Sections 5-9 and 5-15, *Byers and Grewal, 1970; Denavit and Kruer, 1971; Denavit and Walsh, 1981*), with their highly correlated initial particle coordinates, recurrences and multibeam instabilities, the computation may be terminated long before this theory would apply. Indeed, it is the intention of the user of a quiet start to postpone the development of thermal fluctuations in simulations of collisionless phenomena.

The existence of simulation models with identifiable particles and classical deterministic dynamics reopens unresolved difficulties of classical statistical mechanics, such as Gibbs's paradox, whose resolution has been obviated by appeal to quantum statistical physics. The unfinished development of classical thermal physics is a legitimate, if not pressing, topic in theoretical physics. Other difficulties, associated with short wavelengths (such as the "ultraviolet catastrophe" and the divergence of point-particle field energies), do not arise in computer simulation models because the number of degrees of freedom is limited by the number of computer variables used to represent the state of the system.

12-5 DERIVATION OF THE KINETIC EQUATION

(a) Velocity Diffusion

We now calculate the effect of a fluctuating field on the velocity distribution of the particles moving through it. First we calculate the variance of the change in a test particle's velocity over a time t_s, due to the accelerations $\mathbf{a}_0^{(1)}$ through $\mathbf{a}_{s-1}^{(1)}$. If this variance is proportional to t_s when t_s is large enough, then a diffusion process is indicated. For leap-frog integration, the change in velocity is

$$\mathbf{v}_{s-\frac{1}{2}}^{(1)} = \Delta t \sum_{r=0}^{s-1} \mathbf{a}_r^{(1)} \tag{1}$$

For the acceleration use the force along the unperturbed orbit. Further, the force field used will be that in the absence of the test particle. (These errors are small compared to the leading terms, and make a contribution to the variance, a quadratic quantity, of higher order than we wish to retain.

However, they both make important contributions to the mean change in velocity, or drag, which we calculate later.) By using the transform of the field we can do the sum over r in (1):

$$\mathbf{v}^{(1)}_{s-\frac{1}{2}} = \frac{1}{m}\int_{\infty}\frac{d\mathbf{k}}{(2\pi)^3}\int_g\frac{d\omega}{2\pi}F(\mathbf{k},\,\omega)\,e^{i\mathbf{k}\cdot\mathbf{x}_s^{(0)}-i\omega t_s}\frac{1-e^{i(\omega-\mathbf{k}\cdot\mathbf{v})t_s}}{e^{-i(\omega-\mathbf{k}\cdot\mathbf{v})\Delta t}-1} \tag{2}$$

For the fluctuations of the force we have

$$\big\langle F(\mathbf{k},\,\omega)\,F(\mathbf{k}',\,\omega')\big\rangle$$
$$= q^2 S(-\mathbf{k})\,S(-\mathbf{k}')\,(EE)_{\mathbf{k},\omega}(2\pi)^4\delta(\mathbf{k}+\mathbf{k}',\mathbf{k}_g)\delta(\omega+\omega',\omega_g) \tag{3}$$

where
$$(EE)_{\mathbf{k},\omega} = \frac{\mathbf{\kappa\kappa}}{K^4}(\rho^2)_{\mathbf{k},\omega} \tag{4}$$

Since the force depends on the continuous variable $\mathbf{x}$, the normal Fourier inverse transform is to be used. The factor $\delta(\mathbf{k}+\mathbf{k}',\mathbf{k}_g)$ rather than $\delta(\mathbf{k}+\mathbf{k}')$ means that the ensemble is not quite uniform for $\mathbf{F}$ (or n). Averages depend periodically on position relative to the grid, as well as on the separation. This is expected since grid points can be equivalent but the situation between and at grid points is different. More on this later.

We now form the ensemble average $\langle\mathbf{v}^{(1)}\mathbf{v}^{(1)}\rangle/2t_s$ which we hope will be a diffusion tensor, i.e., independent of t_s as $t_s\to\infty$. To do so we multiply (2) by itself with $\mathbf{k}$ and ω changed to $\mathbf{k}'$ and ω', so that two transform integrals can be made into one multiple integral. Then ensemble average, use (3) and do the integrals over $\mathbf{k}'$ and ω' to obtain the variance

$$\left\langle\frac{\mathbf{v}^{(1)}_{s-\frac{1}{2}}\mathbf{v}^{(1)}_{s-\frac{1}{2}}}{2t_s}\right\rangle = \frac{q^2}{2m^2}\int\frac{d\mathbf{k}}{(2\pi)^3}\int_g\frac{d\omega}{2\pi}\sum_{\mathbf{p}}S(-\mathbf{k})\,S(\mathbf{k_p})\,(EE)_{\mathbf{k},\omega}e^{-i\mathbf{p}\cdot\mathbf{k}_g\cdot\mathbf{x}_s^{(0)}}$$
$$\cdot\frac{1}{t_s}\frac{1-e^{i(\omega-\mathbf{k}\cdot\mathbf{v})t_s}}{e^{-i(\omega-\mathbf{k}\cdot\mathbf{v})\Delta t}-1}\frac{1-e^{-i(\omega-\mathbf{k_p}\cdot\mathbf{v})t_s}}{e^{i(\omega-\mathbf{k_p}\cdot\mathbf{v})\Delta t}-1} \tag{5}$$

We are not interested in distinguishing between the diffusion for particles according to their present position relative to the grid, so we average $\mathbf{x}_s^{(0)}$ over one cell, which eliminates all but the $p=0$ term:

$$\left\langle\frac{\mathbf{v}^{(1)}_{s-\frac{1}{2}}\mathbf{v}^{(1)}_{s-\frac{1}{2}}}{2t_s}\right\rangle = \mathbf{D}(\mathbf{v},\,t_s) \tag{6}$$

$$\equiv \frac{q^2}{2m^2}\int_{\infty}\frac{d\mathbf{k}}{(2\pi)^3}\int_g\frac{d\omega}{2\pi}S^2(\mathbf{k})\,(EE)_{\mathbf{k},\omega}$$
$$\cdot\frac{\Delta t^2}{t_s}\frac{\sin^2\frac{1}{2}(\omega-\mathbf{k}\cdot\mathbf{v})t_s}{\sin^2\frac{1}{2}(\omega-\mathbf{k}\cdot\mathbf{v})\Delta t}$$

Consider the ω integration for large t_s. The last factor is a peak of width t_s^{-1}. If the spectrum $(EE)_{\mathbf{k},\omega}$ varies little across the peak (meaning roughly that t_s is larger than the field correlation time), then there is a contribution

$$\mathbf{D}(\mathbf{v}) = \frac{q^2}{2m^2} \int_\infty \frac{d\mathbf{k}}{(2\pi)^3} S^2(\mathbf{k}) (\mathbf{E}\mathbf{E})_{\mathbf{k}, \mathbf{k} \cdot \mathbf{v}} \tag{7}$$

which is independent of t_s and therefore is diffusion-like. There is another likely contribution from the frequencies of oscillation of the plasma where the spectrum will be large; this represents oscillation of nonresonant particles in the wave field and approaches the constant

$$\left\langle \mathbf{v}_{s-\frac{1}{2}}^{(1)} \mathbf{v}_{s-\frac{1}{2}}^{(1)} \right\rangle_{\text{nonres}}$$

$$= \frac{2q^2}{m^2} \int \frac{d\mathbf{k}}{(2\pi)^3} \int_g \frac{d\omega}{2\pi} S^2(\mathbf{k}) (\mathbf{E}\mathbf{E})_{\mathbf{k}, \omega} \frac{(\Delta t / 2)^2}{\sin^2 \frac{1}{2}(\omega - \mathbf{k} \cdot \mathbf{v}) \Delta t} \tag{8}$$

in our stationary ensemble. This contribution may be troublesome in attempts to measure diffusion in computer experiments since it decreases slowly, proportional to t_s^{-1}. Eventually $\mathbf{D}(\mathbf{v}, t_s)$ approaches $\mathbf{D}(\mathbf{v})$, which we therefore call the diffusion tensor. (On the other hand, one cannot wait so long that the particles are deflected by more than a wavelength from the unperturbed paths so that our linearization breaks down. Both conditions are satisfied when there are enough particles to make the fluctuations low in amplitude.)

(b) Velocity Drag

If we evaluate the ensemble-average change in particle velocity using the field along the zero-order orbit and neglecting the influence of the selected ("test") particle on the rest of the plasma, we get zero. To obtain two contributions to the drag, we correct first one of these approximations and then the other.

In the first case we are to find the mean change in velocity of a particle moving through given fields described by a fluctuation spectrum. $\mathbf{a}^{(1)}$ is the acceleration given by the field along the zero-order orbit. From 9-3(1), the resulting deflection in the particle position is

$$\mathbf{x}_r^{(1)} = \Delta t^2 \sum_{r'=0}^{r-1} (r - r') \, \mathbf{a}_r^{(1)} \tag{9}$$

The difference between the acceleration along the perturbed path and $\mathbf{a}^{(1)}$ is

$$\mathbf{a}_r^{(2)} = \mathbf{x}_r^{(1)} \cdot \frac{\partial}{\partial \mathbf{x}_r^{(0)}} \, \mathbf{a}_r^{(1)} \tag{10}$$

Although $\mathbf{a}^{(1)}$ averages to zero, $\mathbf{a}^{(2)}$ does not, because of course $\mathbf{a}^{(1)}$ and $\mathbf{x}^{(1)}$ are correlated. There results a mean change in particle velocity

$$\mathbf{v}_{s-\frac{1}{2}}^{(2)} = \Delta t \sum_{r=0}^{s-1} \mathbf{a}_r^{(2)} \tag{11}$$

After going through similar steps as with the diffusion, one finds (Problem 12-5a)

$$\left\langle \frac{\mathbf{v}_{s-\frac{1}{2}}^{(2)}}{t_s} \right\rangle = \frac{\partial}{\partial \mathbf{v}} \cdot \mathbf{D}(\mathbf{v}, t_s) \tag{12}$$

where $\mathbf{D}(\mathbf{v}, t_s)$ is the same expression as in (6). Again as t_s becomes large this approaches a constant, so that the net effect is like a drag acceleration

$$\mathbf{a}_{\text{fluct}} = \frac{\partial}{\partial \mathbf{v}} \cdot \mathbf{D} \tag{13}$$

If we combine the diffusion and drag found so far in a Fokker-Planck description of the slow evolution of the velocity distribution, we obtain

$$\frac{\partial f}{\partial t} = \frac{\partial}{\partial \mathbf{v}} \cdot \left[-f \frac{\partial}{\partial \mathbf{v}} \cdot \mathbf{D} + \frac{\partial}{\partial \mathbf{v}} \cdot f \mathbf{D} \right] \tag{14}$$

$$= \frac{\partial}{\partial \mathbf{v}} \cdot \mathbf{D} \cdot \frac{\partial f}{\partial \mathbf{v}} \tag{15}$$

Since the fluctuation spectrum in $\mathbf{D}$ [(7) with (4)] does not need to be that of 12-3(7) but could be the result of turbulence in a weakly-unstable plasma, we have an alternative derivation of the quasilinear theory particle equation (in the slow growth limit).

The remaining source of drag is due to the distortion of the surrounding plasma by the test particle. It is adequate to treat the plasma as a Vlasov gas and make the test particle move at constant speed. Averaging as usual over one cell, we obtain (Problem 12-5b)

$$\mathbf{F}(\mathbf{v}) = \int_{\infty} \frac{d\mathbf{k}}{(2\pi)^3} q^2 S^2 \frac{\mathbf{\kappa}}{K^2} \operatorname{Im} \frac{1}{\epsilon(\mathbf{k}, \mathbf{k} \cdot \mathbf{v} + i0)}$$

$$= -\int_{\infty} \frac{d\mathbf{k}}{(2\pi)^3} \frac{q^2 \mathbf{\kappa} S^2}{K^2 |\epsilon(\mathbf{k}, \mathbf{k} \cdot \mathbf{v})|^2} \operatorname{Im} \epsilon(\mathbf{k}, \mathbf{k} \cdot \mathbf{v} + i0) \tag{16}$$

This result, but not (13), applies to time integration schemes other than leapfrog (Problem 12-5d).

(c) The Kinetic Equation

Combining (7), (13), (15), and (16) with 12-3(11) into a Fokker-Planck equation for $f(\mathbf{v})$ we obtain the computer simulation plasma analogue of the Balescu-Guernsey-Lenard kinetic equation:

$$\frac{\partial f}{\partial t} = \frac{\partial}{\partial \mathbf{v}} \cdot \pi \frac{\omega_p^4}{n_0} \int \frac{d\mathbf{k}}{(2\pi)^3} \frac{S^2(\mathbf{k})}{|\epsilon(\mathbf{k}, \mathbf{k} \cdot \mathbf{v})|^2} \frac{\mathbf{\kappa}\mathbf{\kappa}}{K^4} \cdot \sum_{\mathbf{p}} S^2(\mathbf{k_p})$$

$$\cdot \int d\mathbf{v}' \, \delta(\mathbf{k} \cdot \mathbf{v} - \mathbf{k_p} \cdot \mathbf{v}', \, \omega_g) \left[\frac{\partial}{\partial \mathbf{v}} - \frac{\partial}{\partial \mathbf{v}'} \right] f(\mathbf{v}) f(\mathbf{v}') \tag{17}$$

This discouragingly lengthy expression can be compared to the real plasma BGL equation (itself quite complex; see *Lenard*, 1960; *Balescu*, 1960;

Guernsey, 1962). Our (17) reduces to their result

$$\frac{\partial f}{\partial t} = \frac{\partial}{\partial \mathbf{v}} \cdot \pi \frac{\omega_p^4}{n_0} \int \frac{d\mathbf{k}}{(2\pi)^3} \frac{1}{|\epsilon|^2} \frac{\mathbf{kk}}{k^4}$$

$$\cdot \int d\mathbf{v}' \delta(\mathbf{k}\cdot\mathbf{v} - \mathbf{k}\cdot\mathbf{v}') \left[\frac{\partial}{\partial \mathbf{v}} - \frac{\partial}{\partial \mathbf{v}'} \right] f(\mathbf{v}) f(\mathbf{v}') \tag{18}$$

in the limit of small Δx and Δt.

The kinetic equation for Lewis' models (Chapter 10) is a special case of a result obtained as above except that no assumptions are made as to periodicity of κ. It is

$$\frac{\partial f}{\partial t} = \frac{\partial}{\partial \mathbf{v}} \cdot \pi \frac{\omega_p^4}{n_0} \int d\mathbf{v}' \int_g \frac{d\mathbf{k}}{(2\pi)^3} \sum_{pp'} \frac{S^2(\mathbf{k_p}) S^2(\mathbf{k_{p'}})}{|\epsilon(\mathbf{k}, \mathbf{k_p}\cdot\mathbf{v})|^2}$$

$$\cdot \sum_q \delta(\mathbf{k_p}\cdot\mathbf{v} - \mathbf{k_{p'}}\cdot\mathbf{v}' - q\omega_g) \frac{\kappa}{K^4} \left[\kappa \frac{\partial}{\partial \mathbf{v}} - \kappa' \frac{\partial}{\partial \mathbf{v}'} \right] f(\mathbf{v}) f(\mathbf{v}') \tag{19}$$

where $\epsilon(\mathbf{k}, \omega)$ is given by 12-2(4) and $\kappa \equiv \kappa(\mathbf{k_p})$, $\kappa' \equiv \kappa(\mathbf{k_{p'}})$. Special cases are the energy-conserving algorithm for which $\kappa = \mathbf{k}$ and the interlaced grid (Problem 12-5c).

For time integration schemes other than leapfrog, such as are used in implicit simulation, there are additional terms (Problem 12-5d).

Our theory does have one advantage over that for a real plasma; there are no difficulties with divergence at large k because the grid smooths out the Coulomb field.

Intermingled here are normal collisions modified by the force smoothing (*Langdon and Birdsall, 1970; Okuda and Birdsall, 1970*) plus all other grid effects such as stochastic heating from the grid noise (*Hockney, 1966, 1971; Hockney and Eastwood, 1981*, Section 9-2). There are few nice approximations to make in interesting cases, but several physical features are shown analytically in the next Section.

PROBLEMS

12-5a Derive (12). *Hints*: Write (10) with (9) as

$$\mathbf{a}_r^{(2)} = \frac{\Delta t}{m^2} \sum_{r'=0}^{r-1} (t_r - t_{r'}) \, \mathbf{F}^{(1)}(\mathbf{x}_{r'}^{(0)}, t_{r'}) \cdot \frac{\partial}{\partial \mathbf{x}_{r'}^{(0)}} \mathbf{F}(\mathbf{x}_{r'}^{(0)} + \mathbf{v}[t_r - t_{r'}], t_r)$$

$$= \frac{1}{m^2} \frac{\partial}{\partial \mathbf{v}} \cdot \Delta t \sum_{r'=0}^{r-1} \mathbf{F}^{(1)}(\mathbf{x}_{r'}^{(0)}, t_{r'}) \, \mathbf{F}^{(1)}(\mathbf{x}_{r'}^{(0)} + \mathbf{v}[t_r - t_{r'}], t_r)$$

Now ensemble average bringing in $(\mathbf{EE})_{\mathbf{k},\omega}$, and average the particle positions relative to the grid as done for (6), to find

$$\mathbf{a}_r^{(2)} = \frac{q^2}{m^2} \frac{\partial}{\partial \mathbf{v}} \cdot \int \frac{d\mathbf{k}}{(2\pi)^3} \frac{d\omega}{2\pi} S^2(\mathbf{k}) \, (\mathbf{EE})_{\mathbf{k},\omega} \, \Delta t \sum_{r'=0}^{r-1} e^{i(\mathbf{k}\cdot\mathbf{v} - \omega)(t_r - t_{r'})}$$

In (11), after dropping terms which do not contribute because of the symmetry $(\mathbf{EE})_{\mathbf{k},\omega} = (\mathbf{EE})_{-\mathbf{k},-\omega}$ and because $\partial/\partial\mathbf{v}$ removes velocity-independent terms, find (12).

12-5b Derive (16), using 12-2(3) and 12-3(2) for a single particle, and the symmetry found in Problem 12-3b.

12-5c Verify that (19) also holds when the electric field is interlaced as in Problem 8-7c, with an appropriate choice for κ.

12-5d Show that (13) is generalized to all time-integration schemes when written as

$$\mathbf{a}_{\text{fluct}} = \frac{q^2}{m^2} \int \frac{d\mathbf{k}}{(2\pi)^3} \frac{d\omega}{2\pi} \mathbf{k} \cdot (\mathbf{EE})_{\mathbf{k},\omega} \; \text{Im} \left[\frac{X}{A} \right]_{\omega - \mathbf{k}\cdot\mathbf{v} + i0} \tag{20}$$

For the leap-frog scheme, there is only the *resonant* contribution as in Problem 9-2b; show this is (13). For noncentered time integration, such as the C and D schemes, there is also a *nonresonant* contribution with $\omega \neq \mathbf{k}\cdot\mathbf{v}$. *Hints:* Express $\mathbf{a}$ in (10) in terms of the transform $F(\mathbf{k}, \omega)$, and use the proportionality of X and A, written as $(X/A)_{\omega-\mathbf{k}\cdot\mathbf{v}}$ as in Chapter 9, to express $x_r^{(1)}$ instead of using (9).

12-6 EXACT PROPERTIES OF THE KINETIC EQUATION

After deriving his kinetic equation, *Lenard* (1960) considers several conservation principles and inequalities which are true microscopically, and the *H*-theorem. His kinetic equation is found satisfactory in all these respects. We begin by making the same checks on our kinetic equation. By using his notation we can avoid duplicating essentially the same manipulations.

The kinetic equation 12-5(20) is

$$\frac{\partial}{\partial t} f(\mathbf{v}, t) = -\frac{\partial}{\partial \mathbf{v}} \cdot \mathbf{J} \tag{1}$$

where the velocity space flux $\mathbf{J}$ is

$$\mathbf{J} = \int d\mathbf{v}' \, \mathbf{Q}(\mathbf{v}, \mathbf{v}') \cdot \left[\frac{\partial}{\partial \mathbf{v}} - \frac{\partial}{\partial \mathbf{v}'} \right] f(\mathbf{v}) f(\mathbf{v}') \tag{2}$$

in which we have from 12-5(20)

$$\mathbf{Q}(\mathbf{v}, \mathbf{v}') = -\pi \frac{\omega_p^4}{n_0} \int_g \frac{d\mathbf{k}}{(2\pi)^3} \sum_{\mathbf{p}\mathbf{p}'} \frac{S^2(\mathbf{k}_\mathbf{p}) S^2(\mathbf{k}_{\mathbf{p}'})}{|\epsilon(\mathbf{k}, \mathbf{k}_\mathbf{p} \cdot \mathbf{v})|^2} \frac{\kappa\kappa}{K^4} \delta(\mathbf{k}_\mathbf{p} \cdot \mathbf{v} - \mathbf{k}_{\mathbf{p}'} \cdot \mathbf{v}', \omega_g) \tag{3}$$

It is convenient to rewrite $\mathbf{Q}$ by making the replacements $\mathbf{p} \to \mathbf{p}'$ and

$$\int_\infty d\mathbf{k} \, g(\mathbf{k}) \to \int_g d\mathbf{k} \sum_{\mathbf{p}} g(\mathbf{k}_\mathbf{p})$$

and employing the periodicity of κ, K, and $\epsilon(\mathbf{k}, \omega)$ with respect to $\mathbf{k}$.

First, particle density is conserved because the collision operator is the velocity divergence of a flux in velocity space which vanishes at large velocities. Thus, when the kinetic equation is integrated over velocity, the collision term can be rewritten using Gauss' divergence theorem as an integral over a velocity-space surface at infinity.

The density f must be positive or zero. Lenard shows that a positive smooth f will not subsequently become negative if the tensor Q is negative (Problem 12-6a). This is true because the quadratic form

$$A \cdot Q \cdot A < 0 \qquad (4)$$

for any A, as Q is a sum and integral of a negative quantity.

Momentum is conserved if

$$Q(v, v') = Q(v', v) \qquad (5)$$

(Problem 12-6b). We verify (5) by interchanging v with v' and p with p' in (3), and then using the δ function and the periodicity of ϵ with respect to frequency to replace its argument $k_{p'} \cdot v'$ by $k_p \cdot v$. Since δ is an even function, we are back to (3) again. Thus momentum is conserved, as it should be since the models conserve momentum exactly.

However, we know that energy is not conserved exactly, and indeed we find we can make no general statement about energy here. (To the order considered the rate of change of total energy is just the rate for kinetic energy.) Lenard's proof of kinetic energy conservation requires that $(v - v') \cdot Q(v, v') \equiv 0$, which is not true here. We leave the question open for the moment.

The H-theorem requires only (4) and (5). We find

$$\dot{H} \equiv \frac{d}{dt} \int dv \, f(v) \ln f(v)$$

$$= \frac{1}{2} \int dv \, dv' \left[\frac{\partial}{\partial v} \ln f - \frac{\partial}{\partial v'} \ln f \right] \cdot Q(v, v')$$

$$\cdot \left[\frac{\partial}{\partial v} \ln f - \frac{\partial}{\partial v'} \ln f \right] f(v) \, f(v')$$

$$< 0 \qquad (6)$$

with equality not possible. Thus, there is no stationary f, not even the Maxwellian! This remains true in one dimension or when either Δx or Δt is separately set to zero. We can say the space-time grid creates entropy even for a plasma which should have the greatest possible entropy for the given density, momentum, and energy. Since H is then an extremum, the only way it can be changing is if a constraint is changing. In this case it must be that the energy is increasing. Indeed, for the Maxwellian case one can show the energy is increasing by the right amount to account for the change in entropy:

$$\frac{1}{2 v_t^2} \frac{d}{dt} \overline{v^2} = -\dot{H}$$

$$= -\frac{1}{2 v_t^4} \int d\mathbf{v} \, d\mathbf{v}' \, (\mathbf{v} - \mathbf{v}') \cdot \mathbf{Q} \cdot (\mathbf{v} - \mathbf{v}') \, f(\mathbf{v}) \, f(\mathbf{v}')$$

$$> 0 \qquad\qquad (7)$$

The H-theorem result provides a likely expression to study further as something which is due solely to the non-physical heating of the model, and which is easily measured in a computer experiment (*Montgomery and Nielson, 1970*). In fact, the total energy is commonly monitored in simulation codes.

For "energy-conserving" models, again following Lenard, we can show that f remains positive and particles are conserved. Since momentum and energy are not conserved microscopically, it is to be expected that they are not conserved by the kinetic equation (except for the energy when $\Delta t = 0$, Problem 12-6c).

A Maxwellian distribution which is not drifting is constant in the $\Delta t = 0$ limit. Otherwise the H-theorem shows that f changes in a fashion which increases entropy ($\dot{H} < 0$).

If f is instantaneously Maxwellian we can say more about the rates of change of momentum, energy, and H. We find that the changes in energy and momentum have no obvious sign separately, but the combination

$$\frac{d}{dt} \frac{1}{2} \overline{(\mathbf{v} - \overline{\mathbf{v}})^2} = -v_t^2 \dot{H} > 0 \qquad\qquad (8)$$

shows that the spread in the drifting frame is increasing as $\overline{v}$ increases: this is how entropy is increased.

When $\Delta t = 0$ we can see that $\overline{\mathbf{v}}$ decreases. Thus, one can say that "collisions with the grid" slow the drift, and the velocity spread increases so as to maintain $\overline{v^2}$ constant.

With finite but small time step we still expect the drift to slow, or rather to move toward the nearest $\mathbf{j} \cdot \Delta \mathbf{x} / \Delta t$, since for such a velocity the grid looks stationary to the plasma.

In each case, the breakdown of a physical property is due to *aliasing* effects, not to "softening" of the $p = 0$ force. The converse need not hold: momentum conservation is readily obtained and conservation of energy in Hamiltonian models is not affected by spatial aliasing.

PROBLEMS

12-6a Show that the collision equations (1), (2) keep f positive if (4) holds. *Hints:* Postulate that f is about to become negative at velocity $\mathbf{v}$, but is $\geqslant 0$ elsewhere. Then $f(\mathbf{v}) = 0$, $\partial f / \partial \mathbf{v} = 0$ and $\partial^2 f / \partial \mathbf{v} \partial \mathbf{v}$ is a non-negative tensor. Thence

$$\frac{\partial f}{\partial t} = -\int d\mathbf{v}' \, f(\mathbf{v}') \, \mathbf{Q}(\mathbf{v}, \mathbf{v}') : \frac{\partial^2 f}{\partial \mathbf{v} \partial \mathbf{v}} > 0$$

12-6b Show that the collision equations (1), (2) conserve momentum if (5) holds. *Hints*: Show that

$$\frac{d}{dt} \int d\mathbf{v} \, \mathbf{v} f = -\int d\mathbf{v} \, \mathbf{v} \frac{\partial}{\partial \mathbf{v}} \cdot \mathbf{J} = -\int d\mathbf{v} \, \mathbf{J}(\mathbf{v})$$

where we integrate by parts. In the integral of (2), the integrand changes sign on interchange of $\mathbf{v}$ and $\mathbf{v}'$ if (5) holds.

12-6c Show that 12-5(19), the kinetic equation for Hamiltonian models, conserves energy exactly, as expected, in the continuous-time limit $\Delta t \to 0$. *Hints*: in (19), keep only the $q = 0$ term. Form

$$\frac{d}{dt} \int d\mathbf{v} \, \frac{v^2}{2} f = -\pi \frac{\omega_p^4}{n_o} \int \frac{d\mathbf{k}}{(2\pi)^3} d\mathbf{v} \, d\mathbf{v}' \sum_{\mathbf{p}\mathbf{p}'} \frac{S^2 S^2}{|\epsilon|^2} \delta(\mathbf{k}_\mathbf{p} \cdot \mathbf{v} - \mathbf{k}_{\mathbf{p}'} \cdot \mathbf{v}') \frac{\mathbf{k}_\mathbf{p} \cdot \mathbf{v}}{K^4} (\cdots) ff$$

by integration by parts. Write an expression which is equal by interchanging $\mathbf{v}, \mathbf{v}'$ and $\mathbf{p}, \mathbf{p}'$; average the two. Lastly, note that the factor $(\mathbf{k}_\mathbf{p} \cdot \mathbf{v} - \mathbf{k}_{\mathbf{p}'} \cdot \mathbf{v}') \, \delta(\mathbf{k}_\mathbf{p} \cdot \mathbf{v} - \mathbf{k}_{\mathbf{p}'} \cdot \mathbf{v}')$, now appearing in the integrand, is zero.

12-7 REMARKS ON THE KINETIC EQUATION

We have derived results for the fluctuations and collisions, including exactly the effects of the space and time differencing. The corresponding results for real plasmas are recovered in the limit of small grid spacing and time step. The collision integrals are examined for several properties which, for physical reasons, should hold exactly. Our kinetic equations fail to retain a physical property in just those cases where the model itself does not. Thus, the defects are not in the kinetic equation but in the models, and these microscopic errors do not average out to zero as one might hope. Furthermore, these nonphysical properties are in qualitative agreement with what the models are observed to do in practice. All this lends credibility to the analysis.

Of course, the results suffer from the same difficulties as for real plasmas with regard to the adiabatic hypothesis at small k, and are similarly limited to stable systems. However, the large k divergences of classical plasma theory are absent.

The results apply equally well to one, two, and three dimensions with the appropriate adjustment to the k integral. Note in particular that the 1-d collision integral does not vanish identically as it does for a sheet plasma. Therefore, when grid effects become important, 1-d collision times will be proportional to $N_D = n\lambda_D$ rather than to N_D^2. Here may be the explanation for the decrease rather than increase in collision time observed by *Montgomery and Nielson* (1970) when Δx was increased above λ_D.

Hockney (1971) has made the interesting experimental observation that, as $v_t \Delta t / \Delta x$ is increased from below to above unity, the ratio of heating time to velocity scattering time decreases rapidly from well above to below unity. We suspect this is due to the change in character of the grid noise when $v_t \, \Delta t > \Delta x$ (Section 12-3(5)), decreasing the velocity drag relative to the diffusion. Careful numerical evaluation of the predictions of this theory in various regimes has not been done.

Even without detailed numerical evaluation, the theory provides guidance for attempts at heuristic estimates. In Chapter 13 we consider speculations by *Abe et al.* (1975), *Hockney and Eastwood* (1981), and others. The theory also suggests what to expect in regimes not yet empirically explored.

Currently, this kinetic theory is helping understand the connection of nonphysical cooling to the use of damped time integration schemes such as are used in implicit simulations.

THIRTEEN

KINETIC PROPERTIES: THEORY, EXPERIENCE, AND HEURISTIC ESTIMATES

13-1 INTRODUCTION

In this Chapter we discuss insights into kinetic behavior of plasma, following the pioneering work of Dawson and others. *Dawson* (1962) showed that $N_D \sim 5-20$ was sufficient to simulate many properties of warm one dimensional plasmas. This last was a revelation to plasma physicists who restricted their thinking to the 3d real world where $N_D = 10^5-10^{10}$ is common. Other papers, by Dawson and by Eldridge and Feix, established the foundations for one-dimension plasma simulation. Nonphysical heating or cooling in 1 and 2d is also discussed here.

13-2 THE ONE-DIMENSIONAL PLASMA IN THERMAL EQUILIBRIUM

(a) The Sheet Model

The early one dimensional plasma models of *Dawson* (1962) and *Eldridge and Feix* (1962, 1963) used charges in the form of thin sheets moving through a uniform, immobile, neutralizing background. As with ES1, the model was periodic. The fields were obtained directly from particle positions, without use of a spatial grid. Runs were restricted to $\sim 100-1000$ sheets.

The initial model, of *Dawson* (1962), shown in Figure 13-2a, consisted of N mobile charged sheets of one sign (say, electrons, $q < 0$) moving in an immobile background of the other sign, uniform over a length L, of charge density $\rho_0 = -qN/L$. The sheets were spaced apart by $\delta = L/N = 1/n$ initially. If unperturbed, the sheets would remain that way. If the ith sheet x_i were displaced from its equilibrium position x_{0i} with $|x_i - x_{0i}| < \delta$, then the field averaged across the sheet $\bar{E}$ would no longer be zero and there would be a force acting to return the sheet to equilibrium, through

$$\frac{d^2}{dt^2} m(x_i - x_{0i}) = q\bar{E}(x_i) \tag{1}$$

$$= -q\rho_0(x_i - x_{0i}) \tag{2}$$

or
$$\frac{d^2}{dt^2}(x_i - x_{0i}) = -\omega_p^2(x_i - x_{0i}) \tag{3}$$

Hence, each sheet, if not crossing another sheet, exhibits simple harmonic motion at the plasma frequency. Displacements which include crossings of the sheets amount either to exchange of equilibrium positions or to perfectly elastic reflection with exchange of velocities; the only difference is in the names of the particles. Implementation is discussed by *Dawson* (1970).

Dawson made a number of checks on his 1962 calculations. First, the conservation of energy was checked. For the 1000-sheet system with $n\lambda_D = 10.5$, the system lost 7 parts in 1000 of its energy during 18 plasma oscillations (2200 time steps). Each sheet was crossed by roughly 2000 other sheets during this time, so that in crossing another sheet, a given sheet lost

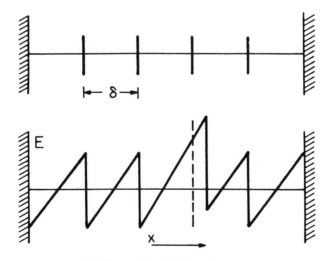

Figure 13-2a Original *Dawson* (1962) model, with thin electron sheets spaced $\delta = 1/n$ apart (in equilibrium) in a uniform positive ion background. The lower part shows $E(x)$ with one sheet displaced.

on the average 3 parts in 10^6 of its energy. It was possible to increase the accuracy at the expense of speed by shortening the time step. However, this did not seem worthwhile.

Second, the motion of a 9-sheet system was reversed and found to retrace its path within an accuracy of one part in 10^3 (all orbits were this accurate) over a period of 6 oscillations.

Third, the drag on a particle was the same in the negative and positive time directions. Thus even though the calculations had a definite direction, forward in time, the results were symmetric in time.

(b) The Equilibrium Velocity Distribution is Maxwellian

The numerical results were obtained by starting the system near thermal equilibrium. To approximate a Maxwellian velocity distribution, the particles were distributed among 16 uniformly spaced velocity groups; the number of particles in a group was proportional to $\exp(-v^2/2v_t^2)$, where v was the velocity of the group. The velocity of a particle i was chosen randomly in the following way. The velocities of all the particles were put on cards (do our present readers know about computer cards?) which were thoroughly shuffled. The resultant deck was used to give the initial velocities of the sheets. All sheets were started at their equally spaced equilibrium positions.

Figure 13-2b shows the time averaged velocity distribution for a system of 1000 sheets with $n\lambda_D = 5.16$. The smooth curve is the theoretical Maxwell distribution obtained by assuming the kinetic energy was equal to

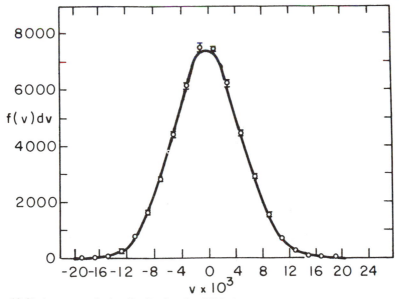

Figure 13-2b Average velocity distribution for 1000 sheet system. The curve is the Maxwellian for the system kinetic energy. The velocity bins were $0.4v_t$ wide. $v_t = 5.1 \times 10^{-3}$, $n_0\lambda_D = 5.1$. *(From Dawson, 1962.)*

the total energy minus the average potential energy (only 8% of the total energy).

The agreement between the numerical results and the theoretical curve is good. However, it is not as good as would be expected if the samples taken at different times were statistically independent. The error in this case would be on the order of $[N(v)\Delta v]^{\frac{1}{2}}$ and is indicated by the error bars. In these calculations the system was sampled three times per plasma period $2\pi/\omega_p$. The *relaxation time* (the time required for a Maxwell distribution to be re-established after a small perturbation) is on the order of $[(2\pi)^{\frac{1}{2}}N_D/\omega_p]$. The system was sampled roughly six times during a relaxation time and thus the expected deviations should be of order $[6N(v)\Delta v]^{\frac{1}{2}}$ rather than $[N(v)\Delta v]^{\frac{1}{2}}$. Fluctuations decreased to the expected statistical fluctuations when the sampling time became comparable with the relaxation time. These results show that the system was constantly fluctuating about its thermal equilibrium state and that agreement with theory is not simply due to the fact that the system was started out with roughly a Maxwell distribution.

(c) Debye Shielding

A charge embedded in a plasma repels charges of like sign and attracts charges of the opposite sign. Thus, around such a charge, a cloud of charge of the opposite sign forms. This cloud contains on the average an equal, but opposite sign of charge to the embedded one. The size of this charge cloud is the Debye length, $\lambda_D \equiv v_t/\omega_p$. Outside this cloud, the system of the particle plus cloud looks neutral. This *Debye shielding* of such an embedded charge was investigated in the single-species charge sheet model.

Debye shielding for a 1000-sheet system with $n\lambda_D = 5.16$ is shown in Figure 13-2c. The points are the average number of particles between 0 and 1, 1 and 2, etc., intersheet spacings from a test sheet. To obtain these averages, the number of sheets within each interval was counted for every tenth sheet. This was repeated at a large number of different times and the average of the whole group found. The shielding amounts to one sheet on the average being absent from a region the size of a Debye cloud which contains many sheets (in this case, $10 \approx 2n\lambda_D$). This small bias can be masked by the fluctuating density in the neighborhood of the test sheet due to random motion of the sheets. Sometimes one finds 12 sheets in the Debye length, other times 8 or 9. The solid curve in Figure 13-2c is the theoretically predicted curve; the points are those obtained from the numerical experiment. The error bars are the statistical uncertainties due to the fact that we have used a finite number of test sheets.

The smooth curve is the theoretical curve obtained from the linearized Debye theory. It is the solution of the linearized form of Poisson's equation,

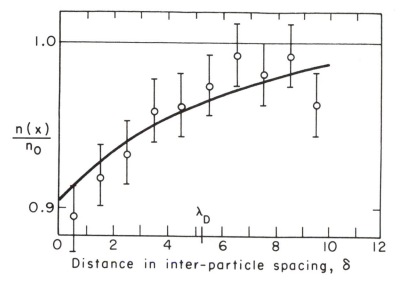

Figure 13-2c Average density of electrons around a test electron sheet at $x = 0$. The curve is the Debye shielding prediction. $n\lambda_D = 5.16$. *(From Dawson, 1962.)*

$$\frac{d^2\phi}{dx^2} = q\,n_0\left[1 - \exp\left(\frac{q\phi}{T}\right)\right] \tag{4}$$

$$\approx \frac{q^2 n_0\,\phi}{T} = \frac{\phi}{\lambda_D^2}$$

where the linearization of the Boltzmann factor assumes $q\phi \lesssim T$. With the boundary conditions at the test charge, *i.e.*, at $x = 0$,

$$E_\pm(0) = -\left(\frac{d\phi}{dx}\right) = \mp\frac{q}{2} \tag{5}$$

the linearized solution is

$$n(x) = n_0\left[1 - (2\,N_D)^{-1}\exp\left(-\frac{|x|}{\lambda_D}\right)\right] \tag{6}$$

$$\phi = \frac{q}{2}\,\lambda_D \exp\left(-\frac{|x|}{\lambda_D}\right) \tag{7}$$

in agreement with 12-2(10). Note that when $N_D \gg 1$, the density depression is small and hence hard to find.

Debye shielding was one of the more difficult quantities to obtain from the machine calculations. This was due to the fact that the statistical error in the density was of the order of $N^{-1/2}$, where N is the number of test charges averaged over. Thus, for the above case, where the maximum depression of the density is 10%, we require 100 cases before the depression equals the fluctuations. To obtain the density depression of 10% accuracy requires 10^4 samples.

Hockney (1971) and *Okuda* (1972) show how to measure λ_D in gridded plasmas from the spatial correlation of the electric field, finding excellent agreement with prediction. *Okuda* (1972) measures both spatial and temporal correlations and spectra, and compares to theory including grid effects.

PROBLEMS

13-2a Sketch $E(x)$, $\phi(x)$, $n_1(x)$ and the pressure force, $\propto (\partial n_1/\partial x)/n_0$ about the stationary test charge.

13-2b Make a table for $N_D = 5, 20, 100$ of the relative depression in n, and of the number of samples needed for the depression to equal the error and to equal the error/10.

(d) Velocity Drag

Consider a homogeneous, one-dimensional plasma with one species of particles plus a neutralizing background. A particle moving through the plasma polarizes it $(v < v_t)$, or excites a plasma oscillation $(v > v_t)$ and is thereby slowed down. It is also accelerated by random electric fields produced by other particles. At time t_0, a group of test particles having nearly the same velocity is selected. At later times, their velocities diffuse into a larger interval. The velocity average of the group decays due both to the polarization or wake and, often overlooked, the fluctuating fields. Equations 12-5(7), 12-5(16), and 12-5(13) describe these contributions.

Consider the drag on a very fast or superthermal sheet in an infinite plasma. Take the velocity of the sheet to be positive. The plasma ahead of such a sheet cannot know of its approach. Thus, there can be no disturbance and hence no electric field ahead of the sheet. However, in going from the negative to the positive side of the sheet the electric field must fall by q as shown in Figure 13-2d. The electric field felt by the sheet is the average of the field on the left and the right, $\bar{E} = q/2$. Its deceleration is

$$\frac{dv}{dt} = \frac{q\bar{E}}{m} = -\frac{\omega_p^2}{2\,n_0} = -\frac{\omega_p v_t}{2\,N_D}. \tag{8}$$

independent of the velocity. The energy is lost to excitation of a wake of plasma oscillations (Figure 13-2d and Problems 13-2a and 13-2b).

Figure 13-2e shows the average absolute velocity as a function of time for two groups of fast particles. The initial velocity for the group represented by the circles is $2.35\,v_t$, while that for the triangles is $-2.35\,v_t$. The particles in these groups are followed in time and the average velocities of the two groups (as functions of time) are recorded. The agreement with (8) is good.

If the system (and hence the code) is time reversible, the drag in the negative time direction should be the same as in the forward time direction. This is found to be the case. At some time t_0, a group of particles with

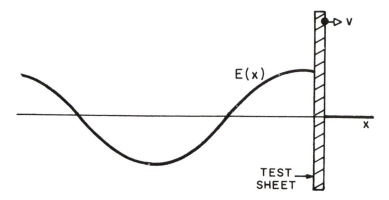

Figure 13-2d The electric field excited by a fast sheet $(v \gg v_t)$ moving to the right. *(From Dawson, 1962.)*

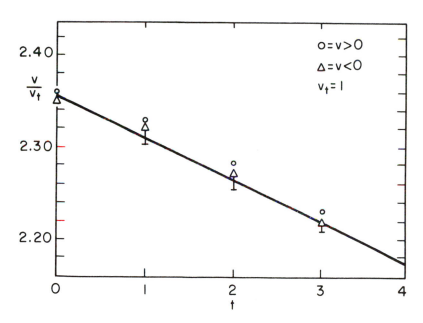

Figure 13-2e Average drag or deceleration on a group of fast particles, starting at $|v| = 2.35\, v_t$ for a 1000 sheet system; $N_D = 10$. The straight line is (8). *(From Dawson, 1962.)*

velocities in the vicinity of 2 v_t is selected. On a plot of the average velocity of the group as a function of $|t - t_0|$, the data for $t - t_0$ positive and for $t - t_0$ negative lie on the same line.

    Eldridge and Feix (1962) calculate the drag and diffusion and obtain agreement with the measurements of *Dawson* (1962). *Eldridge and Feix* (1963) presents theory and measurements of drag and diffusion, as well as considerable physical insight.

The polarization drag is obtained from the linearized Vlasov equation, as in Problem 12-5b. Ignoring grid effects, 12-5(16) yields for one dimensional plasma

$$a_{pol} = \frac{q^2}{m} \int_{-\infty}^{\infty} \frac{dk}{2\pi} \, \text{Im} \, \frac{1}{k\epsilon \, (k, kv+i0)}. \tag{9}$$

Noting that $\epsilon \, (k, kv+i0)$ can be written in the form $\epsilon = 1 + [X(v) + i \, \text{sign}(k) \, Y(v)]/k^2\lambda_D^2$, the integral can be rewritten to find (Problem 13-2c)

$$a_{pol} = \frac{q^2}{m} \int_0^{\infty} \frac{da}{2\pi} \frac{Y(v)}{[a + X(v)]^2 + Y^2(v)} \tag{10a}$$

$$= -\frac{q^2}{2\pi m} \arctan \frac{Y}{X} \tag{10b}$$

where

$$X + i \, Y = \int \frac{dv'}{v - v' + i0} \, v_t^2 \, \frac{\partial f}{\partial v'}$$

For a Maxwellian,

$$X = 1 - \left[\frac{v}{v_t}\right]^2 + \frac{1}{3}\left[\frac{v}{v_t}\right]^4 - \cdots$$

and

$$Y = \left[\frac{\pi}{2}\right]^{1/2} \frac{v}{v_t} \exp\left[\frac{-v^2}{2v_t^2}\right]$$

For small velocities, $v \lesssim v_t$, we find

$$a_{pol} = -\frac{1}{2(2\pi)^{1/2}} \frac{q^2}{m} \frac{v}{v_t} \left[1 + \left[\frac{1}{2} - \frac{\pi}{6}\right]\frac{v^2}{v_t^2} + \cdots\right] \tag{11}$$

$$\approx -\frac{v}{2(2\pi)^{1/2}} \frac{\omega_p}{N_D}$$

For high velocities we find (Problem 13-2d)

$$a_{pol} \to \mp\frac{q^2}{2m} = \mp\frac{\omega_p^2}{2 n_0} \tag{12}$$

as $v/v_t \to \pm\infty$, in agreement with (8).

As in Chapter 12, we assume that the distribution function of the test particles $f(v, t)$ obeys an equation of the Fokker-Planck type:

$$\frac{\partial f}{\partial t} = \frac{\partial}{\partial v}\left[-a_{pol} f - a_{fluct} f + \frac{\partial}{\partial v}(Df)\right] \tag{13}$$

$$= \frac{\partial}{\partial v}\left[-a_{\text{pol}} f + D \frac{\partial f}{\partial v}\right]$$

where the second form follows using $a_{\text{fluct}} = \partial D/\partial v$ from 12-5(13). The diffusion D is obtained by detailed balance: when $f(v)$ is the Maxwell distribution, it does not change with time. So

$$D(v) = -\frac{v_t^2 \, a_{\text{pol}}(v)}{v} \tag{14}$$

For small velocities, we find from (12),

$$D = \frac{v_t^2}{2(2\pi)^{1/2}} \frac{\omega_p}{N_D} \tag{15}$$

$$a_{\text{pol}} + a_{\text{fluct}} = -\frac{\pi v}{6(2\pi)^{1/2}} \frac{\omega_p}{N_D}$$

Note that a_{fluct} is only 5% of a_{pol} in this case. At small velocities, D is easier to measure than the drag. These expressions, first obtained by *Eldridge and Feix* (1962), agree with the measurements of *Dawson* (1962).

(e) Relaxation Time

For a fast sheet, we have deceleration as

$$\frac{1}{v_t} \frac{dv}{dt} = -\frac{\omega_p}{2 N_D} \tag{16}$$

and for a slow sheet, we have

$$\frac{1}{v} \frac{dv}{dt} = -\frac{\omega_p}{2 N_D} \frac{(\pi/2)^{1/2}}{3} \tag{17}$$

where $(\pi/2)^{1/2}/3 = 0.42$, and

$$\frac{D}{v_t^2} = \frac{\omega_p}{2 N_D} \frac{1}{(2\pi)^{1/2}}$$

Hence, we see that there is a time τ

$$\tau \equiv \frac{2 N_D}{\omega_p} \tag{18}$$

that is significant for slowing of either fast or slow particles. τ gives roughly the length of time for a fast or slow particle to change its velocity significantly, which may also be thought of as a randomization time or (to quote Dawson) the time for the plasma to forget the state it was in. If we want two measurements of the velocity distribution to be statistically independent, then we make them at intervals separated by at least τ. As $N_D \geqslant 10$ is usually used, intervals $\geqslant 20/\omega_p$, *i.e.*, greater than 3 plasma cycles, should be used.

For small departures from a Maxwell distribution, the system relaxes to equilibrium in a time roughly equal to the time it takes a slow particle to be stopped, or the length of time it takes a group of particles with a definite velocity to spread out into a Maxwell distribution. There is, of course, no unique relaxation time. However, (18) is a reasonable measure of relaxation time, for a selected group of particles, in a Maxwellian, to acquire the full Maxwellian.

PROBLEMS

13-2a Calculate the wake of a fast sheet, Figure 13-2d, and the drag (8), by considering that a sheet's equilibrium position x_{0i} is displaced a distance $\delta = 1/n$ to the left when the fast sheet crosses it. Show that the energy (kinetic plus electric) density of the wake oscillations is $q^2/2$. The rate of increase of wake energy, as more sheets are brought into oscillation, is $q^2 v/2$, in agreement with (8). Show that the amplitude of the density oscillation in the wake is smaller than the unperturbed density by a factor $\geq N_D$.

13-2b Show analytically that the electric field in front of and behind a fast sheet is

$$
E(x_s + x) = \begin{cases} 0 & x > 0 \\ -q \cos \dfrac{\omega_p x}{v} & x < 0 \end{cases}
\tag{19}
$$

as shown in Figure 13-2d. The sheet is moving to the right; x_s is its instantaneous position. *Hints:* Use 12-2(3) with 12-3(2) specialized to $\Delta x, \Delta t \to 0$ and a single particle. For ϵ use the cold plasma result $\epsilon(k, \omega) = 1 - \omega_p^2/(\omega + i0)^2$. Fourier inversion leads to

$$
E(x_s + x) = \int \frac{dk}{2\pi i} \frac{q}{k} \frac{e^{ikx}}{1 - \omega_p^2/(kv + i0)^2}
\tag{20}
$$

after the trivial integration over ω. The apparent pole at $k = 0$ has vanishing residue $-(i0)^2/\omega_p^2$. The "$i0$" makes clear which poles are enclosed when the contour is closed on an infinite semicircle above or below the real k axis.

13-2c Derive (10). *Hints:* In (9) separate out the integral over $(-\infty, 0)$, substitute $k \to -k$, and add to the integral over $(0, \infty)$.

13-2d Derive (12). *Hint:* As v/v_t is increased, X passes through zero, after which the relevant branch of the inverse tangent in (10b) is $\pi/2 < \arctan \leqslant \pi$.

13-3 THERMALIZATION OF A ONE-DIMENSIONAL PLASMA

As we have seen, the average motion of a sheet moving through a plasma is reduced due to the drag it feels. At the same time it is accelerated by the random fields present in the plasma. The velocity acquired from these random accelerations gives rise to a diffusion or spreading of the velocities of a group of particles. The slowing down and random acceleration compete against each other. Ultimately, they produce the stable Maxwellian distribution. Only for this distribution does the diffusion or spreading to higher velocities exactly balance the slowing down of the particles.

Here we consider the evolution of the distribution functions due to drag and diffusion, as described by a kinetic equation which (in 1d) does not take the system to thermal equilibrium, that is, all species Maxwellian at the same temperature. Next we describe the much slower approach to thermal equilibrium. Finally, we mention additional features due to Δx and Δt effects.

(a) Fast Time-Scale Evolution

Specialized to one-dimension and neglecting Δx and Δt, the kinetic equation 12-5(17) may be simplified. For later use, we generalize it to multiple species. For species a,

$$
\frac{\partial f_a}{\partial t} = \frac{\partial}{\partial v} \int dv' \sum_b \pi \, n_b \, q_b^2 \, \frac{q_a^2}{m_a} \int \frac{dk}{2\pi} \frac{\delta(kv - kv')}{k^2 |\epsilon(k,kv)|^2}
\tag{1a}
$$

$$
\cdot \left[\frac{1}{m_a} \frac{\partial}{\partial v} - \frac{1}{m_b} \frac{\partial}{\partial v'} \right] f_a(v) f_b(v')
$$

$$
= \frac{\partial}{\partial v} \sum_b \frac{1}{2} n_b \, q_b^2 \, \frac{q_a^2}{m_a} \, Q(v) \left[\frac{f_b}{m_a} \frac{\partial f_a}{\partial v} - \frac{f_a}{m_b} \frac{\partial f_b}{\partial v} \right]
\tag{1b}
$$

where

$$
Q(v) = \frac{1}{\chi_i} \arctan \frac{\chi_i}{\chi_r}
\tag{2}
$$

and $\epsilon = 1 + X_r + i \, X_i$ includes contributions from all species (Problem 13-3a).

For a single species plasma, (1b) (good to first order in the small parameter $1/N_D$) predicts that the diffusion in velocities balances the drag and hence there is *no evolution to a Maxwellian*. One may give a simple physical argument why this should be so. Let two particles interacting through a short range force in one dimension, have velocities v_1 and v_2 before the encounter and $\tilde{v}_1$ and $\tilde{v}_2$ after the encounter. Two quantities are conserved for an isolated encounter, the energy $[\tfrac{1}{2} m(v_1^2 + v_2^2)]$ and the momentum $[m(v_1 + v_2)]$. There are only two choices for $\tilde{v}_1$ and $\tilde{v}_2$ which conserve these quantities; first, no change in velocities, $\tilde{v}_1 = v_1$ and $\tilde{v}_2 = v_2$, or, second, an interchange of velocities, $\tilde{v}_1 = v_2$ and $\tilde{v}_2 = v_1$. In either case the number of particles with a given velocity is not changed. One might expect that in a plasma this simple two-isolated-particle-collision argument might not apply since many particles are interacting at the same time due to the long range of the forces. However the (present) theory assumes that all interactions are weak such that, even though there are many simultaneous collisions, they do not interfere with each other so that their effects are simply additive. Thus, the theory predicts *no change* in the distribution function.

This argument applies to a single-species plasma. A more general result follows from (1b) (Problem 13-3b):

$$\frac{\partial}{\partial t} \sum_a n_a \, m_a \, f_a = 0 \qquad (3)$$

This constraint inhibits equilibration between species.

To further understand the evolution of particle velocities in a single species plasma, let us label some of the particles as species a while the rest are species b. For species a we might select all the particles whose initial velocities are in a small interval, or are negative. Equation (1b) describes the evolution of their velocities. At the same time, species b evolves such that $n_a \, f_a + n_b \, f_b$ remains constant.

(b) Slow Time-Scale Evolution

In reality, the simultaneous encounters do affect each other and there is some change in the distribution function. The rate of relaxation to Maxwellian was measured by *Dawson* (1964). The problem investigated was that of the time development of a velocity distribution which initially had the square profile shown in Figure 13-3a. After a short time (about $1/\omega_p$) of rapid adjustment that rounds off the corners of the distribution, the velocity distribution evolves very slowly.

Experimentally, Dawson obtained the distributions after $t = 0$ by counting the numbers of particles in a small velocity intervals (bins), of size $\Delta v = <v^2>^{1/2}/10$ and assigning this number to the center of the velocity interval. (One could be more accurate by using, say, linear interpolation to the two nearest velocity coordinates.) The distributions to be shown are short time averages of all 1000 sheets, made from about six samples taken over a period of a plasma oscillation; this averaging gives relatively smooth $f(v)$ by taking out rapid fluctuations (which are discussed later).

Figure 13-3a shows $f(v)$ for $N_D = 2.5, 5, 10$, and 20 at times $6/\omega_p$, $21/\omega_p$, $41/\omega_p$, and $81/\omega_p$ after initiation of the experiment. Except for the first time, these times correspond roughly to the time required for a group of test particles in a Maxwellian to relax to the full Maxwellian, as noted in the previous section. As readily seen, $f(v)$ tends to retain its initial shape as N_D increases, even with the sampling times chosen proportional to N_D. Hence, the relaxation time for a non-Maxwellian is *not* proportional to N_D and must increase more rapidly than N_D. Actually, most of the variation away from the initial distribution shown here was produced during $1/\omega_p$, caused by setting up particle correlations. After this transient, each $f(v)$ changed only slightly.

The complete relaxation is shown in Figure 13-3b. The first frame shows $f(v)$ at about $\tau = 2N_D/\omega_p$, the previous estimate of relaxation time, and the last shows $f(v)$ very close to its final Maxwellian shape.

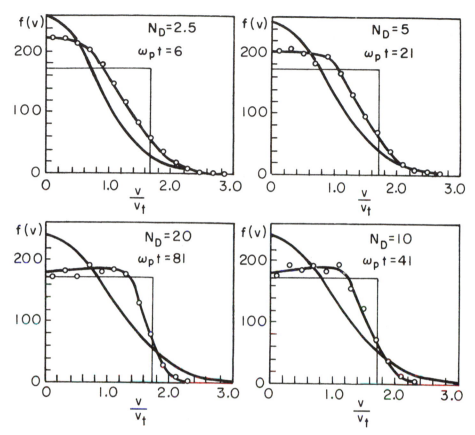

Figure 13-3a Velocity distributions of 1000 sheets for various values of N_D, at times $\omega_p t = 6, 21, 41, 81$. These intervals correspond to about $2\tau = 4N_D/\omega_p$ (except for 6), about twice the time required for test particles in a Maxwellian to relax to the full Maxwellian. Obviously this time does not apply here. *(From Dawson, 1964.)*

Dawson (1964) measured a relaxation time as follows. Note that $f(0)$ in Figure 13-3b creeps up to its final value, as plotted in Figure 13-3c for $N_D = 7.5$ (each point being an average, as noted earlier). The straight line through the points is obtained from least squares fit of the data. The time at which $f(0, t)$ intersects $f(0)$ of its Maxwellian is taken to be the relaxation time τ. The relaxation times are shown as a function of N_D^2 in Figure 13-3d and fit very nicely to

$$\tau = 10 \, N_D^2 \tag{4}$$

Dawson's interpretation is that the simultaneous interaction of three particles gives rise to the relaxation, since the relaxation time due to two-particle interactions would be proportional to $n\lambda_D$, if it did not cancel out.

Dawson observes also that the distribution function undergoes rapid random fluctuations about a mean distribution which gradually drifts toward a

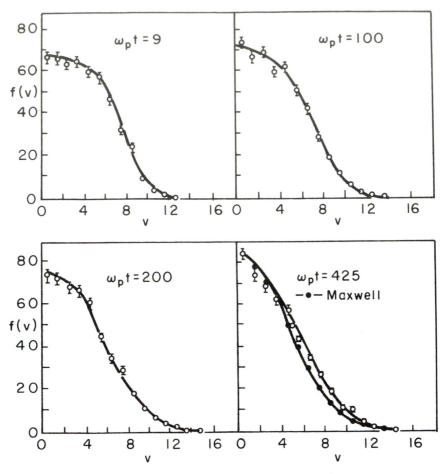

Figure 13-3b Velocity distributions for $N_D = 5$. *(From Dawson, 1964.)*

Maxwellian. The fluctuations in the number of particles with velocities in a small range about zero are measured for $N_D = 20$ for which negligible relaxation took place. Resulting from the constant exchange of energy between the electric field and the particles, the rapid fluctuations are such that they produce very little systematic change. Figure 13-3e is a plot of the number of times $f(0)$ fell between 145 and 215 in bins of length 5; the Gaussian shown fits the data well, typical of completely random fluctuations about a mean value. This shows that the fact that the distribution relaxes slowly results from a very subtle balance and thus the calculation provides an important test of the kinetic theory of plasmas.

Montgomery and Nielson (1970) use the Boltzmann H function 12-6(6) as a measure of thermalization. They find a relaxation time $\propto N_D^2$ in one dimension, versus $\propto N_D$ in two dimensions.

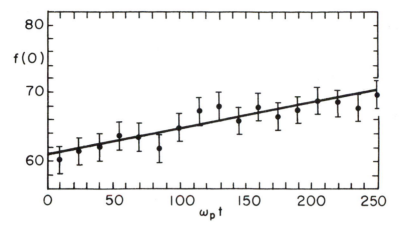

Figure 13-3c Value of $f(v)$ for $v = 0$ as a function of time, for $N_D = 7.5$. The straight line is a least square fit of the data. For this value of N_D, the relaxation time was found to be $\omega_p \tau = 536$ (when $f(0)$ reached the value for a Maxwellian). *(From Dawson, 1964.)*

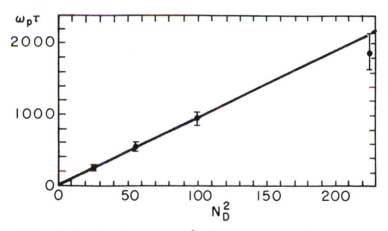

Figure 13-3d Plot of relaxation times versus N_D^2. *(From Dawson, 1964.)*

Virtamo and Tuomisto (1979) found that an operator of the form

$$\frac{\partial f}{\partial t} = \frac{1}{\tau'} \frac{\partial}{\partial v} \left[\frac{v}{\langle v^2 \rangle} f + \frac{\partial f}{\partial v} \right] \tag{5}$$

fit the relaxation they observed, thereby providing yet another way to define and measure a relaxation time τ'. Their simulation used a code like ES1, with parameters $n\lambda_D = 8$, $L = 256 \lambda_D = 1000\Delta x$, and $\omega_p \Delta t = 2\pi/25$. Assuming that $\tau' \propto (n\lambda_D)^2$, their measurements fit $\omega_p \tau' = 28.6 (n\lambda_D)^2$. Further, with the initial square distribution and $\partial f/\partial t$ from (5), Dawson's method of measurement gives $\tau = \tau'/2.6$, in good agreement with the ratio of the measurements, $10/28.6$.

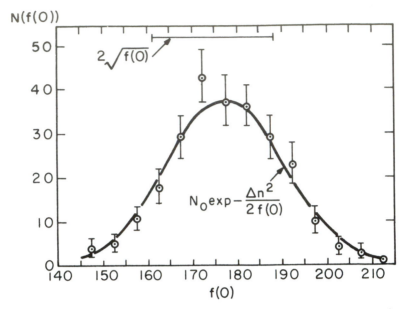

Figure 13-3e Distribution of the fluctuations of $f(0)$ about the mean. $N_D = 20$. *(From Dawson, 1964.)*

(c) Effects of Space and Time Aliasing

When the effects of space and time differencing are taken into account, we found in Section 12-6 that no distribution, not even the Maxwellian, remains constant. Therefore, for large values of N_D, we expect the evolution time in one dimension to scale as N_D/ω_p, while, for small N_D, the evolution time scales as N_D^2/ω_p.

In two dimensions, *Montgomery and Nielson* (1970) find that the thermalization proceeds more slowly when $\lambda_D/\Delta x$ is decreased below unity, perhaps due to cutoff of wavelengths $\sim \lambda_D$. In contrast, thermalization in one dimension is accelerated when $\lambda_D < \Delta x$, an observation consistent with the increased aliasing and the prediction that thermalization in one dimension on the time scale N_D/ω_p is due solely to aliasing.

PROBLEMS

13-3a To derive (1a), first generalize 12-5(17) to multiple species. First review Section 12-3(a) to see how to generalize the numerator and denominator of 12-3(7), the fluctuation spectrum. Note that D and $\mathbf{a}_{\text{fluct}}$ are proportional to q_a^2/m_a^2, while $a_{\text{pol}} \propto q_a^2/m_a$.

13-3b Demonstrate (3) from (1b). Show that conservation of total momentum and kinetic energy (and other moments) follow trivially.

13-4 NUMERICAL HEATING OR COOLING

(a) Self Heating in One Dimension

Measurements of self heating have been made in 1d particle simulations using one species. It is found, for thermal plasmas, that the temperature $T = mv_t^2$ increases linearly with time. Here, a *self-heating* time τ_H is defined as the time for T (or v_t^2) to double. *Abe et al.* (1975) discuss the energy increment due to the random force fluctuations caused by the nonphysical force δF due to the grid (Section 8-3); H. Abe points out (private communication) that their 1975 work leads to

$$\omega_p \tau_H \propto \frac{1}{\eta^2} \left[\frac{\lambda_D}{\Delta x} \right]^2 N_D \tag{1}$$

where η^{-2} is a measure of the goodness of the model, increasing as the order of weighting increases. *Peiravi and Birdsall* (1978), for the ranges $0.1 \lesssim \omega_p \Delta t \lesssim 0.6$ and $0.5 < \lambda_D / \Delta x < 10$, find that the self heating time is maximized for NGP using $(v_t \Delta t / \Delta x) \approx 3/2$ (not a sensitive parameter) with

$$\omega_p \tau_H \approx 3 \left[\frac{\lambda_D}{\Delta x} \right]^2 (N_D + N_C) \tag{2}$$

for CIC-PIC using $(v_t \Delta t / \Delta x) \approx 1/2$ with

$$\omega_p \tau_H \approx 600 \left[\frac{\lambda_D}{\Delta x} \right]^2 (N_D + N_C) \tag{3}$$

and for QS, also using $(v_t \Delta t / \Delta x) \approx 1/2$ with

$$\omega_p \tau_H \approx 4000 \left[\frac{\lambda_D}{\Delta x} \right]^2 (N_D + N_C) \tag{4}$$

where $N_D + N_C = n(\lambda_D + \Delta x)$. These 1d results agree well with the 2d results of *Hockney* (1971). The self heating time is increased by attenuating at short wavelengths; *Peiravi and Birdsall* (1978) found $\omega_p \tau_H$ increased roughly as $(k_{max}/k_{last})^s$ for truncation at k_{last}, with $s = 1, 2, 3$ for zero-, first-, and second-order spline weighting, respectively. Hence, one may use many cells with strong spatial smoothing (large k_{max}/k_{last}) and obtain large τ_H, even with $v_t \Delta t / \Delta x$ much larger than 1/2.

(b) Cooling Due to Damping in the Particle Equations of Motion

Nonphysical cooling is observed in codes using damped equations of motion, both explicit (*Adam et al.*, 1982) and implicit (*Barnes et al.*, 1983b). In the Lenard-Balescu collision operator corresponding to damped time integration we find two spurious terms due to phase errors associated with damping. One is a nonresonant contribution to the polarization drag $\mathbf{a}_{pol}$.

The other is a spurious nonresonant contribution to dynamical friction $\mathbf{a}_{\text{fluct}}$.

In the kinetic theory here we neglect the effects of the spatial grid. In the generalization of the collision operator of Section 12-5 to damped time integration, the velocity diffusion is not altered in any interesting way, but the velocity drag terms are.

The polarization drag 12-5(16) due to anisotropic polarization of an unmagnetized plasma by a test particle, is

$$\mathbf{a}_{\text{pol}} = \frac{q^2}{m} \int \frac{d\mathbf{k}}{(2\pi)^3} \frac{\mathbf{k}}{k^2} \operatorname{Im} \frac{1}{\epsilon(\mathbf{k}, \mathbf{k} \cdot \mathbf{v})} \tag{5}$$

$$= -n_0 \frac{q^4}{m^2} \int \frac{d\mathbf{k}}{(2\pi k)^3} \frac{\mathbf{k}}{k^2} \int d\mathbf{v}' \frac{f(\mathbf{v}')}{|\epsilon(\mathbf{k}, \mathbf{k} \cdot \mathbf{v})|^2} \operatorname{Im} \left(\frac{\mathbf{X}}{\mathbf{A}} \right) \Bigg|_{\mathbf{k} \cdot (\mathbf{v} - \mathbf{v}') + i0}$$

in which ϵ is given in terms of $(\mathbf{X}/\mathbf{A})$ by 9-8(1), and where $(\mathbf{X}/\mathbf{A})_{\omega - \mathbf{k} \cdot \mathbf{v} + i0}$ is the ratio of Fourier amplitudes of $\mathbf{x}^{(1)}$ and $\mathbf{a}^{(1)}$ resulting from the finite-difference equation of motion (and would be $-(\omega - \mathbf{k} \cdot \mathbf{v} + i0)^{-2}$ for exact integration). The meaning of the term "$+i0$" is that $(\cdots)_{\omega + i0}$ is understood to mean the limit of $(\cdots)_{\omega + i\gamma}$ as γ approaches zero through *positive* values. Normally $\operatorname{Im} (\mathbf{X}/\mathbf{A})_{\omega - \mathbf{k} \cdot \mathbf{v} + i0} = -\pi \delta'(\omega - \mathbf{k} \cdot \mathbf{v})$ which leads to the resonant Landau contribution, but here $\operatorname{Im} (\mathbf{X}/\mathbf{A})_{\omega - \mathbf{k} \cdot \mathbf{v} + i0}$ is *also* nonzero for $\omega - \mathbf{k} \cdot \mathbf{v} \neq 0$ due to phase errors associated with numerical damping.

The other part of the drag, the "dynamical friction" $\mathbf{a}_{\text{fluct}}$ is also expressed in terms of $\operatorname{Im} (\mathbf{X}/\mathbf{A})$ in 12-5(20), with the thermal fluctuation spectrum from 12-3(7). As a check on the theory, we can easily verify that the expressions (5) and 12-5(20), with the thermal spectrum 12-3(7), together conserve momentum of the overall distribution of particles.

These results are used in the Fokker-Planck equation describing the evolution of the velocity distribution function $f(\mathbf{v})$,

$$\frac{\partial f}{\partial t} = \frac{\partial}{\partial \mathbf{v}} \cdot \left[-f \mathbf{a}_{\text{pol}} - f \mathbf{a}_{\text{fluct}} + \frac{\partial}{\partial \mathbf{v}} \cdot f \mathbf{D} \right] \tag{6}$$

where $\mathbf{D}(\mathbf{v})$ is the diffusion tensor 12-5(7). Because the resonant parts of (5) and 12-5(20) cancel with continuous time, (6) conserves energy. The rate of cooling due to the nonresonant part (of numerical origin) is:

$$\frac{d}{dt} \text{KE} = \int d\mathbf{v} \; \mathbf{v} \cdot \mathbf{a} \; f(\mathbf{v})$$

$$= -\frac{1}{2} n_0 \frac{q^4}{m^2} \int \frac{d\mathbf{k}}{(2\pi)^3} \frac{1}{k^2} \int d\mathbf{v} \, d\mathbf{v}' \, f(\mathbf{v}) \, f(\mathbf{v}') \tag{7}$$

$$\cdot \left[|\epsilon(\mathbf{k}, \mathbf{k} \cdot \mathbf{v})|^{-2} + |\epsilon(\mathbf{k}, \mathbf{k} \cdot \mathbf{v}')|^{-2} \right] \mathbf{k} \cdot (\mathbf{v} - \mathbf{v}') \operatorname{Im} \left(\frac{\mathbf{X}}{\mathbf{A}} \right) \Bigg|_{\mathbf{k} \cdot (\mathbf{v} - \mathbf{v}')}$$

where it is now to be understood that the resonant part of $\operatorname{Im} (\mathbf{X}/\mathbf{A})$ is dropped. For the C_1 equation of motion scheme, Section (9-8(a)),

$$\text{Im}\left[\frac{\mathbf{X}}{\mathbf{A}}\right]_{\omega} = c_1 \Delta t^2 \sin \omega \Delta t$$

while for the D_1 scheme, Section (9-8(b)),

$$\text{Im}\left[\frac{\mathbf{X}}{\mathbf{A}}\right]_{\omega} = \frac{\Delta t^2 \sin \omega \Delta t}{5 - 4 \cos \omega \Delta t}$$

In both cases, if the spread in particle velocities is less than $\pi/k_{\max}\Delta t$ then the integrand is always positive so only cooling results.

For these schemes with 3rd order damping, the factor $\mathbf{k}\cdot(\mathbf{v} - \mathbf{v}')\,\text{Im}\,(\mathbf{X}/\mathbf{A})_{\mathbf{k}\cdot(\mathbf{v}-\mathbf{v}')}$ in the integrand is proportional to $[\mathbf{k}\cdot(\mathbf{v} - \mathbf{v}')\Delta t]^2$ for small values. With first-order damping, this approaches a nonzero constant instead (*Cohen et al.*, 1982b). Other implementations of third order schemes, *i.e.*, *Barnes et al.* (1983b), produce different phase errors and hence different cooling rates. Quantitative calculations of cooling rates based on this kinetic theory have not yet been carried out.

(c) Heuristic Estimates

Since the self heating is due solely to space and time aliasing effects, it is natural to attempt heuristic estimates in which a self heating rate is calculated from δF (Chapter 8), the part of the force due to $p \neq 0$ terms, and some estimate of correlation time. This is done by *Abe et al.* (1975), and *Hockney and Eastwood* (1981, p. 250 ff.). However, it is difficult to construct reliable estimates of heating (or cooling), since it results from a slight imbalance in the competition between drag and diffusion.

An indication of this difficulty is found when such estimates are applicable with equal plausibility to special cases for which exact results are known. For example, the method of *Abe et al.* (1975), applied to the energy-conserving models (Chapter 10), finds nonzero self heating even in the limit $\Delta t \to 0$.

Since the results in Chapter 12 *do* reproduce qualitative features such as conservation laws, more reliable estimates and scaling laws might be obtained by approximate evaluation of the expressions derived from the kinetic theory.

13-5 COLLISION AND HEATING TIMES FOR TWO-DIMENSIONAL THERMAL PLASMA

Hockney (1971) provided a useful service to electrostatic simulations by making 73 long runs in 2d2v with a thermal (Maxwellian) plasma, varying particle size and shape and Δx and Δt, over wide ranges. From these runs, he found *slowing-down times, self heating times,* and *fluctuation levels,* part of which are given here. The reader is referred to his original article and to

Hockney and Eastwood (1981).

In the laboratory, we expect $\nu_{collision}/\omega_p$ and fluctuation levels $\propto 1/N_D$. In simulations, we expect the same but modified by the effects of the mesh. In the laboratory, there is no self heating. In simulations, we find that there is a slow self-heating $\propto t$, disappearing as $\Delta t, \Delta x \to 0$.

Hockney's model is a spatially uniform plasma in 2d2v, using a five-point finite difference Laplacian and a two-point gradient. The model is doubly periodic, with zero magnetic field, initially loaded with an equal number of thermal ions and electrons ($T_e = T_i$ and mass ratio $m_i/m_e = 64$). The particle positions were selected by random numbers, with uniform distribution in x and y. Particle velocities were selected by using the inverse error function where $|v| < 3 v_t$ for $v_t^2 \equiv T/m$. The electron orbit changes due to collisions were followed; the ions were followed but the results were not used [ion time scales were $(m_i/m_e)^{1/2}$ to m_i/m_e longer]. At $t > 0$ the components parallel and perpendicular to the original direction were measured, $v_\parallel(t)$ and $v_\perp(t)$, with deflection angle $\phi(t)$. The increase of the kinetic energy of each particle from its initial value, was followed,

$$h(t) \equiv \frac{1}{2} m \left[v^2(t) - v^2(0) \right] \tag{1}$$

Four characteristic times, τ_ϕ, $\tau_{v\perp}$, τ_s, τ_{th} are found, due to collisional effects, existing in lab or simulation plasmas. These times are defined by

$$\langle \phi^2(\tau_\phi) \rangle^{1/2} = \frac{\pi}{2} \qquad \tau_\phi \equiv \text{deflection time} \tag{2}$$

$$\langle v_\perp^2(\tau_{v_\perp}) \rangle^{1/2} = \langle v_\parallel(0) \rangle \qquad \tau_{v_\perp} \equiv v_\perp \text{ time} \tag{3}$$

$$\frac{d}{dt} \langle \dot{v}_\parallel(t) \rangle = \left. \frac{\langle v_\parallel(t) \rangle}{\tau_s} \right|_{t=0} \qquad \tau_s \equiv \text{slowing down time} \tag{4}$$

$$\langle h^2(\tau_{th}) \rangle^{1/2} = \frac{1}{2} m v_t^2 = \frac{1}{2} T \qquad \tau_{th} \equiv \text{thermalization time} \tag{5}$$

The angled brackets indicate average of the enclosed quantity, taken over the assembly of electrons and ions separately,

$$\langle \alpha(t) \rangle \equiv \frac{1}{N} \sum_{i=1}^{N} \alpha(t)$$

Hockney uses τ_s as a measure of collisional effects. With $q_i = q_e$, these times are comparable; when $q_i \gg q_e$, τ_ϕ and $\tau_{v\perp}$ for the electrons are reduced due to dominance of their nearly elastic scatter on the ions.

A fifth time τ_H, the heating time, defined by

$$\langle h(\tau_H) \rangle = \frac{1}{2} m v_t^2 = \frac{1}{2} T \tag{6}$$

is a measurement of the failure of conservation of energy in the computer model. Hockney's measurements of $h(t)$ for zero and first-order weighting (NGP and CIC) are shown in Figure 13-5a. Note that increasing the order

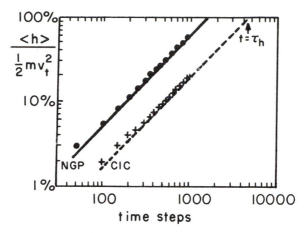

Figure 13-5a Typical results obtained for measurement of the heating time. Note that growth of kinetic energy is roughly linear with time. *(From Hockney, 1971.)*

of the weighting (taking the sharp edges off the particle which reduces the coupling of aliases) lengthens τ_H by 3. An important observation by Hockney is that $\langle h(t) \rangle$ increases linearly in time, which implies stochastic origin. An average electron initially has energy $(\frac{1}{2}T + \frac{1}{2}T)$, as does an ion, so that the system energy density is $n_e T + n_i T$ or $2n_e T$; hence, when $t = \tau_H$ at $\Delta(\mathrm{KE})_e = \frac{1}{2}T$, the system energy is up by 25 per cent, in 2d.

The simulation results to follow are found to depend on the parameter N_C, defined by Hockney to be

$$N_C \equiv n\,[\lambda_D^2 + (R\Delta x)^2] \tag{7}$$

where the second term is the number of cloud centers per cloud. Hockney uses $R = 1, 1,$ and 2 for NGP, CIC, and QS, respectively, from best fits to his data. These choices also imply an effective particle radius given roughly by the location of the maximum interparticle force. The form of (7) implies that when $R\Delta x > \lambda_D$, the particle radius is more important than the Debye length in collisions and fluctuations, a consequence of using finite-size particles.

The *collision time* τ_s is found by least square fitting to be

$$\frac{\tau_s}{\tau_{pe}} = \frac{N_C}{K_1} \tag{8}$$

where $K_1 = 0.98 \pm 0.20$, for 73 runs with $0.25 \leqslant N_C \leqslant 43$, $0.12 \leqslant R\Delta x/\lambda_D < 32$. $K_1 \approx 1$ implies that

$$\frac{\nu_{\mathrm{coll}}}{\omega_{pe}} \approx \frac{1}{2\pi N_C} \tag{9}$$

Even the extreme cases where $\lambda_D/\Delta x \approx 1/8$ which is weakly unstable due to aliasing, R. W. Hockney notes (private communication) that the empirical fit to (8) is satisfactory provided the effect of the particle width is included, as it is in N_C.

The *electric field fluctuations* are found to be (cgs units)

$$\frac{\langle E_x^2 \rangle / 8\pi}{nmv_t^2} = \frac{K_2}{N_C} \tag{10}$$

with $K_2 = 0.12 \pm 0.04$. In order to check this measurement, Hockney estimates from plasma theory that

$$\frac{\langle E_x^2 \rangle / 8\pi}{nmv_t^2} = \frac{1}{16\pi} \ln \left[\frac{1 + (k\lambda_D)_{max}^2}{1 + (k\lambda_D)_{min}^2} \right] \frac{1}{n\lambda_D^2} \tag{11}$$

In an $m \times m$ mesh, neglecting alias contributions, $(k\lambda_D)_{max} = \pi\lambda_D/\Delta x$, $(k\lambda_D)_{min} = 2\pi\lambda_D/m\Delta x$, and using $\lambda_D/\Delta x = 6$ (meaning $N_C \approx N_D$), one finds

$$\frac{\langle E_x^2 \rangle / 8\pi}{nmv_t^2} = \frac{0.112}{n\lambda_D^2} \tag{12}$$

which agrees well with (10). Roughly speaking,

$$\frac{\text{PE}}{\text{KE}} \approx \frac{1}{8N_C} \tag{13}$$

The heating time is found to be strongly dependent on Δx and Δt, with the same dependence on N_C as τ_s. Contours of constant τ_H/τ_s (Figure 13-5b) display the dependence on Δx and Δt. The ratio is the number of collision times in a heating time, which nearly always should be a large number unless both times exceed the duration of the simulation. As the heating increases linearly with t, and there is a 25% error in total energy at $t = \tau_H$, then there is 2.5% error at $\tau_H/10$; this allows us to design thermal plasma experiments with a known error. Note that τ_H/τ_s for CIC is roughly 16 times larger than that for NGP and that values for CIC for typical $\omega_p\Delta t$

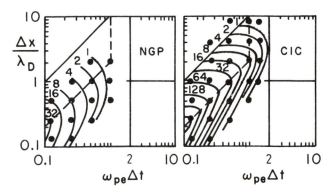

Figure 13-5b Contours of constant τ_H/τ_s on the Δx versus Δt plane for NGP and CIC weighting. *(From Hockney, 1971.)*

(≈ 0.2) and $\lambda_D/\Delta x$ (≈ 1.0) give $\tau_H/\tau_s \approx 64$, which is quite comforting. A possible cause of the decrease of τ_H/τ_s as $v_t \Delta t/\Delta x$ increases is suggested in Section 12-7.

For purposes of design Hockney devised the Δx, Δt plot shown in Figure 13-5c. The region to the right of $\omega_{pe}\Delta t = 2$ is unstable for a cold plasma and $\omega_{pe}\Delta t = 1.62$ for a thermal plasma (move the boundary line a bit). For $\lambda_D/\Delta x \lesssim 0.2$ ($\Delta x/\lambda_D \gtrsim 5$), the non-physical Δx instability becomes important, so add that "boundary." For $v_t \Delta t > \Delta x$ many particles cross a cell in Δt, defining the lower bound shown. His *optimum path* shown is $v_t \Delta t = \Delta x/2$ out to $\omega_{pe}\Delta t = 1$.

Using the τ_H/τ_s values on the optimum path, Hockney then simplified the data to Figure 13-5d. The heating time dependence is found to be

$$\left[\frac{\tau_H}{\tau_s}\right]_{\text{opt}} = K_4 \left|\frac{\lambda_D}{\Delta x}\right|^2 \tag{14}$$

These results are striking, with τ_H/τ_s for CIC ($K_4 \approx 40$) about 20 times longer than for NGP ($K_4 \approx 2$).

There is an apparent contradiction between NGP and CIC having the same τ_s but twenty times different τ_H. Indeed, *Montgomery and Nielson (1970)* found that an NGP model relaxed ten times as fast as a CIC model; if the relaxation to a Maxwellian is due solely to binary collisions, then simulations, with different weighting (NGP and CIC) but the same collision times τ_s should show no difference in relaxation. Hockney points out that there is no contradiction because in the simulation, it is only the error in the fields

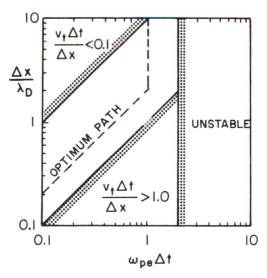

Figure 13-5c The parameter plane used in the study of the heating time. The shaded regions are undesirable and the dotted line shows the optimum path on which $(\omega_{pe}\Delta t)_{\text{opt}} \equiv \min[\Delta x/2\lambda_D, 1]$. *(From Hockney, 1971.)*

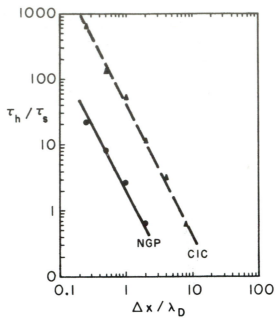

Figure 13-5d Heating to collision time along the optimum path $(\omega_{pe}\Delta t)_{opt} = \min[\Delta x/2\lambda_D, 1]$ as a function of $\Delta x/\lambda_D$ for the NGP, and CIC models. *(From Hockney, 1971.)*

which causes the stochastic heating.

Hockney et al. (1974) published more information on heating times, including more sophisticated particle shapes and using 9 points in the Poisson solver. The newer article gives $K_4 = 100$ for CIC; the reason put forth is that greater care was taken to establish thermal equilibrium before the heating rate measurement was taken. Increasing the Poisson solver from 5 to 9 points does nothing for CIC. Using a QS weighting and a 9 point Poisson solver, with an effective particle radius $R = 1.8$, K_4 increases to 150; adding a potential correction term, with effective $R = 3$, they found $K_4 = 3000$. The latter model takes about twice as long to compute as does standard CIC; however it has 30 times larger K_4 and, for $\lambda_D/\Delta x = 1$, five times larger τ_s/τ_{pe}, hence 150 times larger τ_H/τ_{pe}, which is quite an excellent figure of merit. The latter models in which the potential is adjusted (usually in k-space) to reduce the effect of the mesh, are referred to as *quiet-particle-mesh models* (QPM) by Hockney.

13-6 UNSTABLE PLASMA

As with most kinetic theory, that presented here assumes velocity distributions which are linearly stable, as indicated by the dispersion relation $\epsilon(k, \omega) = 0$. However, it has recently been shown that large-amplitude turbulence can develop in linearly stable one-dimensional plasmas (*Berman et al.,* 1982). A relative drift velocity between ions and electrons, although below the linear stability threshold, provides free energy for a *nonlinear* instability. Detailed measurements of correlations in (x, v) phase space, possible only in computer simulation, disclose "clumps" and density holes. Such diagnostics facilitate successful comparison with theory (*Berman et al.*, 1983; *Dupree*, 1983). The authors believe that these observations, not accessible via perturbative theories, raise questions regarding the relevance of these standard perturbative procedures.

PART

THREE

PRACTICE

PROGRAMS IN TWO AND THREE DIMENSIONS: DESIGN CONSIDERATIONS

This Part introduces the more complicated codes that are used in two and three dimensions and provides some design procedures for codes in all dimensions. While adding realism, the steps from one spatial dimension to two or three increases the complexity of programs and computer running times. We now need 2d and 3d particle movers, particle weightings, field solvers and diagnostics.

Chapter 14 provides details on algorithms for electrostatic programs in two and three dimensions.

Chapter 15 describes simulations including self-consistent electromagnetic fields (as an extension of Chapters 6 and 7), and approximations, such as the Darwin models.

Chapter 16 presents techniques used for loading particles, that is, choosing initial values for $(\mathbf{x}, \mathbf{v})_i$ and for injection and absorption during the run. Also included is the implementation of an external electric circuit.

FOURTEEN

ELECTROSTATIC PROGRAMS IN TWO AND THREE DIMENSIONS

14-1 INTRODUCTION

Simulation in two and three spatial dimensions is much more realistic than in one dimension. Indeed, many problems demand use of more than one dimension in order to obtain useful physics. Hence, this chapter and the next display the steps beyond those already given for 1d, first for electrostatic models and then for electromagnetic models. First, let us estimate costs and complexity.

As in one dimension, electrostatic simulations may be done in 2d, 3d using Coulomb's law directly, having no spatial grid. Particle-particle calculations may be done using the force between charges i and $j \propto 1/r_{ij}$ and $1/r_{ij}^2$. This approach is at its best modeling an isolated plasma, with no boundaries (no image charges). For general plasma models, periodic or bounded, however, there is seldom need to observe behavior at small particle separation or at λ less than λ_D in neutral plasma, and hence, little need to employ the extra computation time needed for particle-particle calculations. Hence, the direct Coulomb law approach is generally replaced by one which obtains the fields from the charge and current densities, using a *spatial grid*. The grid is made fine enough to observe the grid quantities at sufficiently small scale as defined by the physics sought, but not at finer scale. Typically in an x-y 2d program, grids are 32×32 (1024 points) up to 256×256 (65,536 points) with E_x, E_y, B_x, B_y, B_z, ρ stored at each point (6 numbers) in a random access memory. In 3d, using a $64 \times 64 \times 64$ grid, (262,144 points) with all components of **E**, **B**, **J**, and ρ present, storage of 2,621,440 numbers in fast memory is required, currently non-trivial!

The jump from observing $f(x, v_x)$ in 1d1v, two phase-space variables, to $f(x, y, z, v_x, v_y, v_z)$ in 3d3v, with six phase-space variables, may require going to some number like $(64)^3$ cells, and 10^4 to 10^7 particles, depending on the problem. Particle descriptions (*i.e.*, $\mathbf{x}_i$, $\mathbf{v}_i$) jump from two to six numbers and there usually is a need for many times more particles. Being able to construct smooth variations of $f(\mathbf{v})$ and $n(\mathbf{x})$ down to cell size, sometimes expensive even in 1d, may be prohibitive in 3d3v; being able to make both $N_D \equiv n\lambda_D^3$ and L/λ_D large may become very expensive. Each additional dimension in phase space *multiplies* the cost in memory and time. Fortunately, as in 1d, each problem has its own needs; many problems are being done in 2d3v at acceptable cost, certainly relative to having only an approximate theory on which to build, for example, a full scale fusion experiment.

It may be that the greater challenge in moving up from 1d is in the added complexity of doing 2d and 3d. For example, consider the added difficulties in the designing the initial loading and later injection, absorption, reflection of particles and in devising diagnostics and post processing that are readily grasped, as well as in making tests for accuracy. Almost all tasks are tougher, possibly much more so than just obtaining larger memories and faster processing.

The description above suggests that simulations should progress through 1d1v to 1d3v to 2d2v to 2d3v in steps, leading up to and supporting full 3d3v simulations.

Let us outline what to do in solving 2d and 3d electrostatic problems. We proceed from $\mathbf{x} \to \rho \to \phi \to \mathbf{E} \to \mathbf{F}$ as follows:

$\mathbf{x}_{\text{particle}} \to \rho_{\text{grid}}$	weighting, any order.
$\rho_{\text{grid}} \to \phi_{\text{grid}}$	put $\nabla^2\phi = -\rho$ into finite-difference form and solve for ϕ_{grid}.
$\phi_{\text{grid}} \to \mathbf{E}_{\text{grid}}$	put $-\nabla\phi = \mathbf{E}$ into finite-difference form and solve for $\mathbf{E}_{\text{grid}}$.
$\mathbf{E}_{\text{grid}} \to \mathbf{F}_{\text{particle}}$	weighting, any order.

Or, if ρ, ϕ and **E** may be represented by a discrete Fourier series, then we may obtain, for example, $\phi(\mathbf{k})$ from $\rho(\mathbf{k})$ either by division by the finite-difference operator $K^2(\mathbf{k})$ or simply by k^2. Or, there may be a combination of a finite difference solution, say, in x, and Fourier series in y.

The steps from the acceleration, $q/m\,(\mathbf{E} + \mathbf{v}/c \times \mathbf{B})$, to $\mathbf{v}_{\text{new}}$ and $\mathbf{x}_{\text{new}}$ are the same as those given in Chapter 4. There the leap-frog time-centered movers are given, with the steps of half acceleration, rotation and half acceleration. These are presented in vector form, usable in 1d1v up through 3d3v.

Particle weightings in 2d, 3d are straightforward extensions of those already used in 1d, ending up with effective particle shapes that are more or less rods and cubes in 2d and 3d, respectively (See Section 8-9). The particle shape factors become $S(\mathbf{x}) = S(x, y, z)$, generally nearly symmetric, but with some (unwanted) anisotropy due to the squareness or cubeness of the finite-size particles and the rectangular mesh. For example, these effects lead to slightly different wave propagation along axes than between axes, with the effects diminishing as the order of the weighting is increased.

The field solvers also are extensions of those used in 1d. We need to know finite-difference forms for $\nabla^2\phi$ and $\nabla\phi$, with some information about accuracy. The Poisson solvers presented are direct (non-iterative), using either discrete Fourier series and/or matrix inversion to go from $\rho(\mathbf{x})$ to $\phi(\mathbf{x})$.

In choosing models for any number of dimensions, simulators have a variety of choices for boundary conditions on potentials and particles, as well as freedom to use symmetries. In 2d and 3d, models may be wholly periodic or wholly bounded or with different boundary conditions along the 2 or 3 coordinates, and boundaries may be irregular with mixed boundary conditions.

Periodic systems are used to represent a part of an infinite plasma. Such systems are necessarily charge neutral in that

$$0 = \int_{\text{period}} \mathbf{E} \cdot d\mathbf{S} = \int_{\text{period}} \rho \, dV$$

As this relation holds for plasma regions that are more than a few Debye lengths across and away from sheaths, periodic systems are widely used. As there is little difference in 1d, 2d and 3d periodic system boundary conditions, and 1d has been treated earlier, the only further discussion necessary is on $k = 0$ considerations.

In a *bounded system* the charges or potentials or fields are specified on all boundaries. This system need not be neutral. If the system is a grounded rectangular box, then it is similar to a 1d system with grounded boundaries at $x = 0$ and L [*i.e.*, $\phi(0) = 0 = \phi(L)$], which is easily handled with either a Fourier sine series or a finite difference form; extension to 2d and 3d is straightforward.

An *open boundary* is taken to mean an interface between the plasma region described by Poisson's equation, $\nabla^2\phi = -\rho$, and a vacuum region described by Laplace's equation, $\nabla^2\phi = 0$. Potentials are to be matched at such boundaries. For example, in 1d, let there be plasma for $x < L$, and vacuum for $x > L$, where the $\phi(x) = Ax + B$. However, the same open boundary in 2d, for ϕ periodic in y, has Laplace solutions like $\exp[-k_y(x-L)]\exp(ik_yy)$, showing *decay* away from $x = L$ for $k_y \neq 0$; the $k_y = 0$ term provides the same solution as for 1d, given above.

Mixed boundary conditions are possible, combining periodic, specified, and open, combining steps given in this chapter.

Thus, we see some distinct differences between the 1d and the 2d, 3d models. This chapter deals with some of the new aspects.

14-2 AN OVERALL 2D ELECTROSTATIC PROGRAM

In this section, we consider a complete program. We then proceed to examine the parts in detail in succeeding sections. The particles have locations x_i, y_i and velocities v_ξ, v_{yi}. The spatial grid, used for obtaining the fields from the particle charge density and current density, has grid points $X_j = j\Delta x$, $Y_k = k\Delta y$, as sketched in Figure 14-2a. The grid size is made small enough to resolve the details deemed necessary and to avoid numerical troubles (with a warm plasma, keeping $\lambda_D > \Delta x/3$, roughly). The particle density is kept large enough to make density variations smooth, implying a few particles per grid cell. The particles themselves are treated as in 1d, as finite-size particles; this physics comes about automatically by weighting the particles to the grid. Zero-order weighting is again nearest-grid-point-weighting, seldom used. First-order weighting is again linear interpolation, here called bi-linear, or area weighting, due to its geometric interpretation, as shown in Figure 14-2b. The weights are given by

$$\rho_{j,k} = \rho_c \frac{(\Delta x - x)(\Delta y - y)}{\Delta x \Delta y}$$

$$\rho_{j+1,k} = \rho_c \frac{x(\Delta y - y)}{\Delta x \Delta y}$$

$$\rho_{j+1,k+1} = \rho_c \frac{xy}{\Delta x \Delta y} \tag{1}$$

$$\rho_{j,k+1} = \rho_c \frac{(\Delta x - x)y}{\Delta x \Delta y}$$

where ρ_c is the charge density uniformly filling a cell (q/area) and the charge position x, y is measured from the lower left-hand grid point (j, k obtained by truncating the particle coordinate x, y). The charge densities at all of the grid points then become the right hand side of Poisson's equation

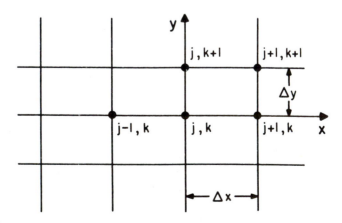

Figure 14-2a Typical two dimensional rectangular grid in x,y.

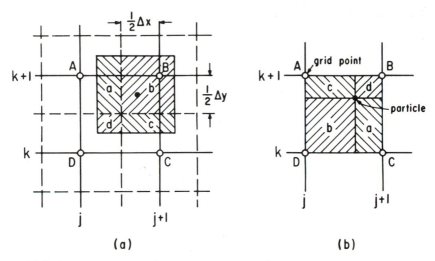

Figure 14-2b Charge assignment for linear weighting in 2d. Areas are assigned to grid points; *i.e.*, area a to grid point A, b to B, etc. as if by NGP, with the particle center located as indicated. (a) is the CIC cloud interpretation. (b) is the PIC bilinear interpolation interpretation.

$$\nabla^2 \phi(x,y) = -\rho(x,y) \tag{2}$$

In finite difference form, this becomes the five-point form,

$$\frac{(\phi_{j-1} - 2\phi_j + \phi_{j+1})_k}{\Delta x^2} + \frac{(\phi_{k-1} - 2\phi_k + \phi_{k+1})_j}{\Delta y^2} = -\rho_{j,k} \tag{3}$$

This set of equations is then solved for all of the potential $\phi_{j,k}$, including use of the appropriate boundary conditions (periodic, bounded, open, other). The **E**'s are obtained from the ϕ's using

$$\mathbf{E}(\mathbf{x}) = -\nabla \phi(\mathbf{x}) \tag{4}$$

in the usual two-point difference form; for E_x, as sketched in Figure 14-2c,

$$(E_x)_{j,k} = \frac{(\phi_{j-1} - \phi_{j+1})_k}{2\,\Delta x} \tag{5}$$

with a similar differencing for E_y. The fields are located at the same points as potentials. Next, the fields are weighted back to the particles; for linear weighting, each field component at the four nearest grid points is weighted to each particle using the same weights as in (1) above. Finally, particles are advanced by the particle mover, from v_{old} to v_{new} and x_{old} to x_{new}, completing one time step, as done in Chapter 4.

Diagnostics are put in where appropriate, with some snapshots now being in 2d, such as contour plots of charge density or potential, which are logical extensions of working in more than one dimension.

The simplest approach is that just given. However, with 2d electrostatic models, there may be good reason to use other coordinates, such as

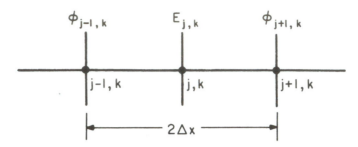

Figure 14-2c Location of $E_{j,k}$ with relation to $\phi_{j,k}$.

cylindrical coordinates r-z or r-θ, as dictated by the physics or the natural boundary conditions, and as a step toward a full 3d model with r-θ-z coordinates. Use of higher or lower order weightings may be called for; these may be worked out from the 1d models given earlier. Section 14-9 treats the Poisson equation and gradient operators for cylindrical r, r-θ, r-z gridding.

The sections following expand upon the various parts given above, such as the accuracy of the Poisson equation finite differencing, better ways to obtain **E** from ϕ, boundary conditions, and weighting and effective particle shape.

14-3 POISSON'S EQUATION SOLUTIONS

We have set up Poisson's equation in ϕ and ρ in a finite-difference form in rectangular coordinates in 14-2(3). For detailed solutions of this form, the reader is referred to introductory work by *Potter* (1973) and the extensive work by *Hockney* (1970) and in *Hockney and Eastwood* (1981, chap. 6). Hockney, working with Buneman and Golub at Stanford University, pioneered fast direct solutions of Poisson's equation in 2d in the early 1960's, making practical the jump from 1d to 2 and 3d simulations, particle or fluid. These references provide a wealth of information that is not repeated here.

Or, we may use Fourier series representations for ϕ and ρ in order to obtain direct solutions, as was done earlier in 1d in ES1; steps for a doubly periodic Poisson solver are given in a later section. We also display in the later sections 2d slab models which involve both Fourier and f. d. e. methods. The models are periodic in y and may be periodic, open, bounded or inverted in x.

In many models, conditions on ϕ or **E** are given at regular boundaries of the plasma or vacuum region such as at $x = 0$, L_x and $y = 0$, L_y. In models with electrodes held at given potentials in the *interior* of the plasma region, the charges on the electrodes are needed. These may be obtained from the *capacitive matrix method*. The details of this method, extensions, and timing are given in *Hockney and Eastwood* (1981, Sec. 6-5-6). The method involves precalculation of a capacity matrix **C** (*e.g.*, see *Ramo, Whinnery, and Van Duzer, 1965,* chap. 5) which relates the potential at the points which are on the interior electrodes and the charge induced on them by the surrounding plasma. First, solve Poisson's equation with no charges on the electrode points; record the difference between the potential found at the electrode points and the desired value. Multiplying this error by **C** gives the negative of the desired surface charge at each electrode point. Then solve Poisson's equation again with this desired surface charge; this solution is correct everywhere, including at the electrodes.

14-4 WEIGHTING AND EFFECTIVE PARTICLE SHAPES IN RECTANGULAR COORDINATES: $S(\mathbf{x})$, $S(\mathbf{k})$, FORCE ANISOTROPY

Assignment of particle charge to neighboring grid points proceeds much as we have already done in 1d, with zero, first, and second order weighting: nearest-grid-point, bilinear, and biquadratic spline, respectively. Our interest here is in the effective shape of the particle (*e.g.*, how anisotropic?), the Fourier representation (*e.g.*, what is the coupling from higher order Brillouin zones?), and the anisotropy of the force. Weighting is done here for rectangular coordinates; weighting for cylindrical coordinates is done in Section 14-10.

Zero-order assignment (NGP) to one grid point is simplest. The nearest grid point is found from the particle position $\mathbf{x} = (x, y)$ by truncating $x + 0.5$, $y + 0.5$ (for $x, y > 0$). The particle charge is assigned to a grid point when the particle center is in the cell Δx by Δy centered on the grid point; when the particle center moves past the edge of the cell, the particle is assigned to the next cell. The particle shape $S(\mathbf{x})$ is a rectangular cloud of height 1 and sides Δx by Δy as sketched in Figure 14-4a. The force between two particles is discontinuous as shown in Figure 14-4b which leads to noise and self-heating, but becomes close to the physical $1/r$ dependence at charge separation of a few cells. The short-range force vanishes, as is true for all weightings. The rectangular shape leads to the anisotropy in force indicated in Figure 14-4c. For these reasons NGP is seldom used.

First-order assignment is linear weighting to the nearest four grid points in two dimensions (hence, bilinear), which is often called *area weighting* due to its geometric interpretation, as shown in Figure 14-2b. Cloud-in-cell (CIC) is the name given by *Birdsall and Fuss* (1969), who considered the particle to be nominally rectangular with the particle fraction in each cell weighted to its cell center. Particle-in-cell (PIC) is the name used by *Harlow* (1964) in fluid velocity weighting, followed by *Morse and Nielson* (1969), who considered the particle to be a point but with charge assigned by linear interpolation to the nearest grid points. The weights are given by 14-2(1). Another form of the weight to the point j, k (for x, y normalized to Δx,

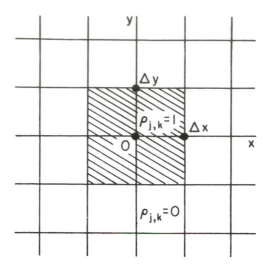

Figure 14-4a Particle shape for zero-order weighting in 2d (NGP), which is uniform as viewed by the grid, of size Δx by Δy, centered on a grid point. *(From Birdsall and Fuss, 1969.)*

Δy) is

$$\Delta x \, \Delta y \, S(x,y) = \begin{cases} (1 - |x|)(1 - |y|) & |x| \leqslant 1, \, |y| \leqslant 1 \\ 0 & \text{otherwise} \end{cases} \tag{1}$$

We may produce the particle shape $S(x,y)$ by measuring the charge assigned to a grid point as the particle moves relative to that point. Hence, (1) provides the particle shape. The particle density contours are shown in Figure 14-4d, indicating an improvement over the flat, rectangular NGP particle (Δx by Δy), yet not quite circular. The force is now piece-wise continuous, as shown in Figure 14-4e, reducing the noise and the self-heating relative to NGP.

The Fourier transform of the particle shape factor, $S(x,y)$ to $S(k_x, k_y)$, is informative. In particular, $S(k_x, k_y)$ indicates the coupling into the fundamental Brillouin zone ($|k_x \Delta x| < \pi$, $|k_y \Delta y| < \pi$) from shorter wavelengths, that is, *aliasing*. $S(\mathbf{k})$ for linear weighting (CIC, PIC) was shown in Figure 11-6c. Note that $S(k_x, 0)$ and $S(0, k_y)$ are the same as in 1d, so that coupling from the $|p_x| = 1$, $p_y = 0$ and $p_x = 0$, $|p_y| = 1$ zones into the fundamental $(0,0)$ zone is much as in 1d; see 8-9 (9) for definition of $\mathbf{p}$. The $(1,1)$ coupling is much smaller, as the $S_{\max}(1,1) \approx 0.15$ [whereas $S_{\max}(0,1) \approx 0.4$]. Note that $S(k_x, k_y)$ is very nearly $S(k)$, $k = (k_x^2 + k_y^2)^{1/2}$ in the $(0,0)$ zone and remains nearly isotropic well into the $(0,1)$, $(1,0)$ and $(1,1)$ zones. Compare this $S(k)$ with that for the "improved dipole" shown in Figure 11-6b, which has larger anisotropy and larger values of $|S|$ well out of the $(0,0)$ zone, hence, larger coupling to aliases.

Second-order assignment (QS, quadratic spline) is to the nearest nine grid points, given by

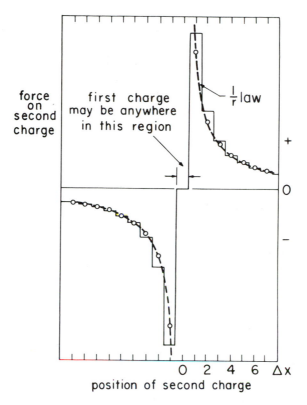

Figure 14-4b Force produced by a charge in the cell at the origin on a second charge at x for zero-order weighting, NGP. The second particle moves parallel to the x-axis. The dashed line shows the physical $1/r$ law for the second particle located at the circles, centers of cells. *(From Hockney, 1966.)*

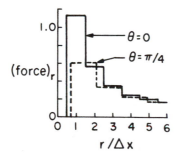

Figure 14-4c Similar to Figure 14-4b, but including the second particle moving at 45° to the x-axis where the peak force is less and the steps are $(2)^{1/2}$ longer. *(From Eastwood and Hockney, 1974.)*

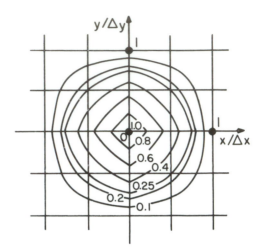

Figure 14-4d Particle density contours for linear weighting in 2d (CIC, PIC). The total particle area is $4\,\Delta x\,\Delta y$. *(From Birdsall and Fuss, 1969.)*

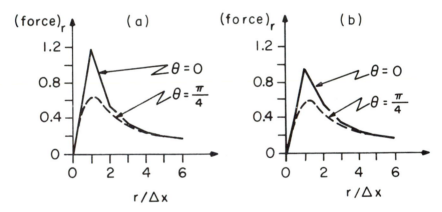

Figure 14-4e (a) Force between two particles for linear weighting (CIC, PIC). A positive charge is located at the origin (a grid point). The force on a positive charge is shown, as this charge moves along the x axis or at $\theta = \pi/4$. The 5 point Poisson stencil was used. (b) Same as (a) but for the 9 point Poisson stencil. *(From Eastwood and Hockney, 1974.)*

$$S(x,y) = S(x)\,S(y) \qquad (2)$$

where $S(x)$ is S_2 of 8-8(2-4). Density contours, profiles and the force are shown in Figure 14-4f. $S(k)$ is given by 8-9(16). There is additional computational effort in weighting the particle to 9 points (beyond the 1 point of NGP, 4 of CIC, PIC) but the particle is now nearly circular. The particle is also larger in area; for $\Delta x = \Delta y$, the NGP particle has area $(\Delta x)^2$, the CIC, PIC particle has area $4(\Delta x)^2$, and the QS particle has area $9(\Delta x)^2$.

The reader is encouraged to obtain $S(\mathbf{k})$ for any new proposed weightings prior to use.

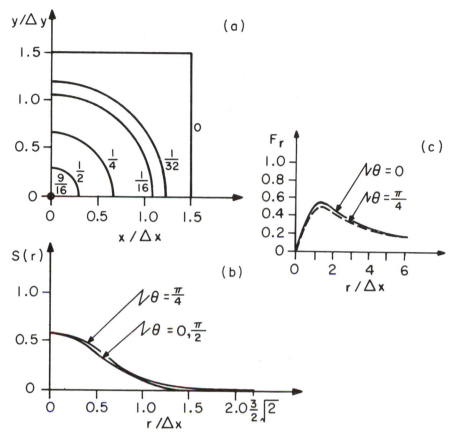

Figure 14-4f (a) Particle density contours for quadratic weighting in 2d (QS). The total particle area is $9\Delta x\Delta y$. (b) Profile of the shape factor $S(r)$ (or density) along the particle radius, along the x or y axis and in between. (c) Force (as in earlier figures) for quadratic weighting in 2d (QS). *(From Eastwood and Hockney, 1974.)*

14-5 DOUBLY PERIODIC MODEL AND BOUNDARY CONDITIONS

As we provide considerable detail on a periodic Poisson solver in 1d, it is appropriate to describe the extension to 2d. We also emphasize the need to consider $\mathbf{k} = 0$ in 2d periodic codes.

(a) Doubly Periodic Poisson Solver

The discrete Fourier transform method is readily applied to a *doubly periodic model*, as done in ZOHAR. The size is $N_x = L_x/\Delta x$ by/$N_y = L_y/\Delta x$ cells. The subroutine names used here are from ZOHAR; the reader may construct these in detail, adding calls to the subroutines RPFTI2 and CPFT used in the transform given in Appendix A. No scratch memory is needed.

The Fourier transform of the charge density and its normalization is RHO (equivalenced to RHOPHI),

$$\rho_{l,m} = \Delta x\, \Delta y \sum_{k=0}^{N_y-1} e^{-ik_y y_k} \sum_{j=0}^{N_x-1} e^{-ik_x x_j}\, \rho_{j,k} \tag{1}$$

where $k_x = 2\pi l/L_x$, $k_y = 2\pi m/L_y$. This is done by subroutine RPPFT, which computes the sine and cosine coefficients, defined as

$$\begin{bmatrix} CS_{l,m} & SS_{l,m} \\ CC_{l,m} & SC_{l,m} \end{bmatrix} = \Delta x\, \Delta y \sum_{j,k} \rho_{j,k} \begin{bmatrix} \cos k_x X_j \sin k_y Y_k & \sin k_x X_j \sin k_y Y_k \\ \cos k_x X_j \cos k_y Y_k & \sin k_x X_j \cos k_y Y_k \end{bmatrix} \tag{2}$$

for $l = 0$ to $N_x/2$ and $m = 0$ to $N_y/2$. One can construct $\rho_{l,m}$ as

$$\rho_{l,m} = (CC_{l,m} - SS_{l,m}) - i\,(CS_{l,m} + SC_{l,m}) \tag{3}$$

For l, m in other quadrants $\rho_{l,m}$ may be found similarly, by using the symmetries of the sine and cosine coefficients. The coefficients are stored in an array RHOK, equivalenced to RHOPHI, according to

$$\text{RHOK}(-L, M) = CS_{l,m}, \qquad\qquad \text{RHOK}(L, M) = SS_{l,m} \tag{4a}$$

$$\text{RHOK}(-L, -M) = CC_{l,m}, \qquad\qquad \text{RHOK}(L, -M) = SS_{l,m} \tag{4b}$$

The potential coefficients obtained from 8-9(17) are

$$\phi_{l,m} = \frac{\rho_{l,m}}{K_{l,m}^2} \tag{5}$$

where
$$K_{l,m}^2 = k_x^2\, \text{dif}^2\frac{k_x \Delta x}{2} + k_y^2\, \text{dif}^2\frac{k_y \Delta y}{2} \tag{6}$$

In a continuum we would have $K_{l,m}^2 = k_x^2 + k_y^2$.

The inverse, performed by RPPFTI, is

$$\phi_{j,k} = \frac{1}{L_x L_y} \cdot \sum_{l=-N_x/2}^{N_x/2-1} e^{ik_x x_j} \sum_{m=-N_y/2}^{N_y/2-1} e^{ik_y y_k}\, \phi_{l,m} \tag{7}$$

The potential and its transform are stored in arrays PHI and PHIK, which are also equivalenced to RHOPHI. Thus the density and potential and their transforms occupy the same memory locations, each replacing the preceeding in the order RHO, RHOK, PHIK, PHI. The use of separate identifiers in the program indicates what quantity is being worked with at any time, and permits convenient indexing to be used for the transforms.

As in the continuum case, a field energy may be computed in either real or transform space:

$$\int dx\, dy\, \frac{\rho\phi}{2} \rightarrow \Delta x\, \Delta y \sum_{j,k} \frac{1}{2} \rho_{j,k}\, \phi_{j,k} \tag{8}$$

$$= \frac{1}{L_x L_y} \sum_{l=-N_x/2}^{N_x/2-1} \sum_{m=-N_y/2}^{N_y/2-1} \frac{1}{2} \rho_{l,m}^{*}\, \phi_{l,m}$$

Note that in this sum a given sine-cosine Fourier coefficient may appear as many as four times. Indexing over l is split into two parts: $l = 0$ and $N_x/2$, and $\pm l = 1$ to $N_x/2 - 1$. Similarly for m.

(b) Periodic Boundary Conditions; k = 0 Fields

Potential and particle boundary conditions for some 2d models are given in earlier sections of this Chapter, such as for the slab models, periodic in y and open or closed or other in x. Connections to external circuits were mentioned in Section 4-11 including how to advance $\mathbf{E}(\mathbf{k}=0)$ in time due to an external current source, and are described further in Chapter 16. Here we add to the discussion in Section 4-11 on the importance of careful treatment of the spatially averaged total current ($\mathbf{k} = 0$).

In dealing with boundary conditions, the component of total current density that is uniform in space ($\mathbf{k} = 0$)

$$\mathbf{J}_{\text{total},\mathbf{k}\,=\,0} = \left[\mathbf{J} + \frac{\partial \mathbf{E}}{\partial t} \right]_{\mathbf{k}\,=\,0} \tag{9}$$

is special. The usual electrostatic periodic model may have one or more of the following properties:

(a) is net neutral with $\rho(\mathbf{k} = 0) = 0$, which may be mistaken to mean $\mathbf{E}(\mathbf{k} = 0) = 0$;
(b) may have no current other than that of the plasma, with $\mathbf{J}_{\text{total}} = 0$ for all $\mathbf{k}$ including $\mathbf{k} = 0$ which may be labelled *open circuit;*
(c) may have periodic ϕ [$\mathbf{E}(\mathbf{k}=0) = 0$] such that $\phi(0) = \phi(L)$, hence may be labelled *short circuit;*
(d) has no net energy flow in or out of the system. Hence, such models may be called undriven, isolated, or closed.

However, we may have interest in systems which call for $\mathbf{k} = 0$ fields, in order to have no "hole" in the spectrum of modes available for nonlinear processes, such as parametric instabilities, or simply to supply a restoring force for $\mathbf{k} = 0$ plasma oscillations (*e.g.*, where all electrons are drifting in the same direction, uncovering ions at one end of the system and electrons at the other, producing a uniform $\mathbf{E}$ in between). For example, $\mathbf{E}(\mathbf{k}=0) = \mathbf{E}_0 \cos \omega_0 t$ may be added (this option is in ES1), independent of the plasma current, called a *driven system*.

The difference between undriven and driven systems has been seen in studies of electrostatic solitons and the two-stream instability. *Morales, Lee, and White* (1974) find relaxation oscillations between a well defined plasma cavity and small amplitude fluctuations for their case B, which corresponds to the undriven system. However, when they change boundary conditions to those of a driven system, they find collapse into a single soliton. *Valeo and Kruer* (1974) report collapse into one or more solitons in their one-dimensional electrostatic simulations of the driven system. *Friedberg and Armstrong* (1968) show that nonlinear stabilization of the linearly unstable modes of the cold two-stream instability occurs before overtaking for their "open circuit" conditions in which average convection current ($k=0$ value) is conserved. But, for their "short circuit" boundary conditions in which the average drift velocity ($k=0$ value) is conserved, they find no nonlinear stabilization.

The lesson is that keeping all or part of $\mathbf{J}_{\text{total}}$ for $k=0$ may make considerable difference in the results.

PROBLEMS

14-5a Write (1) for small Δx, Δy, as a double integral.

14-5b Show that eight of the coefficients in (2) are zero so need not be stored in RHOK. Show that the number of independent elements in RHOK is exactly $N_x N_y$.

14-6 POISSON'S EQUATION SOLUTIONS FOR SYSTEMS BOUNDED IN x AND PERIODIC IN y

We consider the methods for solving Poisson's equation for plasma slab systems that are periodic in y but bounded in x, such as one might encounter in the simulation of a nonuniform plasma as in Figure 14-6a. We solve Poisson's equation with boundary conditions given on the two sides of the system at $x = 0, L_x$. Typical boundaries are *closed* (potential variation in y specified) or *open* (vacuum out to $|x| \rightarrow \infty$).

Inside the simulation volume, $\rho \neq 0$, the potential is a solution to Poisson's equation. If the simulation volume is bounded on one (or both) sides by a vacuum, $\rho = 0$, extending indefinitely, where the potentials are solutions of Laplace's equation, then the solution to Poisson's equation within the simulation volume is to be matched to the appropriate (decaying) solution of Laplace's equation outside the simulation volume. This matching does not mean setting up outgoing waves only in the the simulation volume; an open boundary reflects electrostatic waves. A method of dealing with such open boundary conditions was developed by *Buneman* (1973). We describe a version used in ZOHAR by *Langdon and Lasinski* (1976) for open boundaries at $x = 0, L_x$ and extended by W. M. Nevins to other conditions.

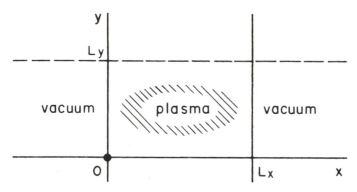

Figure 14-6a 2d plasma slab model, periodic in y, with period L_y, open at $x = 0$, L_x.

A discrete Fourier transform is made on the charge density in the periodic direction y, as

$$\rho_{j,m} = \Delta y \sum_{k=0}^{N_y-1} \rho_{j,k} \exp\left[-i\,\frac{2\pi m}{N_y}\,k\right] \tag{1}$$

where j labels the grid points in x, k labels those in y, m labels the Fourier components in y, and N_y is the number of grid points in one period in y ($N_y = L_y/\Delta y$). Using the usual 2d five-point Laplacian operator, Poisson's equation takes the form

$$\phi_{j+1,m} - 2d_m\,\phi_{j,m} + \phi_{j-1,m} = -\rho_{j,m}\,\Delta x^2 \tag{2}$$

for $j = 0$ to N_x, where

$$d_m = 1 + 2\left[\frac{\Delta x}{\Delta y}\sin\frac{\pi m}{N_y}\right]^2 \tag{3}$$

Hereafter Δx^2 and m are suppressed. To solve (2) we add boundary condition information.

First, let the right and left boundaries be electrodes held at fixed potentials $\phi_0 = V_0$ and $\phi_{Nx} = V_L$, with no y variations. For $m = 0$, $d_0 = 1$ which makes (2) the same as in a 1d electrode-bounded system for which a solution is given in Appendix D, Problem Da. For $m \neq 0$, $d_0 \neq 1$ so then we use the standard Gauss elimination for a tridiagonal matrix, also given in Appendix D.

Second, let the right and left boundaries be open. Outside of the simulation volume, where the charge density vanishes, the solutions to (2) for the potential are found to be (for $m \neq 0$),

$$\phi_j = a\,r^j + b\,r^{-j} \tag{4}$$

where r is the larger root of the quadratic equation,

$$r - 2d + \frac{1}{r} = 0 \qquad (5)$$

(the other root is r^{-1}). The solution to Poisson's equation inside the simulation volume is set equal to the vacuum solution which does not diverge as $|j| \rightarrow \infty$. At the open boundary on the right side of the system, this matching produces

$$a = 0 \quad b = \phi_{N_x} r^{N_x} \quad \phi_j = \phi_{N_x} r^{N_x - j} \quad j \geqslant N_x \quad \phi_{N_x + 1} = r^{-1} \phi_{N_x} \qquad (6)$$

where N_x labels the last grid point at which charge is collected on the right side of the simulation. At the open boundary on the left side of the system, this matching produces

$$b = 0 \quad a = \phi_0 \quad \phi_j = \phi_0 r^j \quad j \leqslant 0 \quad \phi_{-1} = \phi_0 r^{-1} \qquad (7)$$

where 0 labels the last point at which charge is collected at the left side. The boundary conditions and the set (2) constitute a tridiagonal system of $N_x + 1$ simultaneous equations for the ϕ_j, solvable as done in Appendix D.

If one applies the usual Gauss elimination of the subdiagonal, then one finds that all but the last of the new diagonal elements are now equal to r. The reason that the solution of this set of equations is simpler is due to a factorization pointed out by *Buneman* (1973), to be developed below. The system of equations (2) may be written out as

$$(8)$$

$$-\phi_{N_x - 3} + 2d\phi_{N_x - 2} - \phi_{N_x - 1} = \rho_{N_x - 2}$$

$$-\phi_{N_x - 2} + 2d\phi_{N_x - 1} - \phi_{N_x} = \rho_{N_x - 1}$$

$$-\phi_{N_x - 1} + r\phi_{N_x} = \rho_{N_x}$$

We have used (5) and (6) in writing the last equation of the set (8). We now eliminate the upper sub-diagonal and obtain a set of first-order difference equations for ϕ, as follows:

$$(9)$$

$$-r^{-1}\phi_{N_x - 3} + \phi_{N_x - 2} = r^{-1}[\rho_{N_x - 2} + r^{-1}(\rho_{N_x - 1} + r^{-1}\rho_{N_x})]$$

$$-r^{-1}\phi_{N_x - 2} + \phi_{N_x - 1} = r^{-1}(\rho_{N_x - 1} + r^{-1}\rho_{N_x})$$

$$-r^{-1}\phi_{N_x - 1} + \phi_{N_x} = r^{-1}\rho_{N_x}$$

The set of equations (9) has the form

$$-r^{-1}\phi_{j-1} + \phi_j = \psi_j \qquad (10)$$

where the source terms ψ satisfy

$$r\psi_j - \psi_{j+1} = \rho_j \tag{11}$$

The open boundary condition on the potential at the right is manifested by the requirement that

$$\psi_{N_x} = r^{-1}\rho_{N_x} \quad \text{or} \quad \psi_{N_x+1} = 0 \tag{12}$$

We can obtain (10) and (11) directly from Poisson's equation (2) by using the factorization of the Laplacian operator pointed out by *Buneman* (1973),

$$-\phi_{j-1} + 2d\phi_j - \phi_{j+1} = (r - e^{+1})(1 - r^{-1}e^{-1})\phi_j \tag{13}$$

where e is a displacement operator on grid quantities defined by

$$e^{\pm1}\phi_j \equiv \phi_{j\pm1} \tag{14}$$

With this factorization, (2) may be written as

$$(r - e^{+1})\psi_j \equiv r\psi_j - \psi_{j+1} = \rho_j \tag{15}$$

where ψ satisfies

$$(1 - r^{-1}e^{-1})\phi_j \equiv \phi_j - r^{-1}\phi_{j-1} = \psi_j \tag{16}$$

These equations are identical to (10) and (11) above.

Now that we have a boundary condition for ψ at the open boundary on the right, we can *march* across the system from right to left using (11) to calculate the source terms ψ for $j = N_x$ down to $j = 0$. The left boundary provides the boundary condition on the potential (7) that we need to start our march back from left to right, calculating the potential from our first-order partial difference equation for the potential (10). The $m = 0$ solution is in Problem 14-6b.

Langdon and Lasinski (1976) use the other choice in the Buneman factoring for ψ_j, namely $(r - e^{+1})\psi_j$, with $\psi_{-1} = 0$, so that they solve for the ψ's from left to right and the ϕ's from right to left. This factoring and marching works well for the open left boundary. Either factoring has the same operations as in the simplified Gaussian elimination.

Lastly, let the right side be open and let the potential be specified at the left boundary. The first few equations of our reduced set (10) may be written as:

$$\begin{aligned}
-r^{-1}\phi_{0,m} + \phi_{1,m} &= \psi_{1,m} \\
-r^{-1}\phi_{1,m} + \phi_{2,m} &= \psi_{2,m} \\
-r^{-1}\phi_{2,m} + \phi_{3,m} &= \psi_{3,m} \\
\cdot \quad\quad \cdot \quad\quad \cdot \\
\cdot \quad\quad \cdot \quad\quad \cdot \\
\cdot \quad\quad \cdot \quad\quad \cdot
\end{aligned} \tag{17}$$

The potential at this boundary $(j = 0)$ is given as a function of y and may be Fourier transformed to obtain the $\phi_{0,m}$. We use these values in (17) or (10) to start our march back from left to right in calculating the ϕ's.

Finally, we perform the inverse Fourier transform

$$\phi_{j,k} = \frac{1}{L_y} \sum_{m=0}^{N_y-1} \phi_{j,m} \exp\left[i \frac{2\pi m}{N_y} k \right]$$ (18)

to obtain the Poisson solution exact to within roundoff error.

PROBLEMS

14-6a Obtain d_m as in (3).

14-6b Note that for $m = 0$ the equations are underdetermined for open left and open right boundaries. Show that a solution of (2) which is symmetric and is independent of the location of the boundaries of the grid is

$$\phi_j = -\frac{\Delta x^2}{2} \sum_{j'} |j' - j| \rho_{j'}$$ (19)

This may be used to determine ϕ_{-1} and ϕ_0, which provides the needed boundary conditions for the solution of (2). Show that (19) also ensures conservation of [kinetic energy $+ \frac{1}{2} \int \rho \phi \, dx$] and momentum even for a non-neutral system.

14-6c Consider a system that has an open boundary at the left and a closed boundary at the right. The requirement that the potential not diverge as $j \to -\infty$ is (7). It is now advantageous to eliminate the lower subdiagonal of the system of equations (2). Find the first order difference equations satisfied by ϕ. What equation does the source term, ψ_j, now satisfy? *Hint:* Use the *Langdon and Lasinski* factoring.

14-7 A PERIODIC-OPEN MODEL USING INVERSION SYMMETRY

In simulating drift waves in a nonuniform slab plasma, as in Figure 14-7a, it is possible to include only one side of the density profile, $x \geqslant 0$, reducing particle and field computer memory and operations by two. The boundary conditions at $x = 0$ must be chosen with care to avoid introducing spurious or undesirable physical effects. Absorbing walls, for example, destroy overall charge neutrality. Reflecting walls may produce sheaths which dominate the physics near the boundary and could overwhelm the much smaller density gradient sustaining the drift waves. *Lee and Okuda* (1978) have proposed boundary conditions which avoid the sheaths, but which introduce non-physical effects, *viz.*, particles change the sign of their charge when they cross the $x = 0$ boundary. *Naitou et al.* (1979) followed their work, demonstrating some numerical instability, and then proposed several new boundary conditions.

A different set of boundary conditions is used here which avoids sheaths. This model involves several concepts:

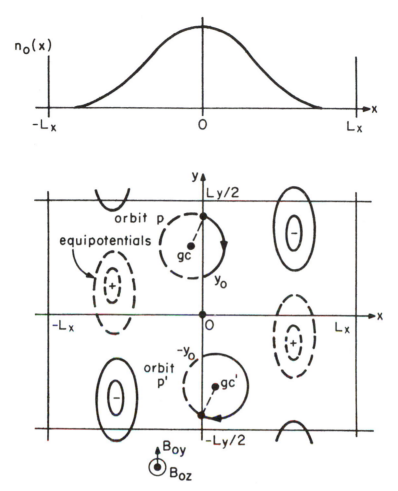

Figure 14-7a 2½d model for inversion symmetry. Particle motion is followed only in the region $0 < x < L_x$ and $-L_y/2 < y < L_y/2$. Potentials and particles have inversion symmetry about $x = 0 = y$, which is an extremum in potential and a zero in field. Equally excited drift waves propagate up and down on the left and right sides of $x = 0$. The result is no wall and no sheath at $x = 0$, as desired. $\mathbf{B_0}$ is a pseudo-vector and inverts into itself across $x = 0 = y$. *(From Nevins et al., 1981.)*

(i) periodic in y (the wave propagation direction)
(ii) open for $x > L_x$
(iii) symmetric about the origin $(0, 0)$.

The latter, termed *inversion symmetry*, was suggested by M. J. Gerver and is required for proper particle and field matching across the plane $x = 0$. Applications are given by *Nevins et al.* (1981). As shown in Figure 14-7a, a particle p going out of the $x = 0$ boundary at $y = y_0$ will be inserted as p' at $y = -y_0$ and $x = 0$, with its velocity (x, y, z components) reversed. Also shown is the inversion of potential $\phi(x,y)$ through the point $(x=0=y)$. Hence the plasma has inversion symmetry through the point $(x=0=y)$. No non-physical effects are present. No sheaths or other special effects occur at $x = 0$ since, according to the physical interpretation, no wall is there at all. One electron drift wave moves up on the left side and another, equally excited, moves down on the right side. B_{0z} and B_{0y} invert into themselves, as **B** is a pseudo-vector. These boundary conditions could also be used at the low density end, $x = L_x$, in which case we would have, in effect, periodic boundary conditions with period $2L_x$. Alternatively, other boundary conditions could be used at $x = L_x$ since, if the density is low enough, very few particles go out the $x = L_x$ boundary and so non-physical or undesirable physical effects are not as important there. For example, *Chen, Nevins, and Birdsall* (1983), in modeling drift waves propagating in y due to ∇n along x, chose to reflect particles at the low density end, $x = L_x$, with no apparent difficulties.

Some care is to be taken in counting charges [*i.e.*, accumulating charges near the grid points, (j,k), by any kind of weighting] near the $x = 0$ plane, due to the symmetry employed. Using linear weighting, a particle at (x,y) with $0 < x < \Delta x$, between y grid points k and $k+1$ is weighted to grid points $(0,k)$, $(0,k+1)$, $(1,k)$ and $(1,k+1)$; its inversion "partner" (image, but same sign) is then at $-\Delta x < x < 0$ between y grid points $-k$ and $-k-1$ and must be counted at $(0,-k)$ $(0,-k-1)$, $(-1,-k)$ $(-1,-k-1)$, as sketched in Figure 14-7b.

The method of solution for the potentials follows the method of Section 14-5, given here in detail. Consider an *open right boundary with inversion symmetry at the left boundary.* The potential and charge density are taken to

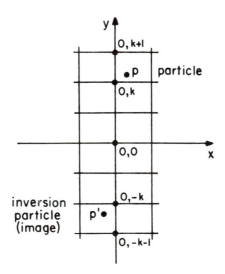

Figure 14-7b Particle p contributes to charge on the boundary $x = 0$, and so does its inversion partner (or image, same sign) p'.

have inversion symmetry about $j = 0 = k$, which means that

$$\phi_{-j,-k} = \phi_{j,k}$$
$$\rho_{-j,-k} = \rho_{j,k} \tag{1}$$

Next, the definition of the discrete Fourier transform 14-6(1) is used to show that

$$\phi_{-j,m} = \phi_{j,m}^{*}$$

$$\rho_{-j,m} = \rho_{j,m}^{*} \tag{2}$$

In particular, we have

$$\phi_{-1} = \phi_{1}^{*}$$
$$\text{Im } \phi_{0} = 0 \tag{3}$$
$$\text{Im } \rho_{0} = 0$$

Hence, Poisson's equation at $j = 0$ may then be written as

$$2d\phi_{0} - 2\,\text{Re } \phi_{1} = \rho_{0} \tag{4}$$

From the first of the set of 14-6(10), we have

$$-r^{-1}\phi_{0} + \phi_{1} = \psi_{1} \tag{5}$$

Solving (4) and (5) for ϕ_{0}, we find

$$\text{Re } \phi_{0} = \frac{\rho_{0} + 2\,\text{Re } \psi_{1}}{r - r^{-1}}$$

$$\text{Im } \phi_{0} = 0 \tag{6}$$

Equation (6) provides the required boundary condition on the potential to start our march from left to right in calculating the potential from 14-6(10).

PROBLEMS

14-7a Is momentum conserved for those particles in the region $0 < x < L_x$? For those in the region $-L_x < x < L_x$?

14-7b Sketch the motion of a particle crossing the $x = 0$ plane from the right for B_{0z} only; for B_{0y} only; for both B_{0z} and B_{0y}.

14-8 ACCURACY OF FINITE-DIFFERENCED POISSON'S EQUATION

For 2d electrostatic programs, the potential $\phi(x,y)$ is obtained from the charge density $\rho(x,y)$ by solving Poisson's equation, 14-2(2, 3). The error in $\nabla^{2}\phi$ (*i.e.*, further terms of the Taylor expansion as in *Collatz* (1966), Table VI, p. 542) is

$$-\frac{(\Delta x)^2}{12}\left[\frac{\partial^4\phi}{\partial x^4}+\frac{\partial^4\phi}{\partial y^4}\right]+\text{higher order terms} \tag{1}$$

In a periodic model when a Fourier series is used, the Fourier coefficients of potential $\phi_{l,m}$ and charge density $\rho_{l,m}$ are related by (see Sections 8-9, 14-5) $K^2(\mathbf{k}_{l,m})$. The difference between the finite difference term $K^2(\mathbf{k})$ and $k^2 = k_x^2 + k_y^2$ depends on $k_x\Delta x$ (as in 1d) and also on the direction of $\mathbf{k}$. The rectangularity of the grid imposed on $\phi_{l,m}$ is seen from the quantity R_5 (for $\Delta x = \Delta y$, small $k\Delta x$), defined by

$$R_5 \equiv \frac{K^2}{k^2} \approx 1 - \frac{(k\Delta x)^2}{24}\left[\frac{k_x^4 + k_y^4}{(k_x^2 + k_y^2)^2}\right]+\text{higher order terms} \tag{2}$$

The bracketed term is 1 along either k axis, and ½ for $k_x = k_y$, but this factor multiplies an already small magnitude correction. At larger $k\Delta x$, the anisotropy in K^2 is readily evident in a contour map of R_5 in which the contours are more square than (the desired) circular, as shown in Figure 14-8a; at a radius of $|k\Delta x| = 2$, R_5 varies ± 8 per cent.

In order to improve $\nabla^2\phi$ (reduce the error term) and K^2/k^2 (making this more nearly 1 out to large $k\Delta x$, and reduce the anisotropy), let us consider what might be gained by using higher-order finite-difference Laplacians; such does *not* necessarily mean a higher-order Poisson f.d.e., a point brought out, *e.g.*, in *Forsythe and Wasow* (1960, p. 195). Nevertheless, let us look at a higher-order stencil for ∇^2 as found in *Collatz* (1966, see Table

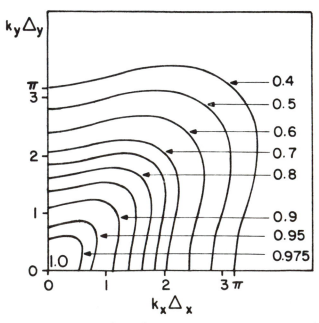

Figure 14-8a Contours of $R_5 \equiv K^2(k)/k^2$ showing anisotropy of the 2d five point finite difference Laplace operator (*i.e.*, contours are not circles).

VI, pp. 542-546); where $\nabla^2 \phi_{a,b}$ occurs, $-\rho_{a,b}$ is inserted. One of these forms, the *nine point* (for $\Delta x = \Delta y$),

$$\frac{1}{(\Delta x)^2} [8 (\phi_{j+1,k} + \phi_{j,k+1} + \phi_{j-1,k} + \phi_{j,k-1}) - 40\phi_{j,k}$$

$$+ 2 (\phi_{j+1,k+1} + \phi_{j-1,k+1} + \phi_{j-1,k-1} + \phi_{j+1,k-1})]$$

$$= \rho_{j+1,k} + \rho_{j,k+1} + 8\rho_{j,k} + \rho_{j-1,k} + \rho_{j,k-1} \quad (3)$$

was used by *Birdsall and Fuss* (1969). The error term in $\nabla^2 \phi$ is now smaller than that given by (1); the charge used is now picked up at 5 points. There is little additional cost needed to compute this form, if one is using a double Fourier transform, as ϕ and ρ are easily picked up at j, k and $j \pm 1$, $k \pm 1$ points. A contrast between the forces derived from 5 and 9 point stencils is given in Figure 14-4e. The end result is a slower decay of K^2/k^2 than in (2), that leads to a more isotropic behavior out to $\theta_x \approx 2$ as we desire; at a radius of 2, R_9 varies by ± 4 per cent.

If we make the step $\rho \rightarrow \phi$ via Fourier series, then we are free to invent and use compensation for the decay and anisotropy of K^2/k^2. Indeed, we may simply obtain $\phi(k)$ from $\rho(k)/k^2$; however, this choice implies a non-local finite-difference algorithm, somewhat defeating this choice, as discussed in Appendix E.

Lastly, lest we focus too closely on the accuracy of just this one step, $\rho \rightarrow \phi$, remember that we seek over-all accuracy in going through the four steps $\mathbf{x} \rightarrow \rho \rightarrow \phi \rightarrow \mathbf{E} \rightarrow \mathbf{F}$. To be sure, the five-point Poisson has quadratic error (generally acceptable) and the nine-point Poisson has less error. However, we have yet to find the error in the form used for the step $\phi \rightarrow \mathbf{E}$. If the latter error compensates for the error in the step $\rho \rightarrow \phi$ using the five point Laplacian (and this may occur), then we may drop the "better" nine-point Poisson.

14-9 ACCURACY OF FINITE-DIFFERENCED GRADIENT OPERATOR

Once ϕ is obtained from ρ, then in 2d, $\mathbf{E}$ is obtained from 14-2(4 and 5). The components E_x and E_y are initially written using the common two-point difference form 14-2(5), with ϕ locations shown in Figure 14-2c. The error in $\partial\phi/\partial x$ is $-1/6(\Delta x)^2 \partial^3\phi/\partial x^3$. For a 2d periodic system, using a Fourier series, the two-point finite-difference gradient operator may be represented by

$$\mathbf{E}(\mathbf{k}) = - i [\hat{\mathbf{x}} k_x \, \text{dif} (k_x \Delta x) + \hat{\mathbf{y}} k_y \, \text{dif} (k_y \Delta y)] \, \phi(kv) \quad (1)$$

The error here is worth investigating further both in magnitude *and* angle; the latter may be more significant than the errors found with the Poisson equation differencing. We start with the two point form, and then introduce the four and six point differencing suggested by *Boris* (1970a), with additional analysis from *Langdon* (1970, unpublished).

The two-point difference form given above produces

$$\mathbf{E}(\mathbf{k}) = - i \, \kappa_2 \phi(\mathbf{k}) \tag{2}$$

The error is found from

$$\frac{\kappa_{x2}}{k_x} = \text{dif}\,(k_x \Delta x) \tag{3}$$

$$\approx 1 - \frac{1}{6}\,(k_x \Delta x)^2 + \frac{1}{120}\,(k_x \Delta x)^4 + \cdots \tag{4}$$

At $k_x \Delta x = \pi/3$ or $\lambda = 6\Delta x$, the error term is 0.17. In 1d, this magnitude error in $\nabla\phi$ could be compensated for by simply multiplying each ϕ_k by $1/\text{dif}\,(k\Delta x)$; in 2d and 3d this simple compensation is not possible, as errors in E_x and E_y have different dependence on $\mathbf{k}$; that is, the corrected ϕ which produces accurate E_x will produce very poor E_y and E_z fields. Changing ϕ affects only the *magnitude* of $\mathbf{E}$; in addition there is a *direction error*, obtained from the sine of the angle between $\mathbf{k}$ and κ_2,

$$A_2 \equiv \sin \theta_2 \equiv \frac{|\mathbf{k} \times \kappa_2|}{|\mathbf{k}||\kappa_2|} \tag{5}$$

$$= \frac{\theta_x \, \theta_y \, |\,\text{dif}\,\theta_y - \text{dif}\,\theta_x\,|}{(\theta_x^2 + \theta_y^2)^{\frac{1}{2}} \, (\theta_x^2 \text{dif}^2\,\theta_x + \theta_y^2 \text{dif}^2\,\theta_y\,)^{\frac{1}{2}}} \tag{6}$$

which is shown in Figure 14-9a. A_2 is 0.3 at $\theta_x \approx 2$, $\theta_y \approx 0.75$ (meaning a direction error of 0.3 radians, 17.5 degrees, which is rather large) and is 0.05 at $\theta_x \approx 1$, $\theta_y \approx 0.5$. At small θ_x, θ_y,

$$A_2 \approx \frac{1}{6} \frac{|k_x k_y|}{k_x^2 + k_y^2} \, |\,(k_y \Delta y)^2 - (k_x \Delta x)^2\,| \tag{7}$$

In a 3d cubic lattice, with $k_x \Delta x$, $k_y \Delta y$, $k_z \Delta z$ all less than $\pi/3$, the maximum two-point direction error is 0.07 radians, a 7 per cent relative error.

Let us move on to a *six-point difference scheme* (in 2d), and *ten-point* (in 3d). Much greater accuracy can be obtained without shifting the potential onto the half-integral grid. By going to a six-point formula (in 2d), one acquires the freedom to choose one parameter after all symmetry requirements are met. This parameter is adjusted to cancel the quadratic terms in the direction error. The resultant difference scheme, using the 2d grid of Figure 14-9b, is

$$E_{x6} = -\frac{1}{3}\left[\frac{1}{2}\frac{\phi_b - \phi_a}{2\Delta x} + 2\frac{\phi_d - \phi_c}{2\Delta x} + \frac{1}{2}\frac{\phi_f - \phi_e}{2\Delta x}\right] \tag{8}$$

The coefficients ($\frac{1}{2}$, 2, $\frac{1}{2}$) are independent of $\Delta x/\Delta y$. Using the e notation

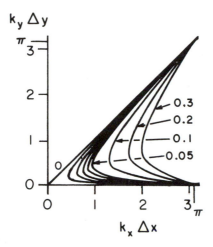

Figure 14-9a Direction error A_2 for two-point gradient difference, with contours of $A_2 = 0$, 0.01, 0.02, 0.03, 0.04, 0.05, 0.1, 0.2, 0.3. The contours reflect about the 45° line ($A_2 = 0$).

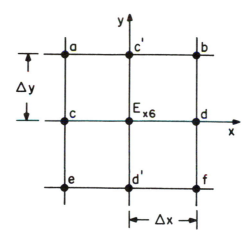

Figure 14-9b Grid used for six-point difference scheme in 2d. E_x is obtained from the 6 points $a-f$. E_y requires using points a, e, c', d', b, f. Hence 8 points are used in $\mathbf{E} = -\nabla\phi$.

defined by

$$e_x^\alpha \, \phi_{j,k,l} \equiv \phi_{j+\alpha,k,l} \tag{9}$$

in 2d, produces

$$E_{x6} = -\frac{1}{3} \frac{e_x^1 - e_x^{-1}}{2\Delta x} \left(\frac{e_y^1}{2} + 2 + \frac{e_y^{-1}}{2} \right) \phi_{j,k} \tag{10}$$

with Fourier representation

$$\kappa_{x6} = k_x \, \mathrm{dif}\, \theta_x \left[\frac{2 + \cos\theta_y}{3}\right] \tag{11}$$

For small θ_x and θ_y, this is

$$\kappa_6 \approx \mathbf{k}\left[1 - \frac{\theta_x^2 + \theta_y^2}{6}\right] + O(k^5) \tag{12}$$

showing no anisotropy up to order k^5, which is very good. The direction error is

$$A_6 = \frac{\theta_x \theta_y \left| \mathrm{dif}\, \theta_y \left[\dfrac{2 + \cos\theta_x}{3}\right] - \mathrm{dif}\, \theta_x \left[\dfrac{2 + \cos\theta_y}{3}\right] \right|}{(\theta_x^2 + \theta_y^2)^{1/2} \left[\theta_x^2 \, \mathrm{dif}^2\, \theta_x \left[\dfrac{2 + \cos\theta_y}{3}\right]^2 + \theta_y^2 \, \mathrm{dif}^2\, \theta_y \left[\dfrac{2 + \cos\theta_x}{3}\right]^2\right]^{1/2}} \tag{13}$$

which is plotted in Figure 14-9c; at $\theta_x \approx 2$, $\theta_y \approx 1$, $A_6 \approx 0.05$. For small θ_x and θ_y,

$$A_6 \approx \frac{1}{180} \frac{\theta_x \theta_y}{\theta_x^2 + \theta_y^2} \left|\theta_y^4 - \theta_x^4\right| \tag{14}$$

This is a large reduction from the two-point and four-point error. The largest error in the region $\max(\theta_x, \theta_y) \leqslant \pi/3$, with $\Delta x = \Delta y$, is now only about 0.002 radians, about factor of 25 smaller than A_2. The improvement is much greater at larger wavelengths because of the $|\theta_y^4 - \theta_x^4|$ dependence.

The six point formula could, of course, be used with any Poisson solver; the direction error would still be decreased. Thus, using the five point

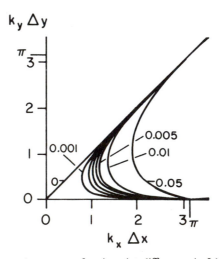

Figure 14-9c Direction error A_6 contours for six point difference in 2d.

Poisson-solving algorithm with the six point difference formula for the acceleration would be equivalent, through quadratic order in θ_x and θ_y, to solving a problem with a modified charge profile to about the same relative accuracy as one would obtain using the original charge profile and the FFT method of Poisson solving.

Keep in mind that there is no reason to seek even more accuracy in this step of the force calculation unless it is shown possible to reduce the comparable errors which arise elsewhere.

In 3d, there are 10 points, with the additional 4 located in the xz plane. The x component is

$$E_{x10} = -\frac{1}{3}\left[\frac{1}{2}\frac{\phi_b - \phi_a}{2\Delta x} + \frac{\phi_b - \phi_a}{2\Delta x} + \frac{1}{2}\frac{\phi_f - \phi_e}{2\Delta x} \right.$$

$$\left. + \frac{\phi_d - \phi_c}{2\Delta x} + \frac{1}{2}\frac{\phi_h - \phi_g}{2\Delta x} + \frac{1}{2}\frac{\phi_j - \phi_i}{2\Delta x} \right] \tag{15}$$

This has a Fourier representation

$$\kappa_{x10} = k_x \text{ dif } \theta_x \left[\frac{1 + \cos\theta_y + \cos\theta_z}{3} \right] \tag{16}$$

The maximum direction error in the cubic lattice A_{10} with θ_x, θ_y, θ_z all smaller than $\pi/3$ is 0.0075 radians.

PROBLEM

14-9a Using higher-order difference operators requiring more than three points along an axis may also reduce errors. For example, let the gradient be obtained from four points along x,

$$-\left[\frac{\alpha(\phi_{j+1} - \phi_{j-1})}{2\Delta x} + \frac{(1-\alpha)(\phi_{j+2} - \phi_{j-2})}{4\Delta x} \right]$$

Show that the Fourier representation is

$$i k_x [\alpha \text{ dif } k_x\Delta x + (1-\alpha) \text{ dif } 2k_x\Delta x]$$

which, for $\alpha = 4/3$ and small $k_x\Delta x$, is

$$i k_x\left[1 - \frac{1}{30}(k_x\Delta x)^4 + \cdots \right]$$

Show that the angle error in 2d for this eight point scheme, for small $k\Delta x$, is $6A_6$, but smaller than A_2 and A_4.

14-10 POISSON'S EQUATION FINITE-DIFFERENCED IN CYLINDRICAL COORDINATES r, r-z, r-θ

Use of cylindrical coordinates is often desired. This section presents a method of obtaining finite difference equations for fields and potentials in

(r), (r,z), and (r,θ) relating $\mathbf{E}$, ϕ, and ρ which has been found powerful in many applications involving partial differential equations with conservation properties (here, conservation of charge and flux). Difficulties with weighting or divergence as $r \rightarrow 0$ are avoided.

This method permits unequal spacing of grid points in r, may be extended to unequal spacing in θ or z, leads to symmetric matrices in ϕ, has no divergence at $r = 0$, and preserves the flux integral exactly. The same method may be applied to spherical coordinates.

Weightings for particles and fields are given in the next section.

(a) r only

In 1d, radius only (no θ, z variations), let the grid quantities be spaced and indexed (using index j) as shown in Figure 14-10a. The particles are cylindrical shells uniform along z. Let the charges of the particles be assigned to the grid points j using a weighting which guarantees charge conservation exactly,

$$\sum_{i=0}^{N} q_i = q_{\text{total}} = \sum_{j=0}^{N_r-1} Q_j \tag{1}$$

Using Gauss' law guarantees flux conservation. Applied at the cylindrical surface $j = \frac{1}{2}$, Gauss' law produces (in rationalized cgs units, or mks with $\epsilon = 1$) the radial electric field from the charges, as

$$Q_0 = 2\pi \, r_{\frac{1}{2}} \, E_{\frac{1}{2}} \tag{2}$$

where $r_{j+\frac{1}{2}}$ is some radius in the interval, such as $\frac{1}{2}(r_{j+1} + r_j)$, and between the $j = 1/2, 3/2$ surfaces,

$$Q_1 = 2\pi \, r_{3/2} \, E_{3/2} - 2\pi \, r_{1/2} \, E_{1/2} \tag{3}$$

and so on. In order to obtain an equation in ϕ, use single cell differencing

$$E_{j+\frac{1}{2}} = -\frac{\phi_{j+1} - \phi_j}{\Delta r_{j+\frac{1}{2}}} \tag{4}$$

where $\Delta r_{j+\frac{1}{2}} \equiv r_{j+1} - r_j$, so that Gauss' law reads

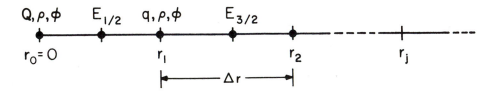

Figure 14-10a Location of grid quantities in a 1d radial grid, index $j = 0, 1, 2, \ldots$. The radial spacing need not be uniform.

$$Q_j = -2\pi\, r_{j+\frac{1}{2}}\frac{\phi_{j+1} - \phi_j}{\Delta r_{j+\frac{1}{2}}} + 2\pi\, r_{j-\frac{1}{2}}\frac{\phi_j - \phi_{j-1}}{\Delta r_{j-\frac{1}{2}}} \tag{5}$$

The charge density is obtained from $\rho_j = Q_j/V_j$. The volumes are $V_j \equiv \pi(\Delta r^2)_j$, where

$$(\Delta r^2)_j = r_{j+\frac{1}{2}}^2 - r_{j-\frac{1}{2}}^2$$
$$(\Delta r^2)_0 = r_{\frac{1}{2}}^2 \tag{6}$$

Equation (5) leads to a three-point finite difference form of Poisson's equation, as

$$\rho_j = -\frac{2\, r_{j+\frac{1}{2}}}{(\Delta r^2)_j}\frac{\phi_{j+1} - \phi_j}{\Delta r_{j+\frac{1}{2}}} + \frac{2\, r_{j-\frac{1}{2}}}{(\Delta r^2)_j}\frac{\phi_j - \phi_{j-1}}{\Delta r_{j-\frac{1}{2}}} \qquad j > 0 \tag{7}$$

$$\rho_0 = -\frac{2\, r_{\frac{1}{2}}}{(\Delta r^2)_0}\frac{\phi_1 - \phi_0}{\Delta r_{\frac{1}{2}}} \tag{8}$$

See Problem 14-9a for the result for uniform Δr. As in all purely radial models, $E_r(a)$ depends only on the net charge enclosed within $r \leqslant a$ and is unaffected by the charges at $r > a$. In the absence of a line charge at the origin, $E_0 = 0$.

(b) $r\text{-}z$

In 2d $r\text{-}z$ coordinates (no θ variations), let the cells appear as in Figure 14-10b, with grid quantities located as shown, with indices j, k. Let Δz be uniform. The particles are rings. For the volume shown by the dashed lines

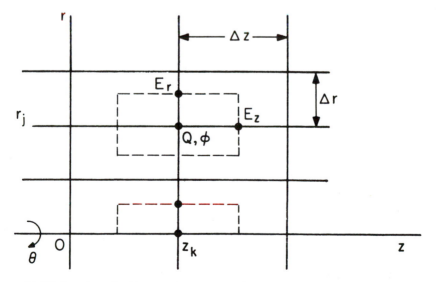

Figure 14-10b Location of grid quantities and Gauss' law volumes in a 2d $r\text{-}z$ grid, with indices j, k.

centered on (r_j, z_k), Gauss' law is

$$Q_{j,k} = 2\pi r_{j+\frac{1}{2}} \Delta z E_{r,j+\frac{1}{2},k} - 2\pi r_{j-\frac{1}{2}} \Delta z E_{r,j-\frac{1}{2},k}$$
$$+ \pi (r_{j+\frac{1}{2}}^2 - r_{j-\frac{1}{2}}^2)(E_{z,j,k+\frac{1}{2}} - E_{z,j,k-\frac{1}{2}}) \qquad (9)$$

At the origin $(j = 0)$,

$$Q_{0,k} = 2\pi r_{\frac{1}{2}} \Delta z E_{r,\frac{1}{2},k} + \pi r_{\frac{1}{2}}^2 (E_{z,0,k+\frac{1}{2}} - E_{z,0,k-\frac{1}{2}}) \qquad (10)$$

The charge density is obtained from $\rho_{j,k} = Q_{j,k}/V_{j,k}$, where

$$V_{j,k} \equiv \pi (\Delta r^2)_j \Delta z \qquad (11)$$

so that $\rho_{j,k} =$

$$\frac{2}{(\Delta r^2)_j} (r_{j+\frac{1}{2}} E_{r,j+\frac{1}{2},k} - r_{j-\frac{1}{2}} E_{r,j-\frac{1}{2},k}) + \frac{E_{z,j,k+\frac{1}{2}} - E_{z,j,k-\frac{1}{2}}}{\Delta z} \qquad (12)$$

Then, using E_r from (4) and E_z as

$$E_{z,j,k+\frac{1}{2}} = - \frac{\phi_{j,k+1} - \phi_{j,k}}{\Delta z} \qquad (13)$$

we obtain a five-point finite-difference form of Poisson's equation in $\phi_{j,k}$, as

$$-\rho_{j,k} = \frac{2r_{j+\frac{1}{2}}}{(\Delta r^2)_j \Delta r_{j+\frac{1}{2}}} (\phi_{j+1,k} - \phi_{j,k}) - \frac{2r_{j-\frac{1}{2}}}{(\Delta r^2)_j \Delta r_{j-\frac{1}{2}}} (\phi_{j,k} - \phi_{j-1,k})$$

$$+ \frac{1}{\Delta z^2} (\phi_{j,k+1} - 2\phi_{j,k} + \phi_{j,k-1}) \qquad j > 0 \qquad (14)$$

$\rho_{0,k}$ is obtained from $Q_{0,k}/V_{0,k}$, which produces

$$-\rho_{0,k} = \frac{2r_{\frac{1}{2}}}{(\Delta r^2)_0} \frac{\phi_{1,k} - \phi_{0,k}}{\Delta r_{\frac{1}{2}}} + \frac{\phi_{0,k+1} - 2\phi_{0,k} + \phi_{0,k-1}}{\Delta z^2} \qquad (15)$$

Again, in the absence of a line charge at $r = 0$, $E_{r,0,k} = 0$.

(c) r-θ

In 2d r-θ coordinates (no z variation) let the cells and grid quantities appear as in Figure 14-10c. The particles are rods. Let $\Delta\theta$ be uniform. For the volume shown by the dashed lines centered on (r_j, ϕ_k), Gauss' law is

$$Q_{j,k} = \Delta\theta (r_{j+\frac{1}{2}} E_{r,j+\frac{1}{2},k} - r_{j-\frac{1}{2}} E_{r,j-\frac{1}{2},k}) + \Delta r_j (E_{\theta,j,k+\frac{1}{2}} - E_{\theta,j,k-\frac{1}{2}}) \qquad (16)$$

where $\Delta r_j \equiv r_{j+\frac{1}{2}} - r_{j-\frac{1}{2}}$. Charge density is obtained from $\rho_{j,k} = Q_{j,k}/V_{j,k}$

where

$$V_{j,k} = \frac{\Delta\theta}{2} (\Delta r^2)_j \qquad (17)$$

so that $-\rho_{j,k} (\Delta r^2)_j$

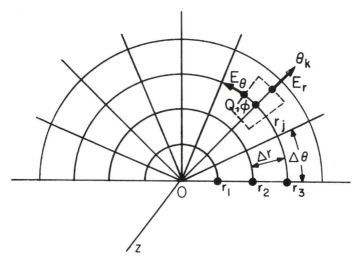

Figure 14-10c Location of grid quantities and Gauss' law volumes in a 2d r-θ grid, with indices j,k.

$$= 2\left(r_{j+\frac{1}{2}} E_{r,j+\frac{1}{2},k} - r_{j-\frac{1}{2}} E_{r,j-\frac{1}{2},k}\right) + \frac{2\Delta r_j}{\Delta\theta}\left(E_{\theta,j,k+\frac{1}{2}} - E_{\theta,j,k-\frac{1}{2}}\right) \quad (18)$$

Using E_r from (4) and E_θ as

$$E_{\theta,j,k+\frac{1}{2}} = -\frac{\phi_{j,k+1} - \phi_{j,k}}{r_j \Delta\theta} \quad (19)$$

we obtain a five-point finite-difference form of Poisson's equation in $\phi_{j,k}$, as

$$-\rho_{j,k}(\Delta r^2)_j = \frac{2r_{j+\frac{1}{2}}}{\Delta r_{j+\frac{1}{2}}}\left(\phi_{j+1,k} - \phi_{j,k}\right) - \frac{2r_{j-\frac{1}{2}}}{\Delta r_{j-\frac{1}{2}}}\left(\phi_{j,k} - \phi_{j-1,k}\right) \quad (20)$$

$$+ \frac{2\Delta r_j}{(\Delta\theta)^2\, r_j}\left(\phi_{j,k+1} - 2\phi_{j,k} + \phi_{j,k-1}\right) \quad j > 0$$

Near the origin, application of Gauss' law is not so obvious. Consider the r-θ grid to be a rectangular j-k grid extending to the origin ($r=0=j$) with charges weighted by some algorithm to produce $\rho_{0,k}$ for each $(0,k)$ point. In this view, for the surface at $r_{\frac{1}{2}}$ Gauss' law relates the total charge at the origin,

$$Q_0 \equiv \sum_k Q_{0,k} = \sum_k \rho_{0,k} V_{0,k} \quad (21)$$

to the field near the origin

$$Q_0 = \sum_k E_{r,\frac{1}{2},k}\, r_{\frac{1}{2}} \Delta\theta \quad (22)$$

The Poisson equation at the origin is given from (22) as

$$Q_0 = -\sum_k \frac{\phi_{1,k} - \phi_0}{\Delta r_{1/2}} \, r_{1/2} \, \Delta\theta \tag{23}$$

Thus the full Poisson equation set is (20) plus (23). The $E_{r,1/2,k}$ are obtained from the $\phi_{1,k}$, ϕ_0 as in (23), and the $E_{\theta,1,k+1/2}$ are obtained from the $\phi_{1,k}$, and $\phi_{1,k+1}$ as in (19).

Because it is difficult to handle the origin well, x-y coordinates are often preferred over r-θ coordinates.

PROBLEMS

14-10a It is asserted that the particle and grid charges have units of:

r only	charge/length, along z
r-z	charge
r-θ	charge/length, along z

Check these assertions.

14-10b Show that (7) for constant Δr, reduces to

$$-\rho_j = \frac{\phi_{j+1} - 2\phi_j + \phi_{j-1}}{(\Delta r)^2} + \frac{\phi_{j+1} - \phi_{j-1}}{2r_j \, \Delta r} \tag{24}$$

with readily recognized relation to the usual finite-difference Laplacian in r, as found, for example, in *Sköllermo and Sköllermo* (1978), and *Sköllermo* (1982), and to the partial differential Laplacian

$$\frac{\partial^2 \phi}{\partial r^2} + \frac{1}{r}\frac{\partial \phi}{\partial r} = \frac{1}{r}\frac{\partial}{\partial r}\left(r\frac{\partial \phi}{\partial r}\right)$$

14-10c In differencing E_θ in the last term of (18), something like $\partial E_\theta/r\partial\theta \approx \Delta E_\theta/r\Delta\theta$, the arc $r_j\Delta\theta$ might be replaced by the chord $2r_j \sin\frac{1}{2}\Delta\theta$. In differencing ϕ in (19), the arc $r_j\Delta\theta$ might be replaced similarly. Using both replacements changes $(\Delta\theta)^2$ in (20) to $(2\sin\frac{1}{2}\Delta\theta)^2$. Show that these replacements handle a uniform field correctly, with potential like $\phi_{j,k} = -E_0 r_j \cos\theta_k$ for $\rho = 0$.

14-11 WEIGHTING IN CYLINDRICAL COORDINATES FOR PARTICLES AND FIELDS

We now propose methods for weighting in cylindrical coordinates r-θ, with methods for r and r-z left to the reader. One method for particle weighting is to weight the charge q_i bilinearly in $(r$-$\theta)$ to the nearest four grid points, similar to that done in rectangular coordinates 14-2(1). Another method is bilinear in (r^2, θ), or *area weighting*, as shown in Figure 14-11a (the r-θ version of 14-2(b)). Let the particle be located at (r_i, θ_i). The fraction of the charge q_i assigned to point A (r_j, θ_k) is (area a)/(areas a + b + c + d) = $f_{j,k}$, leading to

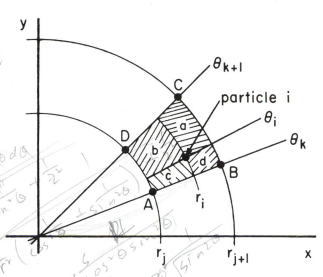

Handwritten marginalia (left margin):

$x = r\cos\theta$
$y = r\sin\theta$

$r = \sqrt{x^2 + y^2}$

$\int r \, dr \, \sin\theta \, d\theta$

Figure 14-11a Particle weighting to r-θ grid points, using area ratio $a/(a + b + c + d)$ for assignment to A, etc., interpreted as area or "r^2" weighting.

$$Q_A = Q_{j,k} = q_i \frac{(r_{j+1}^2 - r_i^2)(\theta_{k+1} - \theta_i)}{(r_{j+1}^2 - r_j^2)(\theta_{k+1} - \theta_k)} = q_i f_{j,k} \tag{1}$$

and the rest follow. Note that these weightings are valid for $r_i < r_1$, where grid points A and D are at the origin. Using the view given at the end of the last section, $Q_{0,k}$ and $Q_{0,k+1}$ are obtained in the same way and hence, Q_0, as needed in 14-10(23).

The weighting of the E_r and E_θ fields to the particles is done in the same manner. However, the fields we have obtained are not at the grid points, but roughly halfway between. One suggestion is that E_r and E_θ be formed at the grid points by an *unweighted average; i.e.*,

$$E_{r,j,k} \equiv \frac{E_{r,j-\frac{1}{2},k} + E_{r,j+\frac{1}{2},k}}{2} \tag{2}$$

Then the bilinear or area weights as above are used to weight these E_r and E_θ to the particle. That is, using Figure 14-11a and (1) we find at particle i,

$$E_r = E_r(A) f_{j,k} + E_r(B) f_{j+1,k} + E_r(C) f_{j+1,k+1} + E_r(D) f_{j,k+1} \tag{3}$$

A second suggestion is to use a *flux weighted average*,

$$r_j E_{r,j,k} \equiv \frac{r_{j+\frac{1}{2}} E_{r,j+\frac{1}{2},k} + r_{j-\frac{1}{2}} E_{r,j-\frac{1}{2},k}}{2} \tag{4}$$

At the origin $r = 0 = j$, we need E_r and E_θ for weighting **E** to those particles having $r_i \leqslant r_1$. One suggestion is to use

$$E_{r,0,k} = E_{r,\frac{1}{2},k} \qquad \text{and} \qquad E_{\theta,0,k+\frac{1}{2}} = E_{\theta,1,k+\frac{1}{2}} \tag{5}$$

...or example, for a uniform vacuum field, as discussed in

...duces

$$_1 E_{r,j+1,k} - r_j E_{r,j,k} = \frac{Q_{j+1,k} + Q_{j,k}}{4\pi} \tag{6}$$

which is a form of $\nabla \cdot E = \rho$. Equation (6) may be integrated to obtain E_r in a cylindrically symmetric system much as using the trapezoidal rule in a linear system. Indeed, the spherically symmetric version of (6) was used (*Langdon,* 1979, unpublished) to obtain E_r in a spherically symmetric system, which replaced FIELDS in ES1.

PROBLEMS

14-11a Check that the weights in (1) sum to unity: $Q_A + Q_B + Q_C + Q_D = q_i$. Repeat for bilinear weighting.

14-11b Obtain (6) from (4).

14-11c Using the "r^2" weights in (1), plot the fraction assigned to r_j as the particle position r_i varies from r_{j-1} to r_{j+1}. What is the implied shape factor $S(r)$? What is the shape factor $S(\theta)$? Repeat for bilinear weighting.

14-12 POSITION ADVANCE FOR CYLINDRICAL COORDINATES

The velocity advance in vector form given in Chapter 4 is usable in rectangular and other coordinate systems. However, the position advance in cylindrical coordinates poses problems as the particle passes close to the origin. For example, with a circular orbit, using $\Delta\theta = (v_\theta \Delta t)/r$ produces very large $\Delta\theta$ as $r \to 0$. One solution, following *Boris* (1970b), is to use v_r and v_θ to make the particle advance in (x,y), and then calculate the new r and θ. The advance in z is the same as for rectangular coordinates. The particle at (r_1, θ_1) is to be moved with v_r and v_θ known at $(t_1 + \Delta t/2)$ as shown in Figure 14-12a, using the x' axis along r_1 (and v_{r_1})

$$x'_2 = r_1 + v_{r_1}\Delta t \qquad\qquad y'_2 = v_{\theta_1} \Delta t$$

producing
$$r_2 = \sqrt{x'_2{}^2 + y'_2{}^2} \geqslant 0 \qquad \theta_2 = \theta_1 + \alpha$$

Next, it is necessary to refer v_2 to the new angle θ_2 from the coordinate rotation (v preserved) prime to double prime, still at time $(t + \Delta t/2)$.

$$\begin{bmatrix} v_r \\ v_\theta \end{bmatrix}_2 = \begin{bmatrix} \cos\alpha & \sin\alpha \\ -\sin\alpha & \cos\alpha \end{bmatrix} \begin{bmatrix} v_r \\ v_\theta \end{bmatrix}_1$$

where $\sin\alpha = y'_2/r_2$, $\cos\alpha = x'_2/r_2$. If a particle stops at the axis $r_2 = 0$,

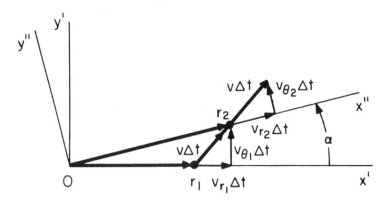

Figure 14-12a Coordinates for position advance in cylindrical coordinates.

then set $\cos \alpha = 1$, $\sin \alpha = 0$, making all momentum radial, as it must be for a particle to stop at the axis. This method provides full second-order accuracy due to the time centering (leap-frog) and reversibility. The cost of avoiding the errors near $r = 0$ is the square root and the coordinate rotation.

PROBLEM

14-12a Show that the cylindrical integrator is time reversible. *Hint*: Change sign of **v** *and* rotate **v** cleverly.

14-13 IMPLICIT METHOD FOR LARGE TIME STEPS

Characteristic time scales for collective phenomena in plasmas encompass many orders of magnitude. Where kinetic effects are crucial, particle simulation methods have been applied very successfully to studies of the nonlinear evolution of plasma phenomena on the faster time scales. However, for both applications and basic studies, there is increasing interest in extending simulation techniques to kinetic phenomena on slower time scales. One approach to modelling long time-scale behavior in such systems is to alter the governing equations in order to eliminate uninteresting high frequency modes. Examples include the electrostatic and Darwin field approximations in particle simulation. Other approaches are subcycling, orbit-averaging and implicit time integration, as already mentioned in Section 9-7.

The highest frequencies in plasmas are the Langmuir frequency ω_p and the electron cyclotron frequency ω_{ce}. The shortest times are the transit time for electrons or light to cross a characteristic distance. In contrast, long time scales can be set by ion inertia, electromagnetic effects, and large spatial scale lengths. The ratios of electron to ion plasma and cyclotron frequencies, and of hydromagnetic to electron transit times, are determined by the small number Zm_e/m_i, where Z is the ionic charge state. Where the dominant forces are from magnetic fields due to currents in the plasma itself, the frequencies are reduced relative to ω_p by at least the ratio c/v_e, where c and v_e are light and electron speeds.

This section describes a method for implementing *implicit time differencing* in particle plasma codes, in which the equations for the time-advanced quantities are constructed *directly* from the particle equations of motion by linearization, rather than by introducing fluid (velocity moment) equations. This section is based on the review chapter by *Langdon and Barnes* (1985).

(a) Implicit Time Differencing of the Particle Equations of Motion

The first major issue is the choice of finite-differenced equations of motion for the particles which have the necessary stability at large time-step and are accurate for the low frequency phenomena to be simulated. We choose not to consider backward-biased schemes with relative errors of order Δt. It is not expensive to achieve relative error of order Δt^2, with error Δt^3 in $\text{Im}\,\omega$, the growth/decay rate.

Several suitable schemes for time-differencing the the particles have been analyzed and applied (*Cohen, Langdon, and Friedman,* 1982). Here, we discuss only the D_1 algorithm (Section 9-8), also called the $\bar{1}$ scheme (*Barnes et al.,* 1983), which can be written

$$\frac{\mathbf{x}_{n+1} - \mathbf{x}_n}{\Delta t} = \mathbf{v}_{n+\frac{1}{2}} \qquad \frac{\mathbf{v}_{n+\frac{1}{2}} - \mathbf{v}_{n-\frac{1}{2}}}{\Delta t} = \bar{\mathbf{a}}_n + \frac{\mathbf{v}_{n+\frac{1}{2}} + \mathbf{v}_{n-\frac{1}{2}}}{2} \times \frac{q\,\mathbf{B}_n}{mc} \qquad (1)$$

where
$$\bar{\mathbf{a}}_n = \frac{1}{2}\left[\bar{\mathbf{a}}_{n-1} + \frac{q}{m}\,\mathbf{E}_{n+1}\,(\mathbf{x}_{n+1})\right] \qquad (2)$$

To check the accuracy of this scheme, we can derive and solve a dispersion relation for harmonic oscillations, analogous to Section 9-8,

$$(\omega_0\,\Delta t)^2 + \left[\frac{2}{z} - \frac{1}{z^2}\right]\frac{(z-1)^2}{z} = 0 \qquad (3)$$

For $\omega_0\,\Delta t \leq 1$, we find (*Cohen, Langdon, and Friedman,* 1982)

$$\pm\,\text{Re}\,\frac{\omega}{\omega_0} = 1 - \frac{11}{24}\,(\omega_0\,\Delta t)^2 + \cdots \qquad \text{Im}\,\frac{\omega}{\omega_0} = -\frac{1}{2}\,(\omega_0\,\Delta t)^3 + \cdots \quad (4)$$

and an extraneous damped mode with $|z| \rightarrow \frac{1}{2}$. For $\omega_0\,\Delta t \gg 1$, the modes are heavily damped, $|z| \rightarrow (\omega_0\,\Delta t)^{-2/3}$.

Equation (1) can be solved exactly for $\mathbf{v}_{n+\frac{1}{2}}$ by adding $\frac{1}{2}\bar{\mathbf{a}}_n\Delta t$ to $\mathbf{v}_{n-\frac{1}{2}}$, doing a rotation, and again adding $\frac{1}{2}\bar{\mathbf{a}}_n\Delta t$. The result is

$$\mathbf{v}_{n+\frac{1}{2}} = \frac{1}{2}\,\bar{\mathbf{a}}_n\Delta t + \mathbf{R} \cdot \left[\mathbf{v}_{n-\frac{1}{2}} + \frac{1}{2}\,\bar{\mathbf{a}}_n\,\Delta t\right] \qquad (5)$$

where the operator $\mathbf{R}$ effects a rotation through angle $-2\tan^{-1}(\Omega\,\Delta t/2)$ where $(\Omega \equiv q\,\mathbf{B}_n/mc)$, and can be written

$$(1 + \theta^2)\,\mathbf{R} = (1 - \theta^2)\,\mathbf{I} + 2\,\boldsymbol{\theta\theta} - 2\,\boldsymbol{\theta} \times \mathbf{I} \tag{6}$$

where $\theta \equiv \boldsymbol{\Omega}\,\Delta t/2$ and $\mathbf{I}$ is the unit tensor. For small $\boldsymbol{\Omega}\,\Delta t$,

$$\mathbf{R} \approx \mathbf{I} - \boldsymbol{\Omega}\,\Delta t \times \mathbf{I} \tag{7}$$

For large $\boldsymbol{\Omega}\,\Delta t$,

$$\mathbf{R} \approx -\mathbf{I} + \frac{2\,\boldsymbol{\Omega}\,\boldsymbol{\Omega}}{2\Omega} - \frac{4\,\boldsymbol{\Omega} \times \mathbf{I}}{\Omega^2 \Delta t} \tag{8}$$

(b) Direct Method with Electrostatic Fields; Solution of the Implicit Equations

The second major issue in implicit codes is the more complicated time-cycle splitting. With explicit differencing, the time-cycle is split between advancing particles and fields; these calculations alternate and proceed independently. However, an implicit code must solve the coupled set of equations 8-5(1, 2) and 14-2(3, 5) with (1, 2) or (3, 4). In all implicit schemes the future positions $\mathbf{x}_{n+1}$ depend on the accelerations $\mathbf{a}_{n+1}$ due to the electric feld $\mathbf{E}_{n+1}$. But this field is not yet known, as it depends on the density ρ_{n+1} of particle positions $\{\mathbf{x}_{n+1}\}$. The solution of this large system of nonlinear, coupled particle and field equations is our task here.

Historically, in the method implemented first for this solution, the fields at the new time level are predicted by solving coupled field and fluid equations, in which the kinetic stress tensor is evaluated approximately from particle velocities known at the earlier time. After the fields are known, the particles are advanced to the new time level, and, if desired, an improved stress tensor is calculated and the process iterated. This moment approach has been described in detail by *Denavit* (1981) and *Mason* (1981).

The next implementation, as done here, is to predict the future electric field $\mathbf{E}_{n+1}$ *directly* by means of a *linearization of the particle-field equations*. One form of this method, its implementation, and some examples verifying its performance, have been outlined by *Friedman, Langdon, and Cohen* (1981). Another form is described by *Barnes et al.* (1983). *Langdon, Cohen, and Friedman* (1983) obfuscate the algorithm with great generality, and consider many important details, such as spatial differencing and filtering, and iterative solution of the implicit equations.

The essence of the *direct method* is that we work directly with the particle equations of motion and the particle/field coupling equations. These are linearized about an estimate (extrapolation) for their values at the new time level $n+1$. The future values of $\{\mathbf{x}, \mathbf{v}\}$ are divided into two parts:

(i) increments $\{\delta\mathbf{x}, \delta\mathbf{v}\}$ which depend on the (unknown) fields at the future time level $n+1$, and

(ii) extrapolations $\{\mathbf{x}_{n+1}^{(0)}, \mathbf{v}_{n+\frac{1}{2}}^{(0)}\}$ which incorporate *all* other considerations to the equation of motion.

The charge density $\rho_{n+1}^{(0)}$ corresponding to positions $\{\mathbf{x}_{n+1}^{(0)}\}$ is collected, as are the coefficients in an expression for the difference $\delta\rho(\{\delta\mathbf{E}\}) = \rho_{n+1} - \rho_{n+1}^{(0)}$ between the densities obtained after integration with $\mathbf{E}_{n+1}^{(0)}$ and with the corrected field $\mathbf{E}_{n+1} = \delta\mathbf{E} + \mathbf{E}_{n+1}^{(0)}$. These comprise the source term in Gauss' law

$$\nabla \cdot \mathbf{E}_{n+1} = \delta\rho(\{\delta\mathbf{E}\}) + \rho_{n+1}^{(0)} \tag{9}$$

This becomes a linear elliptic equation, for $\delta\phi$ or ϕ_{n+1}, with non-constant coefficients.

The care with which we express the increment $\{\delta\rho\}$ is a compromise between complexity and strong convergence (*Langdon, Cohen, and Friedman, 1983; Barnes et al., 1983*). If necessary, $\delta\rho$ may be evaluated rigorously as derivatives of 8-5(1), ["strict differencing" (*Langdon, Cohen, and Friedman, 1983, sec. 4*)], or as simplified difference representations of 8-5(1) [as in *Langdon, Cohen, and Friedman, 1983, sec. 3.4; Barnes et al., 1983*] for each species.

(c) A One-Dimensional Realization

The direct implicit method is illustrated in the following one-dimensional unmagnetized electrostatic example. The position x_{n+1} of a particle at time level t_{n+1}, as given by an implicit time integration scheme, can be written as

$$x_{n+1} = \beta\Delta t^2\, a_{n+1} + \tilde{x}_{n+1} \tag{10}$$

where $\beta > 0$ is a parameter controlling implicitness and is ½ for the D_1 algorithm; $\tilde{x}_{n+1}$, the position obtained from the equation of motion with the acceleration a_{n+1} omitted, is known in terms of positions, velocities and accelerations at time t_n and earlier. Eliminating $v_{n+\frac{1}{2}}$ between (1) and (2) we find

$$\tilde{x}_{n+1} = x_n + v_{n-\frac{1}{2}}\Delta t + \frac{1}{2}\bar{a}_{n-1}\Delta t^2 \tag{11}$$

In its most obvious form, which we adopt for this example, the direct implicit algorithm is derived by linearization of the particle positions relative to $\tilde{x}_{n+1}$, that is, $E_{n+1}^{(0)} = 0$ and therefore $x_{n+1}^{(0)} = \tilde{x}_{n+1}$.

At the grid point located at $X_j \equiv j\,\Delta x$, the charge density $\tilde{\rho}_{j,n+1}$ is formed as in 8-5(1) by adding the contribution of the simulation particle at positions $\{\tilde{x}_{i,n+1}\}$

$$\tilde{\rho}_{j,n+1} = \sum q_i\, S(X_j - \tilde{x}_{i,n+1}) \tag{12}$$

If we expand S in 8-5(1) with respect to position, then

$$\delta\rho_{n+1} = -\sum_i q_i \, \delta x_i S'(X_j - \tilde{x}_{i,n+1}) \tag{13}$$

with $\delta x_i = x_{i,n+1} - \tilde{x}_{i,n+1}$ and $S'(X) \equiv dS/dX$. In terms of E_{n+1}, the particle acceleration is obtained from 8-5(2) evaluated at $\tilde{x}_{n+1}$,

$$m_i a_{i,n+1} = q_i \, \Delta x \sum_i E_{j,n+1} S(X_j - \tilde{x}_{i,n+1}) \tag{14}$$

Writing (13) as

$$\delta\rho = -[\nabla \cdot \sum q \, \delta x \, S(x - \tilde{\mathbf{x}}_{n+1})]_{x=X_j} \tag{15}$$

we see that (13) is a finite-difference representation of

$$\delta\rho = -\nabla \cdot (\tilde{\rho} \, \delta\mathbf{x}) \tag{16}$$

Hence, we have the elliptic partial differential equation

$$-\tilde{\rho} = \nabla \cdot \{[1 + \chi(\mathbf{x})] \nabla\phi\} \tag{17}$$

where $\chi(\mathbf{x}) = \beta\tilde{\rho}(\mathbf{x})(q/m)\Delta t^2$ summed over species, *i.e.*, $\chi = \beta(\omega_p \Delta t)^2$. Because of the similarity of (17) to the field equation in dielectric media, we call χ the *implicit susceptibility*. Where $\omega_p \Delta t$ is large, the regime we wish to access, note that $\chi \gg 1$ is dominant in the right-hand-side of (17).

With the extrapolated charge density $\tilde{\rho}_{j,n+1}$ and a reasonable finite-difference representation of the linearized implicit contribution $\delta\rho = -\partial(\chi E)/\partial x$, the field equation in one dimension is

$$\tilde{\rho}_{j,n+1} = \frac{(1 + \chi_{j+\frac{1}{2}})E_{j+\frac{1}{2},n+1} - (1 + \chi_{j-\frac{1}{2}})E_{j-\frac{1}{2},n+1}}{\Delta x} \tag{18}$$

Two representations of $\chi_{j+\frac{1}{2}}$ used here are

$$\chi_{j+\frac{1}{2}} = \frac{\chi_j + \chi_{j+1}}{2} \tag{19a}$$

or

$$\chi_{j+\frac{1}{2}} = \max(\chi_j, \chi_{j+1}) \tag{19b}$$

where

$$\chi_j = \Delta t^2 \sum_s \left[\beta \, \tilde{\rho}_{j,n+1} \frac{q}{m} \right]_s \tag{20}$$

is a sum over all species index s [*Langdon, Cohen, and Friedman*, 1983, eqs. (28a, b)]. In both (18) and (20), $\tilde{\rho}_{j,n+1}$ is given by (12).

From E at half-integer positions, form

$$E_{j,n+1} = \frac{E_{j-\frac{1}{2},n+1} + E_{j+\frac{1}{2},n+1}}{2} \tag{21}$$

In terms of $E_{j,n+1}$, the particle acceleration is evaluated at $\tilde{x}_{n+1}$ using (14). This algorithm is the shortest implicit scheme we have seen, and is robust in the test problem reported by *Langdon and Barnes* (1985).

(d) General Electrostatic Case

We return to the multidimensional case, possibly including a magnetic field imposed by external currents, showing the calculational steps to be performed starting with (1) and (2). We begin by restating the method in a more general form. The extrapolated charge density $\rho_{n+1}^{(0)}$ is evaluated as in 8-5(1), but from positions $\{x_{n+1}^{(0)}\}$ obtained from the equation of motion with a_{n+1} given by $E_{n+1}^{(0)}$, which is a guess for E_{n+1}. This charge density does *not* correctly correspond to the field $E_{n+1}^{(0)}$; that is, $\nabla \cdot E_{n+1}^{(0)} \neq \rho_{n+1}^{(0)}$. We wish to calculate an improved field E_{n+1} with which the particles are re-integrated to positions $\{x_{n+1}\}$, whose charge density ρ_{n+1} does satisfy

$$\nabla \cdot E_{n+1} = \rho_{n+1} \tag{22}$$

To this end we rewrite (22) as

$$\nabla \cdot \delta E_{n+1} - \delta \rho_{n+1} = \nabla \cdot E_{n+1}^{(0)} - \rho_{n+1}^{(0)} \tag{23}$$

where $E_{n+1} = E_{n+1}^{(0)} + \delta E_{n+1}$, and similarly for ρ_{n+1}; $\delta \rho_{n+1}$ is due to the increments $\{\delta x\}$ in the particle positions, which in turn are due to the difference between E_{n+1} and $E_{n+1}^{(0)}$.

Using (17) and the equation of motion, we express $\delta \rho_{n+1}$ as a linear functional of δE_{n+1}. In the general case, the increments $\{\delta x, \delta v\}$ are evaluated by linearization of each equation of motion (*Langdon, Cohen, and Friedman*, 1983; *Barnes et al.*, 1983) about position $x_{n+1}^{(0)}$; here, we have

$$\delta x_{n+1} = \delta v_{n+1/2} \Delta t \tag{24}$$

and

$$\delta v_{n+1/2} = \frac{q \, \Delta t}{2 \, m} \, \delta E_{n+1} (x_{n+1}^{(0)}) \qquad \text{(unmagnetized)} \tag{25}$$

or

$$\delta v_{n+1/2} = T \cdot \frac{q \, \Delta t}{2 \, m} \, \delta E_{n+1} (x_{n+1}^{(0)}) \qquad \text{(magnetized)} \tag{26}$$

where $T \equiv [I + R_n (x_{n+1}^{(0)})]/2$ which follows from (5).

With (26), the implicit term $\delta \rho = - \nabla \cdot (\rho \, \delta x_{n+1}^{(0)})$ in (23) is seen to be

$$\delta \rho = - \nabla \cdot (\rho_{n+1}^{(0)} \, \delta x) = - \nabla \cdot \left[\left[\sum_s \frac{\rho_{n+1,s}^{(0)} \, q_s \Delta t^2}{2 \, m_s} T_s \right] \cdot \delta E \right] \tag{27}$$

$$= - \nabla \cdot (\chi \cdot \delta E) \tag{28}$$

The summation is over *species* s, not each particle. If only the electrons are implicit, only they appear in (27). In this case, the terms in the summation require only a knowledge of the electron ρ [in addition to the *net* ρ used on the right side of (23)]. In general, it is sufficient to accumulate $\rho_{n+1,s}^{(0)}$ separately from the species with differing q/m. This requires more storage, but *no more computation* than for an explicit code.

The implicit susceptibility

$$\chi = \sum_s \frac{\rho^{(0)}_{n+1,s} \, q_s \, \Delta t^2}{2 \, m_s} \, \mathbf{T}_s \qquad (29)$$

is a *tensor* due to the rotation **R** induced by **B**.

We now have everything needed to write an equation for $\mathbf{E} = -\nabla \delta \phi$. On substituting our expressions for $\rho^{(0)}_{n+1}$ and $\delta \rho$ into the field equation (23) we have our electrostatic implicit field equation

$$\nabla \cdot [(1 + \chi) \cdot \nabla \, \delta \phi_{n+1}] = \nabla \cdot \mathbf{E}^{(0)}_{n+1} - \rho^{(0)}_{n+1} \qquad (30)$$

This is an elliptic field equation whose coefficients depend directly on particle data accumulated on the spatial grid in the form of an effective linear susceptibility. The rank of the matrix equation is determined by the number of field quantities defined on the zones; it is independent of the number of particles and normally is much smaller.

This formalism guides successful implementation of spatial smoothing (*Langdon, Cohen, and Friedman*, 1983; *Barnes et al.*, 1983). If spatial smoothing, denoted by the operator $\hat{S}$ is to be applied to ρ and ϕ on the grid, then $\hat{S}$ must be included in χ if the field solution is to take this into account. In some applications, this has been essential. Inconsistent smoothing has consequences important to linear stability.

The field $\mathbf{E}_{n+1}$ is evaluated at positions $\{\mathbf{x}^{(0)}_{n+1}\}$ in 8-5(2) in integrating the particles to their final positions $\{\mathbf{x}_{n+1}\}$. The error resulting from this approximation, and from the linearization of $\delta \rho$, introduces a possible limitation on Δt that depends on field and density gradients; see *Langdon and Barnes* (1985) and Section 9-7(b). Nonetheless, useful results have been made with $\omega_{pe} \Delta t$ up to 10^3.

After advancing each particle to position $\mathbf{x}_{n+1}$, one can immediately calculate its $\tilde{\mathbf{x}}_{n+2}$ and its contribution to $\tilde{\rho}_{n+2}$. In this way, only one pass through the particle list is required per time step, an advantage when the particles are stored on a slower memory device such as rotating magnetic disk.

Particle boundary conditions in implicit codes can be complex. Particle deletion or emission at a surface depends on $\mathbf{E}_{n+1}$; therefore the particle boundary conditions enter into the implicit field equations. For electromagnetic fields it appears that methods used in explicit codes can be adapted; for example, the outgoing wave boundary conditions of *Lindman* (1975) have been implemented by using implicit differencing of his boundary wave equations by J. C. Adam and A. Gourdin (private communication).

The reader may consult *Langdon and Barnes* (1985) and references therein for many more details.

14-14 DIAGNOSTICS

During the computer runs in 1d, we make *snapshots* of the gridded quantities (ρ, ϕ, E_x) versus coordinate x, plots of distribution functions

$f(v_x)$ versus v_x and the phase-space plots of particle v_x versus x, v_y versus x, and v_x versus v_y. From time to time and at the end of a run we make *history plots* of various quantities versus time t, such as potential, kinetic, thermal, drift energies, or gridded quantities at some value of x. The objects are to observe the physics, checking against linear and non-linear theories, and to apply numerical checks, such as conservation of energy or of some component of momentum, or some other particle or field quantity which is expected to remain invariant.

For 2d and 3d runs, we do much the same but with graphics having x-y or x-y-z or r-θ-z spatial coordinates, requiring more complicated plotting routines. $\phi(x)$ versus v_x in 1d may become $\phi(x,y)$ versus x,y, either as a three-dimensional plot (using say ϕ versus x for many separate values of y, a perspective plot) or as a two-dimensional plot (coordinates x,y) with equipotential *contours*. Also, for a 2d system periodic in y, we may want to look at $\phi(x)$ for a particular wavenumber k_y as a function of x or t. For phase space plots of, say, particle velocities $v_\perp$ and $v_\parallel$, we may convert particle locations into contours of constant phase-space densities, $f(v_\perp, v_\parallel)$ (using, for example, a linear weighting with some smoothing); these may originate inside a magnetic loss-cone and then flow (drift, diffuse) across the loss-cone boundary as the run progresses due to scattering and/or instability. Another variation, using high resolution printers (say, 1000×1000 lines), is to print density of dots to reflect density of particles, making half-tone plots; *e.g.*, in a 3×3 dot region giving 10 levels from white to black.

Where we seek evidence of particles being trapped in a wave, we need to relate the particle $(\mathbf{x}, \mathbf{v})$ to some phase of the wave. In 1d with wave propagation only along x, plots of particle v_x versus x for cold beams (or for marker particles initially in beams), trapping of particles in a wave (formation of phase-space vortices) is very clear, as seen in Chapter 5 projects. In 2d, for example, with a single (or dominant) wave propagating in (x,y) at some $k_\perp$ ($\mathbf{B}$ is $\hat{z}B_0$), *Cohen et al.* (1983) use single-particle plots of $v_\perp$ versus relative gyrophase ψ over some interval in time (ψ is defined by $\theta - \int \omega_{ci} dt'$ where $\tan \theta = -v_y/v_x$). They note that the "orbit of an ion gyrating and bouncing in the magnetic field in the absence of fluctuating electric fields would trace out an ellipse in the (x, v_x) plane, a circle in the (v_x, v_y) plane, and a point in the $(\psi, v_\perp)$ plane .... An ion interacting adiabatically with a wave has a regular, small-amplitude excursion superposed on its orbit. An ion interacting strongly with the wave fields has a large and possible stochastic excursion;" their Figure 8 shows wave trapping and stochasticity, over an interval of about 8 gyroperiods. N. Otani (1983, private communication) has made similar snapshot plots but of many particles where ψ is phase relative to the local transverse wave phase [arccos $(\mathbf{v}_\perp \cdot \mathbf{B}_\perp / v_\perp B_\perp)$, where $\mathbf{B}_\perp$ is that of the wave] and found evidence of particle trapping.

A 1d technique for following, say, potential ϕ in x and t is successive offset *overlays* of $\phi(x)$ at times $t_1 < t_2 < t_3$, etc.. In 2d, following ϕ in x, y, and t by offset overlays is confusing, unless done by making *movies*, in this case with, say, contours of ϕ in (x, y) at successive times. The eye readily picks up wave motion in (x, y), much as watching waves break on a beach. Indeed, such movies are a great aid in sorting out, for example, some rapidly moving but small amplitude waves along one direction from longer term evolution along another direction of larger, more nearly stationary potentials (or fields). Putting, say, $\phi(x, y)$ plots on the same movie frame with ion and electron position (x, y) and velocity (v_x, v_y) plots can provide very convincing evidence of the dominant or essential physics, while the eye filters out the lesser effects (see *Hockney,* 1966 and *Hockney and Eastwood,* 1981, Figure 9-12 for such a frame). On occasion, the eye picks up effects from movies that are barely perceptible (or easily missed) from viewing separate plots. Lastly, movies made in *several colors* can be extraordinarily dramatic in presenting very complex behavior.

The use of a small number of *test particles* placed judiciously at $t = 0$ and followed in time in $(\mathbf{x}, \mathbf{v})$ can be very educational. In an early movie of the diocotron instability of a magnetized ring of electrons (*Birdsall and Fuss,* 1967, unpublished), one particle was also followed separately. The movie showed formation of vortices in (x, y), giving the impressions that most electrons may become trapped in such; however, while the single particle was observed to fall into vortex motion, it then moved on, fell again, moved on, and so on, all the while drifting toward the axis. Indeed, the vortices coalesced finally into one large vortex centered on the axis. The movie taken by itself could have been misleading.

Energy, power, and rate of work are quadratic quantities of use in understanding the physics being done, and may also be useful in numerical checks; hence, diagnostics may include such, with some care, as will now be shown. Let us calculate the rate of change of kinetic energy (KE) due to the work done by $\mathbf{E}$ on $\mathbf{J}$ over a volume V, first using electrostatic energy density as $\frac{1}{2}E^2$ and then as $\frac{1}{2}\rho\phi$ in order to see whether there are differences. The rate of change is

$$-\frac{d(\text{KE})}{dt} = -\int_V \mathbf{E} \cdot \mathbf{J} \, d\mathbf{x} \tag{1}$$

then put in $\nabla\phi$ as

$$-\frac{d(\text{KE})}{dt} = \int_V \nabla\phi \cdot \mathbf{J} \, d\mathbf{x} = \int_V (\nabla \cdot \phi\mathbf{J} - \phi\nabla \cdot \mathbf{J}) \, d\mathbf{x} \tag{2}$$

then use the equation of continuity, Poisson's equation and Gauss' law to obtain

$$-\frac{d(\text{KE})}{dt} = \int_S \phi\mathbf{J} \cdot d\mathbf{S} - \int_V \phi\frac{\partial}{\partial t}(\nabla^2\phi) \, d\mathbf{x} \tag{3}$$

which then becomes

$$-\frac{d\,(\mathrm{KE})}{dt} = \int_S \phi \left[\mathbf{J} - \nabla \frac{\partial \phi}{\partial t} \right] \cdot d\mathbf{S} + \frac{d}{dt} \int_V \frac{1}{2} (\nabla \phi)^2 \, d\mathbf{x} \qquad (4)$$

For the $\frac{1}{2}\rho\phi$ density, we use similar steps, but keep ρ (use continuity, but not Poisson's equation) to obtain

$$-\frac{d\,(\mathrm{KE})}{dt} = -\int_S \phi \, \mathbf{J} \cdot d\mathbf{S} + \frac{d}{dt} \int_V \frac{1}{2} \rho \phi \, d\mathbf{x} \qquad (5)$$

In (4) and (5), remember that V is defined in (1) as the volume enclosing the particles of interest. The difference between the results (4) and (5) may be understood as follows. The integral of $\frac{1}{2}\rho\phi$ in (5) need be only over the volume where ρ exists, yet includes the change of $\frac{1}{2}E^2$ energy outside of V. The integral over $\mathbf{S}$ in (4) of $-\phi\nabla(\partial\phi/\partial t)$ (a power term) represents the rate of flow of $\frac{1}{2}E^2$ energy across $\mathbf{S}$. Or, in (4), with $\frac{1}{2}E^2$ density, the surface integral is power density $\phi \mathbf{J}_{\mathrm{total}}$ (including the displacement current density in $\mathbf{J}_{\mathrm{total}}$), where in (5) with $\frac{1}{2}\rho\phi$ density, it is $\phi \mathbf{J}$ (only the particle current density). For simulation, a diminution of particle KE (which is $-d(\mathrm{KE})/dt$) is expected to be balanced by an increase in field energy, allowing for a numerical check on energy balance. For systems that are periodic or grounded (ϕ on $\mathbf{S}$ is zero), the surface (power) integrals vanish so that only the energy densities of $\frac{1}{2}\rho\phi$ or $\frac{1}{2}E^2$ need to be computed. For more general systems, the choice of energy density as $\frac{1}{2}\rho\phi$ or $\frac{1}{2}E^2$ then governs the choice of surface integrands using power densities as $\phi \mathbf{J}_{\mathrm{particle}}$ or $\phi \mathbf{J}_{\mathrm{total}}$, one of which must be computed. For theorems on momentum conservation, see the work of *Decyk* (1982). Equation (5) was derived using (6) which does not always hold. The potential due to external charge must be handled separately; see Section 10-4 and *Decyk* (1982).

PROBLEM

14-14a Fill in all of the steps needed to obtain (4) and (5). For the latter, provide a proof that

$$\int_V \left[\phi \frac{\partial \rho}{\partial t} - \rho \frac{\partial \phi}{\partial t} \right] d\mathbf{x} = 0 \qquad (6)$$

as also used in Section 10-4, below 10-4(4).

14-15 REPRESENTATIVE APPLICATIONS

Plasma journals and books regularly include articles and chapters on plasma behavior using particle simulation. We call attention to a representative few, in 2d and 3d, electrostatic.

(a) Diffusion Across B

Diffusion of plasma across a magnetic field is well studied in electrostatic particle simulations in 2d and 3d. *Taylor and McNamara* (1971) find agreement between theory and simulation for 2d guiding center diffusion; in the high magnetic field limit ($r_{\text{Larmor}} < \lambda_D$, $\omega_p < \omega_c$) they find the diffusion coefficient has the Bohm ($1/B$) variation, much larger than the usual ($1/B^2$), with weak dependence on system size; *Christiansen and Taylor* (1973) add further corroboration. The subject is carried further by *Joyce and Montgomery* (1973) on the nature of a 2d guiding center plasma, finding formation of quasi-stable spatially inhomogeneous states. *Hsu, Joyce, and Montgomery* (1974) followed the thermal relaxation of an initially spatially uniform 2d plasma in a strong uniform B field, without the guiding center approximation, finding, for $(\omega_{ce}/\omega_p) > 4$, that $\tau_{\text{relaxation}}$ is proportional to $(n_0\lambda_D^2)^{1/2}$ and ω_{ce}/ω_p. *Okuda and Dawson* (1973) and *Dawson, Okuda, and Rosen* (1976) find three distinct regions of particle guiding center diffusion across B. For small values of B (large ω_p/ω_{ce}), the diffusion coefficient $D_\perp$ is found to be proportional to $1/B^2$ as expected from binary collision theory. For intermediate values, $D_\perp$ is nearly constant. For large B ($\omega_p/\omega_{ce} \leqslant \frac{1}{2}$), $D_\perp \propto 1/B$ is found, as is expected for the Bohm diffusion. They go on to compare transport in 3d for closed and nonclosed magnetic field lines; the results for the former are similar to that in 2d; the results for the latter follow $D_\perp \propto 1/B^2$ to much smaller ω_p/ω_{ce}, with interesting increases for what amount to rational rotational transforms, of interest in tokamaks and stellarators. *Kamimura and Dawson* (1976) extended the above work to including magnetic mirrors, finding significant enhancement of convective transport. *Tsang, Matsuda, and Okuda* (1975) observed cross-field diffusion in a model toroidal **B** field in 3d, finding enhancement of electron diffusion due to the toroidal field. The role of electrostatic Bernstein modes in transport (and other effects) is studied by *Kamimura, Wagner, and Dawson* (1978). The crossed field heat transport, which may differ from particle transport, is studied by *Naitou, Kamimura, and Dawson* (1979), and *Naitou* (1980). *Dawson* (1983) summarizes the work of the UCLA group and their collaborators.

(b) Instabilities

Growth of instabilities, from the linear stage through to nonlinear saturation or stabilization, is very well studied using particle simulations. As there are large numbers of plasma instabilities, there are many articles. The usual presentation includes linear theory, the simulations to check such, followed by observations to determine the mechanism of saturation (*e.g.*, particle trapping, quasilinear diffusion, modification of the zero-order distributions, etc.), and other large amplitude effects (such as particle and heat transport). Representative of such studies is the work on the lower-hybrid drift instability, first in 1d (*Chen and Birdsall,* 1983) where it was found that stabilization came by ion trapping for fixed drifts, or by current relaxation for non-fixed drifts. This was followed by simulation in 2d (*Chen, Nevins, and Birdsall, 1983*) which showed that nonlocal effects become important after early growth (when the 1d local theory holds), with a coherent mode structure; stabilization is found to be by local current relaxation due to both ion quasilinear diffusion and electron $\mathbf{E} \times \mathbf{B}$ trapping. In a different limit, $\omega_{ce} \ll \omega_p$, *Biskamp and Chodura* (1973) examine the importance of a weak magnetic field on the nonlinear behavior and turbulence of the current-driven ion sound instability, showing the difference between 1d and 2d turbulent spectra. In additional work on this instability, *Dum, Chodura, and Biskamp* (1974) in 2d find that the dominant saturation mechanism is quasilinear rather than nonlinear.

(c) Heating

It is desirable, if possible, to take advantage of instabilities, to achieve, say, heating. One example of this is the detailed understanding of the modified two-stream instability (or lower-hybrid two-stream instability) provided by *McBride et al.* (1972) in 1d and 2d theory and simulations; they show that this instability can be a very important turbulent heating mechanism, starting from a small electron-ion relative drift velocity. It is also possible to drive this instability in a thermal plasma, for example, by applying an oscillating $\mathbf{E}_0$ applied normal to a steady $\mathbf{B}_0$ and achieve strong ion heating as shown by *Chen and Birdsall* (1973). Such particle simulations provide insight into the heating process unavailable by other means.

FIFTEEN

ELECTROMAGNETIC PROGRAMS IN TWO AND THREE DIMENSIONS

15-1 INTRODUCTION

In this chapter we introduce the use of the *complete electromagnetic fields,* following our earlier work using only the static fields. Now we have two (and sometimes three) components of **E**, **B**, and **J**, as well as the scalar ρ; we might also use the vector and scalar potentials **A**, ϕ along with **E** and **B**. There is much greater variety of collective interaction, both physical and, occasionally, nonphysical. Electromagnetic programs are more complicated than electrostatic programs, and generally tougher to use and more expensive.

Major motivations for EM programs have been simulation of interaction of intense laser light with hot plasmas, instabilities, shocks, etc. in magnetized plasmas, and for intense electron-beam-plasma studies. The algorithms chosen for discussion here were developed for such applications. The primary reference used in the early part of this chapter is *Langdon and Lasinski* (1976) on the ZOHAR code. Additions include more on alternative approaches such as Darwin (nonradiative) codes, and extensions such as hybrid particle-fluid codes.

15-2 TIME INTEGRATION OF THE FIELDS AND LOCATION OF THE SPATIAL GRIDS

The fields are integrated forward using their time derivatives as given by Faraday's law and the Ampere-Maxwell law:

$$\frac{\partial \mathbf{B}}{\partial t} = -c \nabla \times \mathbf{E} \tag{1}$$

$$\frac{\partial \mathbf{E}}{\partial t} = c \nabla \times \mathbf{B} - \mathbf{J} \tag{2}$$

As done in earlier chapters, these equations are written in rationalized c.g.s. (or Heaviside-Lorentz) form which eliminates almost all occurrences of factors of 4π during problem design and physical interpretation of the results.

Putting the time derivatives of $\mathbf{B}(\mathbf{E})$ on the left-hand-side, with the $\mathbf{E}(\mathbf{B})$ field occurring on the right-hand side, suggests use of a *leap-frog scheme* for the time integration, as sketched in Figure 15-2a. The figure also indicates the time stepping of the current density $\mathbf{J}$ particle position $\mathbf{x}$ and velocity $\mathbf{v}$ as well as correction potential $\delta\phi$; the potentials $\mathbf{A}$, ϕ (as alternatives to $\mathbf{E}$, $\mathbf{B}$) are also shown. Such centered time differencing is accurate to second order.

With two space dimensions, the fields may be divided into transverse electric TE and transverse magnetic TM sets. All spatial variation, and therefore $\mathbf{k}$, is in the x, y plane. The TM fields, with $\mathbf{k} \cdot \mathbf{B} = 0$, have components E_x, E_y, B_z. The TE fields, $\mathbf{k} \cdot \mathbf{E} = 0$, have components E_z, B_x, B_y. These sets are uncoupled, as seen by writing out the components of their Maxwell equations. The relative spatial locations of components may be chosen so as to provide centered spatial differencing; these locations are given in Figure 15-2b, for the TM fields (sufficient for 2d) and TE fields (needed for 2½d, 3d). A complete field grid is shown in Figure 15-2c, with the TM and TE fields aligned so as to make the code indexing and boundary

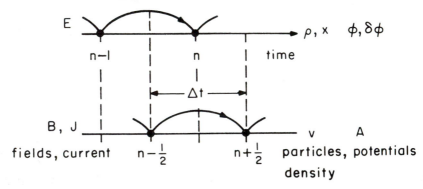

Figure 15-2a Temporal layout of field and particle quantities used in the leapfrog integration of Maxwell's equations. $\mathbf{E}$ is advanced, then $\mathbf{B}$ is advanced in time. New values overwrite old values; none are retained more than one time.

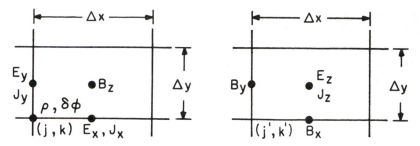

Figure 15-2b Locations of the TM and TE field components on an x, y spatial grid, chosen to allow centered spatial differencing of (1)-(2). As the TM and TE sets are uncoupled, the relative location of (j, k) and (j', k') is not fixed.

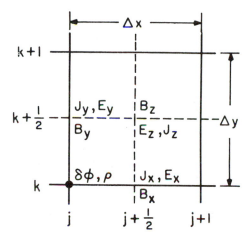

Figure 15-2c Spatial layout on the field grid of the 2d TM fields (E_x, E_y, B_z) and their source terms (ρ, J_x, J_y, $\delta\phi$) as they appear during their integration forward in time via the differenced Maxwell equations. The particle grid is the same as for the charge density. In the 2½d case, the TE fields are added; B_x and B_y are collocated with E_x and E_y, while E_z and J_z are collocated with B_z. *(From Langdon and Lasinski, 1976.)*

conditions analogous. In some applications the TE fields remain zero (Problem 15-6c), and so need not be computed or stored.

We are now ready to difference Maxwell's equations explicitly in a very simple way whose accuracy is second order in space and time, much like *Buneman* (1968), *Boris* (1970b), and *Morse and Nielson* (1971). Specifically the time derivative becomes

$$(\partial_t E_x)_{j+\frac{1}{2},k}^{n+\frac{1}{2}} \equiv \frac{E_{x,j+\frac{1}{2},k}^{n+1} - E_{x,j+\frac{1}{2},k}^{n}}{\Delta t} \tag{3}$$

where $E_{x,j+\frac{1}{2},k}^{n} \equiv E_x([j + \frac{1}{2}]\Delta x, k\Delta y, n\Delta t)$, etc. Let the spatial differences ∂_x and ∂_y be defined analogously. The gradient ∇ becomes ∂_x. What makes this notation helpful later is that these operators, applied to fields defined on our space-time grid, *commute*. Therefore the difference equations can be manipulated in the same ways as the similar-appearing differential equations.

The differenced Maxwell equations, (1)-(2) are, for the TM components,

$$(\partial_t B_z)_{j+\frac{1}{2},k+\frac{1}{2}}^n = -c \, (\partial_x E_y - \partial_y E_x)_{j+\frac{1}{2},k+\frac{1}{2}}^n \tag{4}$$

$$(\partial_t E_x)_{j+\frac{1}{2},k}^{n+\frac{1}{2}} = (c \, \partial_y B_z - J_x)_{j+\frac{1}{2},k}^{n+\frac{1}{2}} \tag{5}$$

$$(\partial_t E_y)_{j,k+\frac{1}{2}}^{n+\frac{1}{2}} = (-c \, \partial_x B_z - J_y)_{j,k+\frac{1}{2}}^{n+\frac{1}{2}} \tag{6}$$

When $B_z^{n-\frac{1}{2}}$ and E^n are known, (4) determines $B_z^{n+\frac{1}{2}}$. The electric field is then advanced similarly. For example, (5) expands to

$$\frac{E_{x,j+\frac{1}{2},k}^{n+1} - E_{x,j+\frac{1}{2},k}^n}{\Delta t} = c \, \frac{B_{z,j+\frac{1}{2},k+\frac{1}{2}}^{n+\frac{1}{2}} - B_{z,j+\frac{1}{2},k-\frac{1}{2}}^{n+\frac{1}{2}}}{\Delta y} - J_{x,j+\frac{1}{2},k}^{n+\frac{1}{2}} \tag{7}$$

The code alternates, first advancing E, then B, as was shown in Figure 15-2a. At each step the new values for a field overwrite the old values in memory. It is not necessary to retain values for any field at more than one time.

PROBLEM

15-2a Obtain the references of this section and the next and sketch the locations of the variables given therein. Compare with those in Figure 15-2c. Explain the differences.

15-3 ACCURACY AND STABILITY OF THE TIME INTEGRATION

One can learn much about the accuracy and stability properties of the scheme in Section 15-2 by seeing how it reproduced plane electromagnetic waves in vacuum. Assuming that the fields are of the form $(\mathbf{E}, \mathbf{B}) = (\mathbf{E}_0, \mathbf{B}_0) \exp(i\mathbf{k} \cdot \mathbf{x} - i\omega t)$ and substituting into the difference equations, we find

$$\Omega \mathbf{B} = c \, \boldsymbol{\kappa} \times \mathbf{E} \tag{1}$$

$$\Omega \mathbf{E} = -c \, \boldsymbol{\kappa} \times \mathbf{B} \tag{2}$$

where $\Omega \equiv \omega \, \mathrm{dif}(\omega \Delta t / 2)$, $\kappa_x \equiv k_x \, \mathrm{dif}(k_x \Delta x / 2)$. In the continuum limit, Ω and $\boldsymbol{\kappa}$ reduce to ω and $\mathbf{k}$. Eliminating $\mathbf{E}$ and $\mathbf{B}$ yields

$$\Omega^2 = c^2 \kappa^2 \tag{3}$$

which, expanded, is

$$\left(\frac{\sin\frac{\omega\Delta t}{2}}{c\,\Delta t}\right)^2 = \left(\frac{\sin\frac{k_x\Delta x}{2}}{\Delta x}\right)^2 + \left(\frac{\sin\frac{k_y\Delta y}{2}}{\Delta y}\right)^2 \tag{4}$$

Obviously ω is real (no damping or growth) if

$$1 > (c\,\Delta t)^2\left[\frac{1}{\Delta x^2} + \frac{1}{\Delta y^2}\right] \tag{5}$$

or $c\,\Delta t < \Delta x/\sqrt{2}$ for $\Delta x = \Delta y$, a *Courant condition*. When this condition is violated, $\sin^2(\omega\Delta t/2)$ exceeds unity for $k_x\Delta x$, $k_y\Delta y$ near π; the ω roots are now complex, with one root giving nonphysical growth which can be very rapid. When condition (5) is satisfied, there are no phase or magnitude errors between **E** and **B**. The errors in the magnitude of ω, in the relative directions of the fields, and in **k** are second order in Δx, Δy and Δt. All these properties are a direct result of the centered differencing in space and time.

A plot of ω versus k is given in Figure 15-3a showing $\omega \leqslant kc$ for $c\,\Delta t \leqslant \Delta x$. Note that at the edge of the fundamental zone, $k_x\Delta x = \pi = k_y\Delta y$, v_{phase} drops as low as $2c/\pi = 0.637c$. Hence, relativistic particles may have $v > v_{\text{phase}}$ at short wavelengths, producing unwanted particle-wave growths, or Cerenkov emission. *Boris and Lee* (1973) and *Haber et al.* (1973) mention noise produced by Cerenkov emission from particles whose velocities exceed the minimum phase velocity. *Godfrey* (1974,1975) and

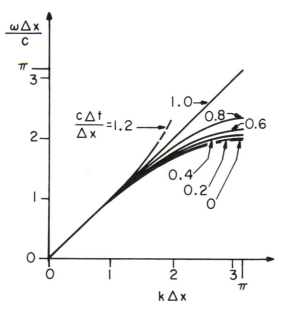

Figure 15-3a Vacuum dispersion solution of Maxwell's equations for finite Δx, Δt, from (4). In one dimension, no dispersion error occurs for $c\,\Delta t/\Delta x = 1.0$, which is marginally stable.

Godfrey and Langdon (1976) examine collective instabilities involving interaction between relativistic electron beams and these slow light waves; this is discussed in Section 15-5. This defect creates interest in algorithms which propagate vacuum light waves at the correct velocity and have other features which also improve stability; (Chapter 6 and Section 15-9b).

PROBLEMS

15-3a Verify the statement in the text following (5) relative to phase and magnitude errors between **E** and **B** and relative directions of **k** and $\boldsymbol{\kappa}$.

15-3b From (4), plot contours of $v_{\text{phase}}/c = 0.9, 0.8, 0.7$ on $k_x \Delta x$ versus $k_y \Delta y$, using $\Delta x = \Delta y$ and $c \Delta t / \Delta x = 0.5$. Comment on whether the contours are circles (phase velocity independent of direction of k) and, if not, where the largest errors are.

15-3c Derive (1), (2).

15-4 TIME INTEGRATION OF THE PARTICLE EQUATIONS

Assume that **E** and **B** are interpolated as in Chapter 14 from the grid fields to the particles at time $t^n = n \Delta t$. We generalize the particle integrator of Section 4-3 to include relativistic effects and varying magnetic field (*Boris, 1970*). For the *relativistic generalization* of 4-3(3), we use $\mathbf{u} \equiv \gamma \mathbf{v}$ rather than **v**.

$$\frac{\mathbf{u}^{n+\frac{1}{2}} - \mathbf{u}^{n-\frac{1}{2}}}{\Delta t} = \frac{q}{m} \left[\mathbf{E}^n + \frac{1}{c} \frac{\mathbf{u}^{n+\frac{1}{2}} + \mathbf{u}^{n-\frac{1}{2}}}{2\gamma^n} \times \mathbf{B}^n \right] \tag{1}$$

where m is the *rest* mass and $\gamma^2 = 1 + u^2/c^2$. In the magnetic field force, centering by time averaging of **u**, as shown, and of **B** (next Section), leaves only γ^n to be specified. Centering of γ^n is easier in the method of *Boris* (1970b). The addition of electric impulses, equations 4-3(7) and 4-3(8), carry over with no formal change,

$$\mathbf{u}^{n-\frac{1}{2}} = \mathbf{u}^- - \frac{q \mathbf{E}^n \Delta t}{2m} \tag{2}$$

$$\mathbf{u}^{n+\frac{1}{2}} = \mathbf{u}^+ + \frac{q \mathbf{E}^n \Delta t}{2m} \tag{3}$$

We substitute these into (1) to obtain

$$\frac{\mathbf{u}^+ - \mathbf{u}^-}{\Delta t} = \frac{q}{2\gamma^n mc} (\mathbf{u}^+ + \mathbf{u}^-) \times \mathbf{B}^n \tag{4}$$

In Section 4-3 we found that (4) results in a rotation of **u** about an axis parallel to **B** through an angle $\theta = -2 \arctan(qB\Delta t / 2\gamma mc)$. The angle is reduced by a factor $\approx \gamma$. Therefore 4-4(11) becomes $\mathbf{t} = q\mathbf{B}\Delta t / 2\gamma^n mc$, with $(\gamma^n)^2 = 1 + (u^-/c)^2$. Since $(\gamma^n)^2 = 1 + (u^+/c)^2$ also, this scheme is time reversible and the overall momentum integration is second-order accurate. This γ^n may be used to calculate a time-centered kinetic energy. On short word-length computers it is best to use the identity $(\gamma - 1)mc^2 = mu^2(\gamma + 1)^{-1}$; the second form of the kinetic energy is far less susceptible than the first to roundoff error.

There is a large class of physical problems in which v_z, E_z, B_x, and B_y are unimportant. If these quantities are zero initially, then the equations of motion show they remain zero (Problem 15-6c). The plane rotation represented by (4) is done very concisely by Buneman's algorithm (Section 4-4),

$$u_x' = u_x^- + u_y^- t \qquad (5)$$

$$u_y^+ = u_y^- - u_x' s \qquad (6)$$

$$u_x^+ = u_x' + u_y^+ t \qquad (7)$$

where $t = -\tan \theta/2 = qB_z \Delta t / 2\gamma^n mc$, $s = -\sin \theta = 2t/(1 + t^2)$.

In the more general 2½d case in which **E**, **B**, and **u** all can have three nonzero components, we use Boris' rotation, generalizing 4-6(10) to 4-6(13),

$$\mathbf{u}' = \mathbf{u}^- + \mathbf{u}^- \times \mathbf{t} \qquad (8)$$

$$\mathbf{u}_+ = \mathbf{u}^- + \mathbf{u}' \times \mathbf{s} \qquad (9)$$

with $\mathbf{t} = q\mathbf{B}\Delta t / 2\gamma^n mc$, $\mathbf{s} = 2\mathbf{t}/(1 + t^2)$. The error in the angle between $\mathbf{u}^-$ and $\mathbf{u}^+$ is $\approx t^3/3 = (\omega_c \Delta t)^3/24$ and does not seem worth correcting in our applications.

In all cases the position is advanced according to

$$\mathbf{x}^{n+1} = \mathbf{x}^n + \mathbf{v}^{n+\frac{1}{2}}\Delta t = \mathbf{x}^n + \frac{\mathbf{u}^{n+\frac{1}{2}}\Delta t}{\gamma^{n+\frac{1}{2}}} \qquad (10)$$

with $(\gamma^{n+\frac{1}{2}})^2 = 1 + (u^{n+\frac{1}{2}}/c)^2$. This step also is reversible and produces a second-order error in the particle orbit.

In many applications, the magnetic field does not significantly affect the ion motion and the ion transverse current is negligible. In such cases some time may be saved by using a mover for the ions in which only the electric field accelerates the ions, relativity is ignored, and only the ion charge density, not the current, is collected. The manner in which the longitudinal current is taken into account is discussed in Section 15-6.

PROBLEM

15-4a Show that $(\gamma - 1)mc^2$ and $mu^2(\gamma + 1)^{-1}$ are both equal to the particle kinetic energy. Consider the susceptibility to roundoff error of the two forms when $u^2 \ll c^2$.

15-5 COUPLING OF PARTICLE AND FIELD INTEGRATIONS

In coupling the particle and field integrations, we have to relate particle and field quantities, which are provided at different locations and times.

The simplest case is the magnetic field, $\mathbf{B}$, given at half-integer times (*i.e.*, $n \pm \frac{1}{2}$) in the field equations, which is needed at integer times (n) for integration of the particle velocity. Since Faraday's law can be used to advance $\mathbf{B}$ to a time ahead of $\mathbf{E}$, one may simply time average $\mathbf{B}$, as

$$\mathbf{B}^n = \frac{\mathbf{B}^{n-\frac{1}{2}} + \mathbf{B}^{n+\frac{1}{2}}}{2} \tag{1}$$

This may be used in the particle mover, 15-4(1) and also in certain diagnostics. In practice, the average does not appear explicitly. To avoid using additional computer storage, the $\mathbf{B}$ integration may be split into two steps (*Boris, 1970, p. 12*). As the last step of the field integration, we advance $\mathbf{B}$ only half way, obtaining

$$\mathbf{B}^n = \mathbf{B}^{n-\frac{1}{2}} - \frac{c\,\Delta t}{2}\, \partial_\mathbf{x} \times \mathbf{E}^n \tag{2}$$

which replaces $\mathbf{B}^{n-\frac{1}{2}}$ in memory. The particles are then integrated, using this $\mathbf{B}^n$ in 15-4(1), As the first step in the following field integration, $\mathbf{B}^n$ is advanced in the same way to $\mathbf{B}^{n+\frac{1}{2}}$.

Specifying the current density, $\mathbf{J}^{n \pm \frac{1}{2}}$, from the particle velocities ($\mathbf{v}$ known at $n \pm \frac{1}{2}$ times) and positions ($\mathbf{x}$ known at n) poses a similar problem in assignment. One method (*Boris, 1970, p. 33*) is to use $\mathbf{v}^{n+\frac{1}{2}}$ times the average of the weights S for the two positions x^n and x^{n+1} producing a time centered $\mathbf{J}^{n+\frac{1}{2}}$. Explicitly, this is

$$\mathbf{J}_\mathbf{j}^{n+\frac{1}{2}} = \sum_i \mathbf{v}_i^{n+\frac{1}{2}}\, \frac{S(\mathbf{X_j} - \mathbf{x}_i^n) + S(\mathbf{X_j} - \mathbf{x}_i^{n+1})}{2} \tag{3}$$

We see that a particle contributes to the current density at its four nearest grid points, just as it does to charge density. (When a particle crosses the cell boundary during advancement of $\mathbf{x}_i^n$ to $\mathbf{x}_i^{n+1}$, more grid points are affected.)

Another way to obtain $\mathbf{J}^{n+\frac{1}{2}}$ is to use weights for the mid-positions, $\mathbf{x}^{n+\frac{1}{2}} = (\mathbf{x}^n + \mathbf{x}^{n+1})/2 = \mathbf{x}_n + \mathbf{v}^{n+\frac{1}{2}}\Delta t/2$ (*Morse and Nielson, 1971, p. 839*). The two methods are similar in computational expense. In ZOHAR, the former was chosen in the hope that it would have better noise properties. However, *Godfrey and Langdon (1976)* show that the use of the average of the weights (3) may be less susceptible to numerical instabilities.

Special consideration must be given to conservation of charge; this is done in the next section. Also, the noise properties of various alternative ρ and $\mathbf{J}$ weighting schemes need separate consideration; this is done in the Section 15-8.

The electric and magnetic fields at the particle positions are obtained by interpolation from the field grid (weighting). In ZOHAR, linear weighting is used. The most obvious thing to do is to interpolate separately from each of the three sets of grid locations shown in Figure 15-2c (*Boris,* 1970; *Morse and Nielson,* 1971; *Palevsky, 1980*). However, this is time consuming for the particle mover. Because the mover accounts for the majority of the computer time, ZOHAR and other codes redefine the fields beforehand to a single set of grid locations. This may be done simply by a spatial average to the ρ grid positions. There are other advantages: The longitudinal part of **E** now is the same as for the momentum-conserving electrostatic codes and the additional smoothing decreases short wavelength noise. Diagnostics are also simplified. The same benefits are realized in an **A**, ϕ code by changing the differencing used to derive **E** and **B** from **A** and ϕ (*Nielson and Lindman,* 1973a, b).

After the particle integration, the averaged fields might be restored to the original field grids by a further spatial averaging. However, this produces severe damping of electromagnetic waves that is unacceptably rapid. Boris observed that the original E_x, for example, can be easily reconstructed from the averaged E_x if the values of E_x at one side are saved before averaging. For B_z, the simplest procedure is to unaverage in x first, along with E_x, then in y along with E_y. In this way we can redefine the fields to a common grid and restore them without adding appreciably to computer storage requirements.

As with the redefinition of field grids, it is advantageous to collect **J** by area-weighting to a single set of grid points, as in (3), collocated with ρ in ZOHAR, and then spatial average to obtain the currents at the locations shown in Figure 15-2c, where they are needed for the field equations. The y component of current (needed at $j, k + \frac{1}{2}$) is then

$$J_{y\,j,k+\frac{1}{2}}^{n+\frac{1}{2}} = \frac{J_{y\,j,k}^{n+\frac{1}{2}} + J_{y\,j,k+1}^{n+\frac{1}{2}}}{2} \tag{4}$$

the x component, needed at $j + \frac{1}{2}, k$ is obtained similarly. No restoration of the unaveraged **J** is needed, of course.

15-6 THE $\nabla \cdot$ **B** AND $\nabla \cdot$ **E** EQUATIONS; ENSURING CONSERVATION OF CHARGE

There are two other Maxwell equations. We now show that the difference equations have the property, as do the continuum equations, that if the divergences of **E** and **B** are correct initially, they remain correct. That is,

$$\partial_t (\partial_x \cdot \mathbf{B}) = \partial_x \cdot (\partial_t \mathbf{B}) = -c\,\partial_x \cdot \partial_x \times \mathbf{E} \equiv 0 \tag{1}$$

Similarly,

$$\partial_t (\partial_x \cdot \mathbf{E} - \rho) = \partial_x \cdot \mathbf{J} - \partial_t \rho \qquad (2)$$

Therefore, if $\mathbf{J}$ and ρ satisfy the continuity equation, Gauss' law remains satisfied if it holds initially.

Neither of the current densities given in the previous section satisfy the continuity equation with ρ calculated by any method which depends only on the present particle locations. This may be seen even in the $\Delta t \rightarrow 0$ limit (Prob 15-6a).

Methods for calculating a charge conserving $\mathbf{J}$ have been developed corresponding to ρ as obtained by zero-order weighting (NGP) (*Buneman*, 1968) and to ρ obtained by first-order weighting (*Morse and Nielson*, 1971, "Method A"). The latter authors found that the noise level in the electromagnetic fields rose in time at an inconveniently rapid rate. The same has been experienced by others. Causes are discussed in Section 15-8. This approach is advisable (if ever) only with complicated curvilinear coordinates for which it is inconvenient to perform the Poisson solution (6) or 15-7(5). Here and in the next Section we show how to use a nonconservative $\mathbf{J}$.

We advance $\mathbf{E}$ using $\mathbf{J}$ from 15-5(3) in Ampere's law, then adjust the longitudinal part of $\mathbf{E}$ in order to correct $\nabla \cdot \mathbf{E}$. Using the uncorrected $\mathbf{J}$ produces an $\mathbf{E}$ which does not, in general, satisfy Gauss' law, $\nabla \cdot \mathbf{E} = \rho$, because of microscopic inconsistencies between $\mathbf{J}$ and ρ due to use of the mesh and weights. Hence, we invent a correction of the form

$$\mathbf{E}' = \mathbf{E} - \nabla \delta \phi \qquad (3)$$

such that

$$\nabla \cdot \mathbf{E}' = \rho \qquad (4)$$

which means that

$$\nabla \cdot (\mathbf{E} - \nabla \delta \phi) = \rho \qquad (5)$$

thus requiring a Poisson solution for $\delta \phi$:

$$\nabla^2 \delta \phi = \nabla \cdot \mathbf{E} - \rho \qquad (6)$$

This correction, due to *Boris* (1970), is computationally convenient and is widely used. The difference form for the Laplacian in (6), consistent with the gradient and divergence operators already used, is the simple five-point operator, ∂_x^2. Although this correction is applied *after* the fields are advanced in time, the overall procedure is time-centered and reversible (Problem 15-6b).

It may be worthwhile to filter ρ spatially before use in (6), in order to boost the medium wavelengths (for better dispersion) and perhaps suppress noise at short wavelengths.

One might fear that the use of solutions of Poisson's equation may permit the propagation at speeds greater than light of information about the particle motion. However, with linear weighting, the contribution of each particle to the source term in (6) is a quadrupole. Therefore the contribution from each particle to the correction fields, $-\nabla \xi$ and $-\nabla \delta \phi$, drop off in a short distance so the correction is very local.

PROBLEMS

15-6a Verify explicitly the lack of continuity mentioned in the second paragraph. One way is to consider a particle which moves in a small circle inside a quarter-cell. After one turn, $\nabla \cdot \mathbf{J}$ has a nonzero average, yet ρ is the same as at the start.

15-6b Show that the asymmetric looking procedure, (3) and (6), produces exactly the same final fields as does 15-7(4) and 15-7(5), which is more obviously a time-centered algorithm.

15-6c Show that if E_z, B_x, B_y, and v_z are all zero at $t = 0$, then they remain zero, and therefore need not be included in the computation.

15-7 A-ϕ FORMULATION

Morse and Nielson's (1971) response to EM noise problems in their $(\mathbf{E}, \mathbf{B})$ code was to develop a Coulomb gauge $(\mathbf{A}, \phi)$ model, their method B. The equations integrated are

$$\nabla^2 \mathbf{A} - \frac{1}{c^2} \frac{\partial \mathbf{A}}{\partial t} = -\frac{1}{c} \mathbf{J}_T \qquad \mathbf{J}_T = \mathbf{J} - \nabla \eta \qquad (1)$$

where η, given by

$$\nabla^2 \eta = \nabla \cdot \mathbf{J} \qquad (2)$$

preserves the gauge $\nabla \cdot \mathbf{A} = 0$. Then

$$\mathbf{E} = -\nabla \phi - \frac{1}{c} \frac{\partial \mathbf{A}}{\partial t} \qquad \mathbf{B} = \nabla \times \mathbf{A} \qquad (3)$$

are used to integrate the particles. The longitudinal $\mathbf{E}$ field is determined from ρ by a Poisson solution; a second Poisson solution is also needed (for the transverse current $\mathbf{J}_T$) each time step.

These fields are in fact the *same* as obtained directly from Maxwell's equations previously in this Chapter. This may be seen by deriving the equations satisfied by the fields (3). Faraday's law 15-2(1) is trivially satisfied, as are $\nabla \cdot \mathbf{B} = 0$ and $\nabla \cdot \mathbf{E} = \rho$. The remaining equation is (Problem 15-7a)

$$c \nabla \times \mathbf{B} - \frac{\partial \mathbf{E}}{\partial t} = \mathbf{J}' \equiv \mathbf{J} - \nabla \xi \qquad (4)$$

where $\xi = \eta - \partial \phi / \partial t$ satisfies

$$\nabla^2 \xi = \nabla \cdot \mathbf{J} + \frac{\partial \rho}{\partial t} \qquad (5)$$

The term $-\nabla \xi$ is simply a correction to $\mathbf{J}$ to force the current $\mathbf{J}'$ to satisfy the continuity equation $\nabla \cdot \mathbf{J}' + \partial \rho / \partial t = 0$, thereby ensuring that Gauss' law, $\nabla \cdot \mathbf{E} = \rho$, remains satisfied. With $\mathbf{J}'$ as the source, the Maxwell equations produce the *same* fields as the A-ϕ equations. With the fields finite-differenced as in Section 15-2, this statement is identically true for the computer implementation. Further, as the longitudinal correction can as well be applied to **E** after, as to **J** before, integration of (4), the **E** and **B** fields here are the same as obtained in the preceeding Section (Problem 15-6b). As well as providing an alternative algorithm, these results show that the noise properties observed by *Morse and Nielson* (1971) were related to the methods of forming **J**, not to the use of potentials.

PROBLEM

15-7a Derive (4). *Hint:* Into the left side of (4), substitute (3), then use the gauge condition and (1).

15-8 NOISE PROPERTIES OF VARIOUS CURRENT-WEIGHTING METHODS

Two numerical methods were used by *Morse and Nielson* (1971) to simulate the Weibel instability. Method A uses the Maxwell equations for **E** and **B**, with a current density constructed so as to satisfy the differenced continuity equation. Method B uses the potentials **A** and ϕ in the Coulomb gauge with an area weighted current density similar to 15-5(3). Method B was found to suffer much less from buildup of noise in the radiation fields. It was found later that the superiority of method B was not due to the use of potentials but due only to the smoother variation of the grid current density as the particles moved through the grid (*Langdon,* 1972). An earlier method due to *Buneman* (1968), motivated similarly to method A, was simpler, but reason was given to expect it to be noisier. This simplified algorithm was almost equivalent to that of *Boris* (1970), therefore, his code was expected to share the good noise properties of Morse and Nielson's method B.

In Section 15-7, we showed that the same fields as in method B may be obtained from the field equations of method A by replacing its current density by the area-weighted, divergence-corrected current **J'**, 15-7(4). (This correction is not necessary in method A because **J** is calculated in a charge-conserving manner.) Alternatively, we could generate the fields of method A from the field equations of method B by using the current density of method A for which the divergence correction vanishes. The only differences between the fields in methods A and B, and also *Buneman's* (1968) method, are due to the differences in the calculation of the current from the particle coordinates; this is where we must look to understand the differences in noise properties.

Noise in the *transverse* radiation fields in EM codes differs from *longitudinal* field noise which is troublesome in electrostatic simulation and which also exists in electromagnetic codes. Electrostatic fluctuations at any time are due only to density fluctuations at that time. The transverse field modes act like harmonic oscillators which are undamped in the absence of lossy media or boundaries and are driven by current fluctuations. Therefore, the radiation field noise energy depends on all past fluctuations; this *noise grows in time* when the energy is started out at much less than the thermal equilibrium level, as observed by *Morse and Nielson* (1971). When studying collective effects one wishes to postpone the thermalization of the radiation fields. Where the problem simulated permits, one may decrease the mode noise level by including radiation field damping.

Incidentally, the blackbody radiation, although classical, suffers from no ultraviolet catastrophe because the fields have no more degrees of freedom than there are field grid points.

To see differences in **J** between the several methods, consider a single particle moving with constant velocity $v \ll c$ parallel to the x axis and passing through the grid points at which J_x is defined, as in Figure 15-2c.

In *Buneman's* (1968) method, $\mathbf{J} = 0$ except for once in about $\Delta x/(v\Delta t)$ time steps, when the particle crosses a cell boundary producing an *impulse* of current density of magnitude $J_x = q/(\Delta y \Delta t)$ corresponding to all of the particle charge moving a distance Δx in that one time step, as sketched in Figure 15-8a(a). The corresponding fields are pulses of radiation. Their spectrum includes large ω and k because the temporal spectrum of **J** includes all frequencies $|\omega| < \pi/\Delta t$ equally (no falling off).

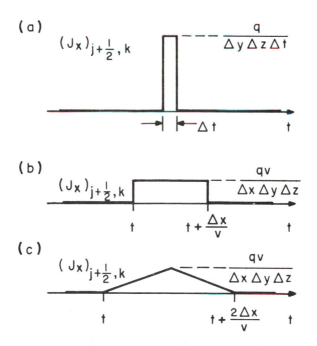

Figure 15-8a (a) Current impulse due to zero-order particle weighting; (b) discontinuous current of method A; (c) continuous current of method B. The noise frequency spectra associated with these fall off as ω^0, ω^{-1}, ω^{-2}, respectively.

In method A, $J = 0$ in this example except at the J_x grid point nearest the particle, where $J_x = qv/(\Delta x \Delta y)$. As the particle moves from cell to cell, **J** varies in a discontinuous piecewise-constant fashion, as sketched in Figure 15-8a(b). Its transform is therefore peaked at low frequencies $\sim v/\Delta x$, and falls off slowly as ω^{-1}.

In method B, or using 15-5(3), the current density $qv/(\Delta x \Delta y)$ is shared by two grid points, with linear weighting according to particle position x. As the particle moves, J varies in a continuous piecewise-linear manner, as sketched in Figure 15-8a(c). Its transform is, therefore, peaked more strongly at low frequency and falls off more rapidly, as ω^{-2}. These smoothness characteristics remain true for more general particle motions.

15-9 SCHEMES FOR $\Delta t_{\text{particles}} > \Delta t_{\text{fields}}$

If we consider a reasonable set of parameters, *e.g.*, $c\Delta t \approx \Delta x/2$, Debye length $\lambda_D = \Delta x/2$, and thermal velocity $v_t = c/20$, then we find that $\omega_p \Delta t = 0.05$. For many applications, this time step is much smaller than is needed for accuracy in moving the particles. In order to relax the time step imposed by the fields, we can take several steps for the fields between particle steps, or employ the known vacuum propagation of the Fourier modes. Another approach is to use an algorithm for the fields which is not subject to the Courant condition; see Section 15-16 and *Nielson and Lindman* (1973a, b).

(a) Subcycling of the Maxwell Equations

Since the particle integration is expensive, it is helpful to advance the particles less often than the fields. In order to illustrate this extension and to summarize a complete time cycle, we outline the operations taken in one particle time step for the case where the fields are integrated twice as often. Superscript n represents the particle step number. Care must be exercised about centering and averaging.

Start with $\mathbf{E}^n$, $\mathbf{B}^n$, $\mathbf{x}^n$, and $\mathbf{u}^{n-\frac{1}{2}}$; proceed as in Figure 15-9a, described by

(0) Average fields $\mathbf{E}^n$ and $\mathbf{B}^n$ to the particle grid.
(1) Advance $\mathbf{u}^{n-\frac{1}{2}}$ to $\mathbf{u}^{n+\frac{1}{2}}$, $\mathbf{x}^n$ to $\mathbf{x}^{n+1}$; form $\mathbf{J}^{n+\frac{1}{2}}$ and ρ^{n+1}.
(2) Average $\mathbf{J}$ to the field grid. Reform $\mathbf{E}$ and $\mathbf{B}$ on the field grid.
(3) Advance $\mathbf{B}^n$ to $\mathbf{B}^{n+1/4}$ using $\mathbf{E}^n$.
(4) Advance $\mathbf{E}^n$ to $\mathbf{E}^{n+\frac{1}{2}}$ using $\mathbf{J}^{n+\frac{1}{2}}$ and $\mathbf{B}^{n+1/4}$.
(5) Advance $\mathbf{B}^{n+1/4}$ to $\mathbf{B}^{n+3/4}$ using $\mathbf{E}^{n+\frac{1}{2}}$.
(6) Advance $\mathbf{E}^{n+\frac{1}{2}}$ to $\mathbf{E}^{n+1}$, using $\mathbf{J}^{n+\frac{1}{2}}$ and $\mathbf{B}^{n+3/4}$.
(7) Advance $\mathbf{B}^{n+3/4}$ to $\mathbf{B}^{n+1}$ using $\mathbf{E}^{n+1}$.
(8) Correct $\nabla \cdot \mathbf{E}^{n+1}$ using ρ^{n+1}.

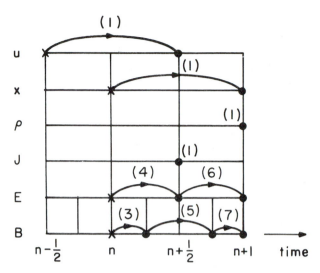

Figure 15-9a Time-stepping for fields advanced twice as often as the particles.

Checking for time centering, we note that steps (3) and (7), and steps (4) and (6) are symmetric. The longitudinal part of E^{n+1} is affected only by $J^{n+\frac{1}{2}}$, and is the same no matter how many time steps are used for the fields in reaching time $n+1$. Therefore the argument used earlier still holds, that the divergence correction does not affect the time centering.

This splitting of the time step can lead to numerical instability. In many applications, the instability reaches very large amplitude over a few thousand time steps. It is possible to restrain this effect by clever averaging from time to time, not a satisfactory solution. Theoretical analysis of subcycling describes this instability accurately, but the underlying cause is seen more simply in problem 15-9a.

(b) Fourier-Transform Field Integration

In order to achieve dispersionless vacuum wave propagation as in Chapter 6, but without the restrictions to one dimension and $\Delta x = c\Delta t$, in some codes the fields are Fourier transformed in space and advanced in time with the correct phase change for each mode (*Haber et al.,* 1973; *Lin et al.,* 1974; *Buneman et al.,* 1980). This can be done by replacing 6-3(4) and 6-3(8) by

$$\pm F = \frac{E \mp \hat{k} \times B}{2} \qquad \hat{k} \equiv k/k \qquad (1a)$$

$$\pm F^{n+1} = e^{\mp ick\Delta t}\left(\pm F^n - \frac{1}{4}J^{n+1/4}\Delta t\right) - \frac{1}{4}J^{n+3/4}\Delta t \qquad (1b)$$

Expressing (1b) in terms of E and B gives

$$\mathbf{E}^{n+1} = \mathbf{E}^n \cos ck\,\Delta t + i\,\hat{\mathbf{k}} \times \mathbf{B}^n \sin ck\,\Delta t - \Delta t\, \frac{\mathbf{J}^{n+1/4}\cos ck\,\Delta t + \mathbf{J}^{n+3/4}}{2} \qquad (2a)$$

$$\mathbf{B}^{n+1} = \mathbf{B}^n \cos ck\,\Delta t - i\,\hat{\mathbf{k}} \times \mathbf{E}^n \sin ck\,\Delta t + \frac{i}{2}\Delta t\,\hat{\mathbf{k}} \times \mathbf{J}^{n+1/4}\sin ck\,\Delta t \qquad (2b)$$

It seems simplest to correct $\nabla \cdot \mathbf{E}^{n+1}$ so that $\mathbf{E}^n$, $\mathbf{J}^{n+1/4}$, and $\mathbf{J}^{n+3/4}$ can include both transverse and longitudinal contributions.

In two dimensions it may be troublesome to collect two current densities. The current density 15-5(3), $\mathbf{J}^{n+1/2} = \frac{1}{2}(\mathbf{J}^{n+1/4} + \mathbf{J}^{n+3/4})$, could be added halfway through the rotation of $^{\pm}\mathbf{F}$:

$$^{\pm}\mathbf{F}^{n+1} = e^{\mp ick\,\Delta t/2}\left[^{\pm}\mathbf{F}^n\,e^{\mp ick\,\Delta t/2} - \mathbf{J}^{n+1/2}\frac{\Delta t}{2}\right]$$

which changes the current contributions in (2a) and (2b) to

$$\mathbf{E}^{n+1} = \mathbf{E}^n \cos ck\,\Delta t + i\,\hat{\mathbf{k}} \times \mathbf{B}^n \sin ck\,\Delta t - \mathbf{J}^{n+1/2}\,\alpha\,\Delta t \cos\frac{ck\,\Delta t}{2} \qquad (3a)$$

$$\mathbf{B}^{n+1} = \mathbf{B}^n \cos ck\Delta t - i\,\hat{\mathbf{k}} \times \mathbf{E}^n \sin ck\Delta t + i\,\hat{\mathbf{k}} \times \mathbf{J}^{n+1/2}\alpha\Delta t \sin\frac{ck\Delta t}{2} \qquad (3b)$$

with $\alpha = 1$. Or, $\mathbf{J}^{n+1/2}$ could be substituted for $\mathbf{J}^{n+1/4}$ and $\mathbf{J}^{n+3/4}$ in (2a) and (2b), yielding (3a) and (3b) with $\alpha = \cos\frac{1}{2}ck\,\Delta t$. The schemes of *Haber et al.* (1973) and *Buneman et al.* (1980), which are exact when $\mathbf{J}$ is constant in time, correspond to $\alpha = (\sin\frac{1}{2}ck\,\Delta t)/(\frac{1}{2}ck\,\Delta t)$. Spatial filtering of $\mathbf{J}$ adds a factor to $\alpha(k)$. These choices agree to order Δt^2 but differ in stability properties for $ck\,\Delta t \geq \pi/2$.

Although these methods are stable for any Δt *in vacuum*, with plasma present they exhibit instability unless $\alpha(\mathbf{k})$ is sufficiently small for $ck\Delta t$ near multiples of π (Problem 15-9a).

PROBLEM

15-9a With $\mathbf{J}$ given by the cold plasma response $(\mathbf{J}^{n+1/2} - \mathbf{J}^{n-1/2} = \omega_p^2\,\Delta t\,\mathbf{E})$, show that the dispersion relation corresponding to the Fourier integration (3ab) is

$$z - 2\cos ck\,\Delta t + z^{-1} = -(\omega_p\,\Delta t)^2\,\alpha\cos\frac{ck\,\Delta t}{2}$$

Show that this can indicate instability in a narrow band of wavenumbers such that the vacuum wave frequency ck is just below $\pi/\Delta t$. Assuming $\omega_p\Delta t \leq 1$, the maximum growth rates are $\mathrm{Im}\,(\omega) = \omega_p^2\Delta t/4$ with $\alpha = 1$, and $\mathrm{Im}\,(\omega) = \omega_p^2\Delta t/2\pi$ with $\alpha = (\sin\frac{1}{2}ck\,\Delta t)/(\frac{1}{2}ck\,\Delta t)$, while $\alpha = \cos\frac{1}{2}ck\,\Delta t$ provides stability (unless $\omega_p\Delta t > 2$).

For $ck\,\Delta t = 2\pi$ and $\alpha = 1$, $\mathrm{Im}\,(\omega) \simeq \omega_p$, while the other two choices for α are stable.

15-10 PERIODIC BOUNDARY CONDITIONS

As with electrostatic codes, much interesting work can be done simulating a system which is periodic in x and y. In this case, boundary conditions offer no conceptual problems, with one possible exception: As formulated so far, the $\mathbf{k} = 0$ component of the electric field is obtained from

$$0 = \mathbf{J}_{\text{total}} = \left(\mathbf{J} + \frac{\partial \mathbf{E}}{\partial t} \right)_{\mathbf{k}=0} \tag{1}$$

and is not zero, in general [case (a)]. Contrast this to the usual electrostatic code in which $\mathbf{E}(\mathbf{k} = 0)$ is zero or a specified function of time (usually meaning driven by an external field), as usually coded [case (b)]. ZOHAR has a switch to select between these options.

One difference is that (a) provides the restoring force for $\mathbf{k} = 0$ plasma oscillations. This means there is *no hole* in the $\mathbf{k}$-spectrum of modes available for nonlinear processes such as parametric instabilities. Plugging this hole makes a drastic difference in a study of self-trapping of light near the critical density (*Langdon and Lasinski, 1983*).

Although this choice arises in electrostatic codes as well, note that electrostatic and electromagnetic codes arrive naturally at different choices.

PROBLEM

15-10a Show that the right-hand side of (1) is $(\rho_0 - m\omega^2/q)\mathbf{v}$, in a uniform plasma, requiring $\omega = \pm\omega_p$ if $\mathbf{v}$ or $\mathbf{E}$ are to be nonzero for $\mathbf{k} = 0$.

15-11 OPEN-SIDED BOUNDARY CONDITIONS

The work-horse particle codes in laser-plasma interaction studies are periodic in y and open-sided, in some sense, on the left and right, as shown in Figure 15-11a. One wants to illuminate the plasma from one side and allow scattered light emerging from the plasma to leave the system. We discuss ways to match the fields to vacuum on the outside of the system and some procedures for dealing with particles which reach a boundary.

(a) The Longitudinal Field

For both ϕ and the correction $\delta\phi$ we wish to obtain a potential solution in the system that is the same as if the grid had been extended to infinity on the left and right, with no charge outside the simulated portion space. This is given in Section 14-8 as worked out by *Buneman (1973)*.

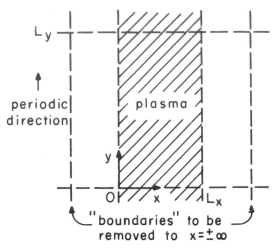

Figure 15-11a Unbounded in x, periodic in y model, with Poisson's equation holding in the plasma, Laplace's elsewhere.

(b) Absorbing Outgoing Electromagnetic Waves in a Dissipative Region

A common way to prevent light leaving the plasma from being reflected at the sides and re-entering the plasma is to place a dissipative region at the sides. The most obvious way to implement this is to introduce a resistive current into the Maxwell-Ampere law. A disadvantage of this approach is that the resistive region must be quite thick in order to avoid both penetration by and reflection of the longest wavelength light.

Some improvement may be obtained by also introducing a magnetic current into the Faraday law, which becomes

$$-c \nabla \times \mathbf{E} = \frac{\partial \mathbf{B}}{\partial t} + \mathbf{J}_m \tag{1}$$

with $\mathbf{J}_m = \sigma_m(\mathbf{x})\mathbf{B}$. This corresponds to a flux of magnetic monopoles. A method equivalent to this is to multiply $\mathbf{B}$ by a factor somewhat less than unity after advancing it in time (*Boris, 1972*). With σ_m equal to the electric conductivity, one finds that a normally incident wave ($k_y = 0$) is not reflected even if σ becomes large in a short distance. However, this is not true for oblique incidence, and we see below that normal incidence is trivial to handle by other means.

J. P. Boris (private communication, 1972) has an interesting method for making the electric resistivity purely transverse. That is, the current does not respond to the longitudinal part of the field, and the current does not itself produce any charge separation. The intent is that electrostatic activity in the plasma not be damped due to resistive currents driven by its fringe fields. The method appears to be equivalent to adding a loop current around each cell proportional to $\nabla \times \mathbf{E}$ in the cell. This is like a *magnetization current*, and is equivalent to adding a current

$$\mathbf{J} = c\nabla \times \mathbf{M} = -\nabla \times [\alpha(\mathbf{x})\nabla \times \mathbf{E}] \tag{2}$$

which has zero divergence and to which $-\nabla\phi$ does not contribute.

The trouble with a dissipative region is that it occupies a lot of memory space. To give good absorption, rather than reflection, for a range of frequencies and angles of incidence, the region must be about a wavelength thick, through which all fields must be defined. *Tajima and Lee* (1981) attempt to optimize their use of such a region. This method is readily adapted to irregularly-shaped boundaries.

(c) A Simple Closure of the Maxwell Equations at the Open Boundaries

A simple boundary condition which solves the problem of closing the differenced Maxwell equations is an option in ZOHAR and other codes. It works well in some applications and illustrates several points applicable to more complicated boundary conditions to be discussed later.

By closure we mean the following: Consider the left side, and assume E_x, E_y, and B_z for $x \geqslant 0$ are to be advanced using the Maxwell equations. When we come to advance $E_{y,0,k+1/2}$ we need $B_{z,-1/2,k+1/2}$, which must be given by some additional condition in order to make the set of equations self-contained.

For a plane wave incident in the $-x$ direction we have $E_y = -B_z$. A time average of E_y and a spatial average of B_z provides the needed extra condition:

$$B_{z,-1/2,k+1/2}^{n+1/2} + B_{z,1/2,k+1/2}^{n+1/2} + E_{y,0,k+1/2}^{n} + E_{y,0,k+1/2}^{n+1} = 0 \tag{3}$$

This is solved for $B_{z,-1/2,k+1/2}^{n+1/2}$ simultaneously with

$$(\partial_t E_y + c\partial_x B_z)_{0,k+1/2}^{n+1/2} = 0 \tag{4}$$

which involves the same quantities. The value of B_z thus obtained is stored in the appropriate column of the B_z array. The interior ($x \geqslant \Delta x/2$) values of B_z are advanced to time $t^{n+1/2}$ with Faraday's law. Then E_x and E_y for $x \geqslant -\Delta x/2$ are advanced using the Ampere-Maxwell law. This algorithm, due to K. H. Sinz (private communication, 1973), is all that is needed for a one-dimensional code and is used in this form in the OREMP code (K. G. Estabrook, private communication).

At the end of the field integration we use the same idea, but without the time average, to obtain

$$B_{z,-1/2,k+1/2}^{n+1} + B_{z,1/2,k+1/2}^{n+1} + 2E_{y,0,k+1/2}^{n+1} = 0 \tag{5}$$

The field averaging then gives all fields for $x \geqslant 0$.

In order to examine errors caused by nonnormal incidence and by the averaging, we consider the reflection of a plane wave incident on the boundary. Take

$$B_z(x, y, t) = B_i\, e^{i(-k_x x + k_y y - \omega t)} + B_r\, e^{i(k_x x + k_y y - \omega t)} \tag{6}$$

for $x \geq \Delta x/2$ and eliminate $B_{z,-\frac{1}{2},k+\frac{1}{2}}$ using (3), then express E_x and E_y in terms of B_z and solve for the ratio

$$\frac{B_r}{B_i} = \frac{c\kappa_x \cos\dfrac{\omega\Delta t}{2} - \Omega \cos\dfrac{k_x \Delta x}{2}}{c\kappa_x \cos\dfrac{\omega\Delta t}{2} + \Omega \cos\dfrac{k_x \Delta x}{2}} \tag{7}$$

At small angles $c\kappa_x \rightarrow \Omega$ and the averages are the largest sources of error,

$$\frac{B_r}{B_i} = \frac{\cos\dfrac{\omega\Delta t}{2} - \cos\dfrac{k_x \Delta x}{2}}{\cos\dfrac{\omega\Delta t}{2} + \cos\dfrac{k_x \Delta x}{2}} \approx \frac{(k_x \Delta x)^2 - (\omega\Delta t)^2}{16} \tag{8}$$

In practice this ranges up to about 0.5%.

When the angle of incidence θ is not so small the largest error comes from assuming $E_y = -B_z$:

$$\frac{B_r}{B_i} \approx \frac{ck_x - \omega}{ck_x + \omega} = -\tan^2\frac{\theta}{2} \tag{9}$$

At 45° the reflection is 17%, or 3% in terms of energy. For many applications this is good enough. If only one angle of incidence is of interest, (3) to (5) can be modified to correspond to that angle.

There is a flaw in all this so far: We have assumed that E is purely transverse. Suppose a point charge is held fixed near the boundary. The steady-state fields are $B_z = 0$ everywhere and $E_y = 0$ on the boundary, which is the same as the electrostatic field of the charge with a conducting boundary. The point charge is therefore attracted toward the boundary, which can be very troublesome in some cases. The correction to $\nabla \cdot E$ has no effect on this. The cure is to subtract $-\partial\phi/\partial y$ from E_y in (3) to (5). Then the steady state fields equal the electrostatic field as obtained by the Poisson solver with open-sided boundary conditions, and there is no force on the charge.

Incidentally, these boundary conditions are nearly equivalent to those used in an early version of the Los Alamos code WAVE (*Nielson and Lindman*, 1973). A difference is that the boundary conditions in the Poisson solver in WAVE are either $\phi = 0$ or $\partial\phi/\partial y = 0$ at the boundary. Thus a charge is attracted to, or repelled by, respectively, image charges outside the system.

In laser plasma interaction studies, one wants a wave such as $B_{z0}\cos(k_0 x - \omega_0 t)$ to propagate in from the left side (say). The simplest way to do this is to add $4B_{z0}\cos\omega_0 t$, evaluated at the appropriate times, to the right hand sides of (3) and (5). The fields of the incoming wave clearly satisfy the modified equations, as do outgoing waves. This procedure generalizes to more complicated input waveforms much more easily than does the common current sheet antenna method.

The boundary conditions on the fields B_x, B_y, and E_z are directly analogous, thanks to their similar spatial locations.

Lastly, we point out that these boundary conditions do not cause $\nabla \cdot \mathbf{E}$ or $\nabla \cdot \mathbf{B}$ to be altered at the sides. To see this, consider that the boundary conditions are used to determine only B_z and E_z at the sides. The fields E_x, E_y, B_x, and B_y are then advanced everywhere by the differenced Maxwell equations, and our remarks in Section 15-6 regarding the preservation of $\nabla \cdot \mathbf{E}$ and $\nabla \cdot \mathbf{B}$, continue to apply.

(d) Boundary Conditions for Waves Incident at (almost) Any Angle

In many problems light is scattered from the plasma in more than one direction or the incoming light is not a simple plane wave. In such cases the simple boundary condition of the last section is not adequate. Here we describe the boundary conditions normally used in ZOHAR to meet these requirements.

Lindman (1975) decomposes a complex wave propagating out of the system into a superposition of plane waves. For each plane wave a relation

$$c\,\frac{\partial A}{\partial x} = C\left|\frac{ck_y}{\omega}\right|\frac{\partial A}{\partial t} \tag{10}$$

holds, where A is the y or z component of the vector potential in the WAVE code, and $C \approx \cos\theta = (1 - c^2 k_y^2/\omega^2)^{1/2}$. The Coulomb gauge condition $\nabla \cdot \mathbf{A} = 0$ is used to determine $\partial A_x/\partial x$. Lindman then regards C as a linear operator involving $\partial/\partial y$ and $\partial/\partial t$, which can be evaluated at the boundary without extrapolation. The form he found to be both stable and accurate was a partial fraction expansion. This concept was adapted to apply directly to the fields by E. Valeo (private communication, 1973), and our further discussion mainly concerns this form as incorporated in ZOHAR. After including incoming light, we use at the left side

$$B_z(0,y,t) + C^{-1}\left(-c\frac{\partial_y}{\partial_t}\right)E_y(0,y,t) = 2B_{z0}(y,t) \tag{11}$$

where B_{z0} is the desired incoming wave at $x = 0$, and

$$C^{-1} \approx \left[1 - c^2\frac{\partial_y^2}{\partial_t^2}\right]^{-1/2} \tag{12}$$

$$= 1 + \sum_n \frac{\alpha_n}{\partial_t^2/c^2\partial_y^2 - \beta_n} \tag{13}$$

In terms of the computer code this means that

$$-B_z + 2B_{z0} = C^{-1}E_y = E_y + \sum_n E_n \tag{14}$$

where E_n is the solution of

$$\left[\frac{\partial_t^2}{c^2} - \beta_n\partial_y^2\right]E_n = \alpha_n\partial_y^2 E_y \tag{15}$$

As in the preceding section, $-\partial_y\phi$ must be subtracted from E_y in these relations, and we must do time and space averages in order to make (14) second order accurate. The only difference between (14) and (3) is the E_n terms; these are known at the same times and positions as E_y, and therefore are time-averaged in the boundary condition.

Considering once again the reflection of a plane wave incident on the boundary, we find

$$\frac{B_r}{B_i} = \frac{c\kappa_x \cos\dfrac{\omega\Delta t}{2} - \Omega C \cos\dfrac{k_x\Delta x}{2}}{c\kappa_x \cos\dfrac{\omega\Delta t}{2} + \Omega C \cos\dfrac{k_x\Delta x}{2}} \tag{16}$$

Neglecting finite difference errors, this becomes

$$\frac{B_r}{B_i} = \frac{\cos\theta - C}{\cos\theta + C} \tag{17}$$

This shows that the reflection in steady state is half the relative error in the expansion $C(ck_y/\omega)$. This is to be kept in mind when choosing the α and β coefficients.

Useful diagnostic information may be obtained from $B_z - B_{z0} \approx (B_z - C^{-1}E_y)/2$, which is the field of the outgoing waves only.

Lindman's coefficients for a three-term expansion are $\alpha = $ (0.3264, 0.1272, 0.0309), $\beta = $ (0.7375, 0.98384, 0.9996472). With so few terms, the amount of computer memory and computation required is much less than with a dissipative region. Because the β's are less than unity, (15) does not require for stability any reduction in time step below what is required by Maxwell's equations.

Lindman also discusses a difficulty with the transient response of the boundary conditions. If the prescribed incoming wave is turned on too suddenly, the fields take a long time to settle into a steady state. If the angle of entry θ is close to 90°, then "too suddenly" may be a very inconveniently long time. He describes a more complicated expansion which improves the transient response. Experiments in ZOHAR with boundary conditions which are other linear combinations of E_x, E_y, and B_z different than (11), also showed poor transient response, to varying degrees. (11) is very satisfactory in this regard. The reasons for the differences in transient response are not fully understood. As *Lindman* (1975) points out, the transient behavior is due in large part to the fact that a general disturbance contains Fourier components with $\omega < ck_y$. These do not propagate away from the boundary and the expansion, (12), cannot approximate the analytic continuation of $(1 - c^2 ky^2/\omega^2)^{-\frac{1}{2}}$ for $\text{Re}\,\omega < ck_y$ and $\text{Im}\,\omega > 0$. Lindman's newer expansion differs in that it does approximate the analytic continuation.

We have observed another problem also caused by behavior of the expansion for $\omega < ck_y$. An instability in the fields E_x, E_y, and B_z occurs when there is a density jump parallel to and near the boundary. The jump supports a surface wave with $\omega < ck$. For some frequency intervals in that range, the expansion C is negative. Thus, the direction of the Poynting flux is reversed, and energy flows into the system to drive up the surface wave. We have had no difficulty except when the distance from the density jump to the boundary is less than about $4c/\omega_p$. Perhaps Lindman's newer expansion would cure the problem. However, all that is necessary to suppress the instability is to have an expansion which remains positive.

(e) Particle Boundary Conditions

Similar in most respects, particle boundary conditions in electrostatic and electromagnetic simulations are discussed in Chapter 16.

15-12 CONDUCTING-WALL BOUNDARY CONDITIONS

In this discussion we have perfectly-conducting walls at $x = 0$ and L_x, and periodicity in y. More general boundary shapes, such as those found in magnetrons, are implemented by *Palevsky* (1980).

(a) Closure of Maxwell's Equations at the Walls

For the time integration of Maxwell's equations we need only the boundary conditions on the tangential components of $\mathbf{E}$

$$E_y = 0 \quad \longrightarrow \quad E_{y,0,k+\frac{1}{2}} = 0 \tag{1}$$

$$E_z = 0 \quad \longrightarrow \quad \frac{E_{z,-\frac{1}{2},k+\frac{1}{2}} + E_{z,\frac{1}{2},k+\frac{1}{2}}}{2} = 0 \tag{2}$$

at $x = 0$ $(j = 0)$, and similarly at $x = L_x$ $(j = N_x)$, where j and k are the x and y grid indices. Shown in Figure 15-12a are the spatial locations where the fields are defined. By averaging E_z we have its boundary condition centered and second order accurate. These conditions are sufficient to obtain closure of the differenced Maxwell equations.

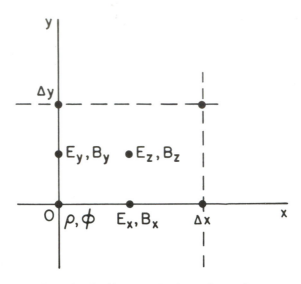

Figure 15-12a The two-dimensional grid at a conducting wall, $x = 0$.

With the Maxwell equations differenced in the interior as described in Section 15-2, the fluxes of tangential components of **B** are conserved exactly:

$$\frac{d}{dt} \int_0^{L_y} dy \int_0^{L_x} dx\, B_y$$

$$\rightarrow \frac{d}{dt} \Delta y \sum_{k=0}^{N_y-1} \Delta x \left[\frac{1}{2} B_{y,0,k+\frac{1}{2}} + \sum_{j=1}^{N_x-1} B_{y,j,k+\frac{1}{2}} + \frac{1}{2} B_{y,N_x,k+\frac{1}{2}} \right] = 0 \qquad (3)$$

$$\frac{d}{dt} \int_0^{L_y} dy \int_0^{L_x} dx\, B_z \quad \rightarrow \quad \frac{d}{dt} \Delta y \sum_{k=0}^{N_y-1} \Delta x \sum_{j=0}^{N_x-1} B_{z,j+\frac{1}{2},k+\frac{1}{2}} = 0 \qquad (4)$$

In addition, the magnetic flux through any surface ($x = $ const) is constant:

$$\frac{d}{dt} \int_0^{L_y} dy\, B_x \quad \rightarrow \quad \frac{d}{dt} \Delta y \sum_{k=0}^{N_y-1} B_{x,j+\frac{1}{2},k} = 0 \qquad (5)$$

From the time varying part of $\nabla \cdot \mathbf{B} = 0$ comes an additional boundary condition, $B_x(x=0,y,t) = 0$ (or perhaps $= B_{x0}(y)$, a function of y but not t, corresponding to a B field established previously by external currents on a longer time scale). This becomes

$$B_x \quad \rightarrow \quad \frac{B_{x,-\frac{1}{2},k} + B_{x,\frac{1}{2},k}}{2} = B_{x0}(y) \qquad (6)$$

This boundary condition is not required in order to close the system of equations.

The total electric field energy is

$$W_E = \frac{\Delta x \, \Delta y}{2} \sum_{k=0}^{N_y-1} \left[\sum_{j=0}^{N_x-1} E_{x,j+\frac{1}{2},k}^2 + \sum_{j=1}^{N_x-1} E_{y,j,k+\frac{1}{2}}^2 \right] \tag{7}$$

The transverse electric field energy

$$W_{\text{trans}} = W_E - W_{es}$$

can be shown to be non-negative as calculated using (7) and (15), below.

(b) Electrostatic Solutions in 2d

For diagnostic purposes we solve $-\nabla^2\phi = \rho$ for the electrostatic potential using the standard 5-point differencing as in Section 14-2. Fourier transforming ρ and ϕ in y produces $\rho_{j,m} \equiv \rho_m(X_j)$ and $\phi_{j,m}$ where m is the harmonic number in y, including $m=0$. Poisson's equation is then a finite difference equation in x, as in Section 14-6. The boundary conditions at the perfectly conducting walls are $E_y(0,y) = 0 = E_y(L_x,y)$. That is, the $x=0=L_x$ walls are equipotentials, $\phi_{0,m} = 0 = \phi_{N_x,m}$ for all $m \neq 0$. For $m=0$ we have

$$\phi_{0,0} - \phi_{1,0} = + (\rho_{0,0} \Delta x + \langle \sigma_0 \rangle) \Delta x \tag{8}$$

$$\phi_{N_x-1,0} - \phi_{N_x,0} = - (\rho_{N_x,0} \Delta x + \langle \sigma_L \rangle) \Delta x \tag{9}$$

where the surface charges on the walls $q_0 \equiv L_y \langle \sigma_0 \rangle$ and q_L are added to the active particle charge densities, weighted to the walls. The surface charge density at the boundary planes varies with y; $\sigma(0,y) = \sigma_0(y)$ and $\sigma(L_x,y) = \sigma_L(y)$; the surface charges in a period in y are, at the left,

$$q_0 = \Delta y \sum \sigma_0(y) \equiv L_y \langle \sigma_0 \rangle \tag{10}$$

and similarly for q_L at the right.

The electrostatic fields E_x, E_y are now obtained at the cell centers, using single-cell differencing of $\phi_{j,k}$, the better to handle charge conservation (Gaussian box centered in point (j,k) with E_x at $j\pm\frac{1}{2},k$, E_y at $j,k\pm\frac{1}{2}$).

Active particles contribute all their charge to $\{\rho_{j,k}\}$, including that part collected at $j=0$ and N_x. The total charge in the system $0 \leqslant x \leqslant L_x$ in a period L_y is

$$Q_{\text{total}} = q_0 + q_L + \Delta x \, \Delta y \sum_{j=0}^{N_x} \sum_{k=0}^{N_y-1} \rho_{j,k} = 0 \tag{11}$$

Q_{total} is zero because the electric field is zero inside the conducting walls, and therefore the electric flux through a surface enclosing Q is zero. External circuit currents transfer charge between q_0 and q_L, which are updated as active particles are created and deleted (Chapter 16). Total charge therefore remains constant (zero), which makes (8) and (9) redundant so that we need another condition in order to determine ϕ uniquely.

The remaining condition needed is a convention for the additive constant in ϕ. We regard the left wall as *grounded:*

$$\phi_{0,k} \equiv \phi_0 = 0, \qquad \phi_{N_x,k} = \phi_L \tag{12}$$

Solution of 14-6(2) with (8) or (9) determines ϕ everywhere, including ϕ_L. This solution can be done by introducing the ($m = 0$ Fourier component of the) field

$$E_{x,j+\frac{1}{2},0} = \frac{\phi_{j,0} - \phi_{j+1,0}}{\Delta x} \tag{13}$$

with which

$$E_{x,j+\frac{1}{2},0} - E_{x,j-\frac{1}{2},0} = \rho_{j,0}\,\Delta x \qquad \text{for} \quad j = 1, \ldots, N_x - 1 \tag{14a}$$

$$E_{x,\frac{1}{2},0} = \rho_{0,0}\,\Delta x + \langle\sigma_0\rangle \tag{14b}$$

This E_x is found in a sweep from left to right using (14a,b); then $\phi_{j,0}$ is found in a second sweep using (13).

The electrostatic field energy is

$$W_{es} = \frac{1}{2}\,q_0\,\phi_0 + \frac{1}{2}\,q_L\,\phi_L + \Delta x\,\Delta y \sum_{j=0}^{N_x}\sum_{k=0}^{N_y-1}\frac{1}{2}\,\rho_{j,k}\,\phi_{j,k} \tag{15}$$

Because of (11), W_{es} is independent of the arbitrary choice (12).

(c) Combined Particle and Field Calculation

For the correction of the longitudinal part of E (Section 15-6) we solve in the same way for the correction potential $\delta\phi$, where the source term in 15-6(12) is the same except that at the walls $\nabla \cdot E$ is calculated as if $E_{x,-\frac{1}{2},k} = E_{x,N_x+\frac{1}{2},k} = 0$. A simple check on this procedure is to apply it twice; the second $\delta\phi$ should be zero (*i.e.*, no further correction).

We can determine the distribution in y of the surface charges q_0 and q_L; these may be used, for example, to control field emission of particles (Chapter 16). At the left wall, Gauss' law over any cell gives

$$\sigma_{0,k} = E_{x,\frac{1}{2},k} - \rho_{0,k}\,\Delta x \tag{16}$$

Gauss' law over a period in y is satisfied at the surface $x_{\frac{1}{2}} = \frac{1}{2}\Delta x$, as we see from

$$\sum_{k=0}^{N_y-1} E_{x,\frac{1}{2},k}\,\Delta y = \sum_{k=0}^{N_y-1} (\rho_{0,k}\,\Delta x + \sigma_{0,k})\,\Delta y \tag{17}$$

$$= q_0 + \Delta x\,\Delta y \sum_{k=0}^{N_y-1} \rho_{0,k} \tag{18}$$

where we have used (10).

As in Section 15-5, for the convenience of the particle integrator and field diagnostics, the fields are averaged in space as in Section 15-5 to obtain fields at the same points as ρ. Before the next integration of Poisson's equation, the original unaveraged fields are restored.

From (1), (2), and (6) we know E_y, E_z, and $B_x = B_{x0}(y)$ at the walls. To form $E_{x,0,k}$ we could *average* $E_{x,+\frac{1}{2},k}$ and $E_{x,-\frac{1}{2},k}$ but, if the latter is zero as when calculating $\nabla \cdot \mathbf{E}$ for the divergence correction, then we cannot represent even the simple case of a charged capacitor with no free charge, in which E_x is uniform. Instead, we use $E_{x,0,k} = E_{x,\frac{1}{2},k}$. We treat B_z similarly. When there is free charge or current in the boundary cells ($\rho_{0,k}$ or $J_{y,0,k}$ nonzero), a number of choices can be made which depend on whether we think we must resolve collective behavior in these cells and how well that can be done. Options for E_x include

(a) $E_{x,0,k} = E_{x,\frac{1}{2},k}$
(b) $E_{x,0,k} = \sigma_{0,k} + \frac{1}{2}\rho_{0,k}\,\Delta x = E_{x,\frac{1}{2},k} - \frac{1}{2}\rho_{0,k}\,\Delta x$
(c) $E_{x,0,k} = \sigma_{0,k} = E_{x,\frac{1}{2},k} - \rho_{0,k}\,\Delta x$

All these reduce to the vacuum case mentioned above when $\rho_{0,k} = 0$. Option (a) changes smoothly as particles are "born" near the wall and move in, while (b) and (c) produce discontinuous change in wall density upon emission of a particle. In a flow which is uniform in y, option (b) results in the correct attraction to the emitting wall for particles in the cells adjacent to the wall, while option (a) produces too little attraction and (c) too much.

15-13 INTEGRATING MAXWELL'S EQUATIONS IN CYLINDRICAL COORDINATES

Boris (1970) describes spatial differencing of Maxwell's equations in r-z coordinates. *Palevsky* (1980) considers both r-z and r-θ coordinates. B. B. Godfrey (private communication) has implemented more general curvilinear coordinates. Here we outline the extension of the electrostatic calculation, Sections 14-10 through 14-12, to an electromagnetic code in r-z coordinates.

First, we introduce a compact notation, useful for the more complicated electromagnetic case, by rewriting the equations of Section 14-10(b). From 14-10(9), the divergence is given by

$$\frac{1}{2}\,\Delta z\,(\Delta r^2)_j\,(\nabla \cdot \mathbf{E})_{j,k} = \Delta z\,(\Delta_r\,r\,E_r)_{j,k} + \frac{1}{2}\,(\Delta r^2)_j\,(\Delta_z\,E_z)_{j,k} \quad (1)$$

where (j,k) are subscripts for (r,z) and the centered Δ_r and Δ_z operators are illustrated by

$$(\Delta_r\,r\,E_r)_{j,k} \equiv r_{j+\frac{1}{2}}\,E_{r,j+\frac{1}{2},k} - r_{j-\frac{1}{2}}\,E_{r,j-\frac{1}{2},k} \quad (2)$$

$$(\Delta_z\,E_z)_{j,k} \equiv E_{z,j,k+\frac{1}{2}} - E_{z,j,k-\frac{1}{2}} \quad (3)$$

Similarly,

$$\Delta r_{j+\frac{1}{2}} \left[\frac{\partial \phi}{\partial r} \right]_{j+\frac{1}{2},k} = \phi_{j+1,k} - \phi_{j,k} \qquad \text{etc.} \qquad (4)$$

Just as the divergence expressions are obtained by application of Gauss' integral theorem to certain volumes, the curl expressions are obtained by application of Stokes' theorem to appropriate line integrals. The line integral of **E** around a rectangle in the r-z plane (Figure 14-10b) gives

$$\Delta r_{j+\frac{1}{2}} \Delta z \, (\nabla \times \mathbf{E})_{\theta,j+\frac{1}{2},k+\frac{1}{2}} = -\Delta z \, (\Delta_r E_z)_{j+\frac{1}{2},k+\frac{1}{2}} + \Delta r_{j+\frac{1}{2}} (\Delta_z E_r)_{j+\frac{1}{2},k+\frac{1}{2}} \quad (5)$$

These difference expressions satisfy $(\nabla \times \nabla \phi)_\theta \equiv 0$ identically (Problem 15-13a).

The r and z components of $\nabla \times \mathbf{E}$ are (Problem 15-13b)

$$\Delta z \, (\nabla \times \mathbf{E})_{r,j+\frac{1}{2},k} = -(\Delta_z E_\theta)_{j+\frac{1}{2},k} \qquad (6a)$$

$$\frac{(\Delta r^2)_{j+\frac{1}{2}}}{2} (\nabla \times \mathbf{E})_{z,j+\frac{1}{2},k} = (\Delta_r r E_\theta)_{j+\frac{1}{2},k} \qquad (6b)$$

These difference expressions satisfy $(\nabla \cdot \nabla \times \mathbf{E})$ identically (Problem 15-13c).

Finally, we often need difference expressions for $\nabla \times (\nabla \times \mathbf{E}) = -(\nabla^2 \mathbf{E}) + \nabla \nabla \cdot \mathbf{E}$. Because ∇^2 operates on the unit vectors $\hat{\mathbf{r}}$ and $\hat{\boldsymbol{\theta}}$, $(\nabla^2 \mathbf{E})_r$ is *not* simply $\nabla^2 E_r$. However, $(\nabla^2 \mathbf{E})_r$ does involve only E_r, so we display this feature in writing the difference equations

$$[\nabla \times (\nabla \times \mathbf{E})]_r = \frac{1}{\Delta z} \Delta_z \left[\frac{\Delta_z E_r}{\Delta z} \right] - \frac{1}{\Delta r} \Delta_r \left[\frac{\Delta_r r E_r}{\Delta r^2/2} \right] + \frac{1}{\Delta r} \Delta_r (\nabla \cdot \mathbf{E}) \quad (7)$$

at position $(j+\frac{1}{2}, k)$, where the first two terms represent $-(\nabla^2 \mathbf{E})_r$. The θ and z components are similar.

In the particle mover, we collect *currents* I_r, I_θ, and I_z, and *charges* Q, rather than *densities* **J** and ρ. As in Section 14-10, this makes it possible to enforce conservation laws. An important subtlety remains: In Section 14-12 we discussed a rotation, applied after advancing the position **x**, which brings the position back into the r,z plane. To avoid an error in the direction of **J**, resulting from the corresponding rotation of **v**, we might use **v** before (after) the rotation in the first (second) part of 15-5(3). In Darwin codes, the collection of $\mathbf{J}_{n+1}$ from an improperly rotated $\mathbf{v}_{n+1}$ can lead to macroscopic accumulation of spurious angular momentum (D. W. Hewett, private communication, 1983).

PROBLEMS

15-13a Show that $(\nabla \times \nabla \phi)_\theta \equiv 0$ by using the commutation property $\Delta_r (\Delta_z \phi) \equiv \Delta_z (\Delta_r \phi)$, which holds on the mesh in Figure 14-10b.

15-13b Derive (6ab). *Hints:* (6a) is from an integral over a segment of a cylinder, divided by $2\pi r$; (6b) is from the difference between the line integral on the outer and inner edges of a disk annulus, divided by 2π.

15-13c Show that $(\nabla \cdot \nabla \times \mathbf{E})_\theta \equiv 0$ by using the commutation property $\Delta_r r (\Delta_z E_\theta) \equiv \Delta_z (\Delta_r r E_\theta)$.

15-14 DARWIN, OR MAGNETOINDUCTIVE, APPROXIMATION

In some applications it is possible to avoid the time-step limitation associated with explicit differencing of Maxwell's equations by altering the field equations so that they do not support wave propagation. A proven approach here is the Darwin, or magnetoinductive, model.

Darwin codes eliminate the Courant restriction $\Delta t \leqslant \lambda/c$ by dropping Maxwell's transverse displacement current term. These "pre-Maxwell equations" eliminate electromagnetic wave propagation while retaining electrostatic, magnetostatic and inductive electric fields. The equivalence of this nonradiative approximation to the Darwin Lagrangian, which retains as much of the electromagnetic interaction as possible without including retardation, is shown by *Kaufman and Rostler* (1971). *Nielson and Lewis* (1976), whose implementation is described here, provide many references for the historical development of these codes.

The field equations in the Darwin approximation are

$$c \nabla \times \mathbf{B} = \mathbf{J}_T = \mathbf{J} + \frac{\partial \mathbf{E}_L}{\partial t} \tag{1}$$

$$c \nabla \times \mathbf{E} = -\frac{\partial \mathbf{B}}{\partial t} \tag{2}$$

Given $\mathbf{E}_n$, $\mathbf{B}_n$, the particles are integrated by explicit differencing to $\mathbf{v}_{n+\frac{1}{2}}$ and $\mathbf{x}_{n+1}$. Extrapolation of $\mathbf{v}_{n+\frac{1}{2}}$ to $\mathbf{v}_{n+1}$ permits collection of $\mathbf{J}_{n+1}$, from which $\mathbf{B}_{n+1}$ is obtained, *e.g.*, by solution of $c\nabla^2 \mathbf{B} = -\nabla \times \mathbf{J}$. Unlike usual electromagnetic codes, the Ampere equation (1) cannot be used to advance $\mathbf{E}_T$ in time. Instead

$$c^2 \nabla^2 \mathbf{E}_T = \frac{\partial \mathbf{J}_T}{\partial t} \tag{3}$$

is used at time level $n+1$. This creates a time-centering problem. To preserve second-order accuracy in time, (3) needs a time-advanced expression for $\partial \mathbf{J}_T/\partial t$. To ensure stability, $\partial \mathbf{J}/\partial t$ is expressed in terms of the advanced $\mathbf{E}$, using moments accumulated from the particles:

$$\frac{\partial \mathbf{J}}{\partial t} = -\nabla \cdot \rho \langle \mathbf{vv} \rangle + \frac{q\rho}{m}\mathbf{E} + \frac{q\rho \langle \mathbf{v} \rangle}{mc} \times \mathbf{B} \qquad (4)$$

summed over species. This leads to an elliptic equation for the advanced fields of form

$$c^2 \nabla^2 \mathbf{E}_T - \omega_p^2(\mathbf{x})\mathbf{E}_T = -\sum \nabla \cdot \rho \langle \mathbf{vv} \rangle + \left(\sum \frac{q\rho \langle \mathbf{v} \rangle}{mc} \right) \times \mathbf{B} + \omega_p^2 \mathbf{E}_L - \nabla \Xi \quad (5)$$

at time level $n+1$. The divergence of this equation, together with $\nabla \cdot \mathbf{E}_T = 0$ determines Ξ. This elliptic equation provides instantaneous propagation of $\mathbf{B}$ and $\mathbf{E}_T$, as is necessary for stability. Iterative methods are usually required to solve (5).

Busnardo-Neto et al. (1977) and *Aizawa et al.* (1980) describe variants of this algorithm.

15-15 HYBRID PARTICLE/FLUID CODES

For applications not requiring a kinetic description of the electrons, codes using a hybrid of particle ions and fluid electrons are indicated. With Darwin and quasistatic approximations, long time scales are accessible as in a fully implicit code but with less noise. Sometimes an energetic minority of electrons is modeled with particles, while the cold ion and electron backround is a fluid. Early models of this type (*Dickman et al.,* 1969; *Byers et al.,* 1974) include the transverse particle current component that is out of the plane of simulation, but treat the remaining particle current and the fluid current very approximately; often the fluid simply neutralizes the particle charge density.

Byers et al. (1978) give a more complete treatment of the fluid and particle currents, in the limit of negligible electron mass. A similar algorithm has been implemented in two dimensions (r-z) by *Mankofsky et al.* (1981) and (x-y) by *Harned* (1982a). From an explicit integration of the fluid electron momentum equation, *Hewett and Nielson* (1978) obtain the transverse fluid current. The longitudinal fluid current is calculated so that the longitudinal particle ion current is cancelled, eliminating Langmuir oscillations. Some plasma phenomena depending on electron inertial effects are retained, *e.g.*, lower hybrid and electron cyclotron waves.

Because of the use of explicit time integration, these codes are unstable at low densities where the Alfvén wave frequency becomes large. In order to treat problems which feature vacuum or low density regions, *ad hoc* interfacing or other stabilization artifices become necessary. *Hewett* (1980) removes this limitation by solving the coupled electron and field equations simultaneously by a noniterated alternating-direction-implicit method.

Applications of hybrid codes are included in Section 15-18.

15-16 IMPLICIT ELECTROMAGNETIC CODES

Implicit fields reproduce electromagnetic wave propagation at long wavelengths ($\lambda \gg c\Delta t$). At short wavelengths, the electrostatic, magnetostatic, and inductive electric fields are retained, as in a Darwin code. Implicit fields can be used with explicit particles.

With implicit particles, Langmuir waves are stabilized at all wavelengths, as in an implicit electrostatic code. The electrostatic fields are accurate for wavelengths longer than the electron transit distance ($v_{te}\Delta t$). These properties make an implicit electromagnetic code attractive *e.g.*, to modeling of intense electron flow which is subject to pinching, Weibel instability (*Brackbill and Forslund*, 1982), and other processes generating magnetic fields which alter the electron flow (*Forslund and Brackbill*, 1982).

Implicit electromagnetic simulation is discussed in detail by *Brackbill and Forslund* (1985), and *Langdon and Barnes* (1985).

15-17 DIAGNOSTICS

We describe a minimal set of diagnostics for understanding the results of laser-plasma interaction simulations. Many are obvious; the need for others is appreciated after experience both with and without them on specific problems. It is these parts of the code which are most frequently changed and should be kept flexible.

(a) Particles

The most familiar particle diagnostic is the phase space scatter plot. Dots are plotted at positions given by two of the particle coordinates, *e.g.* u_x and u_y. Often these plots make direct connection to theoretical descriptions; particle trapping in waves is a well-known example.

While conceptually simple, implementation is complicated by not having all the particles in memory at once. If a phase space plot is to be made on the next time step, then ZOHAR scans the particles for minima and maxima, to see if plot limits must be expanded. The other problem concerns making more than one frame or running several plot channels; we specify an offset as well as a plotting interval. Thus, one plot is made at steps 0, 100, 200, . . . ; another may be made at steps 49, 99, . . . , etc. Usually the slight difference in time of plotting does not hinder comparisons.

An elaboration is to plot one linear combination of particle coordinates versus another, skipping particles not satisfying two linear constraints. For example, plotting $u_x - u_y$ versus $x - y$ for particles with $a < x < b$ displays trapping in waves propagating at 45° in a slice of the system. *Morse and Nielson* (1971) uncover trapping in simulations of Weibel instability by plotting v_y versus y for each of three groups of particles which are sorted according to their canonical momenta $v_x + qA_x/mc$, a constant in their one-dimensional (y) simulation.

Also handled by the same subroutine are plots of $f(q)$ and contours of $f(q_1, q_2)$, where q, q_1, and q_2 are any particle phase space coordinates and q may be u^2 (these are projections over the other coordinates). This includes, for example $f(u_x)$, $f(u^2)$, and $\rho(x, y)$. In the open-sided version, the transform $\rho(x, k_y)$ is often used, and we also plot statistics on particles leaving the system.

(b) Fields

Obvious diagnostics are contour plots for B_z, E_z, and ϕ. We also have plots of (E_x, E_y), (B_x, B_y), and $-\nabla\phi$ consisting of an array of little arrows whose directions and lengths indicate the vector value at that point in space.

Fourier mode energies for B_z, E_z, and ϕ are indicated by an array of vertical lines whose lengths are proportional to the logarithm of the mode energies, over a range of 10^4. This points out modes which should be watched more closely by other means.

In the open-sided version, plots of $B_z(x, k_y)$, etc., for specified y modes aid both in interpretation of results and separating a signal from other effects or from noise.

(c) Histories

Many quantities are saved each time step to provide history plots. Most of these are energies and momenta for fields and particle species. Also saved are specified field probes, mode energies, and mode amplitudes, *e.g.*, for $E_z(x, k_y)$. The latter include E_z and B_z at $x = 0$ for the outgoing waves only, computed as per Sec. 15-11d. From time to time during a run, the code plots the saved quantities, and others derivable from them. For more careful study, an interactive postprocessor ZED reads the history file, performs operations such as calculating frequency spectra, and makes plots.

(d) Remarks

Usually it is not possible to foresee exactly which plots will be decisive in a computer run. Compared to an electrostatic code, there are many times more quantities to monitor. In order to get what is needed the first time, much thought is given to a judicious choice of diagnostics, but we prefer to err on the side of inclusion rather than omission. The result is a little like a telephone directory: a lot of data, much of which will not be useful, but you want to have it all because you do not know in advance what you will need.

For such reasons, we believe this type of computing cannot be seriously pursued without access to a high resolution and high volume graphical output device, such as a cathode-ray tube and camera. Mechanical plotters are too slow, especially for contours and phase space plots. Impact "printer plots" are too coarse, too often hiding or distorting valuable detail. Microfiche is the most compact and easily handled format.

Sometimes one must resort to numerical printouts. Again the telephone directory analogy holds, and microfiche is the best available format.

15-18 REPRESENTATIVE APPLICATIONS

This section is a partial survey of applications of electromagnetic simulation codes. It is not a complete bibliography, but it does indicate some of the diversity of physical problems and computer models. Many additional papers are references in papers cited here, others are cited in Chapters 6 and 7. Some important application areas, such as ionospheric, interplanetary and astrophysical plasmas, free-electron lasers, and collective accelerators, are neglected. We first mention a few papers in various subjects, then discuss some topics in more detail.

Weibel instability and related filamentation of electron flow, an early application of electromagnetic codes (*Morse and Nielson, 1971; Davidson et al., 1972; Lee and Lampe, 1973*), is now well handled by implicit codes (*Brackbill and Forslund, 1982*).

Dissipation phenomena in collisionless shocks are discussed by *e.g. Biskamp and Welter* (1972) and *Forslund et al.* (1972).

Instabilities in whistlers and ion cyclotron waves are discussed by *Hasegawa and Birdsall* (1964) and *Ossakow et al.* (1972a, b).

"Plugging" of end-losses and r.f. heating by means of low frequency radio waves are clarified by *Ohsawa et al.* (1979).

Electron flow in microwave devices is a non-neutral plasma. Electromagnetic codes model magnetrons (*Palevsky, 1980*) and other devices (*Kwan, 1984*).

Collisionless tearing instability, and associated heat transport, are treated by *Katanuma* (1981) and *Katanuma and Kamimura* (1980).

(a) Interaction of Intense Laser Light with Plasmas

This topic has been primarily driven by the intent to ignite significant thermonuclear burn using the very high energy fluxes available from large lasers. Usually, the plasma is also produced by the laser, and therefore the spatial dependence of the plasma density and temperature are related to the intensity and wavelength of the laser light. We simulate a section of the plasma corona encompassing critical density (*i.e.*, where the electron plasma frequency ω_{pe} equals the laser frequency ω_0), down to a few percent of critical density.

First a digression on units: For laser-plasma interaction problems in ZOHAR we set c, the laser frequency ω_0, and $-q_e/m_e$ to unity in the code.

This prescribes a set of units in which numbers in the code are easily related to ratios of physical interest. Since now $\omega_{pe}^2 = -\rho_e$ in rationalized cgs units, density is measured in units of the critical density. For the potential, $\phi = 1$ means $-q_e\phi = m_e c^2$, $i.e.$, $511\,\mathrm{keV}$. The cyclotron frequency, $\omega_{ce} = |q_e|B_z/(m_e c) = B_z$, so that unit B_z means $\omega_{ce} = \omega_0$ which occurs near 100 megagauss for laser wavelength $1.06\,\mu\mathrm{m}$. The species charge is chosen to give the correct total charge, and the mass is obtained from q/m. We divide system energies, etc., by L_y before plotting, so that with our other normalizations the diagnostics do not vary with changes in N_y, L_y, N_e, etc., unless the physics has changed. Other groups achieve similar convenience $e.g.$ by normalizing the cgs electric field to $(4\pi n m_e c^2)^{\frac12}$.

Collisionless absorption near the critical density is due largely to "resonance" absorption, in which part of the light energy tunnels past the classical turning point to excite a plasma wave at critical density. This wave accelerates electrons to high energies. The plasma wave pressure can drastically alter plasma flow through the critical density point, leading to density profile steepening which, in turn, affects the amount of absorption and the temperature of the heated electrons. This nonlinear interplay was demonstrated in simulations by *Forslund et al.* (1975), and *Estabrook et al.* (1975). Magnetic field generation accompanying resonance absorption, with and without collisional effects, are discussed by *Adam et al.* (1982a) and references therein. Magnetic field generation and resulting lateral transport of heat are modeled in an implicit code by *Forslund and Brackbill* (1982).

Two high-frequency instabilities are possible in the underdense plasma region: (a) $2\omega_p$ *decay* of the incident wave into two electron plasma waves near the quarter-critical density region; (b) *Raman scattering*, the decay into an electron plasma wave and an electromagnetic wave, at or below quarter-critical density. *Stimulated Thomson* (or *Compton*) scattering is a variant of Raman, occurring at high temperatures or low densities, in which the longitudinal perturbation is not a plasma wave (because of Landau damping) but is coherent scattering on the electrons. In addition, *stimulated Brillouin scattering*, the decay into a sound wave and an electromagnetic wave, is possible throughout the underdense region. Other processes, such as filamentation, are not discussed here. These processes are named for analogous processes occurring in liquids and gasses.

Early simulations in one dimension of Raman and Brillouin scatter are reported by *Forslund et al.* (1973) and *Kruer et al.* (1973). Raman and Brillouin *sidescatter*, in which the decay wave propagates obliquely to the line of the incident light, is modeled in 2½d simulations by *Klein et al.* (1973), *Ott et al.* (1974), *Biskamp and Welter* (1975), and *Langdon and Lasinski* (1976, 1983).

ZOHAR simulations of the $2\omega_p$ instability (two-plasmon decay) showed nonlinear saturation of this instability by its generation of short-wavelength ion fluctuations and local density profile steepening (*Langdon and Lasinski,* 1976; *Langdon et al.,* 1979). Both features were later observed experimentally by Thomson scattering diagnostics (*Ebrahim et al.,* 1979; *Baldis and Walsh,* 1983).

Aizawa et al. (1980) simulated expansion of laser produced plasma into an externally-generated magnetic field.

(b) Reversed-Field Configurations; Pinches

Hybrid codes have been used to model field-reversed configurations generated by injection of electrons (*Byers et al.,* 1974), ions (*Mankofsky et al.,* 1981; *Harned,* 1982b), and neutral beams (*Byers et al.,* 1978). *Buneman et al.* (1980) simulated a helical instability in a Z-pinch, accessible only to their *three*-dimensional code. An unexpected self-generated toroidal magnetic field was found in simulations by *Hewett* (1984) of theta pinches, and was later observed in experiment.

15-19 REMARKS ON LARGE-SCALE PLASMA SIMULATION

We conclude by discussing matters strongly affecting the practicality of two-dimensional, electromagnetic simulation.

It has often been stated that to obtain "collisionless" behavior, the collision times must be longer than the length of the run. Fortunately this is overly pessimistic. What appears to be closer to the truth is that collision times should exceed, *e.g.*, instability exponentiation times and trapping times. Particularly with the open-sided code, where wave energy and thermal particles are replenished, the whole run can usually be much longer than the latter times.

One might expect to need enough particles to give a good representation of the velocity distribution in each Debye square. With three velocity components ($2\frac{1}{2}$d), the total number of particles required would then be completely out of reach. However, for our applications, details of the distribution in v_z are not of concern; only enough particles to represent the first few v_z moments were needed. Usually no more particles were needed than for a problem with two velocity components (2d). Similarly, for a longitudinal plane wave propagating in the x direction, details of the v_y distribution are not required; one needs enough particles in an area of the order of a square wavelength to represent the v_x distribution well. Since the same tends to be true for each of several superimposed waves traveling in various directions, the particle density required is much lower than one might infer from experience in one dimension. Another way to say this is that one needs good statistics in projections of the distribution, and not necessarily in the full phase space. (Of course, for some phenomena in magnetic fusion, such as cyclotron harmonic waves, the dimensionality of the relevant projection is not as low as in the example here, and many times more particles may be needed for such problems.) This economy is a main reason that particle codes can compete successfully with Vlasov codes in multidimensions.

Caution is required, but one can be paralyzed by a conservative attitude into missing profitable applications.

Finally, some remarks on costs and code efficiency: Due to extensive use on expensive computer systems, it is often true that the cost of the applications of an electromagnetic code exceeds the cost of employing several people. It is then economically advisable to tune the code to the machine, *e.g.* by careful rewriting of sections of the code into the machine's native language using an assembler. In ZOHAR for example, the particle mover (advancing $\mathbf{x}, \mathbf{v}$ and collecting $\rho, \mathbf{J}$) was converted to machine language very early in the code's development, and again when the code moved to a newer computer. The resulting cost savings to the applications of the code repaid this effort in a few months. On the other hand, the particle boundary conditions are changed frequently and are written in FORTRAN, as are almost all the rest of the physics aspects of ZOHAR.

SIXTEEN

PARTICLE LOADING, INJECTION; BOUNDARY CONDITIONS AND EXTERNAL CIRCUIT

16-1 INTRODUCTION

Placing particles in $\mathbf{x}, \mathbf{v}$ at $t = 0$ and creating or removing particles during a run are the main subjects of this Chapter. The placement involves starting with prescribed densities in space $n_0(\mathbf{x})$ and in velocity $f_0(\mathbf{v})$ and generating particle positions and velocities $(\mathbf{x}, \mathbf{v})_i$. The formal process for this is called *inversion of the cumulative density*.

Another subject is the handling of bounded systems having wall charges, net current, and external circuit elements. The latter requires simultaneous solution of Kirchhoff's circuit equations.

In studying plasmas which have very large spatial extent, we usually model only a portion of the plasma, up to some length L (*i.e.*, L_x by L_y by L_z); we usually invoke periodicity along one or more coordinates, where applicable, and make the particles re-entrant or reflected at the periodic boundaries, running with a constant number of particles. In such models the *initial conditions* on the particle distribution function $f(\mathbf{x}, \mathbf{v}, t = 0)$ are very important in determining the later behavior of the system. Hence, it is desirable to load the particles carefully at $t = 0$. Some steps in wide use for inverting given densities $n_0(\mathbf{x})$ and $f_0(\mathbf{v})$ into particle positions and velocities $(\mathbf{x}, \mathbf{v})_i$ are given in this Chapter.

In studying bounded plasma systems, where particles may be created or removed at boundaries during the run, the *boundary conditions* on particles, such as particle flux, $\Gamma = v_{\text{normal}} f_0(x_{\text{wall}}, \mathbf{v}, t)$, are very important in determining the continuing behavior of the system. For some studies, starting with the system empty or full of particles may lead to the same result after some time. For other studies, establishing an equilibrium $f_0(\mathbf{x}, \mathbf{v})$ at $t = 0$, even approximately, may be very important, for example, in studying the stability of this f_0; starting with an empty system and then injecting particles may never produce an $f(\mathbf{x}, \mathbf{v})$ near the desired f_0. Some methods that are in use for creating fluxes at walls are presented in this Chapter.

16-2 LOADING NONUNIFORM DISTRIBUTIONS $f_0(\mathbf{v})$, $n_0(\mathbf{x})$; INVERSION OF CUMULATIVE DISTRIBUTION FUNCTIONS

The physical problem initial conditions usually specify densities $f_0(\mathbf{v})$ and $n_0(\mathbf{x})$ in simple forms, like $\exp(-v^2/2v_t^2)$ or $(1 + \psi \sin kx)$. These forms must be *inverted* in order to obtain $(\mathbf{x}, \mathbf{v})_i$ of the particles. In this section, we look at some general ideas about placing particles in phase space.

Suppose that we wish to place particles so as to form a distribution function (or, density) $d(x)$, from $x = a$ to $x = b$. Let $d(x)$ be given $a \leqslant x \leqslant b$, either analytically or numerically. Form the *cumulative distribution function*,

$$D(x) \equiv \frac{\int_a^x d(x')dx'}{\int_a^b d(x')dx'} \tag{1}$$

where

$$D(a) = 0 \qquad D(b) = 1 \tag{2}$$

and

$$\frac{dD(x)}{dx} = \frac{d(x)}{\int_a^b d(x')\,dx'} \tag{3}$$

We see that equating $D(x_s)$ to a *uniform distribution* of numbers R_s, $0 < R_s < 1$ will produce the x_s corresponding to the distribution $d(x_s)$. (The reader may verify this by sketching $D(x)$ and $D'(x)$.) For example, let

$$d(x) = d_0(1 + \epsilon x) \tag{4}$$

so that

$$D(x) = \frac{(x - a) + \frac{\epsilon}{2}(x^2 - a^2)}{(b - a) + \frac{\epsilon}{2}(b^2 - a^2)} \tag{5}$$

Let R_s be ten numbers 0 to 1 ($s = 0$ to 9), like 0.05, 0.15, . . . , 0.95 (or, a random set); this places the first particle at x_1, which is obtained by solving $D(x_1) = 0.05$, the second at x_2, obtained by solving $D(x_2) = 0.15$, and so on. The simulator may choose to solve these quadratic equations (Problem 16-2a), or may integrate (1) numerically in fine steps (until 0.05 is reached, then 0.15, etc.) which must be done in examples where $d(x)$ is not integrable explicitly. A fast approximation to the integration method is used in ES1 subroutine INIT to create a quiet start velocity distribution, for any one-dimensional distribution; see Section 3-7. See also *Hockney and Eastwood* (1981, secs. 10-3-2 and 10-4-1).

PROBLEM

16-2a Suggest means for identifying and discarding the spurious root in (5), that is, $D(x_s) = R_s$. Show that solving this equation for $1/x_s$ leads to

$$x_s = a + \frac{R_s L (2 + \epsilon L)}{1 + \sqrt{1 + 2\epsilon L (2 + \epsilon L)}}$$

where $L = b - a$. This form avoids dividing by zero for $\epsilon = 0$ and is more accurate.

16-3 LOADING A COLD PLASMA OR COLD BEAM

Cold systems are simple; however, there are some pitfalls, so we start our design examples here. Use of linear weighting is assumed.

A cold uniform plasma may be put together in several ways. The obvious and easy way is to put the electrons in as particles, uniformly spaced, several to a cell, and the ions as a stationary uniform background. [See Section 5-4 for discussion of producing unwanted spikes in density, the Kaiser Wilhelm effect.] With linear or higher-order weighting, any small excitations in displacement (or velocity) excite plasma oscillations; with zero-order weighting, NGP, no field develops unless the excitation is large enough to drive an electron across a cell (or mid-cell) boundary, producing a field. A next step is to treat the ions as particles, which requires that q_i/m_i be given.

A cold beam, with all particles given the same drift speed v_0 (and most probably with an immobile neutralizing background), is numerically unstable in a periodic model. That is, as shown in Part Two, Section 8-12, the cold beam is heated non-physically due to aliasing until $\lambda_D/\Delta x$ reaches about 0.05 (for $\lambda_B/\Delta x \geqslant 0.3$) and then the system becomes stable, although noisy. Hence, one may choose to let the instability grow and eventually quench itself, or one may start off with a mild amount of velocity spread ($v_t \ll v_0$ but with $\lambda_D/\Delta x \geqslant 0.05$) and far less noise than if the growth had been allowed.

16-4 LOADING A MAXWELLIAN VELOCITY DISTRIBUTION

Maxwellian or Gaussian distributions in $\mathbf{v}$ are considered, of form $\exp(-v^2/2v_t^2)$. Such distributions occur in many parts of physics, usually as equilibrium distributions. Hence, we need methods for *inverting* Gaussians to positions or velocities. A useful general reference is Chapter 3 in *Hammersley and Handscomb* (1964).

A normalized thermal distribution is shown in Figure 16-4a. Most of the particles are in the region out to $v = 3v_t$ (99 percent, in fact, in $2v$) so that seldom do we need to place particles beyond 3 or $4v_t$.

Let us find the $(\mathbf{x}, \mathbf{v})_i$ to use with a spatially uniform plasma with isotropic Gaussian $f_0(\mathbf{v})$. The cumulative distribution function for the speed $v = |\mathbf{v}|$

$$R_s(0 \rightarrow 1) = F(\mathbf{v}) = \frac{\int_0^v \exp\left[-\frac{v^2}{2v_t^2}\right] d\mathbf{v}}{\int_0^\infty \exp\left[-\frac{v^2}{2v_t^2}\right] d\mathbf{v}} \tag{1}$$

is set equal to a set of uniformly distributed numbers R_s, varying from 0 to 1, in order to obtain the v's.

For a one-dimensional thermal distribution, the integral over $f(v)$ cannot be done explicitly, but is done numerically, as in INIT in ES1, to produce a "quiet Maxwellian," with thermal velocity v_{t2}; see Section 3-6.

For a two-dimensional isotropic thermal distribution (which is 2v, involving v_x, v_y, or speed $v = (v_x^2 + v_y^2)^{1/2}$ and angle $\theta = \arctan v_y/v_x$); d$\mathbf{v}$ is

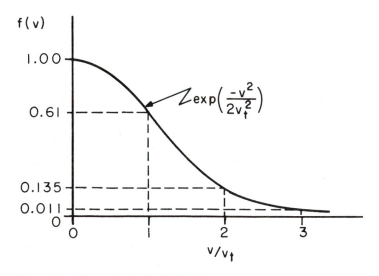

Figure 16-4a Normalized thermal velocity distribution.

$2\pi v \, dv$ so that these integrals can be done explicitly. The inversion for speeds v obtained in terms of R_s gives (new $R_s = 1 - $ old R_s)

$$v_s = v_t \sqrt{-2 \ln R_s} \tag{2}$$

Another set of uniform numbers R_θ is chosen for the θ's, over the range 0 to 2π, $\theta_\theta = 2\pi R_\theta$. An alternative is to use the method of Box and Muller (*Hammersley and Handscomb*, 1964, p. 39), multiplying $\sqrt{-2 \ln R_s}$ by cosine and sine of argument $(2\pi R_\theta)$ to produce normal deviates in independent pairs, that is (for our purposes) Gaussians in v_x and v_y. The cosine and sine lookups may be replaced by the technique of von Neumann (see *Hammersley and Handscomb*, 1964, p. 37), using uniform number sets R_1, R_2,

$$\cos\theta = \frac{R_1^2 - R_2^2}{R_1^2 + R_2^2} \qquad \sin\theta = \pm \frac{2R_1 R_2}{R_1^2 + R_2^2} \tag{3}$$

to generate these for uniformly distributed angles θ; pairs with $R_1^2 + R_2^2 > 1$ are rejected; the sign of $\sin\theta$ is chosen at random, or use R_1 in the interval $(-1, 1)$.

One sample is

$$v_{s,\theta} = v_t \sqrt{-2 \ln R_s} \, \cos 2\pi R_\theta \tag{4a}$$

and a second is, with the same R_s, R_θ

$$v_{s,\theta} = v_t \sqrt{-2 \ln R_s} \, \sin 2\pi R_\theta \tag{4b}$$

Another solution is to work directly with a uniform set of random numbers, $R_1, R_2, \ldots, R_M$, between 0 and 1, generating a random normal (Maxwellian, Gaussian) distribution in

$$v_M = v_t \left[\sum_{i=1}^{M} R_i - \frac{M}{2} \right] \left[\frac{M}{12} \right]^{-1/2} \tag{5}$$

as found in *Abramowitz and Stegun* (1964, pp. 952-953) or *Hammersley and Handscomb*, (1964, p. 39). This method is also used in INIT in ES1, to produce a *random Maxwellian*, with $M = 12$ ($v < v_{\max} = v_t\sqrt{3M} = 6v_t$), with thermal velocity v_{t1}. A check was made on this method by Neil Maron, (LLNL, unpublished) obtaining N velocities with N equal to 200, 400, ..., 102400 as shown in Figure 16-4b; note that the simple check on $\langle v \rangle$ (which should be zero, but is more like $30/N$), is not outstanding; in the same check $\langle v^2 \rangle$ and $\langle (v - \langle v \rangle)^2 \rangle = \langle v^2 \rangle - \langle v \rangle^2$ varied a few percent from unity for all N. The calculated $v_{\max} = v_t\sqrt{3M}$ ($= 4.24 v_t$ for $M = 6$) is quite improbable and the $v_{\max}$ observed was substantially smaller. Note that $f(v)$ will decay faster than a Gaussian at large v/v_t, decreed by having $v_{\max}/v_t \lesssim 3$ or 4.

Let the particle *positions* which go with the velocities as found above, be chosen to produce a plasma which is uniform in **x** space. If the $\mathbf{v}_i$ chosen are in any sense ordered, then their corresponding $\mathbf{x}_i$ are to be scrambled. One solution is to choose the positions in 1d, $0 < x_i < L$, that go with the 1v

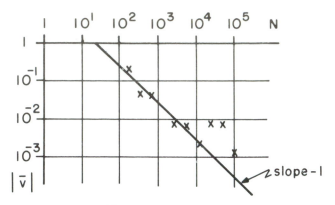

Figure 16-4b The average velocity $\langle v \rangle$ obtained using the random number routine of (5) for $M = 6$, done N times (*i.e.*, for N particles). The desired value is zero.

speeds in (4) or (5), be LR_x where R_x is a random set; alternatively, the R_x may be a *bit-reversed* set as described in the next section on *quiet starts*.

Obviously, it is desirable that $R_s, R_\theta,$ and R_x be uncorrelated. Using random number generators for the R's tends to produce unwanted bunching (*i.e.*, correlations) in x, v_x, v_y space.

The methods described above are rather crude and may be modified in use. Our methods do not arrange particles in **x** and **v** so as to produce Debye shielding, which means that the program will run a while in order to produce the shielding, (about $\tau_p/4$ to $\tau_p/2$). One fine tuning to be used after employing random R's is to correct the lower order moments of the distribution. *Morse and Nielson* (1971) take the particles in each cell and give them an increment to bring their total momentum to what is desired, which for a Maxwellian is zero. *Gitomer* (1971) goes one step further, making sure that the second moment is also correct, using averages over a few cells. Even when these and other modifications have been put in, the simulation will have a fluctuation level $\langle E^2 \rangle / nmv_t^2 \propto (N_C + N_D)^{-1}$ which will generally be much larger than that in a laboratory plasma because $(N_C + N_D)_{\text{simulation}} << N_{D,\text{laboratory}}$ (N_C is number of particles per cell). When weak effects are to be observed, including the low-level linear beginnings of instabilities, other solutions must be used. One such solution is the *quiet start,* as described in the next section.

PROBLEMS

16-4a Show that, if the integral in the numerator of (1) is from v to ∞, that (2) changes but the distribution of velocities is the same.

16-4b Show in detail how to apply (1) in order to obtain v_x, v_y, and v_z values which are uncorrelated. There are three integrals and (perhaps) three sets of R's.

16-5 QUIET STARTS: SMOOTH LOADING IN x-v SPACE: USE OF MIXED-RADIX DIGIT-REVERSED NUMBER SETS

As noted in the previous section, using the usual uniform random number sets to place $(\mathbf{x}, \mathbf{v})_i$ tends to produce computer plasmas with fully developed fluctuation levels, which are larger than laboratory levels; this level of noise may be acceptable for some modeling, but may prevent observing low level physics, which may be desired. In addition, while the desired density variations may be reproduced on a gross scale, the moments of the densities may be poor on a fine scale. For example, over system length L, a Gaussian $f_0(v)$ may be well given, but down to any $L/20$, even the first several moments of $f_0(v)$ may vary considerably from the prescribed values, when using uniform random sets. Hence, it is desirable to improve on the use of uniform random sets, which is done in several ways. The key is loading $\mathbf{x}, \mathbf{v}$ phase space as smoothly as possible. The method is called *quiet starts* and is attributed to J. A. Byers (*Byers and Grewal, 1970*) whose method used multiple beams, discussed in the next section.

A 1d1v "quiet Maxwellian" is mentioned in the previous section and in Section 3-6 as used in subroutine INIT in ES1. Let us consider the detailed steps. The distribution is integrated numerically in small steps to produce $F_0(v)$ which is set equal to $(i+\frac{1}{2})/N$, $i = 0, 1, \ldots, N-1$ to produce the desired function $f_0(v)$. This process is setting $R_s = (i+\frac{1}{2})/N$ in 16-4(1) and inverted to produce the desired v_i. This inversion produces an ordered sequence of v_i's. If the x_i's are chosen similarly, $x_i = [(i+\frac{1}{2})/N]L$, then the full Gaussian extends over L, but is not Gaussian locally. Hence, the x_i's need to be scrambled in order to produce nearly the Gaussian down to, say, $L/20$. That is, we seek a scrambling set, more uniform than the usual random numbers, especially one that works well for small to moderate N. To date, we have had good success with *bit-reversed numbers* (see *Hammersley and Handscomb,* 1964, p. 33), which is now described.

Let there be N particles to be distributed, numbered $i = 0$ to $N-1$, with a 2v isotropic Maxwellian in $v_r \equiv |v|$, and uniform in x. The v_r are obtained from 16-4(1), with R_i set equal to $[1-(i+\frac{1}{2})/N]$, producing 16-4(2) as

$$v_{ri} = v_t \left[-2 \ln \frac{i + \frac{1}{2}}{N} \right]^{\frac{1}{2}} \tag{1}$$

The x_i and θ_i are obtained from

$$x_i = L_x R_{x,i} \qquad 0 \leqslant R_{x,i} < 1 \tag{2}$$

$$\theta_i = 2\pi R_{\theta,i} \qquad 0 \leqslant R_{\theta,i} < 1 \tag{3}$$

where the R's mix the particles in x and θ. Following *Hammersley and Handscomb* (1964), we let $R_{x,i}$ be the result of radix-two digit reversal *bit reversing*, plus zero; the base-two fraction is obtained by mirroring the base-

Table 16-5a Base two bit-reversed fractions

		$R_{x,i}$	
decimal	base-2	base-2 fraction	decimal
0	0	0.0	0
1	1	0.1	$1/2 = 0.5$
2	10	0.01	$1/4 = 0.25$
3	11	0.11	$3/4 = 0.75$
4	100	0.001	$1/8 = 0.125$
5	101	0.101	$5/8 = 0.625$

Table 16-5b Base three bit-reversed fractions

		$R_{\theta,}$	
decimal	base-3	base-3 fraction	decimal
0	0	0.0	0
1	1	0.1	$1/3 = 0.3333$
2	2	0.2	$2/3 = 0.6666$
3	10	0.01	$1/9 = 0.1111$
4	11	0.11	$4/9 = 0.4444$
5	12	0.21	$7/9 = 0.7777$

two index and making it a fraction, as in Table 16-5a. The velocity angles are mixed using radix-three digit reversal, *i.e., trinary reversing*, as in Table 16-5b. The choice of N is open, with $N = 2^a 3^b$ suggested, meaning that one or the other last subsequence is incomplete, with, we think, little harm.

Tests of random, bit-reversed, and Fibonacci numbers for uniformity of scrambling were done by H. S. Au-Yeung, Y.-J. Chen, and C. K. Birdsall (private communications, 1980, 1981). The inability to fill x,y space uniformly with N points with x_i chosen uniformly and y_i chosen by scrambling, was measured by a method suggested by J. Wick (private communication, 1979); random scrambling gave a measure $\propto 1/\sqrt{N}$; the other two gave a far better measure, $3/N$. They also tested for the ability to produce the first three moments of a Gaussian $f(v)$ down to small regions in x, again finding that bit-reversed and Fibonacci numbers were much better than random. *Denavit and Walsh* (1981) compare these three analytically, with graphs, and in simulations; they also provide additional references.

16-6 QUIET START: MULTIPLE-BEAM AND RING INSTABILITIES AND SATURATION; RECURRENCES

An example is to set up a 1d1v Maxwellian distribution using M beams each with N particles ($M, N \gg 1$) uniformly spaced δv apart ($v = 0$, $\pm \delta v$, $\pm 2\delta v$, . . . , $\sim 3v_t$) with both charge q and mass m diminishing as

$$q(v) = q(n) = q(0) \exp\left(-\frac{v^2}{2v_t^2}\right) = q(0) \exp(-\alpha n^2) \qquad (1)$$

q/m is kept the same for all particles. For a uniform spatial distribution, the phase space appears as shown in Figure 16-6a. This system is ordered, with no fluctuations (at least none at low frequencies and long wavelengths). Hence, one may excite waves well below what would have been the fluctuation level had a random loading been used. For example, one may observe Landau damping decay over many decades (see *Byers,* 1970, fig. 1), which is almost impossible to do using random loading without use of an immense number of particles.

Although attractive, there are several problems that may restrict the use of this kind of loading. Suppose that a particular wavelength λ is purposely excited at $t = 0$ to some small amplitude. There is a recurrence time $\tau_r = \lambda/\delta v$ when the $(n + 1)^{th}$ beam will have slipped past the nth beam by $\lambda = 2\pi/k$ so that the initial system will have been reconstituted. At τ_r a jump in E^2 to a level larger than E^2 at $t = 0$ was observed by *Byers* (1970, mentioned on p. 500) followed by further decay and jumps. The jumps are shown in Figure 16-6b, from *Denavit* (1980). The growth observed after τ_r is exponential, toward the thermal fluctuation level which would have occurred with a fully random start, saturating near the thermal level when the beam velocity widths reach δv. This growth is physical, due to the many-beam instability (discussed below) growing at many harmonics of the initial k (if restricted to the initial k, the bounce peaks grow at the calculated rate; private communication from Denavit, 1980). The recurrence time τ_r may be made large by making λ large and δv small. Suppose that we wish to view 10 cycles of oscillation ($\omega_{real} \approx \omega_p$) and the accompanying decay ($\omega_{imag} = \gamma_{Landau}$) out to $t = t_1$; let there be $N = 60$ beams between $\pm 3v_t$ (so that $v_t/\delta v = 10$); this requires that $\lambda > 2\pi\lambda_D$ or $k\lambda_D < 1$, which is the

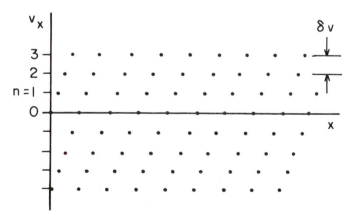

Figure 16-6a Ordered phase space, the trade-mark of the quiet start. For a Gaussian velocity distribution, q and m fall off as $\exp(-v^2/2v_t^2)$, with q/m the same for all particles.

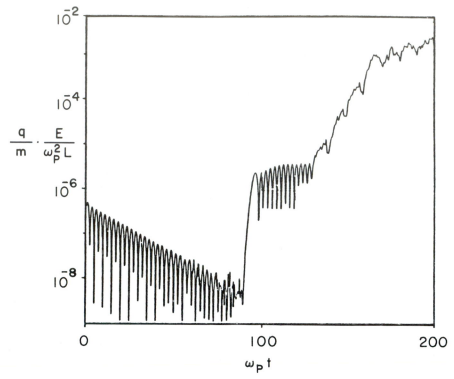

Figure 16-6b Evolution of the electric field for Landau damped electron oscillations with discrete beams having initial uniform velocity intervals, $\delta v = v_t/6$. The oscillation is initially excited at $k\lambda_D = 0.377$. The estimated recurrence time is $\tau_r = 2\pi/k\,\delta v = 100/\omega_p$, very close to that observed. (L is the system length.) *(From Denavit, 1980.)*

range of interest.

A second problem is that almost any system of beams is physically unstable due to the interactions among the beams. We already are well acquainted with a two-stream instability; the same kind of mechanism works for many streams, so that starting from round-off excitation, the whole system will go unstable. This result is predicted by *Dawson* (1960) and observed and identified by many; see *Denavit* (1972 and 1974).

Applications of the quiet start to Maxwellian distributions (always considered stable!) are especially neatly shown in 1d1v simulations by *Gitomer and Adam* (1976). For equally spaced beams the largest activity occurs among the beams near $v = 0$ (these beams have the most charge) as shown in Figure 16-6c. The field energy grows exponentially as shown in Figure 16-6d, with a low growth rate, $\gamma \approx 0.05\omega_p$ The growth saturates at a level of $\epsilon_0\langle E^2\rangle/nmv_t^2) \approx 0.0005$, which is below the level of 0.003 obtained with a fully random Maxwellian velocity initialization, corresponding to the theoretical thermal fluctuation level $\propto 1/N_D$. The start is indeed quiet, roughly 20 orders of magnitude below thermal, 200 dB down. Thus, much as with the

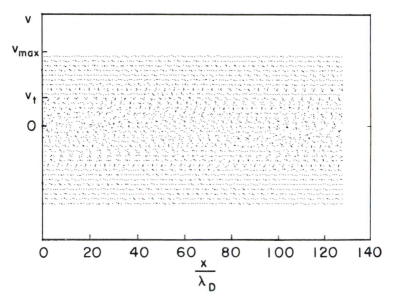

Figure 16-6c Phase-space plot after growth to saturation $t = 74\tau_e$ for 128 beams equally spaced in velocity at $t = 0$ with q and m of each beam weighted to produce a Maxwellian envelope. 16,384 total particles, with every fourth particle plotted. Time step $\omega_p \Delta t = 0.25$; cell size $\Delta x = \lambda_D$; plasma length $L = 128\lambda_D$; periodic system. Beam spacing $\delta v = 0.0419 v_t$, $v_{max} = \pm 2.66 v_t$. Initial spacing in x was uniform. *(From Gitomer and Adam, 1976.)*

nonphysical instability met earlier in the cold beam model, or the thermal model with $\lambda_D/\Delta x$ too small, the physical instability grows and then stabilizes near some form of thermal equilibrium but possibly with larger or smaller than thermal fluctuations.

A second modeling is to use many equally weighted beams unevenly spaced in v so as to produce a Gaussian distribution. The q and m are the same for all particles. The beam spacings are to be obtained from inverting the cumulative distribution function with the smallest spacing at $v = 0$ and largest spacing at the last v (say, $3v_t$). The jumps seen earlier are not expected here because the beam velocity spacing is uneven. However, the many-stream physical instability still occurs. This also is simulated by *Gitomer and Adam* (1976). Here the major interaction is among those beams with the largest velocity separation, on the tail of distribution, as shown in Figure 16-6e. There is exponential growth in field energy at small growth rate and saturation level of 0.002, slightly below the thermal level of 0.003 noted earlier, as shown in Figure 16-6f.

In application, one may choose either many-beam model. If interactions on the tail are of interest with $v_{phase} \gg v_t$, then the first model with uniform δv is recommended. If interactions having $v_{phase} \ll v_t$ are of interest, then the unevenly spaced model is recommended. Such simulations usually are successful if they have instabilities which have growth rates

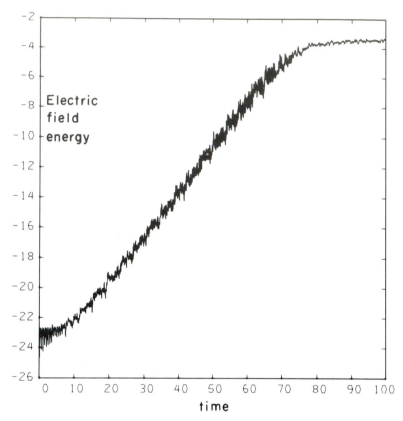

Figure 16-6d Logarithm of electric field energy (normalized by the initial total energy) versus time for equally spaced beams at $t = 0$, as in Figure 16-6c. *(From Gitomer and Adam, 1976.)*

$\gamma > \gamma_{\text{many beams}}$ and also have Reω substantially different from that of the many-beam instabilities; the latter quality allows separation of effects using Fourier analysis in time. If there is concern over unequal q's and m's as used in the uniform beam model, then the uniform q, m model is recommended.

The many-beam growth may be viewed as growth between pairs of streams; the largest growth rate occurs between the pair near $v = 0$ for equal spacing (see Figure 16-6c), and between the fastest pair for equal weighting (see Figure 16-6f). Hence, one might choose to add a thermal spread to either of these pairs (or even to all pairs) in order to prevent their growth. *Denavit and Kruer* (1971, 1980) using 100 equally weighted beams to simulate a warm two-stream instability found that, by staggering in velocity the fastest two beams, the unwanted spurious many-beam oscillations occurred later in time (than with no staggering) when presumably the next pair, with smaller δv spacing and smaller growth rate, caused the spurious oscillations. Using 1600 beams with no staggering for the same model, they found no

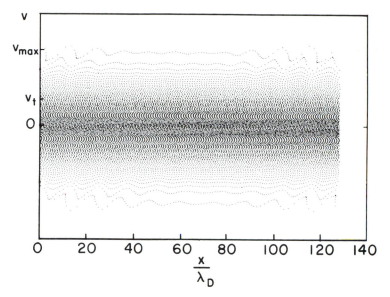

Figure 16-6e Phase-space plot after growth to saturation, $t = 52\tau_e$ for 128 equally weighted beams, unequally spaced in v at $t = 0$ to produce a Maxwellian envelope. Other parameters same as in Figure 16-6c. *(From Gitomer and Adam, 1976).*

spurious oscillations during their runs. Y. J. Chen (private communication, 1980) repeated this for equally weighted beams, Maxwellian distribution and found that warming the last stream merely started the field energy off at a slightly higher level (than in Figure 16-6e) and that saturation simply occurred a little earlier in time, with similar results for adding a $v_t \approx \delta v$ to all beams. She also found growth of the fastest pair occurred even with only one particle per beam in a periodic model; however, with 16,384 particles, her growth rate was $(128/16,384)^{1/2} \approx 1/10$ that of Gitomer and Adam, Figure 16-6e. When she added excitation at finite amplitude, she also observed Landau damping followed by jumps, but smaller jumps than in Figure 16-6b.

Denavit (1972) presented an approach combining particle and distribution function methods, with reconstruction or smoothing every N_s time steps. *Matsuda and Crawford* (1975) showed that this method prevented the many-beam instability (with equally spaced beams) for smoothing with $N_s \leqslant 16$ steps (see their fig. 2), with excellent energy spectra and Landau damping up to $k\lambda_D \approx 0.5$; they commented that this smoothing alone does not prevent the jump recurrence (Figure 16-6b), but that such did not pose problems for their small or large amplitude simulations.

The quiet start idea may be applied to other distributions. In simulating the Dory-Guest-Harris instability in 2d2v, C. K. Birdsall and D. Fuss (private communication, 1969) set up a distribution of magnetized ions in a cold ring in velocity space using velocities chosen from random numbers; they observed growths $\sim t^{1/2}$, which implies growth due to stochastic processes.

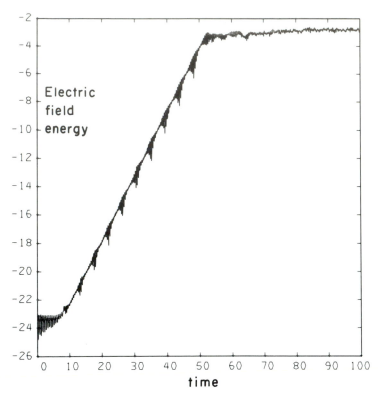

Figure 16-6f Field-energy growth for Maxwellian with equally weighted beams, as in Figure 16-6c. *(From Gitomer and Adam, 1976).*

They then copied the quiet start using wholly ordered ring distributions used in 1d2v by *Byers and Grewal* (1970), with 64 equally-spaced spokes to a ring in each λ. Their simulation succeeded immediately, being able to produce the expected exponential wave growths of many modes over a tremendous range in potential energy (10^{24}, 240 dB), allowing measurement of complex ω to within 1 percent or better of theory (smaller growth rates were observed when noisy) and to observing larger saturation values $(E^2)_{max}$ than with random and noisy starts.

Let us consider again loading a Gaussian velocity distribution in two velocity coordinates, v_x and v_y, which is isotropic, independent of $\theta \equiv$ arctan (v_y/v_x). Our choice here is to load uniformly in angle θ on evenly spaced *spokes* and on *rings* in v_x, v_y space as shown in Figure 16-6g. The radii of the rings are obtained from a set or uniform numbers R_s using the cumulative distribution function 16-4(1) or 16-5(1) and the spoke angles are chosen uniformly over 2π. The many-ring and spoke loading of a *magnetized Maxwellian plasma* in 2v has been analyzed for stability by J.-S. Kim, N. Otani, and B. I. Cohen (private communication, 1980). They derived and solved the dispersion relation for electrostatic flute modes ($\mathbf{k} \cdot \mathbf{B} = 0$) for a

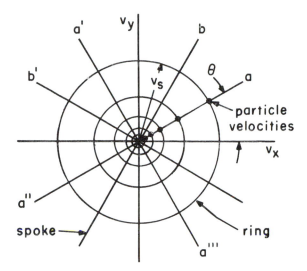

Figure 16-6g Gaussian spaced rings and uniformly spaced spokes for forming a two-dimensional (in $2v$) Maxwellian velocity distribution. Particles placed on a may also be placed on a'' to ensure zero net velocity, or on a, a', a'', and a'''.

Maxwellian velocity distribution made from $N_{v\perp}$ rings and $N_{v\theta}$ spokes (equally spaced gyrophase angles, $\theta = \arctan v_y/v_x$), that is, $N = N_{v\perp}N_{v\theta}$ particles. In contrast to the multibeam unmagnetized Maxwellian model (which is always unstable), the multi-ring-spoke magnetized Maxwellian can be made stable by using a large number of rings and spokes. They found that the effect of the spokes becomes negligible in the dispersion relation for

$$N_{v\theta} > 2 \left| \frac{kv_\perp}{\omega_c} \right|_{\max} \approx 6 \left| \frac{k_{\max}v_t}{\omega_c} \right| \qquad (2)$$

where $v_{\perp\,\max}$ is taken to be $3v_t$ and $k_{\max}$ is the last $k_\perp$ kept. With (2) satisfied, the velocity distribution may be treated as $f(v_\perp)$, rings only, for which Kim and Otani determined the stability boundary, which is roughly

$$N_{v_\perp} > \frac{1}{3} \left| \frac{\omega_p}{\omega_c} \right|^2 \qquad (3)$$

Their parameters at instability threshold are roughly

$$\frac{\omega}{k_\perp v_t} \approx \frac{1}{3} \qquad v_{\text{phase}} \approx \frac{v_t}{3} \qquad \frac{k_\perp v_\perp}{\omega_c} \approx \frac{2}{3} \left| \frac{\omega_p}{\omega_c} \right|^2 \qquad \frac{\omega}{\omega_c} \approx \frac{2}{9} \left| \frac{\omega_c}{\omega_p} \right|^2 \qquad (4)$$

In applying the above to two species, when studying electron Bernstein normal modes with particle electrons, use $(\omega_p/\omega_c)^2$ as that of the electrons; when studying the ion Bernstein modes, take $(\omega_p/\omega_c)^2$ to be $(\omega_{pi}/\omega_{ci})^2$ $/[1 + (\omega_{pe}/\omega_{ce})^2]$. Cohen found, by meeting the stability conditions with sufficient numbers of spokes and rings $[N > 3(k_{\max}v_t/\omega_c)\,(\omega_p/\omega_c)^2]$, and with either spatial replication or scrambling, that a stable quiet start was achieved; this is very desirable for studying the drift cyclotron instability. Note that N_{min} is not at all small for typical magnetic fusion problems $[k_{max} v_t / \omega_c \approx 30, (\omega_p / \omega_c)^2 \approx 1000]$.

The bottom line, however, is that the mixed-radix digit-reversed approach of the previous section, with one particle per beam or per ring, and one spoke per ring, works very well.

16-7 LOADING A MAGNETIZED PLASMA WITH A GIVEN GUIDING CENTER SPATIAL DISTRIBUTION $n_0(x_{gc})$

This section is another example of how one may proceed to load a non-uniform plasma, and is primarily a set of reminders.

Suppose that we wish to load a guiding center distribution $n_{gc}(x_{gc})$ as shown in Figure 16-7a(a). Let the plasma be uniform in y, allowing loading (of gc's) in rows, as shown in Figure 16-7a(b). Then, at each guiding center (or over some group of gc's), we need to put in the desired velocity distribution. That is, for ions, with $f(\mathbf{v}_\perp) = f(v_\perp, \theta)$, the particle $\mathbf{x}$ and $\mathbf{v}$ are given by

$$x = x_{gc} - a_i \sin \theta \tag{1}$$

$$y = y_{gc} + a_i \cos \theta \tag{2}$$

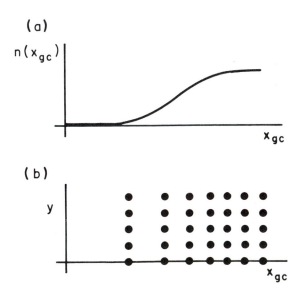

Figure 16-7a (a) Guiding center density desired. (b) Location of guiding centers to produce (a) with uniformity in y.

$$v_x = v_\perp \cos\theta \qquad (3)$$

$$v_y = v_\perp \sin\theta \qquad (4)$$

as shown in Figure 16-7b. a_i is the ion gyroradius $|v_\perp/\omega_{ci}|$ where $\omega_{ci} = (q/m)_i B_{0z}$. For electrons, replace a_i with $-a_e$.

As $n_{\text{particle}}(\mathbf{x}) \ne n_{\text{gc}}(\mathbf{x}_{\text{gc}})$, we need the relation between these two densities, as theory may provide *particle n* and we wish to load n_{gc}. Let the ions have a Maxwellian velocity distribution, with an ion *guiding center* density as follows:

$$n_{\text{gc}}(x_{\text{gc}}) = 1 + \Delta_i \cos k_0 x_{\text{gc}} \qquad (\Delta_i \text{ is a } chosen \text{ factor}) \qquad (5)$$

where
$$x_{\text{gc}} = x + \frac{v_y}{\omega_{ci}} \qquad (6)$$

Then, ignoring the v_z dependence, the ion particle density is

$$n_i(x) = \int f(x, \mathbf{v}) \, d\mathbf{v} = \int F(\mathbf{v}) \, n_{\text{gc}}(x_{\text{gc}}) \, d\mathbf{v}$$

$$= \int F_\perp(v_\perp) \, n_{\text{gc}}\left(x + \frac{v_y}{\omega_{ci}}\right) dv_x \, dv_y \qquad (7)$$

Electrons are taken to be cold and, requiring charge neutrality, we set their particle density equal to that of the ions

$$n_i(x) = n_e(x) = 1 + \Delta_e \cos k_0 x \qquad (8)$$

and find that this requires

$$\Delta_e = \Delta_i \exp\left[-\frac{k_0^2 v_{ti}^2}{2\omega_{ci}^2}\right] \qquad (9)$$

The instructions are to load the ion guiding centers using (5) and the electrons according to (8), with Δ_e obtained from Δ_i using (9). For $\Delta_i = -1$, n_{gc} and n are as shown in in Figure 16-7c.

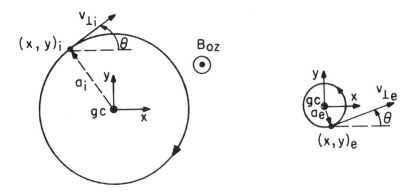

Figure 16-7b Ion and electron particle (x,y) and guiding center (gc) locations.

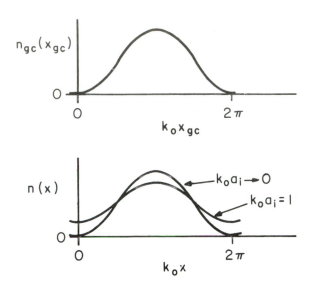

Figure 16-7c Guiding center density variation, $\Delta_i = -1$, and particle density, for Maxwellian ions.

Next, let the ion velocities be distributed in a cold ring, *i.e.*, use

$$F_\perp(v_\perp)\, 2\pi v_\perp\, dv_\perp = \frac{1}{2\pi v_{\perp 0}}\, \delta(v_\perp - v_{\perp 0})\, 2\pi v_\perp\, dv_\perp \tag{10}$$

Then we can write that

$$n_i = \int f(\mathbf{v})\, d\mathbf{v} = \int F_z(v_z)\, \frac{1}{2\pi v_{\perp 0}}\, \delta(v_\perp - v_{\perp 0})\, 2\pi v_\perp\, dv_\perp\, dv_z \tag{11}$$

which (invoking charge neutrality and using cold electrons) leads to

$$\Delta_e = \Delta_i \int_{-v_0}^{v_0} dv_y\, \frac{\exp(ik_0 v_y/\omega_{ci})}{\pi(v_{\perp 0}^2 - v_y^2)^{1/2}} = \Delta_i\, J_0\!\left(\frac{k_0 v_{\perp 0}}{\omega_{ci}}\right) \tag{12}$$

Again let $\Delta_i = -1$; the spatial distributions are shown in Figure 16-7d. Here, as $k_0 a_i$ increases past 2.405 (the first zero of J_0), the particle density becomes flat and then reverses slope.

Naitou et al. (1980) present similar results, allowing both density and temperature gradients, taking these in Fourier transform, which is like using a set of k_0's here. Their loading instructions (which they called "modified guiding center loading," MGCL) are the same as are given here. Their results show no unwanted potential due to charge separation.

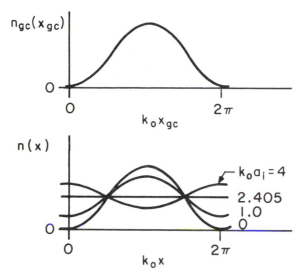

Figure 16-7d Same as Figure 16-7c, except that ions are distributed on a cold ring in velocity space. This allows $n(x)$ to reverse from $n_{gc}(x_{gc})$ for $k_0 a_i > 2.405$.

16-8 PARTICLE INJECTION AND ABSORPTION AT BOUNDARIES; FIELD EMISSION, IONIZATION, AND CHARGE EXCHANGE

It is often desirable to mimic a plasma device which has end walls that continually emit or absorb electrons or ions or neutrals. Or, it may be desirable to study a relatively small but active part of a larger but quiescent plasma, say, by matching the two parts at a common plane; let the quiet part emit a half-Maxwellian $f(v)$ there and allow some kind of return particle reflection also. Let us call these models *axially bounded plasmas*. Phase space for a 1d model is shown in Figure 16-8a.

The device might be a small *thermionic converter* with one hot wall emitting electrons and one cooler absorbing wall, or a *Q-machine* with one or two bounding hot plates emitting electrons and ions, with the plasma guided (transversely contained) by a strong axial magnetic field. The active/quiet matching might be used to study a part of a large *magnetic fusion device*, such as the end cell of a *mirror device*, matched to a center cell. Other examples include those from astrophysics (space-charge limited emission from neutron stars) and ionosphere physics (formation of double layers).

Particle injection may be done by several methods, usually similar to the techniques for initial particle loading. For example, at a wall we may be given the desired injection velocity distribution $f(v_x)dv_x$ or the flux distribution $v_x f(v_x)dv_x = \Gamma_x(v_x)$ particles in the element dv_x per unit time, where

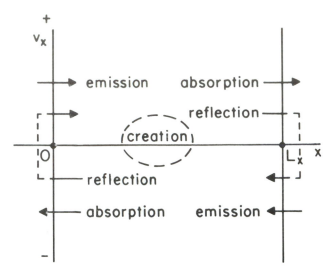

Figure 16-8a Phase-space for a 1d axially bounded plasma, showing some possible particle boundary conditions.

x is the direction normal to the wall. We then calculate the cumulative flux distribution function $F(v_x)$ by integrating $v_x f(v_x)dv_x$ over v_x, set this equal to a uniform set of numbers (random, bit-reversed, other), and solve for the v_x's (uncorrelated with their order, indices). The process is followed in Figure 16-8b. These velocities might be generated as needed or stored in an array (say, 1024 at a time) and used up as the particles are injected. The usual physical or laboratory parameters are flux or current densities (particles or Coulombs per second per unit area) which are translated into a computer parameter of the number of particles injected per time step. These particles are placed in the active simulation region at positions $x = R v_x \Delta t$, where $0 < R < 1$ is a new random number for each particle, tending to fill in the fan between $x = 0$ and $x = v_x \Delta t$. It should be clear that any method used to set up initial x-v_x loading of $f(v_x)dv_x$ can be adapted for particle injection, loading $v_x f(v_x)dv_x$. With injection, creation, and absorption, the program must have some particle index housekeeping, as the number of particles active in the system is not constant.

Particle absorption at the end walls may mean connection to external circuit elements, that is, conversion from plasma current to external circuit current. For example, as a plasma electron or ion passes into the end wall, it may be accumulated as part of the wall charge and be deleted from the list of active particles. The total wall charge also includes that due to the external circuit, as is discussed in the next section.

Numerous other options for particle handling at left and right plasma boundaries are in use. For example, in ZOHAR, incoming particles are controlled by option

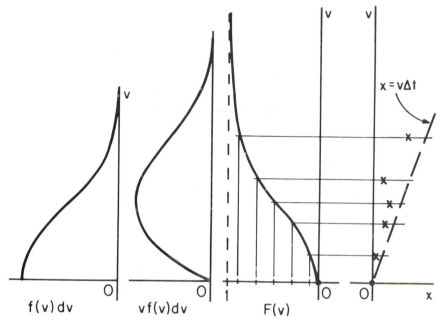

Figure 16-8b Loading *particle* and *flux* distribution functions at an emitting wall, $x = 0$.

$I = 0$ do not emit;
$I = 1$ emit;
$I = 2$ emit, if this reduces $|\langle\sigma_0\rangle|$, $|\langle\sigma_L\rangle|$.

Injection is with thermal and drift velocities, as described earlier. Exiting particles are controlled by option

$E = 0$ delete;
$E = 1$ conditionally delete or reflect specularly;
$E = 2$ conditionally delete or return with a new thermal velocity.

Conditional control is

$F = 0$ return particle if this will reduce $|\langle\sigma_0\rangle|$, $|\langle\sigma_L\rangle|$;
$F > 0$ return a fraction F, chosen randomly.

For example, to send all particles back in with new thermal velocities, use $E = 2$, $F = 1$; this is used, *e.g.*, for electrons with fixed ions. Or, at a boundary with plasma leaving at greater than ion thermal velocity, use $E = 0$ for ions, $E = 1$ or 2 for electrons ($E = 0$ for electrons draws plasma out quicker because because the boundary goes negative). Or, at a boundary with incoming plasma, use $I = 1$, $E = 0$ for ions; for electrons use $I = 2$, $E = 0$ to obtain only incoming thermal particles. In each case we keep a tally of particle energies and any other property of interest.

Ideally, *field emission* of particles with charge q occurs at a wall (say, $x = 0$) when qE_x at the surface is "sufficiently" positive. For space-charge limited flow, the normal electric field E_x is to be zero. However, E_x cannot be controlled in the field solve as can E_y etc.; rather, E_x is controlled *indirectly*, through the emission algorithm. In the simplest case, we model field emission of one species by emitting a particle of charge q at location y on the wall if $q\sigma(x = 0, y)$ where σ is the surface charge density, is positive and still will be positive after emission. We may give the particle a small inward velocity v_x; in order to generate a smooth charge distribution we then set $x = R\,v_x\Delta t$, where the R's are random numbers distributed on the interval $(0, 1)$. In one dimension, this condition is sufficient to specify the number of particles to be emitted for the current time step. Problems can arise when both electrons and ions are to be field emitted. For example, imagine there is a high-frequency component to E_x at the wall, due perhaps to an electromagnetic wave in the device. On alternate half-cycles, ions are emitted. If they are not pushed back into the wall during the other half-cycles, then ions will accumulate near the wall, along with net electron emission to neutralize them. As a result of this build-up of plasma density, eventually $\omega_{pe}\Delta t$ may become large enough to induce instability.

Ionization within the active simulation region may be done. The probability of an energetic electron creating an electron-ion pair is needed, along with the resulting electron-ion pair energy and energy loss of the original electron. There is no net change in charge. *Charge exchange* in the volume may be done, with a fast ion striking a neutral atom, leaving a fast neutral atom and a slow ion. The electrical process simply means a drop in ion velocity. The coding needs a probability. There is no change in charge. This process is important in neutral beam plasma studies. *Photo excitation and ionization* and other processes may be included, up to the ingenuity of the simulator.

16-9 PARTICLE AND FIELD BOUNDARY CONDITIONS FOR AXIALLY BOUNDED SYSTEMS; PLASMA DEVICES

In this section we treat in detail models with plane conducting boundaries at $x = 0, L_x$. The charge and field boundary conditions are first worked out electrostatically in 1d. The extension to 2d with periodicity in y is done in Section 15-12(b). The bounding planes may emit and absorb electrons and ions. The planes may be connected to external circuit elements, so it is shown how to solve the circuit equations along with those used for particle moving.

The model is shown in Figure 16-9a with principal axis along x, with grid index j; the normal coordinate is y, with grid index k.

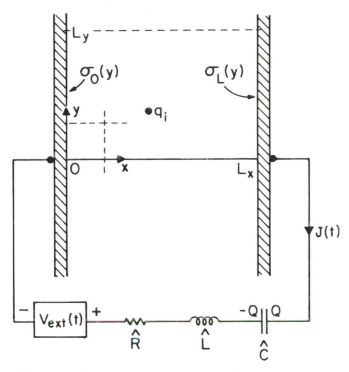

Figure 16-9a Axially bounded plasma model, with external circuit. The charges in the system are the plasma particles q_i, the left and right plane surface charges σ_0, σ_L, and the external capacitor charge Q.

(a) Charge and Field Boundary Conditions in 1d

The new physics here is associated with the surface charge densities at the bounding planes. These are related to the boundary fields as

$$\sigma(0) = \sigma_0 = E_0 \tag{1a}$$

$$\sigma(L_x) = \sigma_L = -E_{N_g} \tag{1b}$$

An algorithm for linear weighting for a particle in the last cell is indicated by the sketch in Figure 16-9b, showing the division between $\rho(L_x)$ and $\rho(L_x - \Delta x)$. Those charges which are moved past the boundary in a time step, $x_i > L_x$, are deposited into $\sigma(L_x)$ (which has units of $\rho \Delta x$) and are deleted from the active charge list. (Other first-cell, last-cell algorithms are possible.) The active charges weighted to the wall are used, for example, in obtaining E a half cell in from the wall, as

$$E_{1/2} = \sigma_0 + \rho_0 \Delta x = -\frac{\phi_1 - \phi_0}{\Delta x} \tag{2a}$$

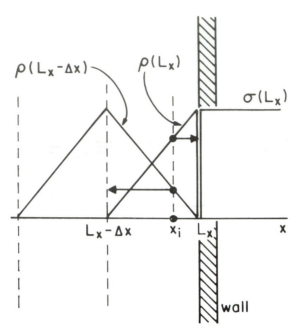

Figure 16-9b A particle at x_i is linearly weighted to neighboring grid planes as indicated by arrows; if the particle is moved past L_x, $x_i > L_x$, then the charge is assigned to $\sigma(L_x)$, the surface charge.

$$-E_{N_g-\frac{1}{2}} = \sigma_L + \rho_{L_x} \Delta x = \frac{\phi_{N_g} - \phi_{N_g-1}}{\Delta x} \tag{2b}$$

To check these results, let $x = 0$ to x be filled uniformly with ρ_{uniform}, making $\rho_0 = \rho_{\text{uniform}}/2$, as ρ_0 (and ρ_{Lx}) collects essentially half the charge in the first (last) cell. Then, analytically

$$E(x) = E_0 + \rho_{\text{uniform}} x$$

so that numerically,

$$E_{\frac{1}{2}} = \sigma_0 + \rho_{\text{uniform}} \frac{\Delta x}{2}$$

which is $\sigma_0 + \rho_0 \Delta x$, as in (2a). To be sure, ρ_0 could be $\frac{1}{2}\rho(0)$, accounting for the half-cell collecting, and can be so used in coding (ρ_0 is a designated *computer* variable, not to be confused with $\rho(0)$ which is *physical*).

Conservation of charge is guaranteed at the bounding plane by applying the integral form of Gauss' law over the pillbox shown in Figure 16-9c, as

$$J_p - J = \frac{\partial \sigma_L}{\partial t} \tag{3}$$

where J_p and J are plasma (or particle) and circuit conduction current densities, respectively. ($J_p - \partial \sigma_L/\partial t = J_p + \partial E_{Ng}/\partial t$ is the total device current,

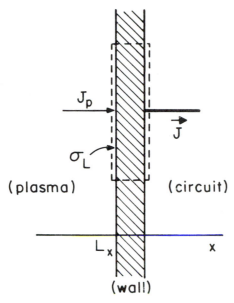

Figure 16-9c The plasma charge current density J_p and external circuit current (density) J differ by $\partial\sigma/\partial t$ at the conducting wall boundary.

equal exactly to J, the circuit current.) Charge deposited at $x = 0$ and $x = L$ by active charges lost from the plasma and by charge flow from the external circuit make up σ, which is needed to obtain the device potentials and fields.

The field equations for ρ, ϕ, E may be written in finite-difference forms, to be solved by common methods.

Similar considerations for 2d charge and field boundary conditions are given in Section 15-12(b).

Instabilities due to boundary conditions as reported by *Swift and Ambrosiano* (1981) do not appear in reasonable applications; their instability developed only when particles crossing a boundary were reintroduced at another boundary at which $q\phi$ is larger (for example, periodic particle boundary conditions with an aperiodic potential!). This is not a procedure one would arrive at from physical considerations, except when there is an external circuit which can add energy to the plasma region.

(b) Solutions with an External Circuit

The behavior of the active charges between the metal bounding walls is readily handled as with earlier models. Deposition of charge on the walls due to the external current is handled by Kirchhoff's circuit equations.

The art of modeling *electron devices in one dimension* may be found, for example in *Birdsall and Bridges* (1966), summarizing the techniques developed and applied successfully from the 1930's on. Implicit in such work are the assumptions:

(a) wholly electrostatic for 1d, with E_x only;
(b) motion only along x which is non-relativistic;
(c) large device (diameter/length) ratio with negligible field variations in y and z;
(d) negligible effects from fringe fields at the edges of real devices of finite diameter;
(e) currents J_x and $\partial E_x/\partial t$ exist, but the magnetic field generated by these currents does not affect the motion (negligible pinch effect, dc or ac);
(f) small enough device diameter so that wave propagation transverse to x is essentially instantaneous;
(g) the walls are equipotentials, with ignorable effects due to wall surface currents.

These assumptions are not all independent; they allow many interesting effects but also rule out much of current interest. Our electrostatic *1d plasma device model* here uses the same assumptions as in the electron device model above.

However, a *2d plasma device model* (with metal plates at $x = 0$, L_x and periodic in y with period L_y, no variations in z) has fewer assumptions (really, restrictions), especially in an electromagnetic version. With a better accounting of currents and magnetic fields, more interesting effects may occur (*e.g.*, propagation in y). A periodic model, wholly specified over $0 \leqslant x \leqslant L_x$, $0 \leqslant y \leqslant L_y$, need have no outside connection; the external (or return) current may flow across the device, uniformly in y, producing changes in the charges on the walls as if due to an $E_x(t)$ uniform in y but with no charge separation or magnetic field in the device. This choice is the same as adding (superposing, due to the linearity of Maxwell's equations) the effects of an external circuit which only imposes charge at the walls (in addition to those produced by the active charges). The total current density averaged over y is

$$\hat{J}_x(x,t) \equiv \frac{1}{L_y} \int_0^{L_y} dy \left[J_x(x,y,t) + \frac{\partial E_x(x,y,t)}{\partial t} \right] - J_{\text{external}} \qquad (4)$$

where the sign of J_{external} is chosen for convenience. This density also is

$$\hat{J}_x(x,t) = \frac{1}{L_y} \int_0^{L_y} dy \, (\nabla \times \mathbf{H})_x = \frac{1}{L_y} \int_0^{L_y} dy \, \frac{\partial H_z}{\partial y} \qquad (5)$$

As H_z is periodic in y, the integral is zero and $\hat{J}_x(x,t) = 0$. Then integrating the $\hat{J}_x(x,t)$ equation across the device in x produces

$$J_{\text{external}} = J_{\text{induced}} - \frac{1}{L_y} \frac{dV}{dt} \qquad (6)$$

where
$$J_{\text{induced}}(t) \equiv \frac{1}{L_x L_y} \int_0^{L_x} \int_0^{L_y} dx \, dy \, J_x(x,y,t) \tag{7}$$

a concept used in electron devices (p. 10 in *Birdsall and Bridges,* 1966). The device potential $V(t)$ in (6) is the line integral of E_x straight across the device and averaged over y,

$$V(t) = \frac{1}{L_y} \int_0^{L_y} dy \left[- \int_0^{L_x} dx \, E_x(x,y,t) \right] \tag{8}$$

taken as $V(t) = \phi(L_x,t) - \phi(0,t)$. Note that the second term in (6) is $-\hat{C}_v \, dV/dt$, $\hat{C}_v \equiv 1/L_x$ (where the carat means capacitance per unit area).

This 2d electromagnetic modeling may be used to make clearer the origin of the assumptions in the 1d model where J_{external} flows outside the device as in Figure 16-9a. Inside the device there is an H_z, but the radial extent (in y) of the 1d device is taken to be small, as implied in assumptions (a, c, d, e, f) so that the integral in (5) also vanishes and (6), (7), and (8) hold. Or, $(\nabla \times \mathbf{H}) \cdot d\mathbf{S}$ may be substituted as $(J_x + \partial E_x/\partial t) \, dy \, dz$ across the device (surface S_1) *plus* $(-J_{\text{external}}) \, dy \, dz$ across the external circuit (surface S_2); taking surfaces $S_1 + S_2$ as covering a *volume*, the total integral across $S_1 + S_2$ vanishes (as the integral of $\nabla \cdot \mathbf{J}_{\text{total}}$ vanishes), which is equivalent to setting $\hat{J}_x = 0$. Hence, (6) holds, and (7) and (8) do not need the averaging in y.

We are now ready to integrate the system forward in time. First, let us treat a simple external circuit, with a resistor, capacitor and voltage source (as in Figure 16-9a) with no inductor, $\hat{L} = 0$. The sequence followed is shown in Figure 16-9d. At time t_n, σ_0 and σ_L are the cumulative result of all charge transfers up to t_n, both of particles to and from the walls *and* of external currents. As the particles are moved to their new positions at time level $n+1$, particles that leave or enter advance the wall charges to

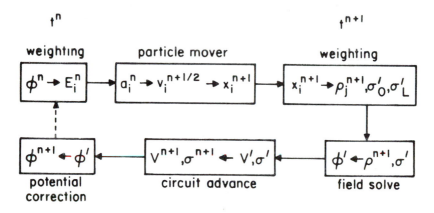

Figure 16-9d Flow diagram for RC external circuit. Primes indicate time intermediate to t_n and t_{n+1}.

provisional values σ_0', σ_L'; that is, at the end of the particle move and weighting, ρ, σ_0', σ_L' include all *particle* contributions up to time t_{n+1} but do not include the wall charge increments and fields due to external circuit current. The field solve at this time produces provisional potentials ϕ_j' and total device voltage V_{n+1}'. The lumped circuit Kirchhoff voltage equation is first order in capacitor charge Q

$$V = V_{\text{ext}} + \hat{R}\, \frac{dQ}{dt} + \frac{Q}{\hat{C}} \tag{9}$$

where $J_{\text{ext}} = dQ/dt$ has been used (both are densities). We use the linearity of the field equations in the device to allow superposition of the voltage V_{n+1}' (due to particles), and external charge flow in the interval t_n to t_{n+1} as

$$V_{n+1} = V_{n+1}' - \frac{Q_{n+1} - Q_n}{\hat{C}_v} \tag{10}$$

Equations (9) and (10) are used to advance Q_n to Q_{n+1} and to obtain V_{n+1}. The second term is like $\Delta Q = J_{\text{ext},n+\frac{1}{2}} \Delta t$. (10) is not time centered; and this both provides stability for all values of $\hat{R}\hat{C}$ decay times, and ensures that $V_{n+1} = V_{\text{ext},n+1}$ for $\hat{R} \to 0$, $\hat{C} \to \infty$. Hence, the circuit equation is written as

$$V_{n+1} = V_{n+1}' - \frac{Q_{n+1} - Q_n}{\hat{C}_v} = V_{\text{ext},n+1} + \hat{R}\, \frac{Q_{n+1} - Q_n}{\Delta t} + \frac{Q_{n+1}}{\hat{C}} \tag{11}$$

(with dQ/dt taken at time $n+\frac{1}{2}$ for stability) to be solved for Q_{n+1} or ΔQ;

$$\Delta Q = \left[V_{n+1}' - V_{\text{ext},n+1} - \frac{Q_n}{\hat{C}} \right] \left[\frac{\hat{R}}{\Delta t} + \frac{1}{\hat{C}_v} + \frac{1}{\hat{C}} \right]^{-1} \tag{12}$$

Note that shorting out $\hat{C}$ (meaning $\hat{C} \to \infty$) is no problem. This equation produces ΔQ and hence V_{n+1}, the correct device voltage. ΔQ is added to σ_0' and subtracted from σ_L' which now correspond to time t_{n+1}. The internal potentials are now updated to t_{n+1} by adding the uniform field produced by ΔQ, $\Delta E_x = -\Delta Q$

$$\phi_j = \phi_j' + \Delta Q\, X_j \tag{13}$$

The fields for the mover are obtained as usual by spatial differencing.

If the longitudinal part of E needs correction (in the EM model) then obtain the correction potential $\delta\phi$ as in Chapter 15. This step and the adjustment (13) can be combined.

Adding in an external inductance $\hat{L}$ requires changes in the sequence used, to that shown in Figure 16-9e. The circuit equations are now second order in Q

$$V_n = V_{\text{ext},n} + \hat{R}\, J_n + \hat{L} \left[\frac{dJ}{dt} \right]_n + \frac{Q_n}{\hat{C}} \tag{14}$$

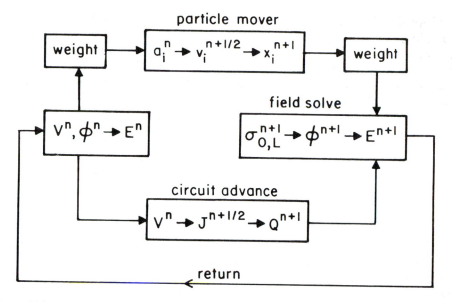

Figure 16-9e Flow diagram for *RLC* external circuit.

$$\left(\frac{dQ}{dt}\right)_{n+\frac{1}{2}} = J_{n+\frac{1}{2}} \tag{15}$$

This set is used to solve for $J_{n+\frac{1}{2}}$ and, hence Q_{n+1}. As a first approach, construct

$$J_n = \frac{J_{n+\frac{1}{2}} + J_{n-\frac{1}{2}}}{2} \tag{16}$$

and

$$\left(\frac{dJ}{dt}\right)_n = \frac{J_{n+\frac{1}{2}} - J_{n-\frac{1}{2}}}{\Delta t} \tag{17}$$

which is recognized as a leap-frog method, time-centered and second-order accurate. Then we obtain the updated current,

$$J_{n+\frac{1}{2}} = \left[V_n - V_{\text{ext},n} - \frac{Q_n}{\hat{C}} + J_{n-\frac{1}{2}}\left(-\hat{R} + \frac{\hat{L}}{\Delta t}\right)\right]\left[\frac{\hat{R}}{2} + \frac{\hat{L}}{\Delta t}\right]^{-1} \tag{18}$$

and the new capacitor charge,

$$Q_{n+1} = Q_n + J_{n+\frac{1}{2}}\Delta t \tag{19}$$

Next, ΔQ is added to σ_0' and subtracted from σ_L', the provisional charges on the walls due to the active charges. Lastly, the field solve is done, with the $(\rho, \sigma_0, \sigma_L)$ at time t_{n+1}. As done in a problem, this approach is stable for $\Delta t \ll \hat{L}/\hat{R}$ and $\hat{R}\hat{C}$ and $(\hat{L}\hat{C})^{\frac{1}{2}}$, hence unusable for $\hat{R}, \hat{L}$, or $\hat{C} \to 0$ or for $\hat{R} \to \infty$ (open circuit). Hence, we seek another method.

As a second approach, write the circuit equation (14) at time $n+1$ (changing Figure 16-9e appropriately) and use the superposition (10). Next, construct the second-order-accurate representations

$$J_{n+1} = \frac{3 J_{n+\frac{1}{2}} - J_{n-\frac{1}{2}}}{2} \tag{20}$$

and

$$\left(\frac{dJ}{dt}\right)_{n+1} = \frac{2 (J_{n+\frac{1}{2}} - J_{n-\frac{1}{2}})}{\Delta t} - \frac{J_{n-\frac{1}{2}} - J_{n-3/2}}{\Delta t} \tag{21}$$

Then we obtain

$$J_{n+\frac{1}{2}} = \left[V'_{n+1} - V_{\text{ext},n+1} - \frac{Q_n}{\hat{C}} + \frac{\hat{L}}{\Delta t} (3 J_{n-\frac{1}{2}} - J_{n-3/2}) + \frac{\hat{R}}{2} J_{n-\frac{1}{2}} \right] \tag{22}$$

$$\cdot \left[\frac{3}{2} \hat{R} + \frac{2\hat{L}}{\Delta t} + \Delta t \left(\frac{1}{\hat{C}_v} + \frac{1}{\hat{C}} \right) \right]^{-1}$$

which is then used to update Q_n to Q_{n+1}, followed by the field solve. This method is stable in several limits and hence, is expected to stable for all values of $\hat{R}, \hat{L}, \hat{C}$, and Δt (Problem 16-9a).

PROBLEMS

16-9a Study the stability of (14) and (15), using (20), (21), and (22). For example, in (22), using Q as the unknown and writing $Q_n = Q z^n$, $Q_{n+1} = Q z^{n+1}$, show that stability is determined by the zeros of

$P(z)$

$$= \left(\frac{1}{\hat{C}_v} + \frac{1}{\hat{C}} \right) z + \frac{\hat{R}}{\Delta t} \left[\frac{3}{2} (z - 1) - \frac{1}{2} \left(1 - \frac{1}{z} \right) \right] + \frac{\hat{L}}{\Delta t^2} \left[2 \left(z - 2 - \frac{1}{z} \right) - \left(1 - \frac{2}{z} - \frac{1}{z^2} \right) \right] \tag{23}$$

being inside the unit circle $|z| = 1$. This polynomial is cubic [not quadratic as expected from (14) and (15), which are second order in Q], producing an extraneous root when $\hat{L} \neq 0$, which is strongly damped.

16-9b Following Problem 16-9a, examine the stability of (14) and (15) using (16), (17), and (18).

FOUR

APPENDICES

Part Four consists of Appendices which complement the main text: the complex fast Fourier transform, compensating and attenuating, digital filtering, direct solution of Poisson's equation, and a discussion on differencing operators.

FAST FOURIER TRANSFORM SUBROUTINES

(a) Complex Periodic Discrete Fourier Transform

Our object is to compute the discrete Fourier transform (DFT)

$$\hat{x}(\hat{t}) = \sum_{t=0}^{N-1} x(t) \; e\left(\frac{t\,\hat{t}}{N}\right) \qquad \text{for } \hat{t} = 0, 1, \ldots, N-1 \qquad (1)$$

where $e(x) \equiv \exp(2\pi i x)$. Sequences x and $\hat{x}$ are complex-valued. Evaluating (1) straightforwardly requires N complex multiplies (6 arithmetic operations) and $N-1$ complex additions (2 arithmetic operations) for each $\hat{t}$, if we assume that the complex exponentials are already computed. Thus the complete sequence $\hat{x}$ requires $8N^2 + O(N)$ operations, when done in this straightforward fashion. Obvious improvements can be made by exploiting special values, symmetries, and periodicity of the exponential. With industry and wit you may be able to evaluate (1) as quickly as a good FFT program, but history says this is unlikely. And if you do, chances are you have reinvented part or all of the FFT algorithm.

We discuss the Sande-Tukey form of the FFT, rather than the Cooley-Tukey form, and follow the notation of *Gentleman and Sande* (1966). Also we restrict ourselves first to the simple and common case in which N is a power of 2.

We evaluate (1) differently for odd and even $\hat{t}$. For even $\hat{t} = 2\hat{b}$:

$$\hat{x}(2\hat{b}) = \sum_{t=0}^{N-1} x(t) \; e\left(\frac{t\,\hat{b}}{N/2}\right) \qquad b = 0, 1, \ldots, \frac{N}{2} - 1$$

$$= \sum_{t=0}^{N/2-1} [x(t) + x(t + N/2)] \; e\left(\frac{t\,\hat{b}}{N/2}\right) \qquad (2)$$

For odd $\hat{t} = 2\hat{b} + 1$:

$$\hat{x}(2\hat{b} + 1) = \sum_{t=0}^{N-1} x(t)\, e\left[\frac{2t}{N}\right] e\left[\frac{t\,\hat{b}}{N/2}\right]$$

$$= \sum_{t=0}^{N/2-1} e\left[\frac{2t}{N}\right][x(t) - x(t + N/2)]\; e\left[\frac{t\,\hat{b}}{N/2}\right] \qquad (3)$$

Note that (2) and (3) are each $N/2$ length DFT's. Because the work to evaluate a length N DFT is proportional to N^2, we have made a gain of about a factor of 2 (for large N) by exchanging $8N^2 + O(N)$ operations for:

(1) Formation of the two sequences in square brackets in (2) and (3), taking $O(N)$ operations, plus
(2) Two DFT's of length $N/2$, taking $2 \times 8(N/2)^2 + O(N)$ operations, or a total of $4N^2 + O(N)$ operations.

The group properties of the exponential which made this reduction possible are:

$$e(x + y) = e(x)\, e(y) \qquad (4)$$

$$e(x + 1) = e(x) \qquad (5)$$

Properties such as these are so rare that there are only a few transforms for which highly efficient algorithms are known.

The next key point is that we can gain another factor of two (nearly) by using the same trick on the two length $N/2$ DFT's in (2) and (3), which leaves four length $N/4$ transforms to be done. This reduction may be repeated $\log_2 N$ times in all, which ends with $N/2$ transforms of length 2 to be done. This is the principle of the fast Fourier transform algorithm. A more careful count shows that the number of operations is proportional to $N \log_2 N$. Also, the roundoff error is smaller than for the straightforward summations in (1). This description of evaluating a DFT in terms of shorter DFT's is an example of a *recursive* definition of an algorithm.

There are still some problems to be resolved before one can write a good FFT computer program. One is the generation of the complex exponentials in the square bracket in (3). A fast and easy way is to store a large table of sines and/or cosines. This ties up a lot of computer memory and is not necessary. Because of the simple order in which the exponentials' arguments arise during a Sande-type FFT, efficient and very compact methods are possible, although few authors have used them. *Singleton* (1967, 1969) generates them all by a multiple-angle recursion method which is designed to avoid roundoff buildup and which requires a short table of $\log_2 N_{\max}$ sines, where $N_{\max}$ is the greatest length DFT the program is asked to do.

Another problem is to minimize the amount of computer memory used during the calculation. Some programs use $2N$ memory cells, in addition to the storage needed for x and $\hat{x}$, and any sine-cosine tables. It is usually most convenient to have the results of each of the $\log_2 N$ stages stored in the same places as was x. We now show how this logistics problem is solved.

Table Aa shows the steps in a FFT of length 8, and shows what is in memory at any time. β_0 and β_4 are calculated from x_0 and x_4 and are stored back in the *same* locations; the others are similar. According to (2) and (3), we must next do length 4 DFT's on the sequences $(\beta_0, \beta_1, \beta_2, \beta_3)$ and $(\beta_4, \ldots, \beta_7)$. These are normally done in parallel, and also are done by splitting each into shorter DFT's of the 4 sequences (γ_0, γ_1), (γ_2, γ_3), etc. which are stored where the β's were. The DFT's of these sequences are (δ_0, δ_1), etc. The δ's are the elements of $\hat{x}$ but not in order. To put them in order, pairs of δ's must be exchanged; again no extra arrays are needed.

(b) Transform of Real-Valued Sequences, Two at a Time

In simulation, we usually transform real data, or inverse transform back to real data. If the sequence x is real-valued and is transformed using the complex FFT subroutine CPFT, then the array i is first filled with zeros. Similarly, there is redundance in the transformed r and i arrays due to the Hermitian symmetry

$$[\hat{x}(\hat{i})]^* = \hat{x}(-\hat{i}) = \hat{x}(N-\hat{i}) \tag{6}$$

which follows from (1). It is possible to FFT a real sequence using roughly half the storage and computation required by this straightforward application

Table Aa Sande-Tukey organization of the FFT.

start	1st level	2nd level	3rd level	sort
x_0	$\beta_0 = x_0 + x_4$	$\gamma_0 = \beta_0 + \beta_2$	$\delta_0 = \gamma_0 + \gamma_1$	$\hat{x}_0 = \delta_0$
x_1	$\beta_1 = x_1 + x_5$	$\gamma_1 = \beta_1 + \beta_3$	$\delta_1 = \gamma_0 - \gamma_1$	$\hat{x}_1 = \delta_4$
x_2	$\beta_2 = x_2 + x_6$	$\gamma_2 = \beta_0 - \beta_2$	$\delta_2 = \gamma_2 + \gamma_3$	$\hat{x}_2 = \delta_2$
x_3	$\beta_3 = x_3 + x_7$	$\gamma_3 = i(\beta_1 - \beta_3)$	$\delta_3 = \gamma_2 - \gamma_3$	$\hat{x}_3 = \delta_6$
x_4	$\beta_4 = x_0 - x_4$	$\gamma_4 = \beta_4 + \beta_6$	$\delta_4 = \gamma_4 + \gamma_5$	$\hat{x}_4 = \delta_1$
x_5	$\beta_5 = e(1/8)(x_1 - x_5)$	$\gamma_5 = \beta_5 + \beta_7$	$\delta_5 = \gamma_4 - \gamma_5$	$\hat{x}_5 = \delta_5$
x_6	$\beta_6 = e(2/8)(x_2 - x_6)$	$\gamma_6 = \beta_4 - \beta_6$	$\delta_6 = \gamma_6 + \gamma_7$	$\hat{x}_6 = \delta_3$
x_7	$\beta_7 = e(3/8)(x_3 - x_7)$	$\gamma_7 = i(\beta_5 - \beta_7)$	$\delta_7 = \gamma_6 - \gamma_7$	$\hat{x}_7 = \delta_7$

$$\beta(a + 2b + 4\hat{c}) = e\left[\frac{\hat{c}(a + 2b)}{8}\right] \sum_{c=0}^{1} e\left[\frac{c\,\hat{c}}{2}\right] x(a + 2b + 4c)$$

$$\gamma(a + 2\hat{b} + 4\hat{c}) = e\left[\frac{a\,\hat{b}}{4}\right] \sum_{b=0}^{1} e\left[\frac{b\,\hat{b}}{2}\right] \beta(a + 2b + 4\hat{c})$$

$$\delta(\hat{a} + 2\hat{b} + 4\hat{c}) = \sum_{a=0}^{1} e\left[\frac{a\,\hat{a}}{2}\right] \gamma(a + 2\hat{b} + 4\hat{c})$$

$$\hat{x}(\hat{i}) = \hat{x}(\hat{c} + 2\hat{b} + 4\hat{a}) = \delta(\hat{a} + 2\hat{b} + 4\hat{c})$$

$$\hat{a}, \hat{b}, \hat{c} = 0 \text{ or } 1.$$

of CPFT (*Cooley, Lewis, and Welch,* 1970) but the algorithm is much more complicated than the alternative in subroutine RPFT2 below. When transform time becomes significant, there are often *many* transforms to do, such as the columns or rows of a two-dimensional array, that can be done in pairs.

Let $z(t) = a(t) + i b(t)$, where a and b are real valued. Then the transforms $\hat{a}$ and $\hat{b}$ can be extracted from $\hat{z}$ using the relations

$$\hat{z}(\hat{t}) = \hat{a}(\hat{t}) + i \hat{b}(\hat{t}) \tag{7}$$

$$\hat{z}(N-\hat{t}) = [\hat{a}(\hat{t})]^* + i [\hat{b}(\hat{t})]^* \tag{8}$$

which follows from (6). This is implemented by RPFT2, which stores Re $\hat{a}$ (Im $\hat{a}$) in the lower (upper) part of the array used for a (Problem Ac). Subroutine RPFTI2 with CPFT implements the inverse.

(c) Sine Transform of Real-Valued Sequences, Two at a Time

The sine transform is

$$\hat{a}(\hat{t}) = \sum_{t=1}^{N-1} a(t) \sin \frac{\pi}{N} t \, \hat{t} \qquad \hat{t} = 0, 1, \ldots, N \tag{9}$$

where $a(0) = a(N) = 0$. This can be performed by extending the definition of $a(t)$ to length $2N$, using sine symmetry

$$a(2N-t) = a(-t) = -a(t) \tag{10}$$

and doing a periodic transform of length $2N$. Instead, we adapt an algorithm of *Cooley, Lewis, and Welch* (1970) to sine transform a pair of sequences a and b, using only one complex periodic transform of length N.

First, we rewrite (9) using (10) as

$$\hat{a}(\hat{t}) = \frac{1}{2} \sum_{t=0}^{2N-1} a(t) \sin \frac{\pi}{N} t \, \hat{t} \tag{11}$$

$$= \frac{1}{2} \sum_{t=0}^{N-1} a(2t) \sin \frac{2\pi}{N} t \, \hat{t}$$

$$+ \left[4 \sin \frac{\pi}{N} \hat{t} \right]^{-1} \sum_{t=0}^{N-1} [a(2t+1) - a(2t-1)] \cos \frac{2\pi}{N} t \, \hat{t} \tag{12}$$

(Problem Ad), which are *periodic* transforms. This result motivates defining the complex periodic sequence

$$z(t) = [a(2t+1) - a(2t-1) - i\,a(2t)] \tag{13}$$
$$+ i[b(2t+1) - b(2t-1) - i\,b(2t)]$$

for $t = 0, 1, \ldots, N-1$, which is transformed using CPFT. The sine transform $\hat{a}$ is extracted from Re $\hat{z}(\hat{t})$, e.g.,

$$4\,\hat{a}\,(\hat{t}) = \text{Re}\,\hat{z}\,(\hat{t}) - \text{Re}\,\hat{z}\,(N-\hat{t}) + \frac{\text{Re}\,\hat{z}\,(\hat{t}) + \text{Re}\,\hat{z}\,(N-\hat{t})}{2\sin\dfrac{\pi}{N}\,\hat{t}} \tag{14}$$

Similarly, $\hat{b}$ is extracted from $\text{Im}\,\hat{z}\,(\hat{t})$.

This algorithm has been used to implement a two dimensional transform of data periodic in y with sine symmetry in x for use on potentials between two plane conducting surfaces.

PROBLEMS

Aa Suppose N is a multiple of 3. Show that the DFT can be evaluated by three DFT's of length $N/3$, by steps similar to those leading to (2) and (3).

Ab It is often said that the δ's are the x's in "bit-reversed" order. To understand this term, rewrite the last column of Table Aa, with the subscripts written in the binary number system, e.g., $\hat{x}_{001} = \delta_{100}$, three digits each.

Ac Show that subroutines RPFT2 and RPFTI2 do what their commentary says they do.

Ad Derive (12). *Hints:* Write (11) as separate sums, over even and odd values of t. Show that the odd sum, and the second part of (12), are both equal to

$$\sum_{t=0}^{N-1} a\,(2t+1)\,\frac{\cos\dfrac{2\pi}{N}\,t\,\hat{t} - \cos\dfrac{2\pi}{N}\,(t+1)\,\hat{t}}{4\sin\dfrac{\pi}{N}\,\hat{t}} \tag{15}$$

(d) Listings for CPFT, RPFT2, and RPFTI2

```
        subroutine cpft(r, i, n, incp, signp)
c   fortran transliteration of singleton's 6600 assembly-coded fft.
c   intended to be of assistance in understanding his code, and in
c   future writing of an fft for another machine.
c   it should be translated into machine code rather than used for
c   production as is because: it is versatile and efficient enough to
c   see lots of use, it benefits greatly from careful hand coding, and is
c   short and simple enough to do quickly.
c   a. bruce langdon, m division, l.l.l., 1971.
c
c   comments below are mostly from 6600-7600 version.
c
c   r       real part of data vector.
c   i       imag part of data vector.
c   n       number of elements (=1,2,4,8...32768).
c   inc     spacing in memory of data (usually 1, but see below).
c   sign    its sign will be sign of argument in transform exponential.
c
c       on entry arrays r and i contain the sequence to be transformed.
c   on exit they contain the transform. input and output sequences are
c   both in natural order (i.e. not bit-reversed scrambled).
c
c       a call to cpft with sign=+1, followed by another call with the
c   first 4 parameters the same and sign=-1, will leave r and i with
c   their original contents times n. the same is true if first sign=-1,
c   and next sign=+1.
c
c       the usefulness of parameter inc may be illustrated by 2 examples:
c       suppose the complex sequence is stored as a fortran complex array
c   z, i.e. real and imaginary parts in alternate memory cells. the
c   separation between consecutive real (or imaginary) elements is 2
c   words, so inc=2. the call might be
c           call cpft(real(z), aimag(z), n, 2, sign)
c   for many compilers one would instead have to do something like
c           call cpft(ri, ri(2), n, 2, sign)
c   where ri is a real array equivalenced to z.
c       suppose one had an array c with dimensions n1, n2. one wants r to
c   be row i1 and i to be row i2. one wants r to
c   is n1 and starting addresses are c(i1,1) and c(i2,1), so use
c           call cpft(c(i1,1), c(i2,1), n2, n1, sign)
c
c   timing, assuming minimal memory bank conflicts:
c       6400 time for n=1024 is 220,000=21.5*n*log2n microseconds.
c       6600 time for n=1024 is  44,500=4.35*n*log2n microseconds.
c       7600 time for n=1024 is   8,300=0.81*n*log2n microseconds.
c
c       a radix 2 fft provokes memory bank conflicts at best, but timing
c   is noticeably worsened when like elements of r and i are in the same
c   bank and/or inc is a multiple of a power of 2. in a worst case on
c   the 7600 the speed was decreased by a factor of 3.
c       thus in the example above, if n1=multiple of 32 one might
c   decide to waste a little memory by increasing n1 to 33, thus
c   decreasing conflicts for transforms over rows or over columns.
c
c   written by r. c. singleton, stanford research institute, nov. 1968.
c   commentary, lrl linkage and other minor changes by a. bruce langdon
c   lawrence radiation laboratory, livermore, april 1971.
c
c   references:
```

```
c     (1) r. c. singleton, 'on computing the fast fourier transform',
c         comm. assoc. comp. mach. vol. 10, pp. 647-654 (1967).
c     (2) r. c. singleton, algorithm 345 'an algol convolution procedure
c         based on the fast fourier transform', comm. acm vol. 12,
c         pp. 179-184 (1969).
c     (3) w. m. gentleman and g. sande, 'fast fourier transforms - for
c         fun and profit', proc. afips 1966 fall joint computer conf.,
c         vol. 29, pp. 563-578.
c
      real r(1), i(1)
      integer signp, span, rc
      real sines(15), i0, i1
c
c     table of sines.
c     these should be good to the very last bit. they are given in octal
c     to prevent an assembler from converting them poorly. they may be
c     obtained by evaluating the indicated sines in double precision and
c     punching them out in octal format (a single precision sine routine
c     is not accurate enough). use the most significant word, rounded
c     according to the least significant word.
c     in this version i do it a lazy way on the first call.
      data sines(1)/0./
      if( sines(1).eq.1. ) go to 1
      sines(1)=1.
      t=atan(1.)
      do 2 is=2,15
      sines(is)=sin(t)
    2 t=t/2.
    1 continue

      if( n.eq.1 ) return

c     set up various indices.
c
      inc=incp
      sgn=signp
      ninc=n*inc
      span=ninc
      it=n/2
      do 3 is=1,15
      if( it.eq.1 ) go to 12
    3 it=it/2
```

```
c     there are 2 inner loops which run over the n/(2*span) replications
c     of transforms of length (2*span). these loops fit into the
c     instruction stack of the 6600 or 7600. one loop is for arbitrary
c     rotation factor angle. the other takes care of the special case in
c     which the angle is zero so that no complex multiplication is needed.
c     this is more efficient than testing and branching inside the inner
c     loop, as is often done. the other special case in which no complex
c     multiply is needed is angle=pi (i.e. factor=i); this is not handled
c     specially. these measures are most helpful for small n.
c
c     the organization of the recursion is that of sande (ref. (3),
c     pp. 566-568). that is, the data is in normal order to start and
c     scrambled afterward, and the exponential rotation ('twiddle') factor
c     angles are used in ascending order during each recursion level.
c     all the sines and cosines needed are generated from a short table
c     using a stable multiple-angle recursion (ref. (1), p651 and ref. (2),
c     pp. 179-180). this method is economical in storage and time, and
c     yields accuracy comparable to good library sin-cos routines.
c     angles between 0 and pi are needed. the recursion is used for
```

```
c    angles up to pi/2; larger angles are obtained by reflection in the
c    imaginary axis (angle:=pi-angle). these pairs of angles are used
c    one right after the other.
c
c       for simplicity, commentary below applies to inc=1 case.
c
c    if truncated rather than rounded arithmetic is used, singleton's
c    magnitude correction should be applied to c and s.
c
10      t=s+(s0*c-c0*s)
        c=c-(c0*c+s0*s)
        s=t
c    replication loop.
11      k1=k0+span
        r0=r(k0+1)
        r1=r(k1+1)
        i0=i(k0+1)
        i1=i(k1+1)
        r(k0+1)=r0+r1
        i(k0+1)=i0+i1
        r0=r0-r1
        i0=i0-i1
        r(k1+1)=c*r0-s*i0
        i(k1+1)=s*r0+c*i0
        k0=k1+span
        if( k0.lt.ninc ) go to 11
        k1=k0-ninc
        c=-c
        k0=span-k1
        if( k1.lt.k0 ) go to 11
        k0=k0+inc
        if( k0.lt.k1 ) go to 10
c    recursion to next level.
12      continue
        span=span/2
        k0=0
c    angle=0 loop.
13      k1=k0+span
        r0=r(k0+1)
        r1=r(k1+1)
        i0=i(k0+1)
        i1=i(k1+1)
        r(k0+1)=r0+r1
        i(k0+1)=i0+i1
        r(k1+1)=r0-r1
        i(k1+1)=i0-i1
        k0=k1+span
        if( k0.lt.ninc ) go to 13
c    are we finished...
        if( span.eq.inc ) go to 20
c    no. prepare non-zero angles.
        c0=2.*sines(is)**2
        is=is-1
        s=sign( sines(is),sgn )
        s0=s
        c=1.-c0
        k0=inc
        go to 11

c
c       arrays r and i now contain transform, but stored in 'reverse-
c       binary' order. the re-ordering is done by pair exchanges.
c       reference for sorting principle is p. 180 and p. 182 of ref. (2).
c
```

```
c     once again, commentary applies to inc=1 case.
c   indices are:
c     ij:=0,1,2...n/2-1 ( a simple counter).
c     ji:=reversal of ij.
c     rc:=reversal of 0,2,4...n/2 (incremented n/4 times).
c   rc is incremented thusly: starting with the next-to-leftmost bit,
c   change each bit up to and including first 0. (the actual coding is
c   done so as to work for any inc>0 with equal efficiency.)
c     for all exchanges ij fits one of these cases:
c       (1) 1st and last bits are 0 (ij,ji even and <n/2), and ij<=ji.
c       (2) one's complement of case (1) (both odd and >n/2).
c       (3) 1st bit 0, last bit 1 (ij odd and <n/2, ji>n/2).
c     the code from label even down to odd is entered with ij even and
c   <=ji. first time thru the complements are done -case (2). second
c   time thru gets case (1). thus a pair of elements both in the first
c   half of the sequence, and another pair in the 2nd half, are
c   exchanged. the condition ij<ji prevents a pair from being exchanged
c   twice.
c     the code from label odd down to increv does case (3).

20      n1=ninc-inc
        n2=ninc/2
        rc=0
        ij=0
        ji=0
        if( n2.eq.inc ) return
        go to 22

c   even.
21      ij=n1-ij
        ji=n1-ji
        t=r(ij+1)
        r(ij+1)=r(ji+1)
        r(ji+1)=t
        t=i(ij+1)
        i(ij+1)=i(ji+1)
        i(ji+1)=t
        if( ij.gt.n2 ) go to 21
c   odd.
22      ij=ij+inc
        ji=ji+n2
        t=r(ij+1)
        r(ij+1)=r(ji+1)
        r(ji+1)=t
        t=i(ij+1)
        i(ij+1)=i(ji+1)
        i(ji+1)=t
        it=n2
c   increment reversed counter.
23      it=it/2
        rc=rc-it
        if( rc.ge.0 ) go to 23
        rc=rc+2*it
        ji=rc
        ij=ij+inc

        if( ij.lt.ji ) go to 21
        if( ij.lt.n2 ) go to 22

        return
        end
```

```
      subroutine rpft2(a,b,n,incp)
      real a(1), b(1)
c   real data, periodic, fourier transform, two at a time.
c
c   interface to complex periodic fourier transform, to do pairs of
c   transforms of real sequences.
c
c   the two sequences are elements 0,inc,2*inc...(n-1)*inc of arrays a,b.
c
c   after a complex periodic fourier transform, with a and b as the
c   real and imaginary parts, rpft2 separates the transforms of a and b
c   and packs them, times 2, back into arrays a and b.
c   thus, the contents of a and b are replaced by twice their transforms
c   by the calls:
c       cpft (a, b, n, inc, sign)
c       rpft2(a, b, n, inc)
c   twice the real parts of the first half of the complex fourier
c   coefficients of a (cosine coef.) are in a(0), a(1)...a(n/2), if
c   inc=1. twice the imaginary parts (sine coef.) are stored in
c   reverse order, in a(n-1), a(n-2)...a(n/2+1). likewise for b.
c
c   no parameter 'sign' is provided for the purpose of changing the sign
c   of the sine coefficients. this may be done with parameter 'sign' of
c   the fourier transform, cpft.
c
c   time required is less than 1/10 of that for cpft.
c
c   should be re-coded in assembly language.
c
c   written by a. bruce langdon, lrl livermore, may 1971.
c
      real ip, im
      inc=incp
      ninc=n*inc
      a(1)=a(1)+a(1)
      b(1)=b(1)+b(1)
      lp=inc
      lm=ninc-lp
      if( lp.ge.lm ) go to 2
1     rp=a(lp+1)
      rm=a(lm+1)
      ip=b(lp+1)
      im=b(lm+1)
      a(lp+1)=rm+rp
      b(lm+1)=rm-rp
      b(lp+1)=ip+im
      a(lm+1)=ip-im
      lp=lp+inc
      lm=ninc-lp
      if( lp.lt.lm ) go to 1
2     if( lp.gt.ninc ) return
      a(lp+1)=a(lp+1)+a(lp+1)
      b(lp+1)=b(lp+1)+b(lp+1)
      return
      end
```

```
      subroutine rpfti2(a,b,n,incp)
      real a(1), b(1)
c  real data, periodic, fourier transform inverse, two at a time.
c
c  interface to complex periodic fourier transform, to do pairs of
c  transforms of real sequences.
c
c  unpacks the cosine and sine coefficients of a and b and combines
c  them so that a + i b is the complex periodic fourier transform of
c  the original sequences. rpfti2 reverses the effect of rpft2, except
c  that a and b are doubled.
c  the calls
c    rpfti2(a, b, n, inc)
c    cpft  (a, b, n, inc, -sign)
c  invert the transform done earlier, except that the arrays have been
c  multiplied by 2*n.
c
c  should be re-coded in assembly language.
c
c  written by a. bruce langdon, lrl livermore, may 1971.
c
      inc=incp
      ninc=n*inc
      lp=inc
      lm=ninc-lp
      if( lp.ge.lm ) return
3     ca=a(lp+1)
      sb=b(lm+1)
      cb=b(lp+1)
      sa=a(lm+1)
      a(lp+1)=ca-sb
      a(lm+1)=ca+sb
      b(lp+1)=cb+sc
      b(lm+1)=cb-sa
      lp=lp+inc
      lm=ninc-lp
      if( lp.lt.lm ) go to 3
      return
      end
```

COMPENSATING AND ATTENUATING FUNCTIONS USED IN ES1

The smoothing function $SM(k)$ in the subroutine FIELDS uses two adjustable parameters a_1 and a_2 given by

$$SM(k) \equiv \exp\left[a_1 \sin^2 \frac{k \Delta x}{2} - a_2 \tan^4 \frac{k \Delta x}{2}\right] \qquad (1)$$

This is used in the product $\rho(k)SM^2(k)$. $a_1 > 0$ is used to cancel for $k \Delta x \leq 1$ the $O(k^2)$ error in the dispersion relation due to $S(k)$, κ and K^2 in the normal algorithms; this is called *compensating* or boosting. $a_2 > 0$ attenuates $\rho(k)$ at short wavelengths, generally called *smoothing*. Using $a_1 = 0 = a_2$ makes $SM(k) = 1$. This Appendix suggests some values to use for a_1 and a_2.

Let the boost given by a_1 at $k \Delta x \leq 1$ be used to compensate for the error in the cold plasma dispersion. If the dispersion is given by the averaged-force result 8-13(3b),

$$\omega^2 \approx \omega_p^2 \left[\frac{k\kappa(k)S^2(k)}{K^2(k)}\right] SM^2(k) \qquad (2)$$

and use the usual formulae,

$$\kappa(k) = k\frac{\sin (k \Delta x)}{k \Delta x}$$

$$S^2(k) = \left| \frac{\sin \frac{k\,\Delta x}{2}}{\frac{k\,\Delta x}{2}} \right|^4$$

$$K^2(k) = k^2 \left| \frac{\sin \frac{k\,\Delta x}{2}}{\frac{k\,\Delta x}{2}} \right|^2 \tag{3}$$

then, using dif $\theta \equiv \sin\theta/\theta$,

$$\omega^2 \approx \omega_p^2 \; \text{dif} \, (k\,\Delta x) \, \text{dif}^2 \frac{k\,\Delta x}{2} \, \exp\left[2a_1 \sin^2 \frac{k\,\Delta x}{2} - 2a_2 \tan^4 \frac{k\,\Delta x}{2} \right] \tag{4}$$

To order $(k\,\Delta x)^2$, with $k\,\Delta x \equiv \theta \ll 1$,

$$\omega^2 \approx \omega_p^2 \left[1 - \frac{\theta^2}{6} + \cdots \right] \left[1 - \frac{\theta^2}{12} \right] \left[1 + a_1 \frac{\theta^2}{2} \right] \tag{5}$$

$$\approx \omega_p^2 \left[1 + \theta^2 \left(-\frac{1}{4} + \frac{a_1}{2} \right) \right] \tag{6}$$

Hence, setting $a_1 = 0.5$ makes $\omega = \omega_p$ to order $(k\,\Delta x)^4$. [Note that choosing a value for a_2 also is unnecessary as $\tan^4(\theta/2)$ is used, not $\tan^2(\theta/2)$.] If we include all spatial harmonics (all aliases), then the cold plasma dispersion with $a_1 = 0$ is given exactly by

$$\omega^2 = \omega_p^2 \cos^2 \frac{k\,\Delta x}{2} \; \text{SM}^2(k) \tag{7}$$

from 8-11(14). For small $k\,\Delta x$ (7) is

$$\omega^2 = \omega_p^2 \left[1 - \frac{\theta^2}{4} + \cdots \right] \text{SM}^2(k) \tag{8}$$

and, hence, is also corrected to order $(k\,\Delta x)^4$ by $a_1 = 0.5$. Plots of

$$\frac{\omega}{\omega_p} \equiv W(\theta) \equiv \cos\frac{\theta}{2} \; \text{SM}(\theta) \tag{9}$$

are given in Figure Ba for $a_1 = 0.5$ and various values of a_2.

Plots of the smoothing applied to the source $\rho(k)$, namely, $\text{SM}^2(\theta)$, are given in Figure Bb for $a_1 = 0$, that is, using just the attenuating factor. The user could choose the smallest wavelength he wishes to include (meaning a value of $k_{max}\Delta x$), locate this on the plot and pick off a value of a_2 that makes $\text{SM}^2(k_{max}\Delta x)$ equal, say, 0.01; the user then ignores all output for $k > k_{max}$, as the source is effectively turned off at larger k.

Sometimes spatial grid effects interfere with the physics. In one-dimension, as in ES1, we can afford to use a very large number of cells (so that $\Delta x \ll \lambda$, a wavelength characteristic of the physics) and then choose

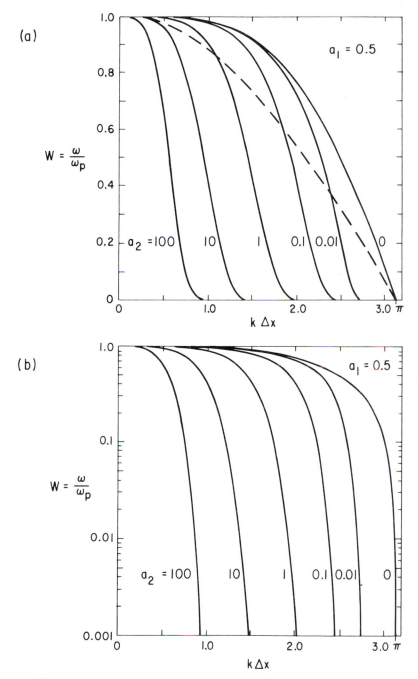

Figure Ba (a) Cold plasma dispersion curves, $\omega(k)/\omega_p \equiv W$ versus $k\Delta x \equiv \theta$ showing compensation and smoothing. The dashed line is $W = \cos\frac{1}{2}k\Delta x$, the cold plasma dispersion for no compensation and no attenuation ($a_1 = 0 = a_2$). The solid lines are for $a_1 = 0.5$ [which makes $W = 1 + O(k^4)$, maximally flat for $k \to 0$] and for various values of a_2 (attenuations). (b) The same, on a log plot.

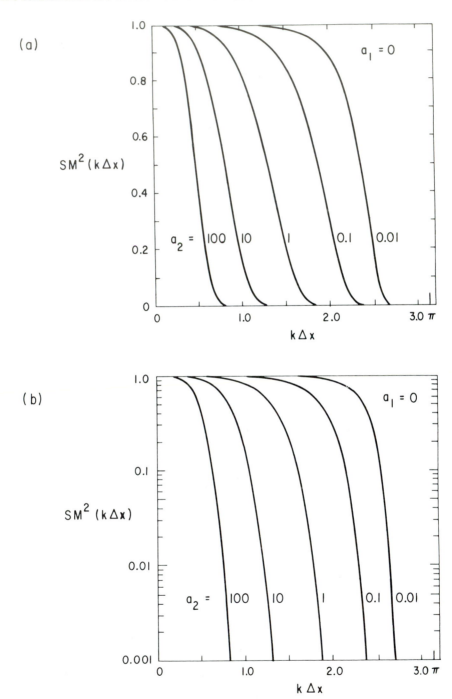

Figure Bb (a) Attenuation of Fourier components of $\rho(k)$ produced by smoothing factor SM^2 versus $k\Delta x \equiv \theta$, for no compensation ($a_1 = 0$), for various values of a_2. (b) Same, log scale.

a_2 so that $\mathrm{SM}(2\pi/\lambda) \sim 0.5$. Thus longer wavelengths are weakly attenuated, while wavelengths $\sim \Delta x$ are strongly suppressed. This usually removes spatial grid effects.

Another simple attenuating factor (not in ES1) which might be used, is

$$\mathrm{SM}_N(\alpha) \equiv \exp\left(-\alpha^N\right) \tag{10}$$

where $\alpha \equiv k/k_{max}$, which is plotted in Figure Bc. Note that:

$$\mathrm{SM}_N(0) = 1$$

$$\mathrm{SM}_N(1) = \frac{1}{e} = 0.368 \qquad\qquad \text{independent of } N$$

$$\mathrm{SM}_0(\alpha) = \frac{1}{e} = 0.368 \qquad\qquad \text{independent of } \alpha$$

$$\mathrm{SM}_{N\to\infty}(\alpha) = \begin{cases} 1, & \alpha < 1 \\ 0, & \alpha > 1 \end{cases} \qquad\qquad \text{step function}$$

$$\mathrm{SM}_N^2(\alpha) = \exp\left(-2\alpha^N\right) \neq \exp\left(-\alpha^{2N}\right) \qquad \begin{array}{l} \textit{i.e. } \text{doubling } N \text{ is } \textit{not} \text{ the} \\ \text{same as squaring } \mathrm{SM}_N(\alpha) \end{array}$$

$$\mathrm{SM}_N(\alpha \to 0) = 1 - O(\alpha^N)$$

Typically, a user would apply SM_N to $\rho(k)$, and then insert his value for k_{max} and use, say, $N = 8$.

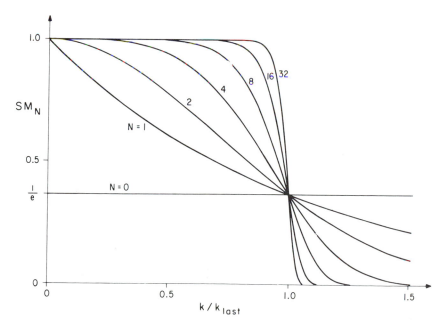

Figure Bc Attenuation factor $\mathrm{SM}_N(\alpha) \equiv \exp\left(-\alpha^N\right)$ for various values of N.

Other smoothing factors are readily constructed. The user is reminded that the sharper he makes the cutoff in k-space (*e.g.*, use of larger values of N in the second attenuator), the more his low-pass filter will emphasize the last harmonic kept. For example, a spatial square wave in $\rho(x)$, passed through a sharp cutoff filter, is returned with a ripple in $\rho(x)$ or $\phi(x)$ at $k = k_{max}$, which is the Gibbs ($\sim 9\%$) overshoot, and a rise length of about $\lambda_{min}/2 = \pi/k_{max}$ in place of the jump. If the sharp cutoff at $k = k_{max}$ is used and attenuation is applied by the Lanczos sigma factor (*e.g. Hamming, 1977*, Chapter 6, 7),

$$\text{SM}_L\left(\frac{k}{k_{max}}\right) \equiv \frac{\sin(\pi k/k_{max})}{\pi k/k_{max}} \qquad k < k_{max} \qquad (11)$$

then the over and undershoot are largely eliminated, but at the expense of resolution, as the rise length doubles to about λ_{min}. It appears that the second attenuator, SM_N of (10), with a moderate value of N (say, $N = 8$), while producing some overshoot and ripple, keeps the long wavelength behavior and a short rise length, near $\lambda_{min}/2$; hence, SM_N is preferable to SM_L.

DIGITAL FILTERING IN ONE
AND TWO DIMENSIONS

Filtering of grid quantities is generally used in simulation in order to

(i) improve agreement with theory [*e.g.*, dispersion $\omega(k)$] at long wavelengths $k\,\Delta x \rightarrow 0$ ($k\,\Delta x \ll \pi$); this is called *compensating or boosting*;

(ii) improve overall accuracy and reduce noise by ignoring source terms at short wavelengths, $k\,\Delta x \rightarrow \pi$, where finite difference algorithms for ∇, ∇^2, etc., become most inaccurate and alias coupling is severe; this is called *attenuating or smoothing*.

In codes using Fourier transforms of spatial grid quantities, the filtering is done directly in k space. Where Fourier transforms of spatial grid quantities are not available, the filtering must be done in x space, and is called *digital filtering*; see *Hamming* (1977) on recursive digital filters and *Collatz* (1966, p. 424). This Appendix is an introduction to digital filtering applied to scalar grid quantities in one and two dimensions. Fourier representations are obtained for the filters proposed, assuming periodic systems, in order to make clear the effect of the filtering.

Let the quantity to be filtered be $\phi(X_j) \equiv \phi_j$, which is known at the grid points, $X_j \equiv j\Delta x$, and is periodic. A simple filtering is done by replacing

$$\phi_j \quad \text{with} \quad \frac{W\phi_{j-1} + \phi_j + W\phi_{j+1}}{1 + 2W} \tag{1}$$

(Caution: the points on the right are always the original values.) Symmetry is required to retain momentum conservation.

The Fourier representation of the filter may be obtained by assuming that we can obtain

$$\phi_{\text{original}}(k) \quad \text{from} \quad \sum_{j=1}^{N} \phi(X_j) e^{ikX_j} \tag{2}$$

Hence, inserting the filtered ϕ of (1) for $\phi(X_j)$ here, as

$$\phi_{\text{filtered}}(k) = \sum_{j=1}^{N} \frac{W\phi_{j-1} + \phi_j + W\phi_{j+1}}{1 + 2W} e^{ikX_j} \tag{3}$$

we obtain (letting $p = j - 1$, $q = j + 1$)

$$\phi_f(k) = \frac{W\sum_{p=0}^{N-1} \phi(X_p) e^{ikX_{p+1}} + \sum_{j=1}^{N} \phi(X_j) e^{ikX_j} + W\sum_{q=2}^{N+1} \phi(X_q) e^{ikX_{q-1}}}{1 + 2W} \tag{4}$$

Recognizing that the assumption of periodicity allows the numbering of grid points to start anywhere, we find the desired result

$$\phi_f(k) = \frac{1 + 2W \cos k\Delta x}{1 + 2W} \phi_0(k) = \text{SM}_W(\theta)\phi_0(k), \quad \theta \equiv k\Delta x \tag{5}$$

The smoothing function $\text{SM}_W(\theta)$ is shown in Figure Ca. Let us consider various values of W. $W > 0.5$ causes $\text{SM}_W(\theta)$ to reverse sign in the first zone, $0 < k\Delta x < \pi$, which is undesirable; hence, filters such as the equally-weighted two and three point averages ($W \gg 1$ and $W = 1$) are not recommended. With $W = 0.5$, $\text{SM}_W(\theta)$ is always positive and goes to zero quadratically as $\theta \to \pi$. Application N times leads to $\cos^{2N}(\theta/2)$ filtering, as would be obtained from single-pass filters with binomial coefficients:

$$\text{three point:} \quad \frac{1}{4}(1, 2, 1) \qquad \to \cos^2 \frac{\theta}{2}$$

$$\text{five point:} \quad \frac{1}{16}(1, 4, 6, 4, 1) \qquad \to \cos^4 \frac{\theta}{2}$$

$$\text{seven point:} \quad \frac{1}{64}(1, 6, 15, 20, 15, 6, 1) \to \cos^6 \frac{\theta}{2}$$

Hence, the filter with $W = 0.5$ is called a binomial filter. Note that binomial smoothing approaches Gaussian smoothing as $N \to \infty$. The choice $W < 0$ produces $\text{SM}_W(\theta) > 1$, useful as a compensation filter. $W = -1/6$ produces compensation that just cancels the attenuation $O(\theta^2)$ of $W = 0.5$; that is,

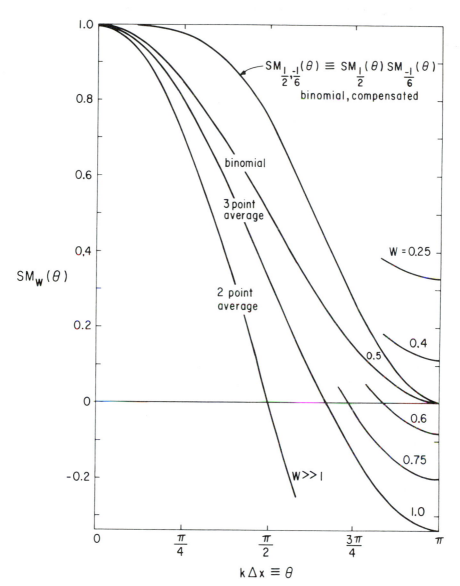

Figure Ca Smoothing function $SM_W(\theta)$ of (5) for various W. The two and three point averages (as well as any $W > 0.5$) produce $SM_W(\theta) < 0$ which alters the physics undesirably. Using first $W = 0.5$, then $W = -1/6$ produces the compensated curve shown.

application of $W = 0.5$ followed by $W = -1/6$ produces $SM_W(\theta) = 1 + O(\theta^4)$ for small θ. This two step filter is equivalent to a five point single filter, with weights $(1/16)\,(-1,\ 4,\ 10,\ 4,\ -1)$.

Using (1) requires keeping a few of the unfiltered values of ϕ, which is not a memory problem in 1d. However, in 2d and 3d, there could be problems with keeping old values of ϕ_j, as the smoothing matrix expands from 3 points in 1d to 9 in 2d and 27 in 3d. Also, it may be simpler to code the smoothing by making more than one pass along each grid row. Hence, it is desirable to see if an algorithm can be found which can be factored. First, sweep forward, replacing all ϕ_j with $\phi_j + U\phi_{j+1}$ and then sweep backward, replacing all ϕ_j with $\phi_j + U\phi_{j-1}$. This requires no scratch memory. The Fourier representation of this convolution process is

$$\text{SM}_U(\theta) = \frac{Ue^{i\theta} + 1}{U + 1} \frac{1 + Ue^{-i\theta}}{1 + U} = \frac{1 + U^2 + 2U\cos\theta}{(1 + U)^2} \tag{6}$$

This is the same as the one-pass $\text{SM}_W(\theta)$ for

$$W = \frac{U}{1 + U^2} \qquad U = \frac{1}{2W} \pm \left[\frac{1}{(2W)^2} - 1\right]^{\frac{1}{2}} \tag{7}$$

For U to be real requires that $-\frac{1}{2} \leqslant W \leqslant \frac{1}{2}$; see Figure Cb. That is, W is restricted to $|W| \leqslant \frac{1}{2}$ in order to produce a factorable form. The binomial form of the smoother, $(1, 2, 1)/4$ with $W = \frac{1}{2}$, corresponds to $U = 1$, with a forward pass of $(0, 1, 1)/2$ and a return pass with $(1, 1, 0)/2$. The corresponding compensator $W = -1/6$ has $U = -3 + 8^{\frac{1}{2}} = -0.171573$, to be applied after the smoother.

Digital filtering in 2d proceeds much as is done in 1d but demands added care in order to make the filtering isotropic and to make the operation efficient in speed and memory. We limit the filter to operation on the grid-point (j, k) and the eight nearest grid points.

The filter consists of replacing $\phi(X_j, Y_k) \equiv \phi_{j,k}$, $X_j \equiv j\Delta x$, $Y_k \equiv k\Delta y$ with weighted values of neighboring points, as

$$\phi_{j,k} \leftarrow \frac{M\phi_{j,k} + S(\text{side terms}) + K(\text{corner terms})}{M + 4(S + K)} \tag{8}$$

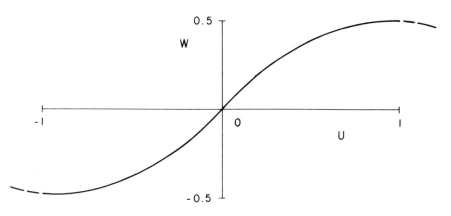

Figure Cb Relation between W and U for one and two-pass digital smoothing filters.

where   side terms $= \phi_{j-1,k} + \phi_{j+1,k} + \phi_{j,k+1} + \phi_{j,k-1}$

corner terms $= \phi_{j-1,k-1} + \phi_{j-1,k+1} + \phi_{j+1,k+1} + \phi_{j+1,k-1}$

and $M \equiv$ middle, $S \equiv$ side, and $K \equiv$ corner weights, as illustrated by the spatial stencil

$$\begin{bmatrix} K & S & K \\ S & M & S \\ K & S & K \end{bmatrix}$$

The Fourier representation of the filter is

$$\phi_{\text{filtered}}(\theta_x, \theta_y)$$
$$= \left[\frac{M + 2S(\cos\theta_x + \cos\theta_y) + 4K(\cos\theta_x \cos\theta_y)}{M + 4(S + K)} \right] \phi_{\text{original}}(k_x, k_y) \quad (9)$$

where $\theta_x \equiv k_x \Delta x$, $\theta_y \equiv k_y \Delta y$. The smoothing function $\text{SM}_{M,S,K}(\theta_x, \theta_y)$ is the bracketed term.

We designate the filters by the values (M, S, K). The hollow 8-point $(0, 1, 1)$ filter (*Hockney*, 1971; *Hockney and Eastwood*, 1981, p. 376) might appear to be a reasonable spatial averaging method; however, its Fourier representation becomes negative over roughly half of the first zone ($0 < \theta_x$, $\theta_y < \pi$). This can lead to nonphysical results, including severe instability (Problem 4-6a, and *Langdon and Birdsall*, 1970).

Application of first the x binomial and then the y binomial spatial stencils

$$\begin{bmatrix} 0 & 0 & 0 \\ 1 & 2 & 1 \\ 0 & 0 & 0 \end{bmatrix} \begin{bmatrix} 0 & 1 & 0 \\ 0 & 2 & 0 \\ 0 & 1 & 0 \end{bmatrix}$$

produces a binomial 9-point filter $(4, 2, 1)$, whose Fourier representation shows excellent attenuation which is nearly isotropic. The stencil is

$$\begin{bmatrix} 1 & 2 & 1 \\ 2 & 4 & 2 \\ 1 & 2 & 1 \end{bmatrix}$$

Each step is a *convolution*, hence, the Fourier representation is the *product* of the θ_x and θ_y representations, *i.e.*, $(\cos\theta_x/2)^2 (\cos\theta_y/2)^2$. In ZOHAR this filter is done in four passes,

$$\begin{bmatrix} 0 & 0 & 0 \\ 0 & 1 & 1 \\ 0 & 0 & 0 \end{bmatrix} \begin{bmatrix} 0 & 0 & 0 \\ 1 & 1 & 0 \\ 0 & 0 & 0 \end{bmatrix} \begin{bmatrix} 0 & 1 & 0 \\ 0 & 1 & 0 \\ 0 & 0 & 0 \end{bmatrix} \begin{bmatrix} 0 & 0 & 0 \\ 0 & 1 & 0 \\ 0 & 1 & 0 \end{bmatrix}$$

forward, backward, up, and down, respectively, which produces the same result without using additional storage. A reasonable compensator is $(20, -1, -1)$; another is $(36, -6, 1)$.

DIRECT FINITE-DIFFERENCE
EQUATION SOLUTIONS

In this Appendix we obtain the solutions to the 1d finite-difference-equation representations of $\nabla^2\phi = -\rho$ and $\nabla \cdot \mathbf{E} = \rho$ begun in the first part of Section 2-5, and the 2d bounded system begun in Section 14-5.

Let Poisson's equation be written as

$$\phi_{j-1} - 2\phi_j + \phi_{j+1} = p_j \tag{1}$$

where

$$p_j \equiv -(\Delta x)^2 \rho_j \tag{2}$$

The problem is to obtain the ϕ_j from the p_j and the boundary and/or auxiliary (e.g., charge neutrality, $\sum p_j = 0$) conditions. Let the system length be L (from $x = 0$ to $x = L$), divided into N cells, as shown in Figure Da. Write the finite-difference-equations once for each j; the 1, -2, 1 coefficients form the square matrix $\mathbf{A}$; the unknowns ϕ form the column matrix $\boldsymbol{\phi}$; the known p_j form the column matrix $\mathbf{p}$. Hence $\mathbf{A}\boldsymbol{\phi} = \mathbf{p}$ is to be solved for $\boldsymbol{\phi}$.

Figure Da Grid space numbering for N cells.

In the *bounded by electrodes* model, $\phi_0 = V_0$ and $\phi_N = V_L$ are given. The interior need *not* be charge neutral. The p_j with $j = 1, 2, \ldots, N-1$ are to be used to obtain the ϕ_j for the same values of j. The matrix equation $\mathbf{A}\boldsymbol{\phi} = \mathbf{p}$ reads

$$
\begin{bmatrix}
-2 & 1 & & & \cdot \\
1 & -2 & 1 & & \cdot \\
& 1 & -2 & 1 & \cdot \\
\cdot & \cdot & \cdot & \cdot & \cdot & \cdot \\
& & & \cdot & 1 & -2
\end{bmatrix}
\begin{bmatrix}
\phi_1 \\
\phi_2 \\
\phi_3 \\
\phi_4 \\
\cdot \\
\phi_{N-2} \\
\phi_{N-1}
\end{bmatrix}
=
\begin{bmatrix}
p_1 - V_0 \\
p_2 \\
p_3 \\
\cdot \\
p_{N-1} - V_L
\end{bmatrix}
\qquad (3)
$$

This set has $N-1$ unknown ϕ_j's and $N-1$ equations which are independent. Hence, any method of solution may be used (*e.g.*, *Potter*, 1973, eq. 4.9, for solution by Gaussian elimination specialized to tridiagonal matrices). See Problem Da for a simple solution.

In a *periodic model*, E and ρ repeat every L; therefore, the system has no net charge, $\langle \rho \rangle = 0$, as noted in Section 4-11,

$$
N\langle \rho \rangle = \sum_{j=0}^{N-1} \rho_j = \sum_{j=1}^{N} \rho_j = 0 \qquad (4)
$$

The matrix equation in ρ_0 through ρ_{N-1}, using periodicity ($\phi_0 = \phi_N$, $\phi_{-1} = \phi_{N-1}$, etc.), now has one unneeded equation; summing as in (4) makes the left-hand side summed also zero which demonstrates that the N equations are not independent. Hence, drop one of the set, the ρ_0 or p_0 equation. This leaves the set

$$
\begin{bmatrix}
1 & -2 & 1 & & & \cdot \\
& 1 & -2 & 1 & & \cdot \\
& & 1 & -2 & & \cdot \\
\cdot & \cdot & \cdot & \cdot & \cdot & \cdot \\
1 & & & \cdot & 1 & -2
\end{bmatrix}
\begin{bmatrix}
\phi_0 \\
\phi_1 \\
\phi_2 \\
\phi_3 \\
\cdot \\
\phi_{N-2} \\
\phi_{N-1}
\end{bmatrix}
=
\begin{bmatrix}
p_1 \\
p_2 \\
p_3 \\
\cdot \\
p_{N-1}
\end{bmatrix}
\qquad (5)
$$

which is the same form as (3), except in the left column. For such sets, see *Temperton* (1975). However, we do not know the value of ϕ_0, nor do we need to know it, because only $\nabla \phi$ is needed to move the particles. Hence, ϕ_0 can be assigned a value (call this the *bias*), the equations solved for ϕ_1, $\phi_2, \ldots, \phi_{N-1}$, and the necessary $\nabla \phi$'s obtained; the easy bias is $\phi_0 = 0$, which zeros the left column of $\mathbf{A}$, making (5) like (3) with $V_0 = 0 = V_L$. Another choice of bias, $\langle \phi \rangle = 0$, corresponds to the ES1 field solution in which $\phi(k = 0) = 0$. The electrostatic energy (periodic model, no wall charges),

$$W_E \equiv \frac{\Delta x}{2} \sum_{j=0}^{N-1} \rho_j \phi_j \qquad (6)$$

is unaffected by the choice of bias; in (6), if the ϕ_j's from (5) are used, which are $\phi_j = \phi_j(\text{true}) + (\phi_0 + \text{constant})$, then

$$W_E = \frac{\Delta x}{2} \sum \rho_j \phi_j(\text{true}) + \frac{\Delta x}{2}(\phi_0 + \text{constant}) \sum \rho_j \qquad (7)$$

$$= \frac{\Delta x}{2} \sum \rho_j \phi_j(\text{true}) \qquad (8)$$

because $\langle \rho \rangle = 0$. To these potentials, obtained from charges inside a period, solutions of the homogeneous equation $\nabla^2 \phi = 0$ may be added. These potentials may be thought of as due to equal and opposite charges at $x = \pm\infty$ (or as due to double layers of charges at $x = 0, L, 2L$, etc.).

Alternatively we might choose to obtain E directly by integrating 2-5(2) from one grid point to the next (the same as using *Gauss' law*), as

$$\int_{x_j}^{x_{j+1}} \frac{\partial E}{\partial x} dx = \int_{x_j}^{x_{j+1}} \rho \, dx \qquad (9)$$

This is, using the trapezoidal rule for the right-hand side,

$$E_{j+1} - E_j = \frac{\rho_{j+1} + \rho_j}{2} \Delta x \qquad (10)$$

Note that E and ρ are given at the *same* grid points; *i.e.*, the E and ρ grids are *not* staggered or interlaced. We need one boundary or other condition on E [as 2-5(2) is a first-order differential equation] for completeness. For example, with a space-charge-limited emitter at $x = 0$, we would choose $E_0 = 0$. For a *periodic model,* we must have $E_0 = E_N$; summing (10) for $j = 0$ to $N-1$ we find (4) again. We can solve for E by assuming, for the moment, $E_0 = 0$ and applying (10). The resulting average field,

$$\langle E \rangle = \frac{1}{N} \sum_{j=0}^{N-1} E_j = \frac{1}{N} \sum_{j=1}^{N} E_j \qquad (11)$$

will *not* be zero. If we wish $\langle E \rangle$ to be zero, we simply subtract the value given by (11) from each E_j. Now the field satisfies (10), and (11) will give $\langle E \rangle = 0$ if evaluated again. A solver similar to this is used by *Denavit and Kruer* (1980), who first solve for E_0, as

$$E_0 = -\frac{\Delta x}{2N} \sum_{s=0}^{N-1} \sum_{j=0}^{s} (\rho_j + \rho_{j+1}) \qquad (12)$$

and then, assured that $\langle E \rangle = 0$, march across the system.

In two dimensional problems, we can set up the finite-difference equations in much the same way. If the configuration permits use of the fast Fourier transform along one direction, the system of transformed equations is tridiagonal and may be solved by Gaussian elimination or other methods.

The Poisson equation from Section 14-5 is

$$\phi_{j+1} - 2d\phi_j + \phi_{j-1} = p_j \tag{13}$$

For a model *bounded by electrodes at fixed potentials* $\phi_0 = V_0$ and $\phi_N = V_L$, then Gaussian elimination is done as follows (from *Forsythe and Wasow, 1960, p.104*). Let the (known) source terms be given by

$$s_1 = p_1 - V_0, \dots, \quad s_n = p_n, \dots, \quad s_{N-1} = p_{N-1} - V_L \tag{14}$$

Forward elimination:

$$w_1 = (-2d)^{-1}; \quad w_{n+1} = (-2d - w_n)^{-1}, \quad n = 1, 2, 3, \dots, N-2 \tag{15}$$

$$g_1 = s_1 w_1; \quad g_{n+1} = (s_{n+1} - g_n)(w_{n+1}), \quad n = 1, 2, 3, \dots, N-2 \tag{16}$$

Back substitution:

$$\phi_{N-1} = g_{N-1}; \quad \phi_n = g_n - w_n \phi_{n+1}, \quad n = N - 2, N - 3, \dots, 1 \tag{17}$$

This procedure works, of course, for $d = 1$, the Poisson equation for 1d, as solved another way in Problem Da .

PROBLEMS

Da In order to solve (3), multiply the last equation by 1, the next to last by 2, etc., and then add all equations to show that

$$\phi_1 = \frac{V_0(N-1) + V_L}{N} - \frac{1}{N} \sum_{n=1}^{N-1} n\, p_{N-n} \tag{18}$$

Hence, ϕ_2 may be obtained from the first equation, ϕ_3 from the second, etc. Rewriting the sum as

$$\sum_{m=1}^{N-1} p_m (N - m) \tag{19}$$

Show that the result for the zero bias periodic model ($V_0 = V_L = 0$) is

$$\phi_1 = \frac{1}{N} \sum_{m=1}^{N} m\, p_m \tag{20}$$

which agrees with *Hockney and Eastwood* (1981, p. 35).

Db Derive (12).

Dc Show that (10) is equivalent to (1) for E obtained from ϕ differenced over $2\Delta x$.

DIFFERENCING OPERATORS;
LOCAL AND NONLOCAL ($\nabla \rightarrow ik$, $\nabla^2 \rightarrow -k^2$)

W. M. Nevins

In solving plasma physics problems on a computer one often converts a differential equation into a finite-difference equation on a spatial (or temporal) grid. In making this conversion, one must choose appropriate differencing operators and be aware of the local and nonlocal effects implied. For example, the first derivative of a function f at the jth grid point might be defined as

$$(\hat{D}_L F)_j = \frac{f_{j+1} - f_{j-1}}{2\Delta x} \tag{1}$$

where the caret means operator and L is used to denote local.

Differencing operators are *linear* operators. Hence, they may be written in the form

$$(\hat{K} f)_i = \sum_{j=1}^{N} K(i-j)f_j \tag{2}$$

where the $K(j)$ are fixed coefficients and the grid points are numbered 1 to N in a system of length L. This set of coefficients $\{K(j)\}$ is the configuration space representation of the operator $\hat{K}$.

Going back to our example (1) we see that the first-derivative operator $\hat{D}_L$ has the configuration space representation

$$
D_L(j) = \begin{cases} \dfrac{1}{2\Delta x} & j = +1 \\[2mm] \dfrac{-1}{2\Delta x} & j = -1 \\[2mm] 0 & \text{otherwise} \end{cases} \tag{3}
$$

This first-derivative operator $\hat{D}_L$ is an example of a *local* operator; *i.e.*, the derivative of a function at the ith grid point involves only the values of the function at *nearby grid points*.

There are two important reasons for expressing derivatives as local operators when working in configuration space. In a continuous space, the derivative of a function is defined locally. Hence, when modeling a continuous system with a discrete system, it is desirable to retain the local character of the derivative. This can be especially true near boundaries or marked internal inhomogeneities. In addition, it is often easier to solve systems of finite-difference equations that contain only local differencing operators.

The Fourier transform is a powerful method for solving linear finite-difference equations. When finite-difference equations are Fourier transformed, the Fourier convolution theorem may be used to show that linear operators like $\hat{K}$ of (3), transform to

$$
(\hat{K} f)_k = K(k) f_k \tag{4}
$$

where f_k is the complex Fourier amplitude, and $K(k)$ is the k-space representation of the operator $\hat{K}$. $K(k)$ is related to the configuration-space representation of the same operator by

$$
K(k) = \sum_{j=1}^{N} K(j) \exp\left[-ik\,\frac{2\pi}{N}j\right] \tag{5}
$$

We take k to be an integer; the corresponding wave number is $k(2\pi/L)$. It follows from (5) that the k-space representation of any linear operator $\hat{K}$ is a periodic function of k with period N.

Equation (5) may be used to obtain the k-space representation of the first-derivative operator $\hat{D}_L$ defined in (1). One obtains

$$
D_L(k) = ik\frac{2\pi}{L}\,\text{dif}\,\frac{2\pi k}{N} \tag{6}
$$

where the diffraction function dif is defined as

$$
\text{dif}(x) \equiv \frac{\sin x}{x} \tag{7}
$$

Once the decision has been made to solve a system of finite-difference equations in k-space, local operators no longer have any computational advantage over nonlocal operators.

It is often suggested that the "exact" expression,

$$D_E(k) \equiv ik \frac{2\pi}{L} \tag{8}$$

be employed as k-space representation of the first derivative. Similarly the "exact" second derivative would be represented by

$$D_E^2(k) = -k^2 \left(\frac{2\pi}{L} \right)^2 \tag{9}$$

Strictly speaking it is not possible to choose these representations for the first and second derivative operators since they are not periodic functions of k. This problem is avoided by requiring that the k-space representations of first and second derivatives be given by (8) and (9) in the first Brillouin zone (i.e., for $-N/2 < k \leqslant N/2$). The k-space representation of these operators is then determined at all other values of k by periodicity.

These representations of the first and second derivatives are widely used [see *Matsuda and Okuda* (1975); *Buneman* (1976)]. It is often stated that no error is introduced into the computation when these "exact" operators are employed, while local operators like (1) do introduce an error. This incorrect conclusion follows from an over-simplified treatment of the error analysis. A correct analysis of the accuracy of differencing operators in a particle simulation is greatly complicated by the fact that the charge density and/or the current density is defined *between* grid points (through the particle phase variables). This leads to aliasing, treated in Chapter 8. In any case, it is clear that a great deal of information is lost when we choose to represent a continuous function (*e.g.*, the charge density or electrostatic potential) by its values on a discrete mesh. When the "exact" differencing operators are employed, this loss of information leads to a very non-local representation of the first and second derivative operators.

Although there is no computational advantage to local operators when working in k-space, the *physics* that the finite-difference equations are modeling often involves local phenomena. Hence, it is desirable to know the configuration-space representation of the differencing operators. It follows from (5) and the orthogonality relation of discrete Fourier transformations, that the configuration space representation of a linear operator is related to its k-space representation by

$$K(j) = \frac{1}{N} \sum_{j=-N/2+1}^{N/2} K(k) \exp \left[ik \frac{2\pi}{N} j \right] \tag{10}$$

Hence, the configuration-space representations of the so-called "exact" derivative operators are given by

$$D_E(j) = \begin{cases} \dfrac{(-1)^j}{j\Delta x} \dfrac{j\pi/N}{\tan(j\pi/N)} & j \neq 0 \\ 0 & j = 0 \end{cases} \tag{11}$$

$$D_E^2(j) = \begin{cases} \dfrac{(-1)^j}{(j\Delta x)^2} \dfrac{-2}{\text{dif}^2(j\pi/N)} & j \neq 0 \\ -\dfrac{\pi^2}{3} \dfrac{1}{\Delta x^2} \left[1 + \dfrac{2}{N^2}\right] & j = 0 \end{cases} \tag{12}$$

It is important to realize that *these "exact" derivatives are very nonlocal operators.* We see from (11) and (12) that the derivative of a function at the jth grid point involves the values of the function at *every other grid point in the system!*

It is instructive to write the "exact" derivative in the configuration space. After regrouping the terms we find

$$(\hat{D}_E f)_i = \sum_{j=1}^{N/2} \frac{f_{i+j} - f_{i-j}}{2j\Delta x} W_1(j) \tag{13}$$

where the weighting function $W_1(j)$ is given by

$$W_1(j) = 2(-1)^j \frac{j\pi/N}{\tan(j\pi/N)} \tag{14}$$

Equation (13) tells us that we may interpret the "exact" derivative as a weighted average of all possible centered differences. The weighting function used in performing this average falls off very slowly with the increase in the interval over which these differences are taken. Hence, the centered difference over half of the system length is very nearly as important in determining the value of the "exact" derivative as is the local centered difference of (1).

Similar considerations show that the "exact" second derivative may be written in configuration space as

$$(D_E^2 f)_i = \sum_{j=1}^{N/2} \frac{f_{i+j} - 2f_i + f_{i-j}}{(j\Delta x)^2} W_2(j) \tag{15}$$

where the weighting function $W_2(j)$ is given by

$$W_2(j) = 2(-1)^j \, \text{dif}^{-2} \frac{j\pi}{N} \tag{16}$$

The weighting functions $W_1(j)$ and $W_2(j)$ are shown in Figure Ea.

In ES1, $\phi(k)$ is obtained from $\rho(k)$ by

$$\phi(k) = \frac{\rho(k)}{K^2(k)}$$

where $K^2(k)$ may be chosen freely in the first Brillouin zone (to produce the

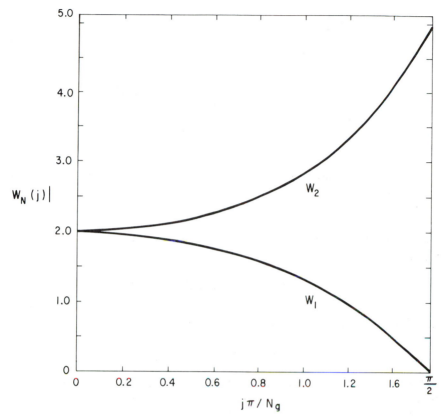

Figure Ea These curves show the absolute value of the weighting functions used in (14) and (16). $|W_1|$ falls off slowly as the differencing interval is increased, while $|W_2|$ actually increases with increasing differencing interval, underscoring the nonlocal nature of the "exact" differencing operators.

desired physics); however, K^2 is actually programmed to be the Fourier representation of the three-point finite-differencing of ∇^2, that is,

$$\frac{\phi_{j-1} - 2\phi_j + \phi_{j+1}}{(\Delta x)^2} \quad \text{produces} \quad K^2(k) = k^2 \, \text{dif}^2 \frac{k \Delta x}{2} \qquad (17)$$

When this value of $K^2(k)$ is inverted by (10), it, of course, reverts to the left-hand-side of (17), the local second derivative with relative error $O(k^2)$.

The lesson here is that there is no perfect choice of a finite-differencing operator. One always has to accept a compromise and choose a differencing operator that is appropriate to the problem being studied. In a nearly uniform system, the nonlocal character of the "exact" derivative is probably less of a handicap, and these expressions for the derivative are a suitable choice. If the system being studied is non-uniform, then it would seem likely that it is important to preserve the local character of the derivatives. Hence, in nonuniform systems a local differencing operator like $\hat{D}_L$ would be a better choice.

This discussion is not meant to emphasize particular choices of either algorithms for ∇ or ∇^2 or particle weightings by themselves; the true tests lie in reproduction of good physics, as in obtaining the correct force or the desired dispersion, which are due to combinations of all choices. Even at best, finite-size-particle methods can produce the correct force only at long range and oscillations and waves at long wavelengths. The chore of the code designer is to produce these effects at least cost in computer time and memory. It appears to us, among the choices

(a) subtracted dipole codes which use heavy smoothing beyond $k\,\Delta x \approx \pi/2$;
(b) Fourier codes ($\nabla \rightarrow ik$, $\nabla^2 \rightarrow -k^2$, with nonlocal derivatives) which use higher-order particle weighting (such as quadratic splines);
(c) linear-weight codes (CIC, PIC) with either local derivatives or their Fourier representations, with smoothing only at large $k\,\Delta x$;

that choice (c) is generally preferable in terms of good accuracy, high speed and maximum use of available grid points.

REFERENCES

Abe, H., J. Miyamoto, and R. Itatani, Grid effects on the plasma simulation by the finite-sized particle, *J. Comput. Phys.* **19**, 134-149, October 1975.

Abramovitz, M. and I. Stegun, *Handbook of Mathematical Functions,* Nat'l Bureau of Standards, Appl. Math. Series 55, U. S. Government Printing Office, Wash., D. C., 1964.

Adam, J. C., A. Gourdin Serveniere, P. Mora and R. Pellat, Effect of collisions on dc magnetic field generation in a plasma by resonance absorption of light, *Phys. Fluids* **25**, 812-814, May 1982a.

Adam, J. C., A. Gourdin Serveniere, and A. B. Langdon, Electron sub-cycling in particle simulation of plasmas, *J. Comput. Phys.* **47**, 229-244, August 1982b.

Aizawa, M., Y. Ohsawa, K. Sato, T. Kamimura, and T. Sekiguchi, Particle simulation studies on behavior or rapidly-expanding high-beta plasma column in a uniform magnetic field, *Japanese Jour. of App. Phys.* **19**, 2211-2227, November 1980.

Albritton, J. R., and A. B. Langdon, Profile modification and hot electron temperature from resonant absorption, *Phys. Rev. Lett.* **45**, 1794-1797, December 1980.

Alder, B., S. Fernbach, and M. Rotenberg, eds., *Meth. Comput. Phys.* **9**, *Plasma Physics,* Academic, New York, 1970.

Auer, P. L., H. Hurwitz, and R. W. Kilb, Low Mach number magnetic compression waves in a collision-free plasma, *Phys. Fluids* **4**, 1105-1121, September 1961.

Auer, P. L., H. Hurwitz, and R. W. Kilb, Large-amplitude magnetic compression of a collision-free plasma. II. Development of a thermalized plasma, *Phys. Fluids* **5**, 298-316, March 1962.

Baldis, H. A., and C. J. Walsh, Growth and saturation of the two-plasmon decay instability, *Phys. Fluids* **26**, 1364-1375, May 1983.

Balescu, R., Irreversible processes in ionized gasses, *Phys. Fluids* **3**, 52-63, February 1960.

Barnes, D. C., T. Kamimura, J.-N. Le Boeuf, and T. Tajima, Implicit particle simulation of magnetized plasmas, *J. Comput. Phys.* **52**, 480-502, December 1983.

Berman, R. H., D. J. Tetreault, T. H. Dupree, and T. Boutros-Ghali, Computer simulation of nonlinear ion-electron instability, *Phys. Rev. Lett.* **48**, 1249-1252, May 1982.

Bernstein, I. B., Waves in a plasma in a magnetic field, *Phys. Rev.* **109**, 10-21, January 1958.

Bers, A. Linear Waves and Instabilities, *Plasma Physics, Les Houches, 1972,* section xxii, Gordon and Breach, New York, 1975.

Birdsall, C. K., Interaction Between Two Electron Streams for Microwave Amplification, Ph. D. Dissertation, Stanford University, Stanford Electronic Research Laboratory Report 36, June 1951.

Birdsall, C. K., Sheath formation and fluctuations with dynamic electrons and ions, *International Conference on Plasma Physics,* Göteborg, Sweden, June 1982.

Birdsall, C. K., and W. B. Bridges, *Electron Dynamics of Diode Regions,* Academic, New York 1966.

Birdsall, C. K., and D. Fuss, Cloud-in-cell computer experiments in two and three dimensions, *Proc. Second Conf. Num. Sim. Plasmas,* Los Alamos Sci. Labs. **LA-3990**, 18-20 September 1968.

Birdsall, C. K., A. B. Langdon, C. F. McKee, H. Okuda, and D. Wong, Theory and experiments for a plasma consisting of clouds interacting with clouds (CIC) with and without a spatial grid, *Proc. Second Conf. Num. Sim. Plasmas,* Los Alamos Sci. Labs. **LA-3990**, 18-20 September 1968.

Birdsall, C. K., and D. Fuss, Clouds-in-clouds, clouds-in cells physics for many-body simulation, *J. Comput. Phys.* **3**, 494-511, April 1969.

Birdsall, C. K., N. Maron, and G. Smith, Cold beam nonphysical instabilities and cures, *Proc. Seventh Conf. Num. Sim. Plasmas,* New York Univ., NY, 2-4 June 1975.

Birdsall, C. K. and N. Maron, Plasma self-heating and saturation due to numerical instabilities, *J. Comput. Phys.* **36**, 1-19, June 1980.

Biskamp, D. and H. Welter, Ion heating in high-mach-number, oblique, collisionless shock waves, *Phys. Rev. Lett.* **28**, 410-413, February 1972.

Biskamp, D. and R. Chodura, Collisionless dissipation of a cross-field electric current, *Phys. Fluids* **16**, 893-901, June 1973.

Biskamp, D., and H. Welter, Stimulated Raman scattering from plasmas irradiated by normally and obliquely incident laser light, *Phys. Rev. Lett.* **34**, 312-316, February 1975.

Blackman, R. and J. W. Tukey, *The Measurement of Power Spectra,* Dover, New York, 1958.

Boris, J. P., The acceleration calculation from a scalar potential, Plasma Physics Laboratory, Princeton University MATT-152, March 1970a.

Boris, J. P., Relativistic plasma simulation-optimization of a hybrid code, *Proc. Fourth Conf. Num. Sim. Plasmas,* Naval Res. Lab., Wash., D. C., 3-67, 2-3 November 1970b.

Boris, J. P., and K. V. Roberts, Optimization of particle calculations in 2 and 3 dimensions, *J. Comput. Phys.* **4**, 552-571, December 1969.

Boris, J. P., and R. Lee, Nonphysical self forces in some electromagnetic plasma-simulation algorithms, *J. Comput. Phys.* **12**, 131-136, May 1973.

Boyd, G., L. M. Field, and R. Gould, Excitation of plasma oscillations and growing plasma waves, *Phys. Rev.* **109**, 1393-1394, February 1958.

Brackbill, J. U., and D. W. Forslund, An implicit method for electromagnetic plasma simulation in two dimensions, *J. Comput. Phys.* **46**, 271-307, May 1982.

Brackbill, J. U., and D. W. Forslund, Simulation of low frequency electromagnetic phenomena in plasmas, in the volume *Multiple Time Scales,* in the series *Computational Techniques,* Academic, New York, 1985.

Brand, L., *Vector Analysis,* Wiley, New York, 1957.

Briggs, R. J., Two-stream instabilities, *Advances in Plasma Physics* **4**, A. Simon and W. B. Thompson, eds., J. Wiley and Sons, Inc., p. 43-78, 1971.

Brillouin, L., *Wave Propagation in Periodic Structures,* Dover, New York, 1953.

Brown, D. I., S. J. Gitomer, and H. R. Lewis, The two-stream instability studied with four one-dimensional plasma simulation models, *J. Comput. Phys.* **14**, 193-199, February 1974.

Buneman, O., Dissipation of currents in ionized media, *Phys. Rev.* **115**, 503-517, August 1959.

Buneman, O., Instability of electrons drifting through ions across a magnetic field, *J. Nucl. Energy, Part C (Plasma Physics)* **4**, 111-117, 1962.

Buneman, O., Time reversible difference procedures, *J. Comput. Phys.* **1**, 517-535, June 1967.

Buneman, O., Fast numerical procedures for computer experiments on relativistic plasmas, *Relativistic Plasmas (The Coral Gables Conference, University of Miami)*, O. Buneman and W. Pardo, eds., Benjamin, New York, 205-219, 1968.

Buneman, O., Subgrid Resolution of Flow and Force Fields, *J. Comput. Phys.* **11**, 250-268, February 1973a.

Buneman, O., Inversion of the Helmholtz (or Laplace-Poisson) operator for slab geometry, *J. Comput. Phys.* **12**, 124-130, May 1973b.

Buneman, O., The advance from 2d electrostatic to 3d electromagnetic particle simulation, *Computer Phys. Comm.* **12**, 21-31, 1976.

Buneman, O., and D. Dunn, Computer experiments in plasma physics, *Science Journal* **2**, 34-43, July 1966.

Buneman, O., C. W. Barnes, J. C. Green, D. E. Nielson, Principles and capabilities of 3d, E-M particle simulations, *J. Comput. Phys.* **38**, 1-44, November 1980.

Burger, P., D. A. Dunn, and A. S. Halsted, Computer experiments on the randomization of electrons in a collisionless plasma, *Phys. Fluids* **8**, 2263-2272, December 1965.

Busnardo-Neto, J., P. L. Pritchett, A. T. Lin, and J. M. Dawson, A self-consistent magnetostatic particle code for numerical simulation of plasmas, *J. Comput. Phys.* **23**, 300-312, March 1977.

Byers, J. A., Noise suppression techniques in macroparticle models of collisionless plasmas, *Proc. Fourth Conf. Num. Sim. Plasmas,* Naval Res. Lab., Wash., D. C., 496-510, 2-3 November 1970.

Byers, J. A., and M. Grewal, Perpendicularly propagating plasma cyclotron instabilities simulated with a one-dimensional computer model, *Phys. Fluids* **13**, 1819-1830, July 1970.

Byers, J. A., J. P. Holdren, J. Killeen, A. B. Langdon, A. A. Mirin, M. E. Rensink, and C. G. Tull, Computer simulation of pulse trapping and pulse stacking of relativistic electron layers in astron, *Phys. Fluids* **17**, 2061-2080, November 1974.

Byers, J. A., B. I. Cohen, W. C. Condit, and J. D. Hanson, Hybrid simulations of quasineutral phenomena in magnetized plasma, *J. Comput. Phys.* **27**, 363-396, June 1978.

Chen, Liu, and C. K. Birdsall, Heating of magnetized plasmas by a large-amplitude low-frequency electric field, *Phys. Fluids* **16**, 2229-2240, December 1973.

Chen, Liu, A. B. Langdon, and C. K. Birdsall, Reduction of grid effects in simulation plasmas, *J. Comput. Phys.* **14**, 200-222, February 1974.

Chen, Liu, and Hideo Okuda, Theory of plasma simulation using multipole expansion scheme, *J. Comput. Phys.* **19**, 339-352, December 1975.

Chen, Y.-J. and C. K. Birdsall, Lower-hybrid drift instability saturation mechanisms in one-dimensional simulations, *Phys. Fluids* **26**, 180-189, January 1983.

Chen, Y.-J., W. M. Nevins, and C. K. Birdsall, Stabilization of the lower-hybrid drift instability by resonant electrons, *Phys. Fluids* **26**, 2501-2508, September 1983.

Chodorow, M., and C. Susskind, *Fundamentals of Microwave Electronics,* McGraw-Hill, New York, 1964.

Christiansen, J. P. and J. B. Taylor, Numerical simulation of guiding center plasma, *Plasma Phys.* **15**, 585-597, 1973.

Cohen, B. I., Theoretical studies of some nonlinear laser-plasma interactions, Ph. D. thesis, University of California, Berkeley, CA, August 1975.

Cohen, B. I., M. A. Mostrom, D. R. Nicholson, A. N. Kaufman, C. E. Max, and A. B. Langdon, Simulation of laser beat heating of a plasma, *Phys. Fluids* **18**, 470-474, April 1975.

Cohen, B. I. and N. Maron, Simulation of drift-cone modes, *Phys. Fluids* **23**, 974-980, May 1980.

Cohen, B. I., T. A. Brengle, D. B. Conley, R. P. Freis, An orbit-averaged particle code, *J. Comput. Phys.* **38**, 45-63, November 1980.

Cohen, B. I., R. P. Freis, and V. Thomas, Orbit-averaged implicit particle codes, *J. Comput. Phys.* **45**, 345-366, March 1982a.

Cohen, B. I., and R. P. Freis, Stability and application of an orbit-averaged magneto-inductive particle code, *J. Comput. Phys.* **45**, 367-373, March 1982.

Cohen, B. I., A. B. Langdon, and A. Friedman, Implicit time integration for plasma simulation, *J. Comput. Phys.* **46**, 15-38, April 1982b.

Cohen, B. I., G. R. Smith, N. Maron, and W. M. Nevins, Particle simulations of ion-cyclotron turbulence in a mirror plasma, *Phys. Fluids* **26**, 1851-1865, July 1983.

Collatz, L., *The Numerical Treatment of Differential Equations,* Springer-Verlag, New York, 1966.

Cooley, J. W., P. A. W. Lewis, and P. D. Welch, The fast Fourier transform algorithm: Programming considerations in the calculation of sine, cosine and Laplace transforms, *J. Sound Vib.* **12**, 315-337, 1970.

Crawford, F. W., and J. A. Tataronis, Absolute instabilities of perpendicularly propagating cyclotron harmonic plasma waves, *J. Appl. Phys.* **36**, 2930-2934, September 1965.

Crume, E. C., H. K. Meier, and O. Eldridge, Nonlinear stabilization of single, resonant, losscone flute instabilities, *Phys. Fluids* **15**, 1811-1821, October 1972.

Davidson, R. C., D. A. Hammer, I. Haber, and C. E. Wagner, Nonlinear development of electromagnetic instabilities in anisotropic plasmas, *Phys. Fluids* **15**, 317-333, February 1972.

Dawson, J. M., Plasma oscillations of a large number of electron beams, *Phys. Rev.* **118**, 381-389, April 1960.

Dawson, J. M., One-dimensional plasma model, *Phys. Fluids* **5**, 445-459, April 1962.

Dawson, J. M., Thermal relaxation in a one-species, one-dimensional plasma, *Phys. Fluids* **7**, 419-425, March 1964.

Dawson, J. M., The electrostatic sheet model for plasma and its modification to finite-size particles, *Meth. Comput. Phys.* **9**, 1-28, B. Alder, S. Fernbach, and M. Rotenberg, eds., Academic, New York, 1970.

Dawson, J. M., Particle simulations of plasmas, *Rev. Mod. Phys.* **55**, 403-447, April 1983.

Dawson, J. M., and T. Nakayama, Kinetic structure of a plasma, *Phys. Fluids* **9**, 252-264, February 1966.

Dawson, J. M., H. Okuda, and B. Rosen, Collective transport in plasmas, *Meth. Comput. Phys.* **16**, 281-325, B. Alder, S. Fernbach, M. Rotenberg, and J. Killeen, eds., Academic, New York, 1976.

Decyk, V. K., Energy conservation theorem for electrostatic systems, *Phys. Fluids* **25**, 1205-1206, July 1982.

Denavit, J., Numerical simulation of plasmas with periodic smoothing in phase space, *J. Comput. Phys.* **9**, 75-98, February 1972.

Denavit, J., Discrete particle effects in whistler simulation, *J. Comput. Phys.* **15**, 449-475, August 1974.

Denavit, J., Collisionless plasma expansion into a vacuum, *Phys. Fluids* **22**, 1384-1392, July 1979.

Denavit, J., Pitfalls in Particle Simulations and in Numerical Solutions of the Vlasov Equation, in *Methoden und Verhaften de Mathematischen Physik* **20**, 247-269, Peter Lang, 1980.

Denavit, J., Time filtering particle simulations with $\omega_{pe}\Delta t \gg 1$, *J. Comput. Phys.* **42**, 337-366, August 1981.

Denavit, J., and W. L. Kruer, Comparison of numerical solutions of the Vlasov equation with particle simulation of collisionless plasmas, *Phys. Fluids* **14**, 1782-1791, August 1971.

Denavit, J., and W. L. Kruer, How to get started in particle simulation, *Comments Plasma Phys. Control. Fusion* **6**, 35-44, April 1980.

Denavit, J. and J. M. Walsh, Nonrandom initializations of particle codes, *Comments Plasma Phys. Control. Fusion* **6**, 209-223, September 1981.

Dickman, D. O., R. L. Morse, and C. W. Nielson, Numerical simulation of axisymmetric, collisionless, finite-β plasma, *Phys. Fluids* **12**, 1708-1716, August 1969.

Dory, R. A., G. E. Guest, and E. G. Harris, Unstable electrostatic plasma waves propagating perpendicular to a magnetic field, *Phys. Rev. Lett.* **14**, 131-133, February 1965.

Drummond, W. E., J. H. Malmberg, T. M. O'Neil, and J. R. Thompson, Nonlinear development of the beam-plasma instability, *Phys. Fluids* **13**, 2422-2425, September 1970.

Dum, C. T., R. Chodura, and D. Biskamp, Turbulent heating and quenching of the ion-sound instability, *Phys. Rev. Lett.* **32**, 1231-1234, 3 June 1974.

Dunn, D. A., and I. T. Ho, Computer experiments on ion-beam neutralization with initially cold electrons, Stanford Electronics Research Laboratory SEL-73-046, Stanford, CA, April 1963.

Dupree, T. H., Growth of phase-space density holes, *Phys. Fluids* **26**, 2460-2481, September 1983.

Eastwood, J. W., and R. W. Hockney, Shaping the force law in two-dimensional particle-mesh models, *J. Comput. Phys.* **16**, 342-359, December 1974.

Ebrahim, N. A., H. Baldis, C. Joshi, and R. Benesch, *Phys. Rev. Lett.* **45**, 1179, 1979.

Eldridge, O. C., and M. Feix, One-dimensional plasma model at thermodynamic equilibrium, *Phys. Fluids* **5**, 1076-1080, September 1962a.

Eldridge, O. C., and M. Feix, Fokker-Planck coefficients for a one-dimensional plasma, *Phys. Fluids* **5**, 1307-1308, October 1962b.

Eldridge, O. C., and M. Feix, Numerical experiments with a plasma model, *Phys. Fluids* **6**, 398-406, March 1963.

Emmert, G. A., R. M. Wieland, A. T. Morse, and J. N. Davidson, Electric sheath and presheath in a collisionless, finite ion temperature plasma, *Phys. Fluids* **23**, 803-812, April 1980.

Estabrook, K. G., E. J. Valeo, and W. L. Kruer, Two-dimensional relativistic simulations of resonance absorption, *Phys. Fluids* **18**, 1151-1159, September 1975.

Estabrook, K., and J. Tull, An 880 ns 1d electrostatic particle mover for the CDC 7600, *Proc. Ninth Conf. Num. Sim. Plasmas,* Northwestern Univ., Evanston, IL, 30 June-2 July 1980.

Feix, M. R., Mathematical models of a plasma, in *Nonlinear Effects in Plasmas,* Gordon and Breach, New York, 1969.

Forslund, D., R. Morse, C. Nielson, and J. Fu, Electron cyclotron drift instability and turbulance, *Phys. Fluids* **15**, 1303-1318, July 1972a.

Forslund, D. W., J. M. Kindel, and E. L. Lindman, Parametric excitation of electromagnetic waves, *Phys. Rev. Lett.* **29**, 249-252, July 1972b.

Forslund, D. W., J. M. Kindel, and E. L. Lindman, Nonlinear behavior of stimulated Brillouin and Raman scattering in laser-irradiated plasmas, *Phys. Rev. Lett.* **30**, 739-743, April 1973.

Forslund, D. W., J. M. Kindel, E. L. Lindman, and R. L. Morse, Theory and simulation of resonance absorption in a hot plasma, *Phys. Rev.* **A11**, 679-683, February 1975a.

Forslund, D. W., and Brackbill, J. U., Magnetic field induced surface transport on laser irradiated foils, *Phys. Rev. Lett.* **48**, 1614-1617, June 1982.

Forsythe, G. E., and W. R. Wasow, *Finite-Difference Methods for Partial Differential Equations,* Wiley, New York, 1960.

Fried, B. D., and S. D. Conte, *Plasma Dispersion Function,* Academic, New York, 1961.

Friedberg, J., and T. Armstrong, Nonlinear development of the two-stream instability, *Phys. Fluids* **11**, 2669-2679, December 1968.

Friedman, A., A. B. Langdon and B. I. Cohen, A direct method for implicit particle-in-cell simulation, *Comments Plasma Phys. Control. Fusion* **6**, 225-236, September 1981.

Gentle, K. W., and J. Lohr, Phase-space evolution of a trapped electron beam, *Phys. Rev. Lett.* **30**, 75-77, January 1973a.

Gentle, K. W., and J. Lohr, Experimental determination of the nonlinear interaction in a one dimensional beam-plasma system, *Phys. Fluids* **16**, 1464-1471, September 1973b.

Gentleman, W. M., and G. Sande, Fast fourier transforms—for fun and profit, *Proc. AFIPS, Fall Joint Computer Conf.* **29**, 563-578, 1966.

Gitomer, S. J., Comments on numerical simulation of the Weibel instability in one and two dimensions, *Phys. Fluids* **14**, 1591-1592, July 1971.

Gitomer, S. J., and J. C. Adam, Multibeam instability in a Maxwellian simulation plasma, *Phys. Fluids* **19**, 719-722, May 1976.

Godfrey, B. B., Numerical Cherenkov instabilities in electromagnetic particle codes, *J. Comput. Phys.* **15**, 504-521, August 1974.

Godfrey, B. B., Canonical momenta and numerical instabilities in particle codes, *J. Comput. Phys.* **19**, 58-76, September 1975.

Godfrey, B. B., and A. B. Langdon, Stability of the Langdon-Dawson advective algorithm, *J. Comput. Phys.* **20**, 251-255, February 1976.

Goldstein, H., *Classical Mechanics,* Addison-Wesley, Cambridge, MA, 1950.

Guernsey, R. L., Kinetic equation for a completely ionized gas, *Phys. Fluids* **5**, 322-328, March 1962.

Haber, I., R. Lee, H. H. Klein, and J. P. Boris, Advances in electromagnetic plasma simulation techniques, *Proc. Sixth Conf. Num. Sim. Plasmas,* Lawrence Livermore Lab., Lawrence Berkeley Lab., Berkeley, CA, 46-48, 16-18 July 1973.

Haeff, A. V., The electron-wave tube—a novel method of generation and amplification of microwave energy, *Proceedings of I. R. E.* **37**, 4-10, January 1949.

Hammersley, J. M., and D. C. Handscomb, *Monte Carlo Methods,* Methuen, London, 1964.

Hamming, R. W., *Numerical Methods for Scientists and Engineers,* McGraw-Hill, New York, 1962.

Hamming, R. W., *Digital Filters,* Prentice Hall, Englewood Cliffs, NJ, 1977.

Harlow, F. H., The particle-in-cell computing method for fluid dynamics, *Meth. Comput. Phys.* **3**, 319-343, B. Alder, S. Fernbach, and M. Rotenberg, eds., Academic, New York, 1964.

Harned, D. S., Quasineutral hybrid simulation of macroscopic plasma phenomena, *J. Comput. Phys.* **47**, 452-462, September 1982a.

Harned, D. S., Kink instabilities in long ion layers, *Phys. Fluids* **25**, 1915-1921, October 1982b.

Harris, E. G., Unstable plasma oscillations in a magnetic field, *Phys. Rev. Lett.* **2**, 34-36, January 1959.

Hasegawa, A., *Plasma Instabilities and Nonlinear Effects,* Springer-Verlag, Berlin, Heidelberg, New York, 1975.

Hasegawa, A., and C. K. Birdsall, Sheet-current plasma model for ion cyclotron waves, *Phys. Fluids* **7**, 1590-1600, October 1964.

Hewett, D. W., A global method of solving the electron-field equations in a zero-inertia-electron-hybrid plasma simulation code, *J. Comput. Phys.* **38**, 378, December 1980.

Hewett, D. W., Spontaneous development of toroidal magnetic field during formation of the field reversed theta pinch, *Nucl. Fusion* **24**, 349-357, March 1984.

Hewett, D. W., and C. W. Nielson, A multidimensional quasineutral plasma simulation model, *J. Comput. Phys.* **29**, 219-236, 1978.

Hockney, R. W., A fast direct solution of Poisson's equation using Fourier analysis, *J. Assoc. Comput. Mach.* **12**, 95-113, January 1965.

Hockney, R. W., Computer simulation of anomalous plasma diffusion and numerical solution of Poisson's equation, *Phys. Fluids* **9**, 1826-1835, September 1966.

Hockney, R. W., Characteristics of noise in a two-dimensional computer plasma, *Phys. Fluids* **11**, 1381-1383, June 1968.

Hockney, R. W., The potential calculation and some applications, *Meth. Comput. Phys.* **9**, 135-211, B. Alder, S. Fernbach, and M. Rotenberg, eds., Academic, New York 1970.

Hockney, R. W., Measurements of collision and heating times in a two-dimensional thermal computer plasma, *J. Comput. Phys.* **8**, 19-44, August 1971.

Hockney, R. W., S. P. Goel, and J. W. Eastwood, Quiet high-resolution computer models of a plasma, *J. Comput. Phys.* **14**, 148-158, February 1974.

Hockney, R. W., and J. W. Eastwood, *Computer Simulation Using Particles,* McGraw-Hill, New York, 1981.

Hsu, J. Y., G. Joyce, and D. Montgomery, Thermal relaxation of a two-dimensional plasma in a d. c. magnetic field Part 2, Numerical simulation, *J. Plasma Phys.* **12**, 27-31, 1974.

Hubbard, J., The friction and diffusion coefficients of the Fokker-Planck equation in a plasma, *Proc. R. Soc. London Ser.* **A260**, 114-126, February 1961.

Hudson, M. K., and D. W. Potter, Electrostatic shocks in the auroral magnetosphere, *Physics of Auroral Arc Formation,* S.-I. Akasofu and J. R. Kan, eds., Geophysical Monograph **25**, American Geophysical Union, Wash. D. C., 1981.

Ishihara, O., A. Hirose, and A. B. Langdon, Nonlinear saturation of the Buneman instability, *Phys. Rev. Lett.* **44**, 1404-1407, May 1980.

Ishihara, O., A. Hirose, and A. B. Langdon, Nonlinear evolution of Buneman instability, *Phys. Fluids* **24**, 452-463, March 1981.

Ishihara, O., A. Hirose, and A. B. Langdon, Nonlinear evolution of Buneman instability. II. Ion Dynamics, *Phys. Fluids* **25**, 610-616, April 1982.

Jackson, J. D., Longitudinal plasma oscillations, *J. Nucl. Energy, Part C (Plasma Physics)* **1**, 171-189, 1960.

Jackson, J. D., *Classical Electrodynamics,* Wiley, New York, 2d. ed., 1975.

Jones, M. E. and J. Fukai, Evolution of the explosive instability in a simulated beam plasma, *Phys. Fluids* **22**, 132-138, January 1979.

Joyce, G. and D. Montgomery, Negative temperature states for the two-dimensional guiding center plasma, *J. Plasma Phys.* **10**, 107-121, 1973..

Jury, E. I., *Theory and Applications of the z-Transform Method,* Wiley, New York, 1964.

Kainer, S., J. M. Dawson, and R. Shanny, Interaction of a highly energetic electron beam with a dense plasma, *Phys. Fluids* **15**, 493-501, March 1972.

Kamimura, T., and J. M. Dawson, Effect of mirroring on convective transport in plasmas, *Phys. Rev. Lett.* **36**, 313, 9 February 1976.

Kamimura, T., T. Wagner, and J. M. Dawson, Simulation study of Bernstein modes, *Phys. Fluids* **21**, 1151-1167, July 1978.

Katanuma, I., Heat transport due to collisionless tearing instabilities, *Jour. of the Phys. Soc. of Japan* **50**, 1689-1697, May 1981.

Katanuma, I. and T. Kamimura, Simulation studies of collisionless tearing instabilities, *Phys. Fluids* **23**, 2500-2511, October 1980.

Kaufman, A. N., and P. S. Rostler, The Darwin model as a tool for electromagnetic plasma simulation, *Phys. Fluids* **14**, 446-448, February 1971.

Klein, H. H., W. M. Manheimer, and E. Ott, Effect of side-scattering instabilities on the propagation of an intense laser beam in an inhomogenous plasma, *Phys. Rev. Lett.* **31**, 1187-1190, November 1973.

Klimontovich, Y. L., *The Statistical Theory of Non-Equilibrium Processes in a Plasma,* MIT Press, Cambridge, MA, 1967.

Krall, N. A., and P. C. Liewer, Low-frequency instabilities in magnetic pulses, *Phys. Rev.* **4**, 2094-2103, November 1971.

Krall, N. A. and P. C. Liewer, Turbulent heating and resistivity in cool-electron θ pinches, *Phys. Fluids* **15**, 1166-1168, June 1972.

Krall, N. A., and A. W. Trivelpiece, *Principles of Plasma Physics,* McGraw-Hill, New York, 1973.

Kruer, W. L. and J. M. Dawson, Sideband instability, *Phys. Fluids* **13**, 2747-2751, November 1970.

Kruer, W. L. and J. M. Dawson, Anomalous high-frequency resistivity of a plasma, *Phys. Fluids* **15**, 446-453, March 1972.

Kruer, W. L., J. M. Dawson, and B. Rosen, The dipole expansion method for plasma simulation, *J. Comput. Phys.* **13**, 114-129, September 1973a.

Kruer, W. L., K. Estabrook, and K. H. Sinz, Instability-generated laser reflection in plasmas, *Nucl. Fusion* **13**, 952-955, November 1973b.

Kwan, T. J. T., High-power coherent microwave generation from oscillating virtual cathodes, *Phys. Fluids* **27**, 228-232, January 1984.

Lamb, S. H., *Hydrodynamics,* Dover, New York, 1945.

Langdon, A. B., Investigations of a sheet model for a bounded plasma with magnetic field and radiation, Ph. D. thesis, Princeton University, Electronics Research Laboratory 41-M257, Univ. Calif., Berkeley, January 1969.

Langdon, A. B., Effects of the spatial grid in simulation plasmas, *J. Comput. Phys.* **6**, 247-267, October 1970a.

Langdon, A. B., Nonphysical modifications to oscillations, fluctuations, and collisions due to space-time differencing, *Proc. Fourth Conf. Num. Sim. Plasmas,* Naval Res. Lab., Wash., D. C., 467-495, 2-3 November 1970b.

Langdon, A. B., Some electromagnetic plasma simulation models and their noise properties, *Phys. Fluids* **15**, 1149-1151, June 1972.

Langdon, A. B., Energy conserving plasma simulation algorithms, *J. Comput. Phys.* **12**, 247-268, June 1973.

Langdon, A. B., Kinetic theory of fluctuations and noise in computer simulation of plasma, *Phys. Fluids* **22**, 163-171, January 1979a.

Langdon, A. B., Analysis of the time integration in plasma simulation, *J. Comput. Phys.* **30**, 202-221, February 1979b.

Langdon, A. B., and J. M. Dawson, Investigations of a sheet model for a bounded plasma field and radiation, *Proc. First Conf. Num. Sim. Plasmas,* College of William and Mary, Williamsburg, VA, 39-40, 19-21 April 1967.

Langdon, A. B., and C. K. Birdsall, Theory of plasma simulation using finite-size particles, *Phys. Fluids* **13**, 2115-2122, August 1970.

Langdon, A. B., and B. F. Lasinski, Electromagnetic and relativistic plasma simulation models, *Meth. Comput. Phys.* **16**, 327-366, B. Alder, S. Fernbach, M. Rotenberg, and J. Killeen, eds., Academic, New York, 1976.

Langdon, A. B., B. F. Lasinski, and W. L. Kruer, Nonlinear saturation and recurrence of the two-plasmon decay instability, *Phys. Rev. Lett.* **43**, 133-136, July 1979.

Langdon, A. B., and B. F. Lasinski, Frequency shift of self-trapped light, *Phys. Fluids* **26**, 582-587, February 1983.

Langdon, A. B., B. I. Cohen, and A. Friedman, Direct implicit large time-step particle simulation of plasmas, *J. Comput. Phys.* **51**, 107-138, July 1983.

Langdon, A. B., and D. C. Barnes, Direct implicit plasma simulation, in the volume *Multiple Time Scales,* in the series *Computational Techniques,* J. U. Brackbill and B. I. Cohen, eds., Academic Press, New York, 1985.

Lee, J. K., and C. K. Birdsall, Velocity space ring-plasma instability, magnetized, Part I: Theory, *Phys. Fluids* **22**, 1306-1314, July 1979a.

Lee, J. K., and C. K. Birdsall, Velocity space ring-plasma instability, magnetized, Part II: Simulation, *Phys. Fluids* **22**, 1315-1322, July 1979b.

Lee, R., and M. Lampe, Electromagnetic instabilities, filamentation, and focusing of relativistic electron beams, *Phys. Rev. Lett.* **31**, 1390-1343, December 1973.

Lee. W. W., and H. Okuda, A simulation model for studying low-frequency microinstabilities, *J. Comput. Phys.* **26**, 139, March 1978.

Lenard, A., On Bogoliubov's kinetic equation for a spatially homogeneous plasma, *Ann. Phys.* **10**, 390-400, July 1960.

Lewis, H. R., Energy-conserving numerical approximations for Vlasov plasmas, *J. Comput. Phys.* **6**, 136-141, August 1970a.

Lewis, H. R., Applications of Hamilton's Principle to the numerical analysis of Vlasov plasmas, *Meth. Comput. Phys.* **9**, 307-339, B. Alder, S. Fernbach, and M. Rotenberg, eds., Academic, New York 1970b.

Lewis, H. R., Variational algorithms for numerical simulation of collisionless plasma with point particles including electromagnetic interactions, *J. Comput. Phys.* **10**, 400-419, December 1972.

Lewis, H. R., A. Sykes, and J. A. Wesson, A comparison of some particle-in-cell plasma simulation methods, *J. Comput. Phys.* **10**, 85-106, August 1972.

Lewis, H. R., and C. W. Nielson, A comparison of three two-dimensional electrostatic plasma simulation models, *J. Comput. Phys.* **17**, 1-9, January 1975.

Lighthill, M. H., *Fourier Analysis and Generalized Functions,* Cambridge University, London, 1962.

Lin, A. T., J. M. Dawson, and H. Okuda, Application of electromagnetic particle simulation to the generation of electromagnetic radiation, *Phys. Fluids* **17**, 1995-2001, November 1974.

Lin, A. T. and N. L. Tsintsadze, Electrostatic parametric instabilities arising from relativistic electron mass oscillations, *Phys. Fluids* **19**, 708-710, May 1976.

Lindgren, N. E., A. B. Langdon, and C. K. Birdsall, Electrostatic waves in an inhomogeneous collisionless plasma, *Phys. Fluids* **19**, 1026-1034, July 1976.

Lindman, E. L., Dispersion relation for computer-simulated plasmas, *J. Comput. Phys.* **5**, 13-22, February 1970.

Lindman, E. L., Free-space boundary conditions for the time dependent wave equation, *J. Comput. Phys.* **18**, 66-78, May 1975.

Mankofsky, A., A. Friedman, and R. N. Sudan, Numerical simulation of injection and resistive trapping of ion rings, *Plasma Phys.* **23**, 521-537, 1981.

Mason, R. J., Implicit moment particle simulation of plasmas, *J. Comput. Phys.* **41**, 233-244, June 1981.

Matsuda, Y., and H. Okuda, Collisions in multi-dimensional plasma simulations, *Phys. Fluids* **18**, 1740-1747, December 1975.

McBride, J. B., E. Ott, J. P. Boris, and J. H. Orens, Theory and simulation of turbulent heating by the modified two-stream instability, *Phys. Fluids* **15**, 2367-2383, December 1972.

Montgomery, D., and C. W. Nielson, Thermal relaxation in one- and two-dimensional plasma models, *Phys. Fluids* **13**, 1405-1407, May 1970.

Montgomery, D., and G. Joyce, Statistical mechanics of 'negative temperature' states, *Phys. Fluids* **17**, 1139-1145, June 1974.

Morales, G. J., Y. C. Lee, and R. B. White, Nonlinear Schrödinger equation model of oscillating-two-stream instability, *Phys. Rev. Lett.* **32**, 457-460, 4 March 1974.

Morse, R. L., and C. W. Nielson, Numerical simulation of a warm two-beam plasma, *Proc. Second Conf. Num. Sim. Plasmas,* Los Alamos Sci. Labs. **LA-3990**, 18-20 September 1968.

Morse, R. L., and C. W. Nielson, Numerical simulation of warm two-beam plasma, *Phys. Fluids* **12**, 2418-2425, November 1969.

Morse, R. L., and C. W. Nielson, Numerical simulation of the Weibel instability in one and two dimensions, *Phys. Fluids* **14**, 830-840, April 1971.

Mynick, H. E., M. J. Gerver, and C. K. Birdsall, Stability regions and growth rates for two-ion component plasma, unmagnetized, *Phys. Fluids* **20**, 606-612, April 1977.

Naitou, H., Cross-field electron heat transport due to high frequency electrostatic waves, *J. Phys. Soc. Japan,* **48**, 608, February 1980.

Naitou, H., T. Kamimura, and J. M. Dawson, Kinetic effects on the convective plasma diffusion and the heat transport, *J. Phys. Soc. Japan* **46**, 258, January 1979a.

Naitou, H., S. Tokuda, and T. Kamimura, On boundary conditions for a simulation plasma in a magnetic field, *J. Comput. Phys.* **33**, 86-101, October 1979b.

Naitou, H., S. Tokuda, and T. Kamimura, Initial particle loadings for nonuniform simulation plasma in a magnetic field, *J. Comput. Phys.* **38**, 265-274, December 1980.

Nevins, W., Harte, J., and Y. Gell, Pseudo classical transport in a sheared magnetic field: theory and simulation, *Phys. Fluids* **22**, 2108-2121, November 1979.

Nevins, W., J. Matsuda, and M. Gerver, Plasma simulations using inversion symmetry as a boundary condition, *J. Comput. Phys.* **39**, 226-232, January 1981.

Nielson, C. W., and E. L. Lindman, An implicit, two-dimensional electromagnetic plasma simulation code, *Proc. Sixth Conf. Num. Sim. Plasmas,* Lawrence Livermore Lab., Lawrence Berkeley Lab., Berkeley, CA, 148-151, 16-18 July 1973.

Nielson, C. W., and H. R. Lewis, Particle-code models in the nonradiative limit, *Meth. Comput. Phys.* **16**, 367-388, B. Alder, S. Fernbach, M. Rotenberg, and J. Killeen, eds., Academic, New York, 1976.

Ohsawa, Y., M. Inutake, T. Tajima, T. Hatori, and T. Kamimura, Plasma paramagnetism in radio-frequency fields, *Phys. Rev. Lett.* **43**, 1246-1249, 22 October 1979.

Okuda, H., Nonphysical instabilities in plasma simulation due to small $\lambda_D / \Delta x$, *Proc. Fourth Conf. Num. Sim. Plasmas,* Naval Res. Lab., Wash., D. C., 511-525, 2-3 November 1970.

Okuda, H., Verification of theory for plasma of finite-size particles, *Phys. Fluids* **15**, 1268-1274, July 1972a.

Okuda, H., Nonphysical noises and instabilities in plasma simulation due to a spatial grid, *J. Comput. Phys.* **10**, 475-486, December 1972b.

Okuda, H., Effects of spatial grid in plasma simulations using higher order multipole expansions, Princeton Plasma Physics Laboratory report PPPL-1355, July 1977.

Okuda, H., and C. K. Birdsall, Collisions in a plasma of finite-size particles, *Phys. Fluids* **13**, 2123-2134, August 1970.

Okuda, H., and J. M. Dawson, Theory and numerical simulation of plasma diffusion across a magnetic field, *Phys. Fluids* **16**, 408-426, March 1973.

Okuda, H., and C. Z. Cheng, Higher order multipoles and splines in plasma simulation, *Computer Phys. Comm.* **14**, 169-176, 1978.

O'Neil, T. M., and J. H. Malmberg, Transition of the dispersion roots from beam-type to Landau-type solutions, *Phys. Fluids* **11**, 1754-1760, August 1968.

O'Neil, T. M., J. H. Winfrey, and J. H. Malmberg, Nonlinear interaction of a small cold beam and a plasma, I, *Phys. Fluids* **14**, 1204-1212, June 1971.

Ossakow, S. L., I. Haber, and E. Ott, Simulation of whistler instabilities in anisotropic plasmas, *Phys. Fluids* **15**, 1538-1540, August 1972a.

Ossakow, S. L., E. Ott, and I. Haber, Nonlinear evolution of whistler instabilities, *Phys. Fluids* **15**, 2314-2326, December 1972b.

Ott, E., W. M. Manheimer, and H. H. Klein, Stimulated Compton scattering and self-focusing in the outer regions of a laser-produced plasma, *Phys. Fluids* **17**, 1757-1761, September 1974.

Palevsky, A., Generation of intense microwave radiation by the relativistic e-beam magnetron (experiment and numerical simulation), Ph. D. thesis, Mass. Inst. of Tech., June 1980.

Panofsky, W. K. H., and M. Phillips, *Classical Electricity and Magnetism,* Addison-Wesley, Reading, MA, 1962.

Peiravi, A., and C. K. Birdsall, Self-heating of 1d thermal plasma; comparison of weightings; optimal parameter choices. *Proc. Eighth Conf. Num. Sim. Plasmas,* Monterey, Calif., PD-9, 28-30 June 1978.

Pierce, J. R., Possible fluctuations in electron streams due to ions, *J. Appl. Phys.* **19**, 231-236, March 1948.

Portis, A. M., *Electromagnetic Fields: Sources and Media,* Wiley, New York, 1978.

Potter, D., *Computational Physics,* Wiley, London, 1973.

Ramo, S., J. R. Whinnery, and T. VanDuzer, *Fields and Waves in Communications Electronics,* Wiley, New York, 1965.

Rayleigh, B., *The Theory of Sound,* Dover, New York, 1945.

Reitz, J. R. and F. J. Milford, *Foundations of electromagnetic theory,* Addison-Wesley, Reading, MA, 1960.

Rostoker, N. Fluctuations of a plasma (I), *Nucl. Fusion* 1, 101-120, March 1961.

Rostoker, N., and M. N. Rosenbluth, Test particles in a completely ionized plasma, *Phys. Fluids* 3, 1-14, January 1960.

Schmidt, G., *Physics of High Temperature Plasmas,* Academic, New York, 1966.

Singleton, R. C., On computing the fast Fourier transform, *Commun. Assoc. Comput. Mach.* 10, 647-654, 1967.

Singleton, R. C., Algorithm 345, an algol convolution procedure based on the fast Fourier transform, *Commun. Assoc. Comput. Mach.* 12, 179-184, 1969.

Sköllermo, A., A Better Difference Scheme for the Laplace Equation in Cylindrical Coordinates, *J. Comput. Phys.* 47, 160-163, 1982.

Sköllermo, A., and G. Sköllermo, A Fourier Analysis of Some Difference Schemes for the Laplace Equation in a System of Rotation Symmetry, *J. Comput. Phys.* ₋₀, 103-114, 1978.

Sommerfeld, A., *Optics,* Academic, New York, 1954.

Stringer, T. E., Electrostatic instabilities in current-carrying and counterstreaming plasmas, *J. Nucl. Energy, Part C (Plasma Physics)* C6, 267-279, May 1964.

Swift, D. W., and J. J. Ambrosiano, Boundary conditions which lead to excitation of instabilities in plasma simulations, *J. Comput. Phys.* 44, 302-317, 1981.

Tajima, T., and Y. C. Lee, Absorbing boundary condition and Budden turning point technique for electromagnetic plasma simulations, *J. Comput. Phys.* 42, 406-412, August 1981.

Tataronis, J. A., and F. W. Crawford, Cyclotron harmonic wave propagation and instabilities, I, perpendicular propagation, *J. Plasma Phys.* 4, 231-248, May 1970.

Taylor, J. B., and B. McNamara, Plasma diffusion in two dimensions, *Phys. Fluids* 14, 1492-1499, July 1971.

Temperton, C. Algorithms for the solution of cyclic tridiagonal systems, *J. Comput. Phys.* 19, 317-323, 1975.

Thomas, V., and C. K. Birdsall, Plasma hybrid oscillations as affected by aliasing, *Proc. Ninth Conf. Num. Sim. Plasmas,* Northwestern Univ., Evanston, IL, PB6, 30 June-2 July 1980.

Tsang, K. T., Y. Matsuda, and H. Okuda, Numerical simulation of neoclassical diffusion, *Phys. Fluids* 18, 1282-1286, October 1975.

Valeo, E. J., and W. L. Kruer, Solitons and resonance absorption, *Phys. Rev. Lett.* 33, 750-753, 23 September 1974.

Vlasov, A. A., *Many Particle Theory and Its Application to Plasma,* Russian original 1950, translation to English, Gordon and Breach, New York, 1961.

Walsh, J. E., and S. S. Hagelin, Van der Pol's equation and nonlinear oscillations in a beam plasma system, *Phys. Fluids* 19, 339-340, February 1976.

Yu, S. P., G. P. Kooyers, and O. Buneman, A time dependent computer analysis of electron-wave interaction in crossed-fields, *J. Appl. Phys.* 36, 2550-2559, August 1965.

AUTHOR INDEX

This index includes only authors of published works.

Abe, 264, 275, 293, 295
Abramowitz, 173, 188, 196, 203, 232, 257, 391
Adam, 95, 204, 293, 384, 396, 397
Aizawa, 380, 385
Albritton, 131
Alder, 156
Ambrosiano, 411
Armstrong, 103, 318
Auer, 64

Baldis, 385
Balescu, 269
Barnes, C., 365, 366, 385
Barnes, D., 205, 212, 293, 295, 340, 341, 342, 343, 344, 345, 381
Benesch, 385
Berman, 301
Bernstein, 127
Bers, 111
Birdsall, 20, 63, 64, 65, 68, 70, 76, 77, 78, 93, 95, 115, 119, 120, 123, 128, 129, 131, 155, 156, 157, 159, 176, 177, 196, 197, 258, 270, 293, 311, 312, 324, 327, 347, 349, 350, 383, 411, 413, 441

Biskamp, 350, 383, 384
Blackman, 160, 188
Boris, 15, 59, 62, 137, 138, 156, 328, 338, 350, 353, 355, 356, 358, 359, 360, 362, 365, 366, 368, 377
Boutros-Ghali, 301
Boyd, 110
Brackbill, 205, 381, 383, 384
Brand, 170
Brengle, 206
Bridges, 76, 77, 411, 413
Briggs, 120
Brillouin, 161
Brown, 169, 229
Buneman, 58, 60, 62, 114, 123, 156, 157, 169, 185, 188, 200, 318, 320, 321, 353, 360, 362, 363, 365, 366, 367, 385, 449
Burger, 156
Busnardo-Neto, 380
Byers, 65, 95, 130, 183, 266, 380, 385, 393, 395, 400

SUBJECT INDEX

469